Modelling nutrient digestion and utilisation in farm animals

Modelling nutrient digestion and utilisation in farm animals

edited by:
D. Sauvant
J. Van Milgen
P. Faverdin
N. Friggens

ISBN: 978-90-8686-156-9
e-ISBN: 978-90-8686-712-7
DOI: 10.3920/978-90-8686-712-7

First published, 2010

Preface

This book contains the papers presented as either posters or oral presentations at the 7th International Workshop on 'Modelling Nutrient Digestion and Utilisation in Farm Animals'. The workshop was co-organised by INRA (the French National Institute of Agricultural Research) and AgroParisTech and was held at AgroParisTech in Paris on 10-12 September 2009. Previous modelling workshops in this series were held at: Grassland Research Institute (1979, Hurley, United Kingdom), Universiy of California (1984, Davis, USA), Lincoln University (1989, Canterbury, New Zealand), Research Center Foulum (1994, Denmark), Capetown (1999, South Africa), Wageningen University (2004, the Netherlands). All workshops were held in conjunction with the International Symposium on Ruminant Physiology.

The major purpose of this workshop was to present the 'state of art' in modelling digestion and nutrient utilization in farm animals. It was attended by scientists, modellers, PhD students, and non-modellers to discuss and exchange new ideas and methods. The contributions covered a large range of topics within modelling activities applied to animal nutrition. The Workshop (and this book) were organised in the following six sessions:

1. Advances in methodological aspects of modelling
2. Modelling feeding behaviour and regulation of feed intake
3. Modelling fermentation, digestion and microbial interactions in the gut
4. Modelling interactions between nutrients and physiological functions: consequence on product quality and animal health
5. Extrapolating from the animal to the herd
6. Modelling the environmental impact of animal production

Approximately 110 participants from 18 different countries made this workshop an exciting event. A total of 212 colleagues contributed as author or co-author to the 67 oral (36) or poster (31) presentations, resulting in a total of 45 papers in this book. This is a considerable increase compared with previous workshops, and illustrates the increasing role of modelling in research programs.

To give an idea of the range of topics covered, a rapid overview of more than 300 keywords of the titles of workshops held since the second workshop (Davis, USA) indicates that the studies presented mainly concerned ruminants (23.8% of the keywords), most of which deal with dairy cattle (14.5%). Fewer studies deal with monogastrics (5.6%). With regard to digestion, rumen function (11.9%) was also more frequently modelled than the digestion processes in monogastrics (1.6%). The dominance of ruminants is mainly the result of the fact that the first 2 workshop essentially focused on these animals. In terms of physiological functions, lactating animals were more frequently studied than growing animals (7.3%). The 3 other major topics of presentation focused on metabolism (11.5%), farm systems and herd management (9.9%) and methodological aspect of on modelling (8.2%). For metabolism, there is an approximate balance between energy (3.3%), protein (3.0%) and mineral studies (2.6%). Surprisingly, only few models presented concern some important topics of research in animal nutrition, this is particularly the case for intake and behaviour (2.0%) and the regulation of metabolism (1.0%). Moreover, disciplines other than nutrition were only rarely considered: pathology (1.3%), genetics (0.7%). Concerning the range of topics in the current workshop, there were no marked changes in comparison with previous workshops. The biggest change was that there was an increase in the proportion of communications focussing on the farm and environmental aspects of animal production (20.5% in the present workshop vs. 12.7% for workshops 5 and 6, and 6.9% for workshops 2, 3 and 4).

We would like to acknowledge the organisations and people who have contributed to the success of the workshop:

- The local scientific committee: Jacques Agabriel (INRA Clermont-Ferrand), Philippe Faverdin (INRA Rennes), Philippe Lescoat (INRA Tours), Daniel Sauvant (AgroParisTech-INRA Paris), Jaap Van Milgen (INRA Rennes)
- The international scientific committee: Jan Dijkstra (NL), Jim France (Canada-UK), John Mc Namara (USA), Dennis Poppi (Australia) and also Roberto Sainz (USA)
- The co-organisers INRA and AgroParisTech, the French Association of Zootechnics (AFZ), which supported us, and ADEPRINA who efficiently contributed to the registration and administrative processes.
- Companies that provided supported: BASF, Evialis, Idena, Inve, Lallemand, Limagrain, Nutreco, Pancosma, Phodé, Provilis, Skretting and Techna.
- My three colleagues Jaap, Nicolas and Philippe who accepted to be the scientific co-editors of this book. They contributed to the reviewing of the articles. Thanks is also expressed towards others colleagues, all co-authors of articles of this book, who also helped us to review some of the articles.
- Our colleagues of the research unit 'UMR INRA-AgroParisTech Physiologie de la Nutrition et Alimentation' (UMR PNA), with a special thanks to our secretary, Marie Paul Poulin, who contributed a great deal to solving the numerous practical aspects that emerged during the preparation and organisation of the congress and the book.

Pr Daniel Sauvant

Chair of the local committee

Table of contents

Part 1 – Advances in methodological aspects of modelling

A generic framework for simulating agricultural production systems

R. Martin-Clouaire and J.-P. Rellier
INRA, UR875 Biométrie et Intelligence Artificielle, 31326 Castanet-Tolosan, France; rmc@toulouse.inra.fr; rellier@toulouse.inra.fr

Abstract

The relevance of simulation approaches to the study and design of agricultural production systems is widely claimed. The methodology and computer software appropriate to such a task have however still not reached the state of a mature technology and are mainly developed in research laboratories. Suitable computer models need to represent the structure and dynamics of the underlying biophysical system together with the coordinated human activities involved in the management of the farm production process. Most existing approaches focus primarily on biophysical processes. This paper outlines the generic framework DIESE especially designed for building and running agricultural production system models. DIESE relies on a rich conceptual basis under the form of an ontology of agricultural production systems. It supports the modelling of the decision process in terms of activities, resources required to realize them, and well-structured constraints bearing on the relevance and feasibility of activities, the interdependencies between them and the restrictions on the uses of resources. Computationally the ontology comes under the form of a C^{++} library. In developing a farm production system model, the ontology acts as a metamodel; implementing a model amounts to particularizing the ontology concepts as required by the domain and then instantiating the corresponding classes to capture the specific aspects of the system to be simulated. A discrete event simulation mechanism realizes the step by step interpretation of the strategy and the progressive execution of the decided activities, which in turn alters the biophysical state that otherwise responds to external factors, e.g. weather, influencing biophysical processes. DIESE is currently used in large modelling projects dealing with various kinds of production such as cash crop, vineyard, pasture-based livestock and pig systems, which attest to the wide scope of applicability of the framework.

Keywords: simulation, ontology, management, plan, activity, resource

Introduction

From a system perspective, an agricultural production system involves at least a biophysical system (composed of land, crops, livestock, etc.), a decision system (the farm manager) and an operating system that implements the decisions using various resources (input, labor, machinery, etc.). The relevance of simulation approaches to understand, evaluate and design such systems is widely claimed, although the methodology and computer software currently used to support such investigation have still not reached the state of a mature technology and are mainly confined to research settings. Suitable computer models need to represent the structure and dynamics of the underlying biophysical system together with the coordinated human activities involved in the management of the farm production processes. Production process improvement involves studying timely interactions among biophysical processes and decision making processes at the farm level while most existing approaches tend to address one at the expense of the other, usually focus primarily on biophysical processes.

Mathematical models of farm management are logically attractive but are of limited practical relevance. In particular, those based on static equilibrium conditions can hardly address challenges created by uncertainty and dynamics. Indeed farm systems are complex and dynamic and farming is conducted under conditions of uncertainty. Performance depends very much on how uncertainty is dealt with. Approaches based on averages do not work either. Indeed aggregate differential equations tend to smooth out fluctuations. It is essential that the model be able to reveal how fluctuations might be amplified and how the system may become unstable to large perturbations. Moreover, individual farms are unique and farmers have significantly different practices and preferences that are objects of study in their own right.

Simulation-based approaches that explicitly incorporate production management processes provide a more promising framework thanks to their ability to grasp realistic situations and issues. The management system model should explicitly represent the decision-making process and the implementation of the technical actions resulting from this process. Agricultural production management deals with how farmers combine land, water, domesticated living things, machinery, commercial inputs, labour, and management skills to produce crop and livestock commodities. Farm management (Dillon, 1979) is the process by which resources and situations are manipulated over time by the manager in trying, with less than full information, to achieve his or her goals that might be competing (increase profits, respond to social objectives, or maintain a way of life). The farmer's management behaviour is observable through the choice and timing of activities involved in the various production aspects. This behaviour results from the situation-dependent implementation of his management strategy. Simply stated a management strategy is a kind of flexible plan coming with its context-responsive adaptations and the necessary implementation details to constrain *in situ* the stepwise determination and execution of the actions.

This paper outlines the generic framework DIESE (Martin-Clouaire and Rellier, 2009) especially designed for building and running agricultural production system models. The capability to represent a farmer's production management behaviour is emphasised. DIESE supports the modelling of the decision process in terms of activities, resources required to realize them, and well-structured constraints bearing on the relevance and feasibility of activities, the interdependencies between them and the restrictions on the uses of resources. Section 2 introduces the ontological basis of the DIESE framework. The main aspects specific to the modelling of management activities are presented in Section 3 and illustrated in Section 4. Dealing with resources in DIESE is addressed in Section 5.

Overview of the production system ontology and DIESE

Basically an ontology (Chandrasekaran *et al.*, 1999) is an explicit and declarative description of the domain we are interested in, that is, the concepts in this domain, the properties of these concepts and the constraints on these properties. In addition to providing a shared vocabulary and sense disambiguation, an ontology enables to reuse pre-formalized concepts and templates that can be particularized, instantiated and then mapped into an executable model interpreted with a discrete event simulation engine.

The ontology of agricultural production systems (Martin-Clouaire and Rellier, 2009) serves as a conceptual meta-model supporting the modeling framework. It contains a number of pre-formalized concepts, templates and mechanisms describing the studied system components at a high level of abstraction. The DIESE modeling framework is an object-oriented modeling tool that implements the ontology through a C++ library of pre-programmed classes and services. It comprises a discrete event simulation engine that enables to emulate the continuous and discontinuous features of the simulated system. The modeling enterprise relies on the ontological concepts, templates and mechanisms by particularizing, instantiating and then mapping these into an executable dynamic model of a specific system. The ontology considerably eases model development and implementation for non-computer-specialists, as the framework helps in clarifying how to organize the expert knowledge about the system of interest in the knowledge base.

The three fundamental concepts of the ontology are: entity, process and event. These represent the structural, functional and dynamic aspects of a system respectively (Rellier, 2005). An entity describes a kind of material or abstract item in the area of interest. The state of a system at a given moment in time is the value of the slots (properties) of the entities it comprises. A process is a specification of part of the behavior of a system, i.e. of the entities composing it. Typically, the process code specifying this behavior includes the use of methods attached to entities affected by the process. A process causes a change in state when a particular event occurs. Thus, events convey the temporality of process triggers. For instance, in a livestock system model, the biophysical part may involve an entity such as *animal-batch* that has slots such as *location* and *composition*. The type of value that the composition slot might take is another entity, *set-of-animal* that has a numerical slot describing its size and slot *representative-animal* describing the typical animal of this set through slots such as *weight*, *intake-capacity* or *intake-amount*. The processes that might affect the state of an individual instantiating *representative-animal* include for instance *growth* and *intake*. An event starting a *growth* process might be created at birth time of an animal.

The DIESE framework includes specific constructs to represent various aspects relevant to the management functions. Fundamental to our conceptual model is the commitment to understand things from a farmer's point of view. To be effective, management behavior must be specified by using constructs and language that are intelligible and conceptually close to those actually used in an agricultural setting. The basic unit of analysis in our approach is work activity, which is a common high level concept in production management. An activity is a purposeful engagement driven by certain needs to achieve a certain purpose. Activities are contextual in the sense that actual circumstances condition their relevance and greatly affect the way the intended objective is achieved. Activities usually involve the use of resources (equipment, labor). Whenever a combination of activities must be undertaken with a view to achieving a pre-conceived result, a plan is needed to express how those composite activities should be coordinated. A work plan is the result of reflection on prior experiences and in anticipation of particular goals and likely occurrences of important events. Because of this, plans are not rigid in the sense of a definite and precise specification of the execution steps. Plans are flexible and adaptable to circumstances. A slightly more formal and encompassing conceptual description is given in the next subsection.

Modeling flexible management

In its simplest form, an activity, which we will call a primitive activity, denotes something to be done to a particular biophysical object or location, e.g. a herd batch or building, by an executor, e.g. a worker, a robot or a set of these. Besides these three components, a primitive activity is characterized by local opening and closing conditions, defined by time windows and/or predicates (Boolean functions) referring to the biophysical states or indicators. An indicator is a contextual piece of knowledge or information invoked, assembled, or structured to substantiate a decision-making step, e.g. appraisal of remaining forage amount on a field to decide withdrawal of the herd from it. The opening and closing conditions are used to determine at any time which activities are eligible (according to the manager's intention) for execution; they play a key role in defining the timing flexibility.

The 'something-to-be-done' component of a primitive activity is an intentional transformation called an operation, e.g. the harvest operation. The step-by-step changes to the biophysical system as the operation is carried out constitute a functional attribute of the operation. These changes take place over a period of time by means of a process that increases the degree of achievement at each step of the operation until it is completed. An operation is said to be instantaneous if its degree of achievement goes from 0 to 1 in a single step, otherwise the operation is durative which implies that its execution might be interrupted. An operation may require resources such as a mower in case of cutting. In addition, the execution of an operation is constrained by feasibility conditions that relate to the biophysical state. Objects on which an operation is carried out can be individual objects, e.g. a field or a set of fields, or objects having numerical descriptors, e.g. an area. Speed is defined as a quantity e.g. number of items or area, which can be processed in a unit of time. The duration of the operation is the ratio of the total quantity to the speed. In order to have the effect realized the operation must satisfy certain enabling conditions that refer to the current state of the biophysical system, e.g. the field to be processed should not be too muddy, muddy being an indicator. The ability to reap the benefits of organizational and timing flexibility depends on execution competence determined by the involved resources (both operation resources and the executor). Careful representation of the resources and their availability might therefore be essential to get a proper understanding of the situation under study.

Activities can be further constrained by using programming constructs enabling specification of temporal ordering, iteration, aggregation and optional execution. To this end, we use a set of non-primitive or aggregated activities having evocative names such as *before*, *iterate*, and *optional*. Others are used to specify choice of one activity among several (*or*), grouping of activities in an unordered collection (*and*) and concurrence of some of them (e.g. *co-start*, *equal*, *include*, *overlap*). Formally a non-primitive activity is a particularized activity. As such it might also be given opening and closing conditions as well as other properties such as a delay between two activities involved in a *before* aggregated activity. In particular, it has a relational property that points to the set of the other activities directly involved in it (or constrained by it). In addition, it is equipped with a set of procedural attributes that convey the semantics of the change in status specific to each non-primitive activity. The opening and closing of a non-primitive activity depend on their own local opening and closing conditions (if any) and on those of the underlying activities. All the activities are connected; the only one that does not have a higher level activity is the plan. In addition to the timing flexibility attached to the opening and closing conditions of its activities a plan is made flexible by the use of composed activities that enable optional execution or choice between candidate activities. Whether an optional activity is executed and which alternative activity is chosen are context-dependent decisions.

Notwithstanding the flexibility of activities it may be necessary to adapt the plan when particular circumstances occur. Indeed a nominal plan conveys the rough course of intended steps to go through under normal circumstances. The specification of when and what changes should be made to a nominal plan is called a conditional adjustment. The trigger for a conditional adjustment is either a calendar condition that becomes true when a specific date is reached, or a state-related condition that becomes true when the current circumstances match this condition. The adjustment can be any change to the nominal plan such as the deletion or insertion of activities. It can also affect the resources used in some activities. Actually a conditional adjustment can also specify a change to be made to conditional adjustments themselves. By this means, the management can be reactive and thus cope with unexpected (though still feasible) fluctuations of the external environment (e.g. drought) and various contingencies.

Examples of activities in grassland-based livestock systems

The example (Martin et *al.*, 2009) considered here concerns the hay-making activity on a farm: cutting the herbage of a grassland plot and, once it is dry enough, storing this new-mown hay. Two primitive activities are involved: cutting and storing. For the cutting activity, the object operated by the cut is a plot, in particular the component herbage, and the executor is the farmer equipped with his tractor and mower. The speed of the cut (the *something-to-be-done* component) is a harvestable area per unit of time. Its effect is the creation of a harvested herbage, the initialization of a drying process on this harvested herbage, and the re-initialization of the herbage component of the plot with its descriptors updated (leaf area index, dry matter, growth cycle age, digestibility, etc.). For the storing activity, the object operated by the storage is the harvested herbage, and the executor is the farmer equipped with his tractor, round-baler and trailer. The speed of the storage is a storable quantity of hay per unit time. Its effect is the crediting of the amount of hay stored in the barn by the harvested quantity minus some losses to the yield associated with the whole hay-making process. Storing of harvested herbage can occur only once cutting is complete. Thus hay-making is a sequence of two primitive activities which can be written:

hay-making = *before* (cutting:
operation: cut with mover
operated object: plot
performer: farmer
storing:
operation: store with tractor, round-baler and trailer
operated object: harvested herbage created in cut
performer: farmer)

The opening of any hay-making activity, and consequently of the cutting activity, has to occur within a particular time range delimited by a minimum and a maximum beginning date. In addition, the opening predicate refers to a threshold on harvestable yield and a given phenological stage for the corresponding herbage, i.e. between stem elongation and flowering, to ensure a compromise between harvested quantity and quality. Once the opening predicate of the hay-making activity has been verified the feasibility conditions attached to the cut operation are examined. These feasibility conditions concern the bearing capacity of the grassland plot, sufficient free space in the barn to store additional hay, and a satisfactory expected air saturation deficit and rainfall in the coming days to ensure proper drying conditions in the field. No closing conditions are specified in this case to ensure completion of the hay-making activity. To summarize the hay-making aggregated activity is represented as follows:

hay-making = *before* (cutting, storing)
earliest beginning date
latest beginning date
opening predicates concerning:
- minimum yield to harvest (an amount expressed in kg / ha)
- earliest phenological stage (stem elongation expressed in degree days)
- latest phenological stage (flowering expressed in degree days)

Farmers seldom make hay on a single field at a time. Typically, they do it on a set of fields that are close together, i.e. that belong to the same islet. This practice may be risky if too many fields are cut and long period rainy weather occurs. A typical risk-limiting attitude is to make small groups of plots and harvest these groups in sequence. Bad weather during drying then harms only the plots in the last group treated. The example of practice considered in this paper, hay-making on the plots of a group can only start if the last hay-making activity executed in the previous group is complete. Moreover a delay is imposed between the hay-making on the different groups of plot (for instance to keep time for daily routine work that cannot be done on the busy days of hay-making). The grouping of activities enables management constraints to be attached to this set, such as the delay between the processing of the groups {Field1, Field2, Field3} and {Field4, Field5}. Using an *and* to make the grouping gives flexibility in the order of execution of the concerned activities, using for instance yield-based preferences. The sequence of hay-making on the two groups of fields can then be written:

before (*and* (hay-making Field 1, hay-making Field 2, hay-making Field 3),
and (hay-making Field 4, hay-making Field 5)
in-between delay = 4 days)

Due, for example, to particular weather conditions in a given year, such a plan might be unachievable. Conditional adjustments of the plan are then necessary to recover a consistent management situation. For instance, in a showery weather period, the farmer might decide to reverse the order of the groups of hay-making activities in the sequence (*before*) to take advantage of the lower drying requirements of herbage on fields 4 and 5. Another adjustment could be the changing of the delay between the processing of the two groups. The above composed activity would then be changed into:

before (*and* (hay-making Field 4, hay-making Field 5),
and (hay-making Field 1, hay-making Field 2, hay-making Field 3)
in-between delay = 3 days)

Resources

As pointed out by Dillon (1979), resource management is the essence of farm management. Basically, in the DIESE framework, a resource is an entity that supports or enables the execution of activities. Typically, the activity executors, the machinery involved and the various inputs (seeds, fertilizer, water, fuel) are resources. Resources are generally in finite supply and have significant influence on when and how activities may be executed. The availability of a resource is restricted by availability constraints that specify the conditions allowing their use or consumption. The constraints are temporal constraints (time windows of availability), capacity-related constraints (the amount available) or state-related constraints. Any resource is possibly constrained with respect to the maximum number of operations supported simultaneously and the maximum number of resources of other types that can be used simultaneously.

There are many types of resources that must be dealt with (Smith and Becker, 1997). A resource can be either consumable (usable only once) or reusable after it has been released. It can be a discrete-state resource (whose availability is expressed by a qualitative state such as *ready* or *not ready*) or a capacity resource (whose availability is characterized by a vector of numerical values expressing a multi-dimensional capacity). We distinguish between single resources and aggregate resources, which are collections of resources.

In a primitive activity, the role of resource is played by the operated object, the operation resources and the executor as well. An operated object is a discrete-state resource that is a part of the biophysical system (an entity or a set of entities of the biophysical system). It is characterized by its ability to be transformed by several operations simultaneously. It may allow several resources to be simultaneously involved in transformations, and several executors to carry out certain transformations simultaneously.

An operation resource is either a discrete state resource (e.g. tools) or a capacity resource (e.g. diesel fuel). It is characterized by its ability to be used simultaneously for several objects acted upon in the biophysical system, to be involved simultaneously in several operations, and to be used simultaneously by several executors.

An executor is a discrete-state resource characterized by its ability to work simultaneously on several objects in the biophysical system, to be involved simultaneously in several operations, to cope with several operation resources used simultaneously in the operations it is engaged in. Another feature of an executor is its work power that has an effect on the speed of the operation and on the requirement of operation resources if the latter are declared proportional to power. An executor is either an individual resource (e.g. a worker) or a labour team (a set of individual workers whose work power is by default the sum of the powers of the individual workers it comprises).

As an illustration, consider a cutting activity having the resource specifications shown in Table 1. The operated object specification refers to a set of spatial entities that are dynamically generated by expanding the entity set specification defining this set. Considering it as a resource is useful in case it is decided to disallow two simultaneous operations on any of these entities. The specification of resources coming with the operation component states that two machines are required: a mower and a tractor. The executor is a person to be selected either from the farmer's sons or his employees.

If we have instances available in each of these classes, we have to consider two alternative allocations. In this example, at the time of allocation, the allocation procedure would return two alternatives {(f1, m1, t2, s2), (f1, m1, t2, e)} if f1 is the only field satisfying the request, m1 and t2 are the mower and tractor that are available, and s2 and e are respectively the second son and the employee that have no duty at that time. It might return a set of only one collection of assignments if either no son or no employee is available. It might of course return no solution at all, meaning that it is impossible to execute the activity immediately.

The use of resources is restricted by various constraints that make resource allocation a tricky combinatorial task. In addition to availability constraints, the ontology makes it possible to specify co-usage restrictions that concern the simultaneous use of a resource in different operations and combined with other resources. These co-usage restrictions are defined as specific entities having a slot whose value is a set (conjunction) of inconsistency conditions defined as cardinality limitations. The restriction called *activity-inconsistency-conditions* stipulates the limitations on the use of some resources for any primitive activity of a given type whereas the one called *resource-sharing-violation-conditions* stipulates for a resource of a given type the restriction on its simultaneous uses with other resources involved in set of primitive activities executed concurrently. Finally, a third type of usage restriction called *activities-resources-inconsistent-commitments* is available in the ontology. It has two slots whose values are a set of *activity-inconsistency-conditions* and a set of *resource-sharing-violation-conditions*; these two conditions must be satisfied in order to meet the requirement expressed by the *activities-resources-inconsistent-commitments.*

Investigating resource allocation means looking at resource bottlenecks and inefficient utilization that have a significant impact on the system performance. Simulation must include the process of

Table 1. Resource requirements in a cutting activity.

What is specified	Specification	Instances of entities or resources[1]
Operated objects	'non-grazing fields greater than 0.5ha'	FIELD: {f1, f2, f3, ...}
Operation resources	'one mower and one tractor'	MOWER: {m1, m2} TRACTOR: {t2}
Executors	'one person from farmer's sons or his employees'	SON: {s1, s2, s3} EMPLOYEE: {e}

[1]Small capitals refer to classes, normal characters refer to existing instances of the class.

dynamic allocation that takes place repetitively as part of the action-oriented decision making. This process implements a dedicated constraint satisfaction solver. At any time, the management strategy can tell what activities are deemed appropriate and the resource allocation must determine among them the combinations that are feasible with respect to the availability and co-usage constraints. As any constraint satisfaction problem, resource allocation may become a highly combinatorial problem depending on the richness of the pool of resources and the flexibility in their usage. When several allocations of resources are possible and, more generally, when several sets of allocated activities are eligible some preferences are used to ultimately select the best one among them and engage execution.

Conclusion

Keating and McCown (2001) already suggested that challenges for farming system modelers are 'not to build more accurate or more comprehensive models, but to discover new ways of achieving relevance to real world decision making and management practice.' In this sense, the DIESE project is the result of consistent efforts to improve the representation of farm management strategies and get closer to the questions raised in practice. Using the integrative conceptual framework that we have described one can develop elaborate simulation models of agricultural production systems. It provides a common structure to help organize and frame monitoring and management activities that can be applied effectively and consistently across the production system. Running such a simulation model under various scenarios of external conditions (weather in particular) helps to give a realistic view of the system's behavior and performance, its sensitivity to external factors and the quality of the tested management strategy as regards robustness and flexibility. We can use this approach to give a clearer meaning to the selection and prioritization of management activities by placing the management process in context.

At this stage, the decision making behavior coming with the DIESE framework lacks an explicit representation of goals. This would become necessary to take into account anticipatory decision making capabilities. Incorporating such capabilities could be done by implementing a Belief-Desire-Intention (BDI) type of decision making architecture (Wooldridge, 2002) in which beliefs express the decision maker current state of knowledge about the production system, intentions are the activities structured in a plan and desires are specifications about target states of the production system.

The complexity of the farmer's management task is not due to the number of components or possible states of the system but rather to the dynamic behavior of the different components which arise from their interactions over time and their dependence on uncontrollable driving factors such as weather. The dynamic complexity relates to human difficulty in dealing consistently with feedback effects, and multiple and delayed consequences of interventions. Much of the information about biophysical system functioning and the cognitive process involved in production management resides in the mental models of farmers where it remains tacit. By using the DIESE framework, one can expect to capture part of this subjective and context-specific knowledge and, in this way, make it an object of scientific investigation. Improving our ability to make this knowledge explicit and usable for formal modeling and learning can have important effects on both research and practice. Researchers are in a better position to build more complete, accurate and insightful models and practitioners can increase their awareness and mastery of organizational and management issues.

DIESE is currently used in large modelling projects dealing with various kinds of production such as cash crop (Cialdella *et al.*, 2009), vineyard (Ripoche *et al.*, 2009), pasture-based livestock (Martin *et al.*, 2009; Chardon *et al.*, 2007) and pig systems (Rigolot *et al.*, 2009), which attest to the wide scope of applicability of the framework.

Acknowledgements

The authors acknowledge the support of the French Agence Nationale de la Recherche (ANR), under grant ANR-06-PADD-017 (SPA/DD project).

References

Chandrasekaran, B., Josephson, J. and Benjamins, V., 1999. What are ontologies and why do we need them. IEEE Intelligent Systems, 14:20-26

Chardon, X., Rigolot, C., Baratte, C., Le Gall, A., Espagnol, S., Martin-Clouaire, R., Rellier, J.-P., Raison, C., Poupa, J.-C. and Faverdin, P., 2007. MELODIE: a whole-farm model to study the dynamics of nutrients in integrated dairy and pig farms. In: Oxley, L. and Kularsiri, D. (eds.), MODSIM 2007 International Congress on Modelling and Simulation, Modelling and Simulation Society of Australia and New-Zealand. pp.1638-1645.

Cialdella, N., Rellier, J.-P., Martin-Clouaire, R., Jeuffroy, M.-H. and Meynard, J.-M., 2009. SILASOL: A model-based assessment of pea (*Pisum sativum* L.) cultivars accounting for crop management practices and farmers' resources. In: Proceedings of Farming Systems Design 2009, Monterey, CA, USA.

Dillon, J.L., 1979. An evaluation of the state of affairs in Farm Management. South African Journal of Agricultural Economics 1:7-13.

Keating, B.A. and McCown, R.L., 2001. Advances in farming systems analysis and intervention. Agricultural Systems 70:555-579.

Martin, G., Duru, M., Martin-Clouaire, R., Rellier, J.-P. and Theau, J.-P., 2009. Taking advantage of grassland and animal diversity in managing livestock systems: a simulation study. In: Proceedings of Farming Systems Design 2009, Monterey, CA, USA.

Martin, G., 2009. Analyse et conception de systèmes fourragers flexibles par modélisation systémique et simulation dynamique. Ph.D. thesis, Université de Toulouse, France.

Martin-Clouaire, R. and Rellier, J.-P., 2009. Modelling and simulating work practices in agriculture. Int. J. of Metadata, Semantics and Ontologies 4:42-53.

Rellier, J.-P., 2005. DIESE: un outil de modélisation et de simulation de systèmes d'intérêt agronomique. Internal report UBIA-INRA, Toulouse-Auzeville. http://carlit.toulouse.inra.fr/diese/docs/ri_diese.pdf.

Rigolot, C., Chardon, X., Rellier, J.-P., Martin-Clouaire, R., Dourmad, J.-Y., Le Gall, A., Espagnol, S., Baratte, C. and Faverdin, P., 2009. A generic framework for the modelling of livestock production systems: MELODIE. In: Proceedings of Farming Systems Design 2009, Monterey, CA, USA.

Ripoche, A., Rellier, J.-P., Martin-Clouaire, R., Biarnès, A., Paré, N. and Gary, C., 2009. Modeling dynamically the management of intercropped vineyards to control the grapevine water status. In: Proceedings of Farming Systems Design 2009, Monterey, CA, USA.

Smith, S.F. and Becker, M.A., 1997. An ontology for constructing scheduling systems, In: Proceedings of the AAAI Spring Symposium on Ontological Engineering, pp 120-129, Palo Alto, CA, USA.

Wooldridge, M., 2002. Introduction to multiagent systems. John Wiley & Sons.

Population based growth curve analysis: a comparison between models based on ordinary or stochastic differential equations implemented in a nonlinear mixed effect framework

A.B. Strathe[1], A. Danfær[2], B. Nielsen[3], S. Klim[4] and H. Sørensen[5]
[1]Department of Animal Health and Bioscience, Faculty of Agricultural Sciences, Aarhus University, Blichers Allé 20, 8830 Tjele, Denmark; abstrathe@ucdavis.edu
[2]Department of Basic Animal and Veterinary Sciences, Faculty of Life Sciences, University of Copenhagen, Gronnegaardsvej 3, 1870 Frederiksberg, Denmark
[3]Department of Genetic Research and Development, Danish Pig Production, Axeltorv 3, 1609 København V, Denmark
[4]Department of Informatics and Mathematical Modelling, Technical University of Denmark, 2800 Kongens Lyngby, Denmark
[5]Department of Mathematical Sciences, University of Copenhagen, Universitetsparken 5, 2100 Copenhagen Ø, Denmark

Abstract

Growth curve analysis is frequently carried out in the study of growth of farm animals because measurements are taken repeatedly over time on the same individual within groups of animals. Growth functions like the Gompertz, Richards and Lopez, which are based on ordinary differential equations (ODE), can be used to model the functional relationship between size or mass and age. Recently, nonlinear mixed effect models have become very popular in the analysis of growth data because these models quantify both the population mean and its variation in structural parameters. The within animal variability is modelled as being independent and normally distributed and thus assumed to be uncorrelated. Systematic deviations from the individual growth curve can occur, which is the consequence of animal intrinsic variability in response to environmental impacts. This leads to correlated residuals. A way to model correlated residuals is to adopt a stochastic differential equation (SDE) model for the within animal variability. The objective of the present work was to compare two methods for population based growth curve analysis, where the individual growth curve was given as a solution to an ODE or a SDE. Data used to compare the two approaches was based on 40 pigs of three genders where body weight (BW) measurements were conducted weekly until 150 kg BW and then biweekly until slaughter. It was found that the population structural parameter estimates provided by the two methods were similar although the ODE models may underestimate parameter standard errors in the presence of serial correlation. Moreover, variance components may be severely overestimated if serial correlation is not addressed properly. In conclusion, application of SDEs model is biologically meaningful as there are many complex factors that affect growth in pigs and thus deviations from the time trend are mainly driven by the state of the pig.

Keywords: pigs, state space models, animal variation

Introduction

Growth curve analysis is frequently carried out in the study of animal growth and it concerns a series of measurements in time of some quantity of interest (e.g. body weight, body composition, length, diameter, etc.). The data will not be independent between time points thus causing a challenging problem in modelling, estimation and interpretation. A large class of growth functions, which are solutions to ordinary differential equations (ODE), can be used to represent the time trend in growth processes, i.e. s-shaped and diminishing return behaviour (Thornley and France, 2007). These are based on biological considerations, meaning that their parameters have a biological interpretation and thus a biologist should be able to draw some meaningful conclusions from the analysis. When growth functions are applied to analysis of population data then two sources of variation must be considered and separated, i.e. within and between-animal variability. Modelling between-animal variation is accomplished by assuming that the growth of all individuals follows the same functional form with parameters varying according to a multivariate normal distribution. Thus, a relevant population attribute (e.g. mature size) is described by a mean level and an associated variance component. This has become a rather standard approach in animal science since the introduction of standard nonlinear mixed model software (Craig and Schinckel, 2001; Kebreab *et al.*, 2007).

The individual's deviation from the expected time course is usually modelled with independent, identically and normally distributed variables (Andersen and Pedersen, 1996; Craig and Schinckel, 2001). However, systematic deviations from the fitted curve may appear thus resulting in serially correlated residuals. Systematic deviations are also the consequence of local fluctuations in the growth rate over time, which are due to environmental and animal intrinsic influences (Strathe *et al.*, 2009). An option is to use stochastic differential equations (SDEs). The main attraction of the SDEs and the primary difference to the corresponding ODEs is the inclusion of a diffusion term, which accounts for model uncertainty. Addressing model uncertainty is biologically meaningful because there are many complex factors affecting intrinsic growth processes, which cannot be explicitly modelled because the mechanism is either not known or may be too complex for inclusion in a model for parameter estimation. With the SDE formulation it is assumed that growth is the result of random processes which are stochastically independent non-over-lapping time intervals. Hence, SDE formulation is a powerful approach to deal with serially correlated residuals that are likely to occur when analysing growth data (Sandland and McGilchrist, 1979; Garcia, 1983).

The objective of the present research is to compare two methods for population based growth curve analysis where the individual growth curves are given as solutions either to an ODE or a SDE.

Material and methods

Animals and feeding

The data used in this analysis was part of a serial slaughter experiment set up to determine growth capacity, body composition as well as energy and nutrient utilization in growing pigs (Danfær and Fernández, unpublished data). The experimental pigs used in that study were crosses of Yorkshire × Landrace sows and Duroc boars. The pigs were fed seven different diets in the corresponding intervals: 4 to 7 weeks, 7 weeks to 25 kg, 25 to 45 kg, 45 to 65 kg, 65 to 100 kg, 100 to 150 kg and 150 kg to maturity. The diets were formulated to supply 110% of the standard nutrient recommandations in Denmark. The pigs were housed individually under thermoneutral conditions and given ad libitum access to feed and water during the entire growth period in order to maximize growth rates. All pigs were weighed at birth, at weaning, weekly until approximately 150 kg BW and then every second week until the time of slaughter. The dataset consisted of 40 animals from 17 litters, which were of three genders (barrow, boar and gilt). The three genders were represented by 11, 16 and 13 pigs, respectively. There were a minimum of 25 and maximum 65 with an average of 36 data points per pig available for the analysis. Thus, a total of 1,431 BW recordings were used. More details are given by Strathe *et al.* (2009).

Candidate differential equations

Thornley and France (2007) have presented a large compilation of mathematical functions, which can be used to study growth. These authors also presented a detailed overview of the bio-mathematical properties of these functions, e.g. point of inflexion. In the present work, three growth functions are chosen and they will be referred to as Gompertz (1825), Richards (1959) and Lopez *et al.* (2000) after the principal sources where they were derived initially. These are given as:

Gompertz: $\frac{dx}{dt} = x \times k \times \log(\frac{W_f}{x})$

Richards: $\frac{dx}{dt} = x \times k \times (\frac{W_f^n - x^n}{n \times W_f^n})$

Lopez: $\frac{dx}{dt} = n \times \left(\frac{t^{n-1}}{k^n + t^n}\right) \times (W_f - x)$

where W_f, k and n are parameters. W_f is interpreted as the mature size (kg) of the pig, k is a rate constant (day^{-1}) in the Gompertz and Richards models, but the age to approximately half mature size (days) in the Lopez model. The state variable is defined by x, i.e. body weight (kg). The initial condition is $x(0) = x_0 = 1.8$ kg, which is the mean birth weight.

The individual stage model with ordinary differential equations

The observation y_{ij} is a vector of body weight measurements for the ith pig measured at the jth time point, N is the number of individuals and n_i is the number of measurements for the i'th individual, i.e.

y_{ij}, $i = 1, \ldots N$, $j = 1, \ldots n_i$

The structural growth model used to describe the within individual data consists normally of solutions to the above ODEs as these have analytical solutions (Thornley and France, 2007), but in order to introduce the state space model framework these are written in ODE form. The ODEs are supplemented by a model of the residual variation that describes the differences between the structural model and the observation, which presents the individual stage model, i.e. the available data is

$$\frac{dx_i}{dt} = f(x_i, t, \theta_i)$$

$$y_{ij} = x_i(t_{ij}) + e_{ij}$$

where for ith individual, x_i is the state of the animal (weight), t is time, t_{ij} is the discrete measurement time points, and θ_i is a vector of structural parameters specific to the individual. The residuals e_{ij} are assumed to be independent, identically and normally distributed with mean zero and variance σ_e^2. Thus, the estimation of model parameters is based on the individual stage likelihood function, which is formed as a product of probabilities for each measurement. The population likelihood function is explained later.

The individual stage model with stochastic differential equations

Models with SDEs extend the usual ODE models by including system noise as an additional source of variation in the individual stage model. This extended model describes the within individual variability in data with two sources of noise: (1) a dynamic noise term, which is part of the system (the animal) such that the value of the process at time t depends on this noise up to the time t, and (2) a measurement noise term, which does not affect the growth process itself, but only its observations. Separation of intra animal variation into two noise components requires additional explanation. The system noise can be thought of as animal intrinsic variability due to the following: (1) animal growth is always embedded in a randomly varying environment (no matter how well the experimental conditions are controlled), (2) growth processes are subjected to external and internal influences that change over time (e.g. shifting diets, sub-clinical diseases, hormonal influences, emotional stress, etc.), which may randomly affect growth. Modelling all these aspects that disturb the animal would produce a very large and complicated model that renders model identifiability and parameter estimation given data. Thus, the equations governing the individual growth can be written as the following stochastic state-space model:

$$dx_i = f(x_i, t, \theta_i)dt + \sigma_w dw_t$$

$$y_{ij} = x_i(t_{ij}) + e_{ij}$$

where x_i, t, t_{ij} and θ_i have already been defined. Note that the state equation is split into two components $f(x_i, t, \theta_i)dt$ and $\sigma_w dw_t$, which are referred to as the drift and diffusion term, respectively. e_{ij} are the independent identically distributed Gaussian measurement errors with mean zero and variance σ_e^2. w_t is a standard Wiener process, which is a continuous time Gaussian process where the increments $(w_{t_2} - w_{t_l})$ between two points in time $(t_l < t_2)$ are $N(0, |t_2 - t_1| I)$ and I is the identity matrix. The Wiener process can be interpreted as the sum of many identically distributed stochastic terms, yielding the difference between the true evolution of the animal state and the evolution described by the drift term $f(\cdot)$. Furthermore, the variance of the Wiener process increases linearly in time, which can be interpreted as a linear increase in the number of stochastic terms contributing to the dynamic noise. The estimation of model parameters is based on the individual stage likelihood function, which is formed as a product of probabilities for each measurement. Due to the assumption of correlated residuals with inclusion of the Wiener process, it is necessary to depend on the previous measurement for defining the probability density of each measurement. If it is assumed that the conditional densities for the states are Gaussian and thus fully described by the state prediction and the state prediction variance for each observation then the method of Kalman filtering may be used. The extended Kalman filter is applied in the present work because the SDE models are nonlinear. In the nonlinear case, the extended Kalman filter is an approximate minimum variance estimate of the evolution of the model states (Øksendahl, 1995). The population likelihood is explained later.

The population stage model

The population stage model describes the between-animal variation, which is presented by the following model for the individual parameters

$$\theta_i = h(\beta, z_i) + b_i$$

where $h(\cdot)$ denotes a population model, which is a function of the fixed effects β and covariates z_i. These may be a set of dummy variables for including categorical factors (e.g. gender) or continuous explanatory variables. This part of the model construction is similar to traditional regression type modelling where the fixed effects are incorporated into the model, reflecting the experimental design. The random effects (b_i) are normally distributed with zero mean and covariance matrix Ω, resulting in a multivariate normal distribution. The individual and population probability density functions define the joint density of the observed population data and random effects, which is formed as a product of the individual and population probability density functions. The population likelihood of the data (the joint likelihood with the random effects integrated out) does not have an analytical solution and various methods can be used to approximate it (Pinheiro and Bates, 1995). In the present work, the first order conditional expectation (FOCE) method is used, which is available in the R-packages nlmeODE and PSM (Tornoe, *et al.*, 2004; Klim *et al.*, 2009). The ODE and SDE models are implemented in those R-packages.

The population model for the Richards model is given here as an example of how the specific growth models were formulated, i.e.

$$W_{f,i} = W_{f(barrow)} z_1 + W_{f(boar)} z_2 + W_{f(gilt)} z_3 + b_i, \quad b_i \sim N(0, \sigma^2_{w_f})$$

$$k_i = k_{(barrow)} z_1 + k_{(boar)} z_2 + k_{(gilt)} z_3$$

$$n_i = n$$

where the three genders have specific values for the mature weight (W_f) and the rate parameter (k) coded with the dummy variables z_1, z_2 and z_3. The Richard and Lopez models include an additional parameter n, which is assumed to be constant across genders and individual pigs. Furthermore, the mature weight is also allowed to vary between animals (b_i), reflecting population variation. Convergence problems for the SDE models occurred if further random effects were introduced.

Variance stabilization with models based on differential equations

Model diagnostics (residual and autocorrelation plots) are employed and visual inspection of the residuals plotted as a function of time is used to decide if variance stabilization is required. Previous analysis of parts of the dataset has shown that the appropriate variance stabilizing function is the square root and thus a transform-both-sides (TBS) approach has been developed (Strathe *et al.*, 2009). The appealing feature about the TBS approach is that the parameter interpretations are unchanged. More specifically, the ODE is transformed and then system noise is added to the transformed ODE. Hence, system noise is assumed to have a constant intensity on the transformed scale. We have recently shown that the (TBS) approach can be generalized to accommodate any type of transformation if the transforming function is continuous, differentiable and has a corresponding inverse function (Strathe *et al.*, 2009). As an example, the approach is applied to the Richards function using the square root transformation and hence the state equation becomes

$$dq_i = \frac{q_i \times k}{2} \times \left(\frac{W_f^n - q_i^{2n}}{n \times W_f^n}\right) + \sigma_w dw_t$$

where dq is the rate of change per unit time in the transformed state variable q ($q = \sqrt{x}$), with derivative q' ($q' = 1/2\sqrt{x}$). All other terms are as defined previously.

Results and discussion

The growth models are represented as state space models although the analysis would be somewhat simpler for ODE models if the analytical solutions are used. However, analytical solutions are not available for the SDE models because the transition probabilities are unknown and thus these are approximated by the extended Kalman filter, giving rise to the state space representation. Hence, the state space representation is chosen due to transparency and compatibility in the present work. When parameter estimation in nonlinear models is conducted, then unique convergence must be checked thoroughly as final estimates should not depend on initial values. This can be done by selecting a good set of initial parameter values and these may be obtained by simple nonlinear least squares method. Second, these initial values are changed multiple times for monitoring the effect of different starting points on final parameter estimates. This hopefully ensures that the nonlinear mixed effect routines have searched the entire parameter space and thus reaches a unique solution for the specific model.

The results of fitting the different models to data are presented in Table 1. Differences between the two methods in terms of parameter estimates for the fixed effects are generally minor. The parameter uncertainty is expressed by the standard errors (SE) of the estimates, which are obtained as the diagonals of the inverse of the empirical information matrix for both methods. A general trend is that the SEs are higher for the SDE method with the largest differences for parameters k and n. The differences between the two methods in terms of SEs for W_f are only minor. Proper estimation of SE is important for calculating confidence limits with suitable coverage and for purpose of hypothesis testing (Jones, 1993).

Model diagnostics is a central part of longitudinal data analysis, and plotting the (scaled) residuals against time reveals if serial correlation is present. In Figure 1 the scaled residuals are plotted against age for Richards ODE and SDE models for nine animals of three genders. The scaled residuals for the ODE models are obtained by dividing the raw residuals (observation – fitted) by residual standard deviation (Pinheiro and Bates, 2000), which scales the residuals in such way that they mainly range between -3 and 3 because they will have the variance of 1. For the SDE models scaled residuals can also be constructed, but it is done in a slightly different way as the one-step prediction error is weighted by the corresponding one-step prediction variance, which can be obtained from the extended Kalman filter (Klim *et al.*, 2009). From Figure 1 it is clear that the Richards ODE

Table 1. Parameter estimates for the three differential equations (Gompertz, Richard and Lopez) expressed as ordinary (ODE) or stochastic differential equations (SDE) and implemented in a nonlinear mixed effect framework. Standard errors are given in parentheses.

Parameter[1]	Gompertz		Richard		Lopez	
	ODE	SDE	ODE	SDE	ODE	SDE
$W_{f(barrow)}$	381	371	415	410	467	471
	(14)	(15)	(17)	(18)	(16)	(23)
$W_{f(boar)}$	456	439	502	472	557	550
	(12)	(15)	(15)	(19)	(13)	(23)
$W_{f(gilt)}$	341	342	356	354	385	378
	(12)	(13)	(15)	(14)	(14)	(17)
$k_{(barrow)}$	9.17e-3	9.62e-3	6.54e-3	7.06e-3	265	267
	(9.8e-5)	(2.6e-4)	(9.5e-5)	(2.7e-4)	(2.8)	(9.2)
$k_{(boar)}$	8.15e-3	8.55e-3	5.97e-3	6.66e-3	292	292
	(7.1e-5)	(1.9e-4)	(7.9 e-5)	(2.3e-4)	(2.6)	(8.7)
$k_{(gilt)}$	9.74e-3	9.45e-3	7.09e-3	7.58e-3	234	231
	(8.7e-5)	(2.34e-4)	(8.6e-5)	(2.6e-4)	(1.9)	(6.9)
n	-	-	-2.58e-1	-2.16e-1	1.98	1.99
			(6.2e-3)	(1.5-e2)	(9.5e-3)	(2.1e-2)
σ_w	-	5.55e-2	-	5.20e-2	-	5.45e-2
σ_e	0.51	3.73e-3	0.40	3.35e-3	0.35	1.82e-3
σ_{Wf}	42.9	30.3	52.3	33.2	48.0	36.0

[1] *Wf* is interpreted as the mature size (kg) of the pig, k is a rate constant (day^{-1}) in the Gompertz and Richards models, but the age to approximately half mature size (days) in the Lopez model.

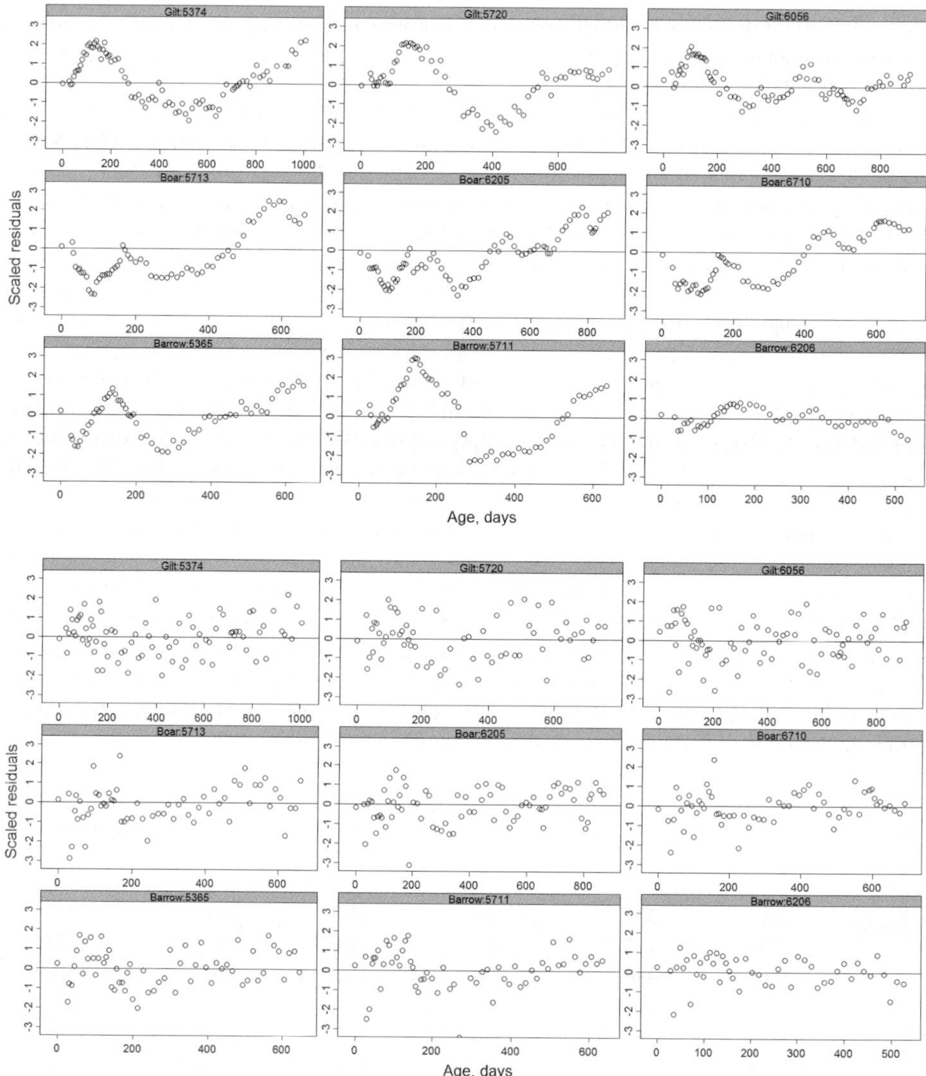

Figure 1. Scaled residuals (i.e. scaled to unity variance) plotted as function of time expressed in days. The top matrix plot displays the performance of the Richards ODE model for nine pigs of three genders. The bottom matrix plot shows the performance of the Richards SDE model for the corresponding animals.

model produces patterned residuals because the residuals are oscillating systematically over time as indicated by long runs of residuals with either positive or negative values. Thus, as stated in the introduction serial correlated residuals are the end result. On the other hand, the SDE formulation of Richards SDE model produces scattered residuals. Moreover, the variance is constant across all ages, which is accommodated by means of the square root transformation.

Table 1 shows that the estimates of the population variation expressed by the variance component σ^2_{Wf} are reduced when SDE models are applied. This was expected because a larger part of the total variation is modelled by the within individual variation, which is consistent with the simulation results reported by Karlson *et al.* (1995). Furthermore, these authors reported severe overestimation of the variance components if serial correlation was not addressed in the statistical models. Hence, if the variance components are of primary interests (e.g. animal breeding) then within individual variation should be considered carefully when analysing longitudinal data. On the other hand, if the structural parameter estimates are of main interest then serial correlated errors are less of an issue, but their SE may be underestimated.

There are important biological and conceptual differences between growth curves based on solutions to ODEs versus SDEs and these differences should be acknowledged. In the SDE growth model the individual growth trajectory is not a smooth curve. In fact, the ODE model assumes that: (1) the mathematical processes generating the observed growth trajectories are smooth (continuous and continuously differentiable) within the considered time frame; (2) the variability of the actual measurements is due to observation error. The present approach results from the hypothesis that the underlying mathematical process is not smooth, but is subjected to random intrinsic perturbations. This system noise represents the complex effect of many factors, each with a small individual effect, which is not explicitly represented in the deterministic part of the model (the drift term of the model). It further implies that an individual animal does not have a fixed growth curve from birth as growth is subjected to disturbances during life time. Assessment of these conceptual differences can be visualized by plotting the individual growth curves as function of time and corresponding expected values for the Richards ODE and SDE, respectively. This is depicted in Figure 2. By conditioning on the previous measurements the SDE model dynamically adapts to changes in growth during the lifetime of the pig – for further discussion see Strathe *et al.* (2009).

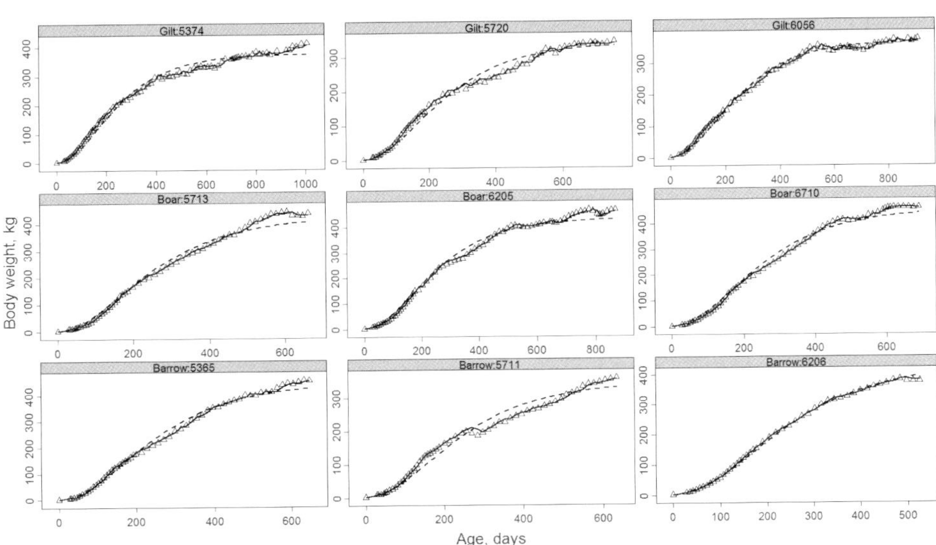

Figure 2. The individual growth curves as function of time and corresponding expected values (Δ) for the Richards ODE (---) and SDE (—), respectively.

Conclusion

Recent development in computational methods and statistical theory allows formulation of stochastic differential equations within the framework of a nonlinear mixed effect model for population based growth curve analysis. Application of this method to pig growth modelling is a powerful method to deal with serially correlated errors arising during analysis of longitudinal growth data. It is shown that growth models based on stochastic differential equations are better alternatives to the traditional ordinary differential equation models as they account for random fluctuations in the modelled growth trajectory. The population structural parameter estimates provided by the two methods are similarly, but the ordinary differential equation models may underestimate parameter standard errors when serial correlation is present. Moreover, variance components may be severely overestimated if serial correlation is not addressed properly. This may have important implications if these are of particular interest.

References

Andersen, S. and Pedersen, B., 1996. Growth and feed intake curves for group-housed gilts and castrated male pigs. Animal Science 63:457-464.

Craig, B.A. and Schinckel, A.P., 2001. Nonlinear mixed effects model for swine growth. Professional Animal Scientist 17:256-260.

Garcia, O., 1983. A stochastic differential equation model for height growth of forest stands. Biometrics 39:1059-1072.

Gompertz, B., 1825. On nature of the function expressive of the law of human mortality, and on a new mode of determining the value of life contingencies. Philosophical Transactions of the Royal Society 115:513-585.

Jones R.H., 1993. Longitudinal data with serial correlation: A state space approach. Chapman and Hall/CRC, 240 pp.

Karlsson, M.O., Beal, S.L., and Sheiner, L.B., 1995. Three new error models for population PK/PD analysis, Journal of Pharmacokinetics and Biopharmaceutics 23:651-672.

Kebreab, E., Schulin-Zeuthen, M., Lopez, S., Dias, R.S., De Lange, C.F.M. and France, J., 2007. Comparative evaluation of mathematical functions to describe growth and efficiency of phosphorus utilization in growing pigs. Journal of Animal Science 85:2498-2507.

Klim, S., Mortensen, S.B., Kristensen, N.R., Overgaard, R.V. and Madsen, H., 2009. Population stochastic modelling (PSM) – An R package for mixed-effects models based on stochastic differential equations. Computer Methods and Programs in Biomedicine 94:279-289.

Lopez, S., France, J., Gerrits, W.J.J., Dhanoa, M.S., Humphries, D.J., and Dijkstra J., 2000. A generalized Michaelis-Menton equation for the analysis of growth. Journal of Animal Science 78:1816-1828.

Øksendal., B. 1995. Stochastic Differential Equations: An Introduction with Applications. 6th edition. Springer.

Pinheiro, J.C. and Bates, D.M., 1995. Approximations to the Log-likelihood Function in the nonlinear Mixed-effects Model, Journal of Computional and Graphical Statistics 4:12-35.

Pinheiro, J.C. and Bates, D.M., 2000. Mixed effects models in S and S-PLUS. Statistics and Computing, Springer, New York, USA, 528 pp.

Richards, F.J., 1959. A flexible growth function for empirical use. Journal of Experimental Botany 10:290-300.

Sandland, R. and McGilchrist, C., 1979. Stochastic growth curve analysis. Biometrics 35:255-271.

Strathe, A.B., Danfær, A. and Sørensen, H., 2009. A new mathematical model for combining growth and energy intake in animals: The case of the growing pig. Journal of Theoretical Biology 261:165-175.

Thornley, J.H.M. and France. J., 2007. Mathematical Models in Agriculture. 2nd ed. CAB Int., Wallingford, UK.

Tornøe, C.W., Agersø, H., Jonsson, E.N., Madsen, H. and Nielsen, H.A., 2004. Non-linear mixed-effects pharmacokinetic/pharmacodynamic modelling in NLME using differential equations, Computer Methods and Programs in Biomedicine 76:31-40.

Regression procedures for relationships between random variables

M.S. Dhanoa[1], R. Sanderson[2], S. Lopez[3], J. Dijkstra[4], E. Kebreab[5] and J. France[6]
[1]North Wyke Research, Okehampton, Devon, EX20 2SB, United Kingdom; dan.dhanoa@bbsrc.ac.uk
[2]Institute of Biological, Environmental and Rural Sciences, Aberystwyth University, Gogerddan, Aberystwyth, Ceredigion, SY23 3EB, United Kingdom
[3]Instituto de Ganadería de Montaña (IGM-ULE), Departamento de Producción Animal, Universidad de León, 24007 León, Spain
[4]Animal Nutrition Group, Wageningen University, Marijkeweg 40, 6709 PG Wageningen, the Netherlands
[5]Animal Science Department, 1 Shields Avenue, University of California, Davis, CA 95616, USA
[6]Centre for Nutrition Modelling, Department of Animal and Poultry Science, University of Guelph, Guelph ON, N1G 2W1, Canada

Abstract

Finding association and relationship among measured random variables is a common task in biological research and regression analysis plays a major role for this purpose. Which type of regression model to use depends on the nature of the predictor variable and on the purpose of the analysis. The most commonly used model ordinary least squares (Type I regression) applies when measurement errors affect only the response variable. The predictor is either without measurement errors or under the control of the investigator. However, if the predictor variable does have measurement errors then an 'errors in both variables' or Type II regression model is more appropriate and takes into account variation in both the response and the predictor variables simultaneously. When repeatability error variances in both variables are known then the maximum likelihood solution (or Deming regression) is appropriate. If the measurement error variance is unknown then a suitable Type II model should be selected. In this respect, Bartlett's three-group method, major axis regression, standard major axis regression (also called reduced major axis), ranged major axis and instrument variable methods are discussed in relation to energy balance studies with cattle.

Keywords: predictor with errors, Type II models, model choice, energy balance

Introduction

Biological variables are generally subject to some degree of measurement error which will affect regression relationships. Attenuation of the estimated regression slope (by a factor of $1/r_{yx}$, where, r_{yx} is the sample correlation), when using ordinary least squares (OLS) regression, a Type I model, is a well known effect. This in turn will lead to bias in derived quantities which are functions of the estimates of the regression intercept and slope, e.g. maintenance energy requirement and efficiency of energy utilisation in animal energy balance studies. There are some situations, however, where OLS is still acceptable. For example, the Berkson case (Berkson, 1950) where the predictor is either fixed or under control of the investigator and the purpose is to predict rather than to describe its association with the *y*-variable. Also, the bias is minimal when correlation is high, e.g. r_{yx}>0.975 (Martin, 2000).

In this communication we aim to illustrate the extent of the OLS slope attenuation problem and suggest some alternative regression procedures generally known as Type II regression models, to minimise the size of the slope-bias.

Material and methods

In OLS a simple linear model $y_i = \hat{\alpha} + \hat{\beta}x_i + \varepsilon_i$ is fitted to a set of bivariate sample values (x_i, y_i). Parameters $\hat{\beta}$ (slope or Δy per unit of Δx) and $\hat{\alpha}$ (intercept on y-axis) are estimated by minimising $\sum\varepsilon_i^2$ and estimates are given by:

$$\hat{\beta}_{OLS} = \sum(y_i - \bar{y})(x_i - \bar{x}) / \sum(x_i - \bar{x})^2 \tag{1.1}$$

$$\hat{\alpha}_{OLS} = \bar{y} - \hat{\beta}_{OLS}\bar{x} \tag{1.2}$$

Some derived quantities such as the value of x for which response $y = 0$, i.e. $-\hat{\alpha}/\hat{\beta}$, have biological significance, e.g. in animal energy balance it is an estimate of the energy requirement for maintenance. In OLS we minimise differences between the fitted model and the observations in a direction that is parallel to the y-axis i.e. assuming measurement errors in only the y-variable. With errors in both the y- and x-variables, adopting the OLS model results in attenuation of the slope estimate with consequences for any derived quantities. In this situation we need to minimise differences between the fitted model and observations in both directions simultaneously using a Type II regression model.

Following McArdle (1988), the observed y value equals $Y + \varepsilon$ and the observed x value equals $X + \delta$ where ε and δ are independent normal measurement errors. If X and Y are random variables (i.e. x and y are measured on a random sample) then they are known to have a structural relationship. Knowledge of σ_ε^2 and σ_δ^2 or the ratio $\lambda = \sigma_\varepsilon^2/\sigma_\delta^2$ is necessary to estimate parameters in these relationships.

Many alternative Type II regression models exist and the choice of the model will depend on whether the ratio of the measurement error variances is known. If the ratio of measurement error variances is known then Maximum Likelihood (ML) solution (or Deming regression) is appropriate. Otherwise alternatives include Bartlett's three-group method, major axis (or orthogonal) regression, standard major axis (or geometric mean functional relationship, least triangles/rectangles or reduced major axis) regression, ranged major axis regression and the instrument variable (or covariate ratio) method.

Based on the ratio of measurement error variances of x- and y-variables, exact ML estimates (Kendal and Stuart, 1966; Madansky, 1959) can be derived as

$$\hat{\beta}_{ML} = \frac{s_y^2 - \lambda_{ML}s_x^2 + \sqrt{(s_y^2 - \lambda_{ML}s_x^2)^2 + 4\lambda_{ML}s_{xy}^2}}{2s_{xy}} \tag{2}$$

where $\lambda_{ML} = \hat{\sigma}_\varepsilon^2/\hat{\sigma}_\delta^2$ and $\hat{\sigma}_\varepsilon^2$ is the estimator of the precision or error variance of a single y-value whilst $\hat{\sigma}_\delta^2$ is the estimator of the error variance of a single x-value (both assumed constant over the range of the data), s_x^2 and s_y^2 being the sample variances of x and y sample values respectively and s_{xy} the sample covariance.

The above ML solution can be written in an alternative form known as the Deming regression model (Deming, 1943) which is used mainly in the context of method-comparison. For Deming regression (like the ML estimator), the ratio of measurement error variances in both x and y variables is required and is assumed constant over the range of data. To calculate the Deming estimate of slope (Cornbleet and Gochman, 1979; Deming, 1943) we have

$$\hat{\beta}_{Deming:y.x} = U + \sqrt{U^2 + (1/\lambda_{Deming})} \tag{3.1}$$

where

$$U = \frac{s_y^2 - \left(1/\lambda_{Deming}\right) s_x^2}{2r_{yx}s_x s_y} \text{ and } s_{xy} = r_{yx}\, s_x s_y$$

Note that λ_{Deming} is the reciprocal of λ_{ML} *i.e.* $\lambda_{Deming} = \hat{\sigma}_\delta^2/\hat{\sigma}_\varepsilon^2$ instead of $\lambda_{ML} = \hat{\sigma}_\varepsilon^2/\hat{\sigma}_\delta^2$.

Thus, by switching the λ ratio and substituting $s_{xy} = r_{yx}s_x s_y$, the Deming formula is the same as that of the ML estimator. Mandel (1964) gives an approximate relationship between Deming slope and the OLS slope as

$$\hat{\beta}_{Deming:\,y.x} = \frac{\hat{\beta}_{OLS:y.x}}{1-(\hat{\sigma}_\delta^2 / s_x^2)} \tag{3.2}$$

If σ_δ^2 can be estimated then corrected least squares (CLS) slope (Snedecor and Cochran, 1980) can be calculated as

$$\hat{\beta}_{CLS} = \hat{\beta}_{OLS}\left(\hat{\sigma}_x^2/\left(\hat{\sigma}_x^2 - \hat{\sigma}_\delta^2\right)\right) \tag{4}$$

where $\hat{\sigma}_x^2$ is the estimator of the sample variance for x.

Use of variance estimates in these formulae can compromise robustness unless care is taken in the estimation of variances (i.e. robust estimates are used). In common with the sample mean, variances also have zero breakpoint.

Martin (2000) gives formulae for iteratively weighted general Deming regression when the measurement error variance is *not* constant over the range of the data and Linnet (1990) addresses the situation where errors are proportional.

ML or the Deming solution give a unique regression line whether $y = f(x)$ or $x = f(y)$ i.e. $y = \alpha_y + \beta_{yx}\, x$ and $x = \alpha_x + (1/\beta_{yx})\, y$, and this is normally a property of a Type II model. However, in situations where the ratio of the measurement error variances, λ, is not known, alternative Type II regression models have to be used. In this respect the Type II regression models used most commonly are as follows:

Bartlett's three-group method

Improving on Wald's two-group method (Wald, 1940), Bartlett (1949) proposed a three-group method for a Type II regression model. In this method matched data for the x- and y-variables are sorted in ascending order according to the rank order of the x-variable. Then the bivariate data are divided into three groups corresponding to smaller, medium and larger ranks. The lower and upper groups are of equal size and the intermediate group comprises the remaining values and can be smaller or larger depending on overall sample size (Davies, 1967). From the estimates of the means of groups 1 and 3 Bartlett's Type II regression slope is calculated as

$$\hat{\beta}_{Bartlett} = \frac{\bar{Y}_3 - \bar{Y}_1}{\bar{X}_3 - \bar{X}_1} \tag{5.1}$$

Here the subscripts indicate the group from which the x and y means are calculated. Any outliers should be eliminated prior to the application of this method because a mean value has a breakpoint of zero in respect of its robustness. One way of ensuring robust estimation is to use medians instead of means since the sample median has the theoretical maximum breakpoint of 50 per cent (Siegel, 1982). In terms of medians Equation 5.1 becomes

$$\hat{\beta}_{Bartlett} = \frac{\mathrm{M}y_3 - \mathrm{M}y_1}{\mathrm{M}x_3 - \mathrm{M}x_1} \quad (5.2)$$

where My and Mx are medians for y and x in the relevant groups (1 or 3), respectively.

The confidence interval (CI) for the Bartlett slope, $\hat{\beta}_{Bartlett}$ (Equation 5.1), is provided by the roots, B_1 and B_2, of the quadratic equation

$$\frac{1}{2}N(\bar{x}_3 - \bar{x}_1)^2(\hat{\beta}_{Bartlett} - B)^2 = t^2(C_{yy} - 2BC_{xy} + B^2C_{xx})/(N-3) \quad (6)$$

where C_{yy}, C_{xx}, C_{xy} are the sums of squares and products *within* the three groups, N is the overall sample size and t is the t-value for a chosen level of confidence interval corresponding to $(N-3)$ degrees of freedom. With this method the property that slope of regression $y = f(x)$ is the reciprocal of slope of regression $x = f(y)$ rarely holds because ranking by x-variable and y-variable invariably produces different groupings of data.

Major (principal) axis or orthogonal regression (MA)

In this method perpendicular distances (which account for data point deviations in both y and x directions) are minimised. Measurement variances of x and y are assumed equal and the measurement scale should be the same. Substituting $\lambda = 1$ in the ML estimator (Equation 2 or 3.1) we get the MA slope estimate

$$\hat{\beta}_{MA} = \frac{s_y^2 - s_x^2 + \sqrt{(s_y^2 - s_x^2)^2 + 4s_{xy}^2}}{2s_{xy}} \quad (7.1)$$

Alternatively,

$$\hat{\beta}_{MA} = \frac{d \pm \sqrt{d^2+4}}{2} \text{ where } d = \frac{(\hat{\beta}_{OLS})^2 - r_{yx}^2}{r_{yx}^2\hat{\beta}_{OLS}} \quad (7.2)$$

The positive root is taken when the correlation is positive and the negative one when negative (Legendre and Legendre, 1998). Alternatively the major principal axis regression (Sokal and Rohlf, 1995) slope can be calculated as

$$\hat{\beta}_{MA} = s_{xy}/\left(\eta - s_y^2\right) \quad (7.3)$$

where

$$\eta = 0.5[s_y^2 + s_x^2 + \sqrt{\left(s_y^2 + s_x^2\right)^2 - 4\left(s_y^2 s_x^2 - s_{xy}^2\right)}]$$

A confidence interval for $\hat{\beta}_{MA}$ may be obtained using either bootstrap or jackknife.

In OLS, slope and CI estimates vary proportionally to the measurement units but this is not the case with MA slope (Legendre and Legendre, 1998). Therefore, it is preferable to make the variables dimensionally homogeneous e.g. transform or standardise the variables.

Reduced major axis regression (RMA)

RMA is also called standard major axis regression or the geometric mean functional relationship and is often referred to as 'y on x' and 'x on y' regression. Unlike MA, with RMA when x and y variables are dimensionally non-homogeneous, results vary with the scale of the variables. If the dimensions are arbitrary, so is the slope estimate. RMA is only used when r_{yx} is significant. RMA, like MA, has the property that the slope of $y = f(x)$ is the reciprocal of the slope of $x = f(y)$. This is important because contrary to OLS, there is no functional distinction between x and y in a type II regression model.

The RMA slope is equivalent to the ratio of the standard deviations in y and x with sign corresponding to that of their correlation or s_{xy} i.e.

$$\hat{\beta}_{RMA} = \pm s_y / s_x \tag{8.1}$$

with sign $\hat{\beta}_{RMA}$ equivalent to that of $\Sigma(y_i - \bar{y})(x_i - \bar{x})$.

Since slopes of MA and RMA are related as follows this is a convenient alternative way to calculate $\hat{\beta}_{MA}$.

$$\hat{\beta}_{MA} = \delta + \sqrt{(\delta^2 + 1)} \text{ where } \delta = 0.5\left(\hat{\beta}_{RMA} - 1/\hat{\beta}_{RMA}\right)/r_{yx} \tag{8.2}$$

The RMA slope is also given by $\hat{\beta} = \sqrt{\hat{\beta}_{yx}/\hat{\beta}_{xy}}$ where $\hat{\beta}_{yx}$ and $\hat{\beta}_{xy}$ are the slope estimates from OLS regression of y on x and x on y, respectively, with intercept $\hat{\alpha} = \bar{y} - \hat{\beta}\bar{x}$. An approximate CI for $\hat{\beta}_{RMA}$ can be calculated (Jolicoeur and Mosimann, 1968) as

$$\hat{\beta}_{RMA}\left(\sqrt{(B+1)} \pm \sqrt{B}\right) \tag{9}$$

where $B = t^2\left(1 - r_{yx}^2\right)/\left(N - 2\right)$ in which t is the two-tailed Student's t with $(N - 2)$ degrees of freedom.

Testing one RMA slope against an assumed or H_0 value (Clarke, 1980) we have

$$\text{var}\left(s_y/s_x\right) = \frac{\sigma_y^2}{\sigma_x^2}\left(1 - \rho_{yx}^2\right)/N \tag{10}$$

Substituting r_{yx} for ρ (the population correlation coefficient) gives the test statistic T_1 with v degrees of freedom and has approximate t-distribution

$$T_1 = \frac{\left|\log\hat{\beta}_{RMA} - \log\hat{\beta}_{H_0}\right|}{\sqrt{\left[\left(1 - r_{yx}^2\right)/n\right]}} \text{ with } n = N - 1 \text{ and } v = 2 + n\Big/\left(1 + \frac{1}{2}r_{yx}^2\right) \tag{11}$$

However, the more common situation is to compare two slopes calculated from different sets of data in which case a similar test can be used, *viz.*

$$T_{12} = \frac{\left|\log\hat{\beta}_{RMA1} - \log\hat{\beta}_{RMA2}\right|}{\sqrt{\left[\left(1 - r_{yx1}^2\right)/n_1 + \left(1 - r_{yx2}^2\right)/n_2\right]}} \text{ with } v = 2 + 2/\left[\text{var}\left(T_{12}\right) - 1\right] \tag{12}$$

where subscripts 1 and 2 indicate the data sets.

Imbrie (1956) gives a standard normal test as

$$T_{12} = \frac{\left|\log\hat{\beta}_{RMA1} - \log\hat{\beta}_{RMA2}\right|}{\sqrt{\left[\hat{\beta}_{RMA1}^2\left(1 - r_{yx1}^2\right)/n_1 + \hat{\beta}_{RMA2}^2\left(1 - r_{yx2}^2\right)/n_2\right]}} \tag{13}$$

Ranged standard major axis regression (RSMA)

Instead of normal standardisation, as employed above in RMA regression, ranged standardisation is preferred in some fields of application such as ecology (Legendre and Legendre, 1998). Unlike MA, ranged standard major axis is suitable for non-homogeneous scales. In this method the y and x variables are standardised by subtracting their respective minimum value and dividing by their respective ranges, i.e.

$$\frac{y_i - y_{\min}}{y_{\max} - y_{\min}} \text{ and } \frac{x_i - x_{\min}}{x_{\max} - x_{\min}} \quad (14)$$

Then the MA slope (Equation 7.1) and its CI are calculated and both are transformed back to the original scale by multiplying by the ratio of y and x variable ranges:

$$\frac{y_{\max} - y_{\min}}{x_{\max} - x_{\min}} \quad (15)$$

After this back transformation the intercept is calculated as usual: $\hat{\alpha} = \bar{y} - \hat{\beta}\bar{x}$. Just like the mean and variance, minimum and maximum values of a sample are also non-robust in the presence of outliers and extreme values since their breakpoint is also zero.

Instrument variable regression or covariate ratio method (IV)

To estimate the causal effect of some variable x on another variable y, an 'instrument' is a third variable z which affects y only through z's effect on x. If such an instrument variable (IV) is available then Type II model slope can be estimated uniquely either as the ratio of IV covariance with y and x (Kuhry and Marcus, 1977)

$$\hat{\beta}_{IV} = \frac{\sigma_{yz}}{\sigma_{xy}} \quad (16)$$

or by two stage least squares, *viz.* regressing y on the fitted values from regression of x on z. Durbin (1954) suggests the use of 'rank order' and/or 2nd/3rd power of the x-variable as suitable instruments. This method is generally quite popular in econometrics and socio-economic sciences mainly because of the availability of suitable instrument variables. It does not appear to work well with biological variables unless the correlation is high (e.g. $r_{y\hat{x}}$>0.75).

Robustness

The above Type II regression models are no more robust than the OLS or Type I models. Use of group means in the Bartlett method leads to non-robustness in the presence of outliers or extreme values. Use of non-robust estimates of sample variances and covariances in ML solution, Deming regression, MA and RMA is the source of non-robustness. Similarly, the use of minimum, maximum and range in the RSMA causes robustness problems. These problems can be minimised using robust estimates of means (e.g. Winsorised means) and variances. Variances can be calculated from the median absolute deviations (MAD) estimates (Analytical Methods Committee, 2001). Also the link between Type II and Type I models can be used to incorporate OLS robust fitting procedures such as the Huber (1981) method and repeated median method (Siegel, 1982).

Results and discussion

For illustration purposes two data sets were used to derive regression relationships using the Type I and the Type II models discussed here. Data set 1 comprised metabolisable energy intake (MEI;

MJ/kg $LW^{0.75}$/d) as the x-variable and milk energy (MilkE; MJ/kg $LW^{0.75}$/d) as the y-variable for dairy cows (mean live weight (LW) = 579 kg) and was extracted from the large data set used in Kebreab *et al.* (2003) by compressing into forty contiguous groups. MilkE data was adjusted for tissue mobilisation and MEI data was adjusted for growth or realimentation. ME_m values were estimated when MilkE = 0. Data set 2 comprised MEI (MJ/kg $LW^{0.75}$/d) as the x-variable and whole body energy retention (ER; MJ/kg $LW^{0.75}$/d) as the y-variable estimated by respiration calorimetry for growing steers (mean LW = 142 kg) (Sanderson, 1992).

The effects of model choice on estimates of efficiency and ME_m are demonstrated clearly in Table 1. Any bias in the estimate of slope feeds into the estimates of the intercept and ME_m since

Table 1. Estimates of slope ($\hat{\beta}$, interpreted as efficiency of metabolisable energy utilisation, k_l or k_g), intercept ($\hat{\alpha}$) and ME_m (MJ/kg $LW^{0.75}$/d) obtained using OLS and various Type II models applied to data sets relating to dairy cows (r_{yx} = 0.6531) and growing steers (r_{yx} = 0.7933).

	Method	λ_{ML}	$\hat{\beta}$	$\hat{\alpha}$	ME_m
Cows	OLS	na	0.4352	-0.1124	0.2583
	ML	0.25[a]	0.4633	-0.1612	0.3480
	ML	0.50[a]	0.4917	-0.2107	0.4285
	ML	0.75[a]	0.5201	-0.2600	0.4999
	ML (= MA)	1.00[a]	0.5478	-0.3082	0.5625
	ML	1.2493[b]	0.5744	-0.3544	0.6170
	ML	1.50[a]	0.5998	-0.3985	0.6644
	ML	1.75[a]	0.6235	-0.4398	0.7053
	ML	2.00[a]	0.6456	-0.4782	0.7407
	Bartlett	unknown	0.4120	-0.0722	0.1752
	RMA	s_y^2/s_x^2	0.6663	-0.5141	0.7716
	RSMA	na	0.6984	-0.5699	0.8160
	IV ($z = x^2$)	na	0.4314	-0.1059	0.2455
	IV (z = rank(x))	na	0.4183	-0.0831	0.1986
Steers	OLS	na	0.3960	-0.1499	0.3784
	ML	0.25	0.4050	-0.1565	0.3863
	ML	0.50	0.4137	-0.1628	0.3936
	ML	0.75	0.4220	-0.1690	0.4004
	ML (= MA)	1.00	0.4300	-0.1748	0.4066
	ML	1.25	0.4376	-0.1804	0.4123
	ML	1.50	0.4448	-0.1857	0.4175
	ML	1.75	0.4516	-0.1907	0.4223
	ML	2.00	0.4581	-0.1955	0.4268
	Bartlett	unknown	0.3731	-0.1330	0.3565
	RMA	s_y^2/s_x^2	0.4992	-0.2257	0.4522
	RSMA	na	0.4744	-0.2075	0.4374
	IV ($z = x^2$)	na	0.3925	-0.1473	0.3752
	IV (z = rank(x))	na	0.3904	-0.1457	0.3733

[a]Assumed $\sigma_\varepsilon^2/\sigma_\delta^2$ value.
[b]Actual value from data; *na* = not applicable.
OLS: ordinary least squares, ML: maximum likelihood, RMA: reduced major axis, RSMA: ranged standard major axis, IV: instrument variable.

$\hat{\alpha}=\bar{y}-\hat{\beta}\bar{x}$. ML slope increases progressively over a series of values for λ_{ML} ranging from there being proportionally more measurement errors in the x-variable to there being more in the y-variable. With the exception of estimates from the Bartlett three-group method and the instrument variable method, all estimates of slope and ME_m are higher than those obtained by OLS. Note that the RMA slope equals the OLS slope divided by the sample correlation r_{yx}, *viz.* 0.6663 = 0.4352/0.6531 and 0.4992 = 0.3960/0.7933. In situations where λ_{ML} is either unknown (Bartlett method) or assumed to be equal (MA) and proportional to data sample variances (RMA), the estimates are quite variable which serves to emphasise the potential implications of model choice. In these examples neither the Bartlett nor the instrument variable method performs well. With the dairy cow data there are substantial differences between the MA, RMA and RSMA estimates and the ML estimates based on the observed λ_{ML}. Therefore, if at all possible, it is recommended that suitable estimates of measurement error variances are obtained.

Conclusion

Much care is needed in the selection of an appropriate regression model (Type I or Type II) when dealing with relationships between random variables. When the functional relationship between two random variables is to be described then a Type II model is appropriate. The choice of Type II model depends on the following three conditions:

1. If errors in the x-variable are negligible or relatively small then ordinary least squares regression (the Type I model) of 'y on x' is appropriate.
2. If errors in the y-variable are always small compared to that of the x-variable then OLS regression of 'x on y' is appropriate.
3. But, if errors in both the y- and the x-variables are not negligible then a Type II model is required. The choice of a Type II model is dependent on the knowledge of the relative magnitude of the measurement error variances from which the ratio $\lambda_{ML} = \sigma_\varepsilon^2/\sigma_\delta^2$ is estimated.

However, OLS can still be used if the purpose of regression analysis is simply to predict the values of the response variable. Similarly, in the Berkson case OLS is justified when the predictor (x-variable) is either fixed or under control of the investigator (Berkson, 1950). Other occasions where OLS application is appropriate is in disciplines where decisions are made on the basis of measured quantities (e.g. econometrics, finance and socio-economics) despite any measurement errors (Johnston, 1972).

References

Analytical Methods Committee, 2001. Robust statistics: a method of coping with outliers. AMC Technical Brief No. 6. Royal Society of Chemistry, London, UK.

Bartlett, M.S., 1949. Fitting a straight line when both variables are subject to error. Biometrics 5:207-212.

Berkson, J., 1950. Are there two regressions? Journal of the American Statistical Association 45:164-180.

Clarke, M.R.B., 1980. The reduced major axis of a bivariate sample. Biometrika 67:441-446.

Cornbleet, P.J. and Gochman, N., 1979. Incorrect least-squares regression coefficients in method-comparison analysis. Clinical Chemistry 25:432-438.

Deming, W.E., 1943. Statistical adjustment of data. John Wiley & Sons, New York, USA, 184 pp.

Davies, O.L. (ed.), 1967. Statistical methods in research and production. Oliver Boyd, London, UK, 396 pp.

Durbin, J., 1954. Errors in variables. Review of the International Statistical Institute 22:23-32.

Huber, P.J., 1981. Robust statistics. Wiley, New York, USA, 308 pp.

Imbrie, J., 1956. Biometrical methods in the study of invertebrate fossils. Bulletin of the American Museum of Natural History 108:215-252.

Johnston, J., 1972. Econometric methods (2nd Edition). McGraw-Hill, Tokyo, Japan, 437 pp.

Jolicoeur, P. and Mosimann, J.E., 1968. Intervalles de confiance pour la pente de l'axe majeur d'une distribution normale bidimensionnelle. Biométrie-Praximétrie 9:121-140.
Kebreab, E., France, J., Agnew, R.E., Yan, T., Dhanoa, M.S., Dijkstra, J., Beever, D.E. and Reynolds, C.K., 2003. Alternatives to linear analysis of energy balance data from lactating dairy cows. Journal of Dairy Science 86:2904-2913.
Kendall, M.G. and Stuart, A., 1966. The advanced theory of statistics. Volume 3. Charles Griffin and Company Limited, London, 552 pp.
Kuhry, B. and Marcus, L.F., 1977. Bivariate linear models in biometry. Systematic Zoology 26:201-2009
Legendre, P. and Legendre, L., 1998. Numerical ecology (2nd English Edition). Developments in environmental modelling, 20. Elsevier, Amsterdam, the Netherlands, 853 pp.
Linnet, K., 1990. Estimation of the linear relationship between the measurements of two methods with proportional errors. Statistics in Medicine 9:1463-1473.
Madansky, A., 1959. The fitting of straight lines when both variables are subject to error. Journal of the American Statistical Association 54:173-205.
Mandel, J., 1964. The statistical analysis of experimental data. John Wiley and Sons, New York, USA, 410 pp.
Martin, R.F., 2000. General Deming regression for estimating systematic bias and its confidence interval in method-comparison studies. Clinical Chemistry 46:100-104.
McArdle, B.H., 1988. The structural relationship: regression in biology. Canadian Journal of Zoology 66:2329-2339.
Sanderson, R., 1992. The effect of supplements of fishmeal and specific amino acids on the growth and efficiency of young steers given grass silage. PhD thesis, University of Reading, UK.
Siegel, A.F., 1982. Robust regression using repeated medians. Biometrika 69:242-244.
Snedecor, G.W. and Cochran, W.G., 1980. Statistical methods (7th Edition). Iowa State University Press, Ames, Iowa, USA, 507 pp.
Sokal, R.R. and Rohlf, F.J., 1995. Biometry (3rd Edition). W.H. Freeman and Company, New York, USA, 887 pp.
Wald, A., 1940. The fitting of straight lines if both variables are subject to error. The Annals of Mathematical Statistics 11:284-300.

Extracting biologically meaningful features from time-series measurements of individual animals: towards quantitative description of animal status

N.C. Friggens[1,2], M.C. Codrea[1] and S. Højsgaard[1]
[1]Faculty of Agricultural Sciences, University of Aarhus, P.O. Box 50, 8830 Tjele, Denmark
[2]INRA UMR Modélisation Systémique de la Nutrition des Ruminants, AgroParisTech, 16 rue Claude Bernard, 75231 Paris, France; nicolas.friggens@agroparistech.fr

Abstract

Our ability to describe differences between animal phenotypes is limited, especially in the context of applying nutritional models to predict performance of different genotypes in different environments. This may reflect the difficulty of measuring animal parameters relative to the ease of measuring feed characteristics. However, recent advances in on-farm technology now provide us with multiple time-series of reliably measured indicators of animal performance and status. This paper presents results from work to develop methods for extracting biologically meaningful features from such time-series that offer the potential for quantification of animal status at the level of the individual. One approach to feature extraction from time-series data is presented using milk yield data as an example. In this approach, the data are reduced to a smooth function using B-splines. These are very flexible in terms of approximating complex patterns, they can also be readily differentiated to provide derivatives that may in themselves be features with biological meaning such as milk yield acceleration. By imposing different degrees of smoothing it is possible to decompose the milk yield curve into components that describe the cow's response to infection or nutritional challenge, as well as describing the phenotypic potential yield function of that cow. Each of these features relate to aspects of the cow's potential and robustness. Features can also be combined across different measures, this is useful if the phenomenon we wish to describe is deemed to be latent with respect to the available measures. Taking mastitis as an example we present a proof of principle for combining time-series measures of different indicators (somatic cell count, electrical conductivity, and an enzyme in milk) to derive a latent quantity; degree of infection. These results are discussed in the context of achieving better descriptions of animal characteristics and their potential for driving nutritional models.

Keywords: B-spline, smoothing, latent process, state-space model

Introduction

Our ability to describe differences between animal phenotypes is somewhat limited, especially in the context of applying nutritional models to predict performance of different genotypes in different environments(Friggens and Newbold, 2007). Currently, the majority of nutritional models assume a standard animal and are not designed to easily accommodate differences in phenotype (or genotype) of the animal being modeled (Bryant *et al.*, 2005). They also tend to be driven by an intake that is input rather than being driven by input parameters describing the animal (Dumas *et al.*, 2008). These shortcomings may reflect the difficulty of measuring animal parameters relative to the ease of measuring feed characteristics.

However, recent advances in on-farm technology now provide us with multiple time-series of reliably measured indicators of animal performance and status. This paper presents results from work to develop methods for extracting biologically meaningful features from such time-series that offer the potential for quantification of animal status at the level of the individual.

The underlying hypothesis is that biologically meaningful descriptions of animal characteristics can be derived from time-series measurements of animal parameters. By biologically meaningful we intend to invoke characteristics which can be considered to be important to the animal, i.e. characteristics that have a fitness value to the animal. Although we value, and select on, the ability to produce, for example, milk there is no *a priori* reason to assume that this measure captures all of the aspects of the animal's ability to contribute to the viability of her offspring. This distinction becomes increasingly important if we wish to describe animal characteristics such as robustness, disease resistance, or ability to obtain resources. These are abilities that are highly relevant to developing sustainable livestock systems in the face of changing environments (Bocquier and Gonzáles-Garcia, 2010).

Methodological approach

From the technical viewpoint it seems clear that it is unlikely that any one on-farm measure captures all of any given biological phenomenon. Instead the phenomenon is likely to be distributed across a number of measures. Conversely, it is likely that any one measure reflects several biological phenomena. We thus need to consider means of feature extraction from these time-series, and also to consider these biological phenomena as being latent processes distributed across multiple time-series measures. As an initial step in exploring the possibilities for deriving such information from time-series measures, this paper presents examples of simple ways to extract features and quantify a latent process. One approach to feature extraction from time-series data is presented using milk yield data as an example (Figure 1). In this approach, the data are reduced to a smooth function using B-splines.

These are very flexible in terms of approximating complex patterns and spline representations of data are usually determined by optimizing a roughness measure (λ), subject to interpolation constraints (for example knot density). They can also be readily differentiated to provide derivatives that may in themselves be features with biological meaning such as milk yield acceleration. This approach has been described and termed functional data analysis by Ramsay and Silverman (2005) who have developed software procedures that facilitate easy implementation. The base equation, denoted f(c), being minimized is:

$$f(c) = \sum_j \left[y_j - x(t_j)\right]^2 + \lambda \int \left[D^4 x(t)\right]^2 dt$$

$$x(t) = c\phi(t)$$

Where: y_j are the observed values of the time-series, x(t) is the smooth function generated from the set of basis B-spline functions ϕ(t) and coefficients c. The roughness penalty, in this case on curvature in the 2nd derivative, is weighted by λ. By imposing different values of λ, different degrees of smoothing are obtained (Figure 1). It is thus possible to decompose the milk yield curve into components that, for example, describe the cow's response to infection or nutritional challenge (λ=100; Figure 1), as well as describing the phenotypic potential yield function of that cow (λ=100,000,000; Figure 1). Each of these features relate to different biological phenomena; the cow's potential and robustness.

It is also relatively easy to introduce an offset to the smoothing. This is relevant in two situations, where there is a likelihood of a systematic bias in random error e.g. if the machine errors are mostly low readings, milk yields resulting from interrupted milkings would be of this type. The second situation is where it is desirable to estimate some form of potential measure e.g. the milk production of the animal when free from biological disturbances such as the impact of diseases or restrictions

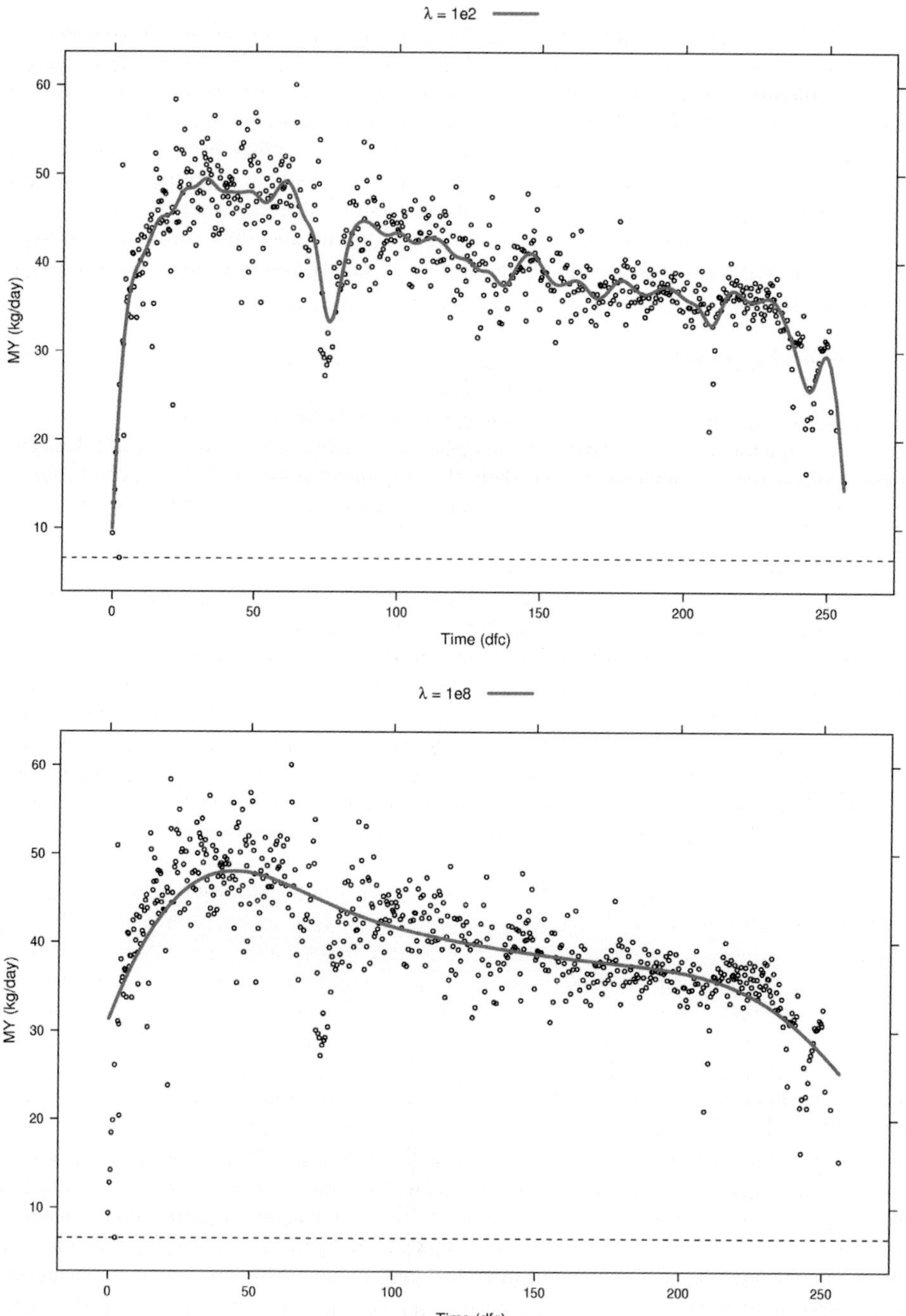

Figure 1. An example of B-spline smoothing of one lactation curve of milk yield (MY) relative to days from calving (dfc), for an individual cow. In the upper panel, a roughness penalty (λ) of 100 is used, in the lower panel a λ value of 100,000,000 is used.

on intake. This provides a useful baseline against which to measure the magnitude of a challenge such as that occurring in response to a mastitis infection. Another example could be the responses to digestive challenges such as acidosis (Desnoyers *et al.*, 2009). Given that the majority of biological disturbances decrease milk yield, there is need to positively offset the smoothed curve that is to be representative of undisturbed phenotypic potential. This can be done in a number of ways, all of which rely on setting up asymmetric ranges for acceptable residuals and then either excluding extreme values or down-weighting their contribution to the sum of squares. An example, using quantile regression with confidence limits of 90 and 60% is shown in the upper panel of Figure 2. The resulting offset curve representing undisturbed phenotypic potential and the much rougher curve that captures biological disturbances is shown in lower panel of Figure 2.

Given that we can identify local biological disturbances, as shown in Figure 2, then we could envisage comparing for different animals the time-series of responses to e.g. disease challenges. Figure 3 shows a schematic representation of this. If we knew that both cows had received the same disease challenge for each of the 3 disease incidences in the above graph, then we would be able to say that the (grey) cow with the greater responses was the more susceptible (the vertical differences between the 2 curves equating to the differences between cows in response to the disease).

However, we do not directly know the size of the disease challenge, especially under naturally occurring infections. The difference in height of response could be due to different susceptibilities and/or to different 'doses'. We need to be able to disentangle susceptibility and size of disease challenge. In general terms, the size of the disease challenge can be seen, at a particular time, t, as being a consequence of the size of disease challenge in the preceding time interval, t-1, and the success of the animal in responding to the disease challenge in the preceding time interval. If we equate some feature of the response profile with disease susceptibility then we can, in principle, model this via a time-dependent linkage between size of challenge and susceptibility as indicated in Figure 4. This type of model is a state-space model and it could allow us to infer the pattern of susceptibility (ability to respond) through time.

A pre-requisite for quantifying susceptibility is that we can develop a credible measure of the size of the disease challenge. In the absence of a definitive direct measure of size of disease challenge we need to use another logic to allow us to convert some of the measures that would classically be seen as response variables into a measure of size of disease challenge or degree of infection (DOI). The logic is as follows. In healthy animals we expect a given measure, e.g. somatic cell count in milk (SCC), to show no systematic time-trend (on the time scale relevant to the development and subsequent cure of an infection). Thus, if we detect a systematic increase in SCC it indicates an increasing DOI (Friggens *et al.*, 2007).

Using a response variable to generate a DOI in this way depends on prior knowledge that the chosen measure is biologically relevant and is not affected by (or can be adjusted for) other non-disease factors. However, using only one measure to give us DOI is equivalent to assuming that this measure is a complete reference for degree of infection. We know from the mastitis literature that no such gold standard exists (Sloth *et al.*, 2003) and it would be risky to base a DOI on a single reference mastitis measure. We can reduce this risk and strengthen our measure of DOI by basing it on a panel of mastitis measures. We assume that there is such a structure as DOI but that it is latent, i.e. not directly observable but reflected to varying degrees in the measures we have. Given this, DOI (size of challenge) can be modelled as a linear combination of the panel of measures. In the following example the features used are concentration of SCC, inter-quarter ratio of electrical conductivity (ECC), and the activity of an immune response enzyme, lactate dehydrogenase (LDH). All of these parameters can be measured automatically on-farm under commercial conditions (Norberg

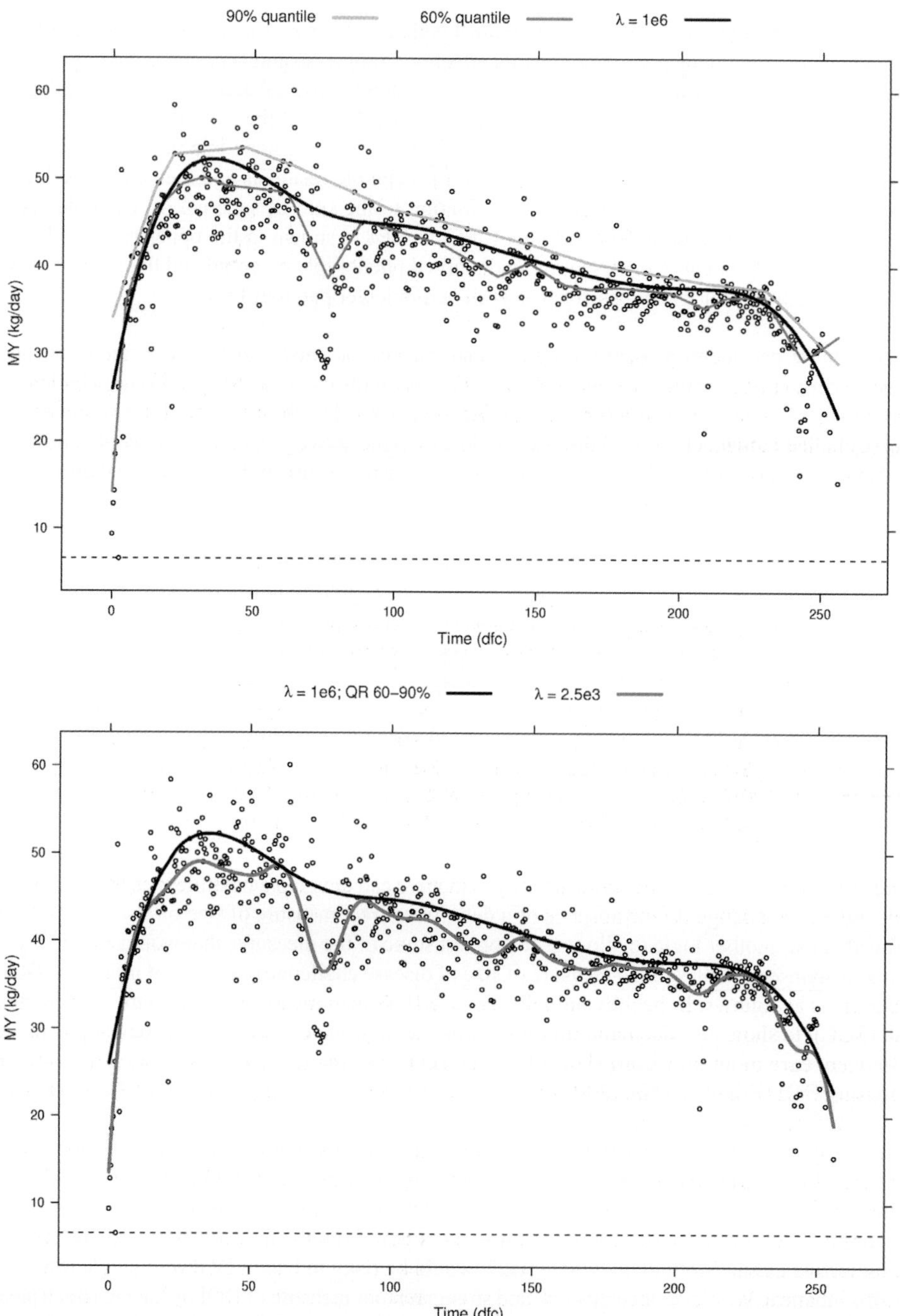

Figure 2. Smoothing with an offset (dark line) based on quantile regression with asymmetric limits of 90 and 60% (light lines) is shown in the upper panel. The offset smoothed phenotypic potential curve (λ = 1,000,000) and biological disturbances curve (λ = 2,500) are shown in the lower panel.

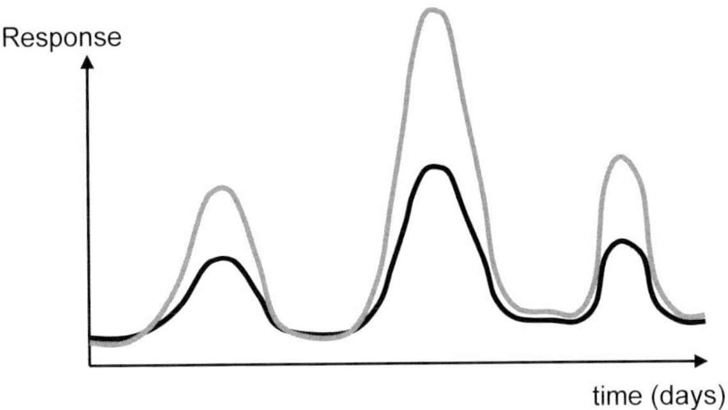

Figure 3. Schematic representation, for two different individuals, of the profile of a biological indicator that increases in response to a challenge. In this scheme, three time synchronized challenges are shown.

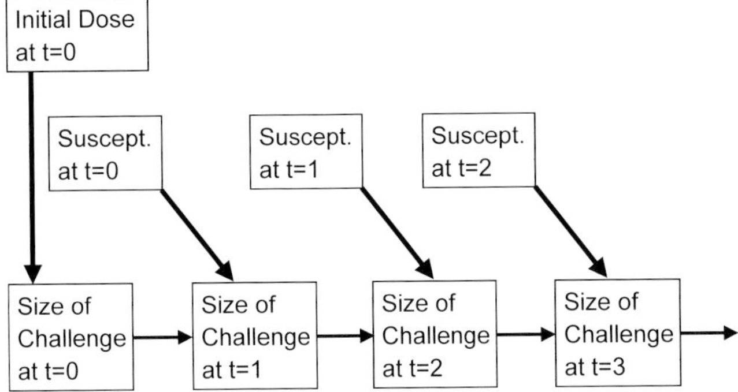

Figure 4. A model of the time-dependent linkage between size of disease challenge and disease susceptibility in a time-series of degree of infection measurements.

et al., 2004, Friggens *et al.*, 2007, Kamphuis *et al.*, 2008). The following example should be seen as a proof of principle for this approach, it is described in greater detail in Højsgaard and Friggens (2010). For each indicator (denoted by the superscript k) at time t_j, $y^k(t_j)$ is assumed to be composed of slowly varying cow specific level $\beta^k(t_j)$, a more rapidly fluctuating short term component $r^k(t_j)$ and an error term $v^k(t_j)$ such that

$$y^k(t_j) = \beta^k(t_j) + r^k(t_j) + v^k(t_j).$$

Onset of mastitis will tend to generate a rapid change in $r^k(t_j)$. To make the assumption that each of the indicators LDH, SCC and ECC are manifestations of the degree of infection operational, it is assumed that each of these indicators is linearly related to a latent variable called $\mathrm{DOIfree}(t_j)$, i.e. that there is a common pattern in these components. This leads to the model assumption

$$y^k(t_j) = \beta^k(t_j) + \lambda^k\,\mathrm{DOIfree}(t_j) + v^k(t_j),$$

where λ^k is a proportionality constant defining the amount of change in indicator $y^k(t_j)$ when $\mathrm{DOIfree}(t_j)$ changes one unit and $v^k(t_j)$ denotes residual variation from this linear relationship. Hence the rapidly varying components $r^k(t_j)$ are assumed to have a common form across all indicators, namely λ^k DOIfree(t_j). Collecting all indicators into vectors gives

$$y(t_j) = \beta(t_j) + \Lambda\, \mathrm{DOIfree}(t_j) + v(t_j),$$

where Λ is a 3×1 matrix of regression coefficients while $v(t_j)$ is the vector of residual terms. It is assumed that $v(t_j) \sim N(0, \Psi)$ where Ψ is a diagonal matrix so that the residuals $v(t_j)$ are uncorrelated. The slowly varying levels $\beta^k(t_j)$ are modelled as:

$$\beta^k(t_j) = \beta^k(t_{j-1}) + w(t_j), \text{ where } w(t_j) \sim N(0,W).$$

This assumes that the change in level $\beta^k(t_j)$- $\beta^k(t_{j-1})$ from t_{j-1} to t_j is a random perturbation with zero mean, and that this perturbation is the same for all indicators k. Alternative ways of modelling the slow moving trend could be considered. Time dependency is also put into the model for DOIfree(t_j). Specifically it is assumed that

$$\mathrm{DOIfree}(t_j) = \{\mathrm{DOIfree}(t_{j-1})\text{-}\mathrm{DOIfree}(t_{j-2})\} + \mathrm{DOIfree}(t_{j-1}) + u(t_j), \text{ where } u(t_j) \sim N(0,U).$$

The term {DOIfree(t_{j-1}) – DOIfree(t_{j-2})} on the right hand side of the equation can be interpreted as the trend in the development of DOIfree at time t_{j-1} while the subsequent term DOIfree(t_{j-1}) gives the level of DOIfree at time t_{j-1}. Therefore this equation postulates a local linear development in DOIfree. The random perturbation $u(t_j)$ will allow the local linear development to vary over time. These equations define a linear state space model (Durbin and Koopman, 2001). Thus, the latent variable DOIfree can be estimated using e.g. a Kalman filter. The quantity DOIfree(t_j) varies freely but from a biological viewpoint it is more relevant to represent the degree of infection (DOI) as a continuum on the interval [0:1] where 0 corresponds to no infection and 1 to full blown mastitis. This can be achieved by using a sigmoid transformation of DOIfree(t_j) to obtain the DOI. An example of a time-series of the original indicators and DOI is shown in Figure 5.

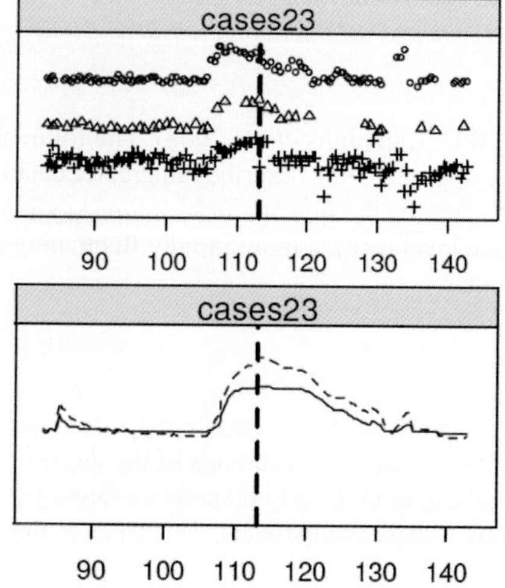

Figure 5. The time-series of measurements of SCC (o), LDH (Δ), and ECC (+) for a cow that was treated for mastitis on day 113 are shown in the upper panel. The corresponding values of DOI_{free} (stipled line) and DOI (solid line) are shown in the lower panel.

The approach was tested using two subsets of cows. These were cows that can – with reasonable certainty – be classified as either having mastitis or being completely free of mastitis (hereafter called healthy controls) within specific time-windows. For each mastitis case, matching control periods were chosen from the pool of cows of the same breed and lactation number class (1, 2, 3+) that had no veterinary treatment for mastitis in that lactation until at least 30 d after the day of lactation of the mastitis case being matched on (Friggens *et al.*, 2007). The chosen control period spanned the same days of lactation as the case being matched on. Mastitis cases occurring after 350 days of lactation were excluded. This resulted in a data set containing 58 true mastitis cases and 71 true healthy control cases. The average DOIfree values for the mastitic and healthy subsets are shown in Figure 6. There were significant differences in DOIfree in a window of 5 d around the treatment/control point.

Although the example considered here relates to a disease challenge, it should be viewed as a proof-of-principle for a more generic approach to characterizing animal status and thus a means to quantify other aspects of an animal's adaptive capacity to a wide range of challenges.

Conclusions

In addition to showing that latent variables such as DOI can be extracted from on-farm time-series measures, this approach also shows that there are considerable gains to be made relative to the classical representations of infection as a binary condition (healthy, sick). Further, given that DOI is a reasonable representation of 'true infection status' then, as shown in Figure 4, we move towards a more robust and operational definition of susceptibility. Thus we add value to time-series measures of biological indicators by feature extraction and combination across measures. We believe that this approach provides a means to improve description of animal states and thereby functional phenotypes.

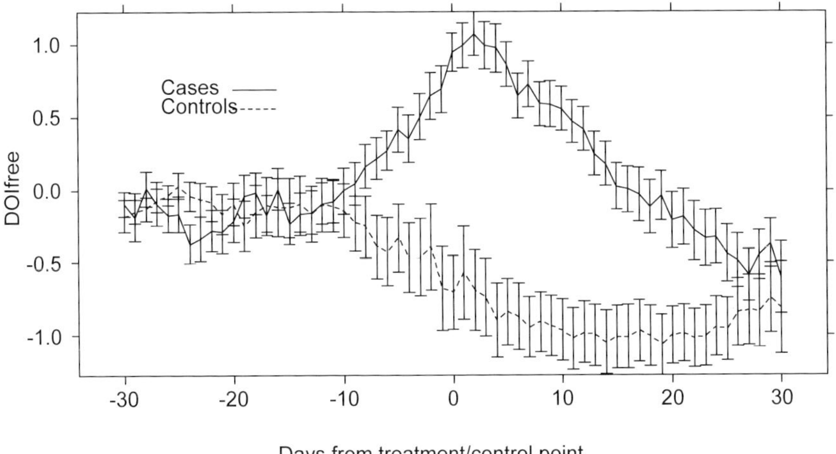

Figure 6. Average DOI_{free} values for mastitis cases (n=58) and control cows (n=71). The bars indicate standard errors. Redrawn from Højsgaard and Friggens (2010).

References

Bocquier, F. and Gonzáles-Garcia, E., 2010. Feeding practices for sustainable ruminant production facing environmental changes and the human food crisis. Animal 4: 1258-1273.

Bryant, J.R., López-Villalobos, N., Holmes, C. and Pryce, J.E., 2005. Simulation modelling of dairy cattle performance based on knowledge of genotype, environment and genotype by environment interactions: current status. Agricultural Systems 86:121-143.

Desnoyers, M., Giger-Reverdin, S., Duvaux-Ponter, C. and Sauvant, D., 2009. Modelling off-feed periods caused by subacute acidosis in intensive lactating ruminants: application to goats. Journal of Dairy Science 92:3894-3906.

Dumas, A., Dijkstra, J. and France, J., 2008. Mathematical modelling in animal nutrition: a century review. Journal of Agricultural Science 146:123-142.

Durbin, J. and Koopman, S.J., 2001. Time Series Analysis by State Space Methods. Oxford University Press, Oxford, UK.

Friggens, N.C., Chagunda, M.G.G., Bjerring, M., Ridder, C., Højsgaard, S. and Larsen, T., 2007. Estimating degree of mastitis from time-series measurements in milk: a test of a model based on lactate dehydrogenase measurements. Journal of Dairy Science 90:5415-5427.

Friggens, N.C. and Newbold, J.R., 2007. Towards a biological basis for predicting nutrient partitioning: the dairy cow as an example. Animal 1:87-97.

Højsgaard, S. and Friggens, N.C., 2010. Quantifying degree of mastitis from common trends in a panel of indicators for mastitis in dairy cows. Journal of Dairy Science 93:582-592.

Kamphuis, C., Sherlock, R., Jago, J., Mein, G. and Hogeveen, H., 2008. Automatic detection of clinical mastitis is improved by in-line monitoring of somatic cell count. Journal of Dairy Science 91:4560-4570.

Norberg, E., Hogeveen, H., Korsgaard, I.R., Friggens, N.C., Sloth, K.H.M.N. and Løvendahl, P.L., 2004. Electrical conductivity in milk: ability to predict mastitis status. Journal of Dairy Science 87:1099-1107.

Ramsay, J.O. and Silverman, B.W., 2005. Functional data analysis,1st. Springer, New York, USA.

Sloth, K.H.M.N., Friggens, N.C., Løvendahl, P.L., Andersen, P.H., Jensen, J. and Ingvartsen, K.L., 2003. Potential for improving description of bovine udder health status by combined analysis of milk parameters. Journal of Dairy Science 86:1221-1232.

Turning a cow into a goat with a teleonomic model of lifetime performance

O. Martin and D. Sauvant
UMR Physiologie de la Nutrition et Alimentation, INRA-AgroParisTech, 16, rue Claude Bernard, 75231 Paris cedex 05, France; olivier.martin@agroparistech.fr

Abstract

Predicting nutrient partitioning throughout lifespan is a major issue in modelling dairy female performance. A general framework has been proposed for dairy cattle. The key concept of this framework is to consider lifetime performance as a trajectory, goal directed towards reproductive success, genetically scaled and potentially altered by environment. The framework is based on two sub-models, namely a regulating model describing a time base pattern of priority for physiological functions and an operating sub-model driving energy partitioning. The objective of this work is to assess the suitability of the model to represent dairy goat lifetime performance. Eight parameters were modified to 'turn a cow into a goat': two parameters were related to foetal growth, three parameters were related to reproductive events timing, two parameters related to the scaling of body weight at maturity and milk potential and a parameter related to labile body mass mobilization. Individual records of the INRA experimental herd were used to assess the model suitability to simulate dairy goat performance. The result shows that the simulated body weight and milk production dynamics over three lactations were globally consistent with the data. Hence, adapting eight parameters from the dairy cow model was sufficient to simulate the performance of a dairy goat. This work opens perspective for the development of a generic animal model.

Keywords: teleonomic model, dairy livestock, lactation, body weight

Introduction

The paradigm in animal nutrition has evolved from the aim of covering requirements to maximize a genetic potential to the aim of predicting multiple performance of a particular genotype in a particular environment (Sauvant, 1992). The prediction of regulations controlling nutrient partitioning, particularly energy, is a major issue in modelling dairy cattle performance. The proportions of energy channelled to physiological functions (growth, maintenance, gestation and lactation) change as the animal ages and reproduces, and according to its genotype and nutritional environment. As a key concept, lifetime performance can be seen as a trajectory, goal-directed towards reproductive success, genetically scaled and potentially adaptable to a large diversity of nutritional challenges.

We have recently proposed (Martin and Sauvant, 2010a,b) a general framework formalizing this concept through the design of a time base pattern of priority for physiological functions (regulating sub-model) driving energy partitioning (operating sub-model). The regulating sub-model describes, throughout the lifespan of dairy cows, the dynamic partitioning of relative priorities to elementary life functions (growth, gestation, lactation, ageing, balance of reserves) and is structured on a teleonomic basis, embodying the idea that the coordination of life functions is goal-directed toward the preservation of life for the reproduction of life form (Bricage, 2002). The teleonomic basis was to consider the animal *as an active biological entity with its own agenda* (Friggens and Newbold, 2007). The model has been parameterized in dairy cows but the approach is meant to be generic and this postulate needs to be tested in other lactating species. According to this perspective, we addressed here the suitability of this model to describe lifetime performance of dairy goats.

Material and methods

Cow model description

A detailed description of the model is given in Martin and Sauvant (2010a,b). Briefly, the proposed conceptual framework is based on the coupling of a regulating sub-model providing driving forces to control the functions of an operating sub-model. The regulating sub-model describes the dynamic partitioning of a female mammal's priority between life functions targeted to growth (G), ageing (A), balance of body reserves (R), and nutrient supply of the unborn (U), newborn (N) and suckling (S) generations. The so-called GARUNS dynamic pattern of priorities defines a teleonomic trajectory goal-directed towards the survival of the individual for the perennity of its life form. The operating sub-model describes energy partitioning between flows involved in changes in body weight and composition, foetal growth, milk yield and composition, and food intake over lifetime and during phases of gestation and lactation of repeated reproductive cycles. It incorporates major concepts described by Van Es *et al.* (1978) and shared by updated dairy energy feeding systems used around the world (e.g. NRC, 2001; Fox *et al.*, 2004; INRA, 2007). The dynamic partitioning of priority controls the dynamic partitioning of energy, which defines a long-term reference pattern of performance. Genetic parameters are incorporated in the model to scale individual performance and to be able to simulate differences within and between breeds. The reference pattern of performance involves a required energy supply which calibrates all the energy flows. Deviations from this reference pattern can occur according to food availability. At any stage of life, energy surplus or deficit is apportioned between energy flows according to reference level of priorities and to the level of animal body reserves.

These adaptative processes to energy supply imply deviations from the reference pattern of performance, in particular that of body reserves acting as a sink for energy surplus and a source in case of energy deficit. In turns, these changes in animal body condition may progressively imply changes in priority. Therefore, the model explicitly involves a homeorhetic drive by way of the tendency to home on to the teleonomic trajectory and a homeostatic control through the apportioning function that maintains an energetic balance in response to nutritional constraints.

The initial model calibration was designed to determine parameter values for a realistic reference pattern of performance in dairy cows and was performed with various data from the literature.

Calibration to goat

A simple model of homeorhetic regulation of major energy flows was published on dairy goats (Puillet *et al.*, 2008). As the current model is much more adaptative, as described above, our specific objective was to evaluate its ability to describe performance of dairy goats. To address this issue, a minimum number of parameters of the regulating and operating sub-models were reset to appropriate values in order to yield simulations of realistic dairy goat milk yield (*MY*, kg/d) and body weight (*BW*, kg) changes from birth to the end of the 3rd lactation. The parameters considered were those initially defined in the cow model as bovine specific parameters used to scale fetal growth dynamics and to time reproductive events or as input parameters (genetic parameters, see Martin and Sauvant, 2010a) used to scale individual body and milk performance.

The two parameters governing fetal growth dynamics were taken from Laird (1966) assuming the available sheep constants to be representative for goats. The two parameters governing reproductive events timing (age at first conception and parturition to conception interval) and the parameter scaling pregnancy length were taken from Puillet *et al.* (2008). The two parameters scaling mature body weight and milk potential were estimated through a fitting procedure achieved on a dataset of individual records (experimental flock of our unit). Additionally, it was also necessary to adapt the parameter scaling the body reserves balancing rate. Parameters scaling milk composition were not considered in the present work and the parameter defined as an individual trait of labile body mass mobilization index was held to the cow value since its change increases only marginally the goodness-of-fit. Finally, the goat version of the model was achieved by changing only 8 parameters among the 52 model parameters.

Implementation

The model was implemented with Modelmaker version 3.0 software (Cherwell Scientific Ltd, 2000) using the Runge-Kutta 4 numerical integration procedure and a fixed integration step of 1 d from t = 0 to t = 1,400 d in goat and t = 1,850 d in cow (approximately 4 yr and 5 yr respectively at the end of the 3rd lactation). The least square procedure of fitting was performed with the simplex algorithm embedded in the optimization tool of Modelmaker version 3.0 software (Cherwell Scientific Ltd, 2000) with default advanced parameters value and 100 convergence steps.

Results

The values of the 8 parameters used to turn the cow model into a goat one are given in Table 1. Modulations of the parameters scaling mature size (W_M) and body reserves changes (b_0) correspond to a 9-fold division relative to the cow values. Considering the three parameters of the operating sub-model, a goat can be seen as 'a very small high yielding cow'.

Model simulations of body weight and milk yield dynamics from birth to the end of the 3rd lactation in goat are plotted in Figure 1 with the individual data used in the global fitting procedure. The goodness of fit is given through reported root mean square errors: *BW*: 1.6 kg (196 records); *MY*: 0.3 kg/d (118 records). Corresponding values obtained on cows with group data were respectively 15 kg and 0.8 kg/d (Figure 2). The model suitability to be adapted to simulate dairy goat body

Table 1. Model parameters to turn the cow model (Martin and Sauvant, 2010a) into a goat model.

Parameter	Description	Unit	Value	
			Cow	Goat
Regulating sub-model				
Pregnancy scaling				
π	Maximal pregnancy length	d	285	150
α	Rate of decay of fetal growth rate	-	0.0111	0.0209
ω	Initial value of fetal growth	kg	$3.5 \cdot 10^{-6}$	$2.4 \cdot 10^{-6}$
Reproductive events timing				
$PCI(0)$	Age at first conception	d	450	240
$PCI(c)$	Parturition to conception interval of cycle	d	120	210
Operating sub-model				
Body mass change rate				
b_0	Balancing body reserves	d^{-1}	1.60	0.17
Individual performance scaling (genetic parameter)				
W_M	Mature non-labile body mass	kg	500	55
v_Y	Milk yield	-	1.0	1.6

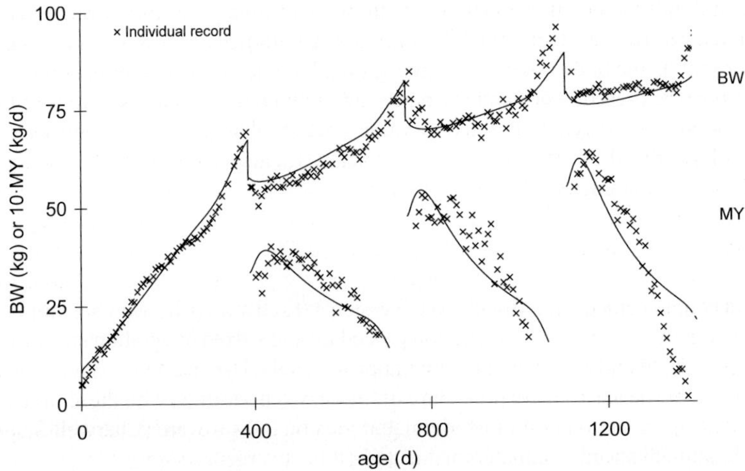

*Figure 1. Model simulations of body weight (*BW*, kg) and milk yield (*MY*, kg/d) from birth to the end of the 3rd lactation plotted with individual data in dairy goats.*

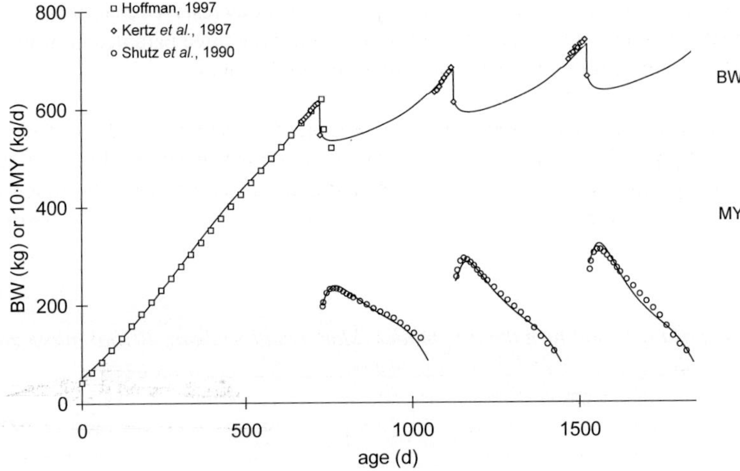

*Figure 2. Model simulations of body weight (*BW*, kg) and milk yield (*MY*, kg/d) from birth to the end of the 3rd lactation plotted with literature date in dairy cows.*

weight changes during growth and reproductive cycles and milk yield during lactation was thus found to be globally satisfactory.

Discussion

The proposed model of lifetime performance initially conceived for dairy 0 provides a basis for predicting nutrient partitioning across different physiological states and genotypes. This work, fully described in Martin and Sauvant (2010a;b), involves (1) a conceptual framework that focuses on priorities for life functions, (2) a holistic view of growth, reproduction and nutrition, (3) a combination of genotype and nutritional effects in the prediction of performance, and (4) scaling parameters allowing the simulation of different animal genotypes.

The ambition of such a model is to be adaptable to different lactating species with the same shared generic framework and species specifications. This first attempt to the application of the model to dairy goats has proved successful and provides a first validation step of the model's generic formalism. In this perspective, the fact that the modulation of only 8 parameters (15% of the overall number of parameters) is sufficient to develop a caprine version is a very promising result. This version is *a priori* much more flexible in function of the environmental challenges than the goat model of Puillet *et al.* (2008). Therefore, this advantage would have to be evaluated on various other contexts on dairy goat.

This result shows the potential of a teleonomic modelling approach to simulate lifetime performance pattern. The concept of priorities among functions is relevant to predicting biological responses throughout lifespan. It provides a conceptual framework to link reproductive strategy, energy allocation and physiology. Hence, this type of modelling approach opens interesting perspectives of collaboration not only in livestock feeding but also in the field of ecology to compare different reproductive strategies in mammals.

However, the work presented here should be performed with a large number of individual records and with other species to explore more precisely its suitability to become a multi-species tool. Such a tool could be in the future used for *in silico* experiments exploring changes in animal performance resulting from the interactions of genetic and environmental factors.

The next step of work will be to carry on the exploration of the suitability of the model for other lactating species but mainly to carry on the development of the cow version of the model to evaluate its accuracy in the prediction of lifetime performance seen as the trajectory of a particular genotype in a particular environment.

Conclusion

An attempt to turn a goat into a cow with a teleonomic model of lifetime performance based on the dynamic partitioning of life function priorities was performed. Results show that 8 parameters were sufficient to adapt the model, suggesting that a goat can be seen as a 'very small high yielding cow'. This work in progress is proposed in the perspective of developing generic animal models.

References

Bricage, P., 2002. Héritage génétique, héritage épigénétique et héritage environnemental: de la bactérie à l'homme, le transformisme, une systémique du vivant. Symposium AFSCET, Evolution du vivant et du social: Analogies et differences. Retrieved July 4, 2008, from http://www.afscet.asso.fr/heritage.pdf.

Cherwell Scientific Ltd. 2000. Modelmaker User Manual, Oxford, England.

Fox, D.G., Tedeschi, L.O., Tylutki, T.P., Russell, J.B., Van Amburgh, M.E., Chase, L.E., Pell, A.N. and Overton, T.R., 2004. The Cornell Net Carbohydrate and Protein System model for evaluating herd nutrition and nutrient excretion. Animal Feed Science and Technology 112:29-78.

Friggens, N.C. and Newbold, J.R., 2007. Towards a biological basis for predicting nutrient partitioning: the dairy cow as an example. Animal 1:87-97.

INRA (Institut National de la Recherche Agronomique), 2007. Alimentation des bovins, ovins et caprins: Besoins des animaux – Valeurs des aliments. Quæ, Paris, France.

Laird, A.K., 1966. Dynamics of embryonic growth. Growth 30:263-275.

Martin, O. and Sauvant, D., 2010a. A teleonomic model describing performance (body, milk and intake) during growth and over repeated reproductive cycles throughout the lifespan of dairy cattle. 1. Trajectories of life function priorities and genetic scaling. Animal, in press: doi 10.1017/S1751731110001357.

Martin, O. and Sauvant, D. 2010b. A teleonomic model describing performance (body, milk and intake) during growth and over repeated reproductive cycles throughout the lifespan of dairy cattle. 2. Voluntary intake and energy partitioning. Animal, in press: doi 10.1017/S1751731110001369.

NRC (National Research Council), 2001. Nutrient requirements of dairy cattle. 7th revisited edition, National National. Academy Press, Washington, DC, USA.

Puillet, L., Martin, O., Tichit, M. and Sauvant, D. 2008. Simple representation of physiological regulations in a model of lactating female: application to the dairy goat. Animal 2:235-246.

Sauvant, D., 1992. Systemic modeling in nutrition. Reproduction Nutrition Development 32:217-230.

Van Es, A.J.H., Vermorel, M. and Bickel, H., 1978. Feed evaluation for ruminants: new energy systems in the Netherlands, France and Switzerland. General introduction. The systems in use from May 1977 onwards in the Netherlands. Livestock Production Science 5:327-330.

Techniques to facilitate metabolic modeling with reference to dairy cows

R.C. Boston[1], P.J. Moate[2], J. Roche[3], R. Dunbar[4] and G. Ward[5]
[1]University of Pennsylvania, Department Clinical Studies, School of Veterinary Medicine, 3800 Spruce St., Philadelphia, PA 19104, USA
[2]Primary Industries Research, Department Primary Industries, Ellinbank 3821, Victoria, Australia
[3]DairyNZ, Private Bag 3221, Newstead, Hamilton 3240, New Zealand
[4]Presbyterian Hospital, Univiversity of Pennsylvania, Philadelphia, 800-789-PENN USA
[5]St. Vincent's Hospital, Deptartment of Endocrinology, 41 Victoria Parade, Fitzroy VIC 3065, Australia

Abstract

Modeling has contributed to the advancement in animal health and animal production with the aid of two types of models; research models and diet formulation models. The former models admit the inclusion of detailed mathematical accounts of digestive and metabolic processes, and the latter deploy extensive mathematical specifications of the nutritional value of feeds in regard to the animal host. Each has been used to help maintain and enhance health, welfare and production of farm animals. Data for models comes from various sources but, until recent times, the detailed investigation of animal responses to metabolic challenges has been limited to empirical quantification. However, we point out that there remain very useful indices and measures afforded from modeling such responses if special care is taken in mathematically teasing out the metabolic information. In this article we describe how this is done, and we demonstrate the methods with reference to models of the glucose and the NEFA responses of the glucose challenge.

Keywords: glucose, non-esterified fatty acids (NEFA), metabolic model, minimal model, model partitioning

Introduction

Metabolic challenge models comprise a class of models that has long been used in research and in clinical diagnosis related to human health conditions (see for example, Boston *et al.*, 2003), but have only recently begun to be used to explore metabolism in farm animals. These models are generally used to assist in the assessment of the ease with which a metabolically challenged system is able to restore its baseline state. Whilst it is not new to examine challenge responses in systems (for example dairy cows, Bauman and Currie, 1980), it is less common to model these responses using dynamic approaches and to assign production and clinical interpretations to the parameters evolving from such models. Because aspects of these models (called minimal models) are not well understood outside the field of human medical research, this paper will discuss critical features of their basis; their purpose, how the models are developed, their application to data from dairy cows, and their limitations. We will take as our example the glucose minimal model developed by Bergman *et al.* (1979), and we will introduce a new model, the NEFA minimal model.

Methods

In what follows, we outline the salient features of models used to describe the critical forces that are invoked when a system is challenged. Our setting is the application of a glucose challenge (typically an injection of glucose (300 mg/kg)) into the blood of a fasted subject, for example a dairy cow. The challenge causes a dramatic increase in blood glucose from its initial baseline or homeostatic state, a sudden spike in plasma insulin concentrations and a sustained depression in plasma NEFA concentrations (see Figure 1).

The quickness with which the blood glucose level returns to baseline provides an indication of the functionality of glucose homeostatic system. To drive the glucose back to baseline, counter-regulatory forces are invoked each contributing in some way to the 'disposition' of the massive glucose 'burden'.

There are three important points regarding the metabolite disposition forces mimicked by metabolic models. First, disposition forces do not act sequentially and cannot be accurately and unambiguously resolved by failing to recognize the concurrent nature of their action. Second, disposition forces as portrayed in the models are not necessarily explicit in that in some instances multiple forces may be compiled into a common component (see for example auto-regulation, Moore *et al.*, 1998). Third, the disposition forces, whether combined or not, may not be an intrinsic, or explicit, part of the model but may need to be teased out of the model using functions of the parameters (see for example insulin resistance).

What are minimal models?

Minimal models are models which describe the temporal fate of key metabolites of a challenged system. They are minimal in the sense that they incorporate the least amount of detail while not compromising their capacity to accurately describe the temporal pattern of the challenge response. Moreover, minimal models generally have the flexibility to admit diverse challenge responses without the accuracy or resolution of the estimates of their parameters being seriously impeded, thus exposing the status of the system in regard to the metabolite of interest.

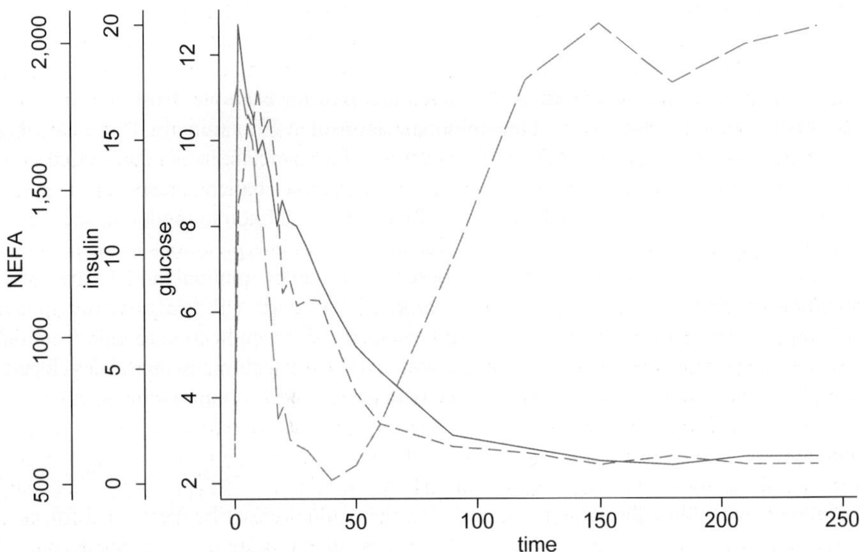

Figure 1. Glucose (mM/l; solid line), insulin (mU/l; dash), and NEFA (μM/l; long dash), patterns following an FSIGT applied to a dairy cow.

How are minimal models developed?

At the heart of developing a minimal model are refinement steps in which: (1) model solutions explain the observed challenge responses accurately and consistently, and (2) model fitting to the observed responses yields identified parameter estimates. They are developed using an iterative sequence of expansion and contraction. We expand the number of compartments and or parameters in the model under development when it is clear that there is not enough flexibility in the model to exhibit the important features of the observed response, and we contract or reduce the number of compartments and or parameters in the model when the parameter estimates from the fitting steps fail to produce identified estimates of the parameters. Of course the principle of simply abandoning metabolic entities from our model in the contraction step is not simply a matter of reaching a point in our model evolution where we decided that mathematical representation of, say, the liver, or kidney, or pancreas should be dropped from our model because some of the model's parameters cannot be identified. Bergman and Cobelli (1980) and Bergman *et al.* (1981) have provided some extremely helpful guides to the contraction steps for the development of a minimal model. If to include a representation of a particular tissue, or organ in our mathematical model, (1) exposes the data fitting process to poorer quality data than other model components for which data exists, (2) substantially increases the complexity of the model *per se* without adding to the model's utility, (3) does not add to, or improve the clarity of the objective of the model development basis, and (4) places burden on the estimation process to accurately and precisely predict important parameters, then there is serious contention regarding the inclusion of the 'additional' organ or tissue.

It is interesting to note that with the glucose minimal model there seems to have been equivocation regarding the role of insulin. Clearly the presence of insulin, in some form or other, is absolutely critical if the glucose minimal model is to make some sense physiologically. However, Bauman and Currie (1980) pointed out that to add insulin as a modeled component of the glucose minimal model would dramatically add to the complexity of the model, and to add the insulin data to the model's estimation database would also seriously hamper the estimation process, especially bearing in mind the questionable quality of insulin data *per se.* Another interesting point that comes up in this context is the balance of weight we wish to place on the data to exposing critical indices of the system versus, possibly complex model structures to expose the same perceived function. In contrast to the glucose minimal model where the data itself easily exposes important systems features, almost all modeling efforts to create an insulin minimal model have placed excessive reliance on the model itself to expose system features (Toffolo *et al.*, 1980). The problem faced was to include insulin in the model and to do this in a way that would not compromise the objective of a minimal model.

Partition analysis: contraction without compromise

An alternative to 'shedding' components of a model is to exercise partition analysis. This approach was employed by Bergman specifically to aid with the development of the glucose minimal model. After eliminating system components having, at most, marginal effects on the dynamics of the key entities under investigation (in the case of the glucose minimal model, we could specifically include glucagon and epinephrine in this category) we are left with the decision of how to best portray the role of important yet un-modeled entities that nevertheless, may have significant influences on the processes and form of the evolving minimal model. Partitioning may suggest that these residual influential entities can be incorporated into the minimal model as linearly interpolated forms, or as an input function. Their patterns are not modeled, thus obviating mathematical and estimation problems, but the influence they have on modeled components is represented computationally as functions defined by lines joining the points in time where their observations exist. Thus, in solving the system of minimal model differential equations, intermediate solutions of these partitioned entities called for by numerical integrators are furnished by interpolations along the lines between the observations. The equations of the minimal glucose model are thus:

$$G`(t) = -G(t) \times (X + Sg) + Sg \times Gb$$
$$X`(t) = -P2 \times X(t) + P3 \times (I - Ib)$$
$$G(0) = G0,\ X(0) = 0$$

Where $G(t)$ [mg/dl,units of all glucose measures] denotes estimated plasma glucose, $X(t)$ [min^{-1}] is insulin action, I [mU/l,units of all insulin measures] is observed plasma insulin, Ib is basal insulin, Gb is basal glucose, $P3$ [$(mU/l)^{-1}.min^{-2}$] is the rate of delivery of insulin to the action space, $P2$ [min^{-1}] is the inverse of the residence time, or persistence, of insulin action, Sg [min^{-1}] is glucose effectiveness, and $SI = P3/P2$ [$(mU/l)^{-1}.min^{-1}$] is insulin sensitivity. Whereas $G(t)$ and $X(t)$, are both modeled, I is not modeled, but rather interpolated.

What is gained from partition analysis in the minimal model?

Because glucose promotes insulin secretion and insulin promotes glucose disposition in a dynamically and tightly coupled feedback mechanism, this creates a very challenging problem for estimating the system parameters. Minor exploratory adjustments to parameters in the wrong direction during fitting insulin and glucose data may lead to the dramatic amplification of estimated responses and hamper estimation convergence. Indeed, it is probably this reason alone that caused the failure of attempts to develop such a model during the twenty-year period before the advent of the glucose minimal model. In the minimal model equations above, rather than modeling the insulin data, we utilize the insulin data *per se*, so that the feedback loop involving insulin and glucose is now opened thus substantially reducing or possibly eliminating problems associated with parameter estimation.

An interesting feature evident in the above set of equations is that insulin is depicted as influencing glucose disposal via the remote insulin pool 'X', which can be thought of as a relatively smoothed, attenuated, and delayed version of insulin observations.

A desirable consequence of the implementation of the partitioning procedure employed here is that the equations above can be fitted explicitly to glucose observations using a simple nonlinear equation estimator. Indeed, using this approach, Boston (unpublished data) showed that estimates of SI (insulin sensitivity) using a standard nonlinear estimator were concordant to over 94% with those from the Minimal Model estimator MinMod Millennium.

The final point to be made regarding the implementation of minimal models relates to the need to anchor model parameters to the mechanisms and entities embedded in the model and to ensure that in themselves they quantify, in a relative sense, the state of the processes they define. In this regard, two critical indices evident in the minimal glucose model are glucose effectiveness (Sg), also referred to as auto-regulation, or the combined effect of glucose mediation of glucose disposal and glucose mediation of hepatic glucose production (Moore *et al.,* 1998), and insulin sensitivity (SI). Each of these is really a function of the rate of disposal of glucose, Rd, with Sg being the rate of change of Rd with respect to glucose, G(t), and SI is the rate of change of Sg with respect to insulin, I(t) (Bergman *et al.*, 1979).

From the above exposition, it can be seen that although the Glucose minimal model is *prima facie*, apparently simple, it nevertheless has the capacity to allow extensive exploration of how various dietary treatments and medication regimens influence the glucose / insulin system. Recently, a new glucose minimal model software, Min Mod Millennium was released (Boston *et al.,* 2003) and it is now used in research and clinical settings in many hundreds of hospitals and Universities around the world. A recent search on Medline revealed that there have been over 1,000 scientific publications concerning or utilizing the glucose minimal model, and over 50 new papers citing it each year. Thus, it is arguable that no mathematical model has had a greater impact on endocrinology in particular and health in general than Bergman's glucose minimal model.

The NEFA minimal model

Whereas the role of glucose in the lactating cow is critical it is not as susceptible to manipulation as in monogastric animals and humans. This is because of the massively dominant demand for glucose by the mammary gland to produce, for example, up to 100 lbs of milk per day per cow around peak production times in modern dairy systems. What is important and potentially subject to manipulation though is the fat and fatty acid status of dairy cows, especially in the immediate post parturient period when large quantities of adipose tissue are mobilized to support milk production (Komoragiri *et al.*, 1998). Accordingly, with a view to providing a tool to quantify the fatty acid status of an individual subject, we decided to create a minimal NEFA (free fatty acid, Boston *et al.*, 2008) model using the methods described above.

As a first step, we reviewed (Moate *et al.*, 2007) the general status of NEFA models and we found that there were no models, minimal or otherwise, that could describe all of the features of the NEFA response to a glucose challenge.

To review very briefly the pattern of NEFA in response to the glucose challenge we see the following (see Figures 1 and 3):

1. Immediately post glucose injection a latent period is observed, where the blood levels of NEFA are fluctuating, to a small degree, unsystematically between oxidative and lipolytic states around a baseline status (this period follows an overnight fast). The latency duration runs from about 5 to 12 or more minutes.
2. Once the cessation of lipolysis is established by insulin's action on adipocyte HSL there follows a precipitous decline in NEFA at a rate of about 10-20% per minute (for dairy cows), for about 35 minutes to a Nadir. Because the challenge follows an overnight fast it is likely that this decline reflects predominantly oxidation.
3. About 5 to 10 minutes prior to the NEFA nadir, the NEFA profile exhibits a 'soft approach' to the point of the minimum (i.e. there is no abrupt change in direction) and then starts to rise again at a rate somewhat slower than the decline. The NEFA rise being slower than its decline is not surprising because the decline does not have to compete with other forces whereas the incline is the sum of lipolysis (and allied inputs) less oxidation (and other outputs). We have found the likely peak lipolysis rate here to be around 450 micromoles/min/l (for dairy cows).
4. At around 80 minutes following the challenge, NEFA levels cross their baseline (or mean baseline where sustained oxidative or lipolytic trends persisted in the latency period) and proceeds to a supra-basal state, in some cases amounting to over twice the baseline levels.

There are thus three or four stages and two or more control points associated with the NEFA response to the glucose challenge and our objective in developing the NEFA model was to produce an accurate mathematical account of the NEFA profile (in its substantial diversity) and to provide model parameters which might serve as a guide to the metabolic state of the cow. The choices in modeling the NEFA response involved a decision as to whether, and how, glucose and insulin might find their ways into the model. It is clear that both will assume critical positions but the question remained would both be needed and would mathematical portrayal of the included entities be needed. The mathematical representation of NEFA within the model was unquestionable. We reflected on the following in making our decisions here:

- Glucose is central to both insulin's role in any energy related control within the body and naturally provides a perfect marker for insulin through the obvious high pharmacologic correlation (Derendorf and Hochhaus, 1995).
- Glucose is exquisitely regulated within the body and only moves as energy signals impose. Thus if switches with energy implications are involved with NEFA movement, glucose will signal and respond, or convey those switches directly.
- There is ample evidence that glucose is a mediator of lipolysis (Park *et al.*, 1990, Qvisth *et al.*, 2004, Bauman and Currie, 1980) and thus, as suggested above, shifts in lipolytic status will be rapidly signaled from glycemic shifts.
- There is extensive evidence of the close relationship between glucose and NEFA in the health status of energy dependent systems. For example, whereas insulin resistance might be thought to be a disease related to the impaired management of glucose (Tonelli *et al.*, 2005), it is now clear that NEFA itself is additionally directly implicated. There is also evidence that much of the control of gluconeogenisis *apropos* of the glucose challenge is associated with NEFA status change signaling as embodied in Bergman's 'single gate hypothesis'.
- Mindful of Bergman's reasoning for not including an insulin sub-model in the glucose minimal model and also taking into account the superior quality of glucose observations compared with those of insulin, we decided to exploit glucose patterns to provide a basis for any specific action of insulin within a NEFA model. Thus a minimal NEFA model capitalizing on glucose observations may be not only more accurate in regard to NEFA predictions, simpler to implement, and produce parameter estimates with higher precision, but also offer a way of bypassing the complex and relatively expensive determination of insulin concentrations to permit inferences on NEFA metabolism in the context of a minimal metabolic modeling approach.

In light of the compelling case for modeling NEFA using glucose observations (in the fashion of partitioned modeling) we present our NEFA minimal model.

$$\frac{dR}{dt} = k_c\{G \times (t) - R(t)\}, \qquad R = R_0, t = 0$$

$$G \times (t) = G(t) - g_s, \qquad G(t) > g_s, \textit{ else } 0$$

$$h(t) = \frac{1}{1 + \dfrac{\Phi}{R(t)}}$$

$$\frac{dNefa}{dt} = S_{ffa}\{1 - h(t)\} - K_{ffa}Nefa(t), \qquad Nefa = Nefa_0, t = 0$$

Here, $R(t)$ [mM/l] represents glucose concentration in a remote glucose pool associated with modification of lipolysis, k_c [min^{-1}] is the delivery and disposition rate of $R(t)$, $h(t)$ is the effect of $R(t)$ on lipolysis rate, g_s is a threshold of blood glucose needing to be present before lipolysis is resumed, $G(t)$ is blood glucose (represented by glucose observations), $G^*(t)$ is blood glucose above the threshold level, S_{ffa} [μM/l·min^{-1}] is the maximum achievable lipolysis rate, K_{ffa} [min^{-1}] is the NEFA oxidation rate, $Nefa(t)$ [μM/l] is the blood NEFA level, and Φ is the value of $R(t)$ at half the peak lipolysis rate.

In Figure 2 we present a graphic of the model, and in Figure 3 we show typical fits of the NEFA model to three dairy cow NEFA profiles following a glucose challenge.

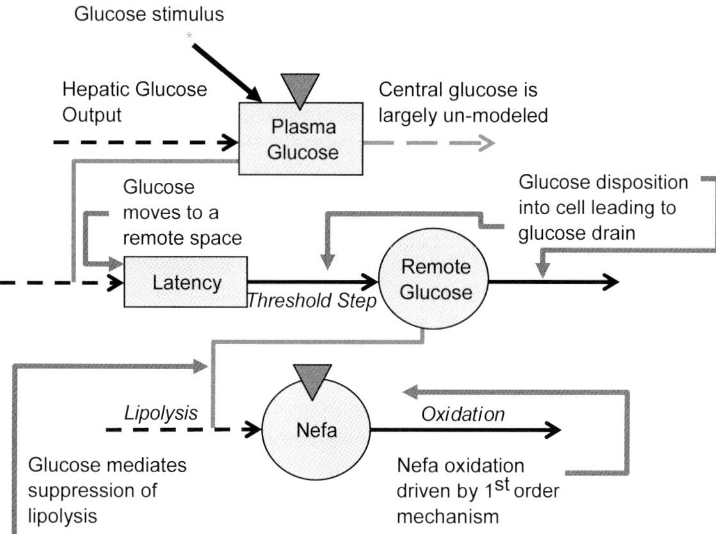

Figure 2. The novel minimal NEFA model in graph form.

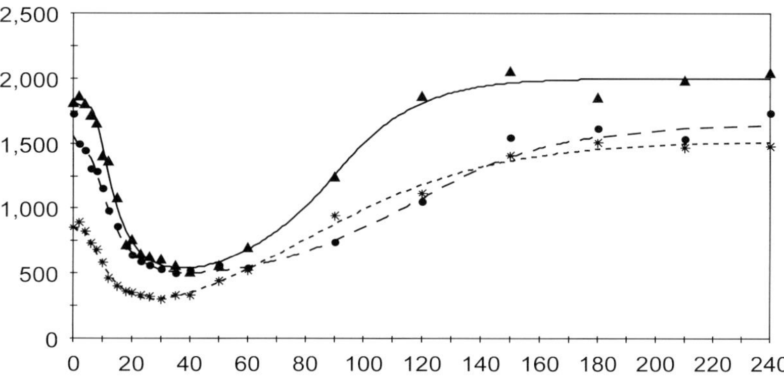

Figure 3. Fits of the minimal NEFA model to dairycow responses from three treatment groups in an energy metabolism study. The triangles, dots and asterisks go with the solid, long-dashed and short-dashed lines respectively. The lines are the NEFA model predictions and the symbols are the observations. NEFA units are μM/l and the time units are minutes after the administration of the glucose injection.

Conclusion

We have explained the principles underpinning the development of a minimal approach to metabolic modeling (Boston and Moate 2008) and we have illustrated their direct application to the glucose minimal model and the development of a new NEFA minimal model. We hope that the NEFA minimal model will provide a working environment for diary scientists to explore the implications of health and productions issues in dairying. We welcome inquiries about the NEFA minimal model and would be delighted to collaborate with researchers in this area.

References

Bauman, D.E., and Currie, B., 1980. Partitioning of nutrients during pregnancy and lactation: a review of mechanisms involving homeostasis and homeorhesis. Journal of Dairy Science 63:1514-1529.

Bergman, R.N., Inder, Y.Z., Bowden, C.R. and Cobelli, C., 1979. Quantitative estimation of insulin sensitivity. American Journal Physiology Endocrinology and Metabolism and Gastrointestinal Physiology 5(6):E667-E677.

Bergman, R.N. and Cobelli, C., 1980. Minimal modeling, partition analysis, and the estimation of sensitivity. Federation Proceedings Vol. 39, No.1 'Theoretical Aspects of Tracer Methods', pp.110-115.

Bergman, R.N., Phillips, L.S. and Cobelli, C., 1981. Physiologic evaluation of factors controlling glucose tolerance in man: measurement of insulin sensitivity and b-cell glucose sensitivity from the response to intravenous glucose. Journal of Clinical Investigations 68:1456-1467.

Boston, R.C., Stefanovski, D. Moate, P.J. Sumner, A.E. Watanabe, R.M. and Bergman, R.N., 2003. MINMOD Millennium: A Computer Program to Calculate Glucose Effectiveness and Insulin Sensitivity from Frequently Sampled Intravenous Glucose Tolerance Test. Diabetes Technology & Therapeutics 5:1003-1015.

Boston, R.C., Roche, J.R. Ward, G.M. and Moate, P.J., 2008. A novel minimal model to describe non-esterified fatty acid kinetics in Holstein diary cows. Journal of Dairy Research 75:13-18.

Boston, R.C. and Moate, P.J., 2008. A novel minimal model to describe NEFA kinetics following an intravenous glucose challenge. American Journal Physiology and Regulatory and Integrative and Comparatives Physiology 294:R1140-R1147.

Derendorf, H. and Hochhaus, G., 1995. Handbook of Pharmocokinetic/Pharmacodynamic Correlation. CRC Press. Chapter 1, 1-35. Pharmacokinetic-Pharmacodynamic Modeling of Reversible Drug Effects by Jurgen Venitz.

Komoragiri, M.V.S., Casper, D.P. and Erdman, R.A., 1998. Factors effecting body tissue mobilization in early lactation dairy cows. 2 Effect of dietary fat on mobilization of body fat and protein. Journal of Dairy Science. 81:169-175

Moate, P.J., Roche, J.R., Chagas, L.M. and Boston, R.C., 2007. Evaluation of a compartmental model to describe non-esterified fatty acid kinetics in Holstein dairy cows. Journal of Dairy Research 74:430-437.

Moore, M.C., Connolly, C.C. and Cherrington, A.D., 1998. Autoregulation of hepatic glucose production. European Journal of Endocrinology 138:240-248.

Park, K.S., Rhee, B.D., Lee, K.U., Lee, H.K., Koh, C.-S. and Min, H.K., 1990. Hyperglycemia per se Can Reduce Free Fatty Acid and Glycerol Levels in the Acutely Insulin-Deficient Dog. Metabolism 39:595-597.

Qvisth, V., Hagstrom-Toft, E., Enoksson, S., Sherwon, R.S., Sjoberg, S. and Bolinder, J., 2004. Combined Hyperinsulinemia and Hyperglycemia, But Not Hyperinsulinemia Alone, Suppress Human Skeletal Muscle Lipolytic Activity. The Journal of Clinical Endocrinology and Metabolism 89:4693-4700.

Toffolo, G., Bergman, R.N., Finegood, D.T., Bowden, C.R. and Cobelli, C., 1980. Quantitative estimation of beta cell sensitivity to glucose in the intact organism: a minimal model of insulin kinetics in the dog. Diabetes, 29:979-990.

Tonelli, J., Kishore, P., Lee, D. and Hawkins, M., 2005. The regulation of glucose effectiveness: how glucose modulates its own production. Current Opinion in Clinical Nutrition and Metabolic Care 8:450-456.

Implementation of a genetic algorithm for optimization within the Cornell Net Carbohydrate and Protein System framework

T.P. Tylutki[1], *V. Durbal*[1], *C.N. Rasmussen*[1] *and M.E. Van Amburgh*[2]
[1]*Agricultural Modeling and Training Systems LLC, 418 Davis Rd, Cortland NY 13045, USA; tom@agmodelsystems.com*
[2]*Department of Animal Science, 272 Morison Hall, Cornell University, Ithaca NY 14853, USA*

Abstract

Models such as the Cornell Net Carbohydrate and Protein System include many non-linear functions. As such, non-linear optimization techniques that converge quickly and efficiently for field application are required. The objective of this paper is to introduce a genetic algorithm for optimization within the CNCPS ver. 6.1 framework. Genetic algorithms are generally categorized as global search heuristics. The genetic algorithm initially seeds the optimization with binary (0,1) representations of potential solutions (chromosomes). It then introduces crossover and mutation rates (set by the user) that automatically force changes in the chromosome combinations by changing the binary coding. Each solution is evaluated against fitness tests (e.g. nutrient and feed constraints). Two types of nutrient constraints have been utilized: soft and hard. A hard constraint forces the solution to be within set ranges. Soft constraints are set to be either equal, or within the range of the hard constraints. As solutions are evaluated, they are compared with soft constraints first. If a solution falls between a soft and hard constraint, a penalty function is applied. Solutions not meeting hard constraints are removed from the solution set. The penalty adds a 'cost' to the solution. If the resulting 'cost' adjusted solution is favorable over other solutions, it is kept within the solution set. This allows for solutions to be evaluated that may be nutritionally acceptable but slightly less then desirable. As an example, given variation in parameter measurements and model variation, it is nearly impossible to say that a 20.9% peNDF solution is different then a 21% peNDF solution; however, the cost of such a solution may be 1-10% different. Genetic algorithms also allow multiple objective functions. In this implementation, least cost or maximum income over feed costs were selected. Evaluations have shown that marginal incomes can be increased 5-10% by simply changing the objective function.

Keywords: nonlinear, objective function, models

Abbreviations used: CNCPS = Cornell Net Carbohydrate and Protein System, CPM = Cornell-Penn-Miner Dairy Formulation Software version 3, ME = metabolizable energy, MP = metabolizable protein, peNDF = physically effective neutral detergent fibre, GA = genetic algorithm, IOFC = income over (minus) feed cost, EO = evolutionary optimization.

Introduction

The commercial application of any model requires the model to be integrated in a user-friendly interface. It must be fast, stable, and contain an optimizer. Optimization has been used in the feed industry for over thirty years. The majority of optimizers used in animal agriculture are based on the simplex method that was developed by George Dantzing in 1947. Simple linear optimization systems assume a constant response slope. These systems have typically been used with the minimization Lagrange function. The typical structure of the problem would be: minimize f(x, y). This simplest form, while providing mathematically correct solutions, typically fails to provide nutritionally correct solutions. Adding constraints (Minimize f(x, y) SUBJECT TO: $i < x < j$, fi(g, y)) places bounds on potential solutions. These bounds can be nutritional, logistically, or experientially based. Regardless of basis, they add the user's bias to solutions; thus the original Lagrange function may be satisfied representing a local solution.

A common perception exists that every formulation problem, if solvable, has one and only one solution. This may be true for certain problems with specific constraints. However, relaxing the constraints can provide additional solutions that may be as correct – illustrating that there is no one perfect solution. This is a difficult concept for most people to accept as they have been trained to expect either a feasible or infeasible solution. Non-linear models highlight the multiple feasible solutions problem. As models become more complex, modelers are relying on non-linear, and in many cases dynamic, approaches. These approaches represent the biological system more appropriately but they introduce field implementation complexities.

The Cornell Net Carbohydrate and Protein System (Tylutki *et al.*, 2008) contains multiple non-linear functions. Attempts have been made to apply curve-peeling approaches to allow linear optimization with mixed success. A more appropriate solution would be the inclusion of a non-linear optimization routine. Others have attempted this (e.g. CPM Dairy ver. 3), again with mixed success (Boston *et al.*, 2000). Newer approaches, such as genetic algorithms, offer flexibility and robustness not historically found in optimization implementations.

The objective of this paper is to introduce the GA concept within the CNCPS format. The approach with an example is utilized.

Genetic algorithm and evolutionary optimization

The core concepts of GAs and EOs is rather simple for animal scientists to understand. A GA approaches the problem as a biological organism where an initial population is developed. In terms of EOs (which are a type of GA), this initial population is referred to as *chromosomes*. The chromosomes contain *genes* that are simple binary representations (0,1) with the chromosome length (number of genes) determined by the complexity of the problem (constraints, objective function, etc.). During the evolutionary process, the chromosomes are crossed (bred) to produce new offspring. These offspring must pass a fitness test to determine if they will remain in the new population or be removed from future crossings. As offspring are evaluated with fitness tests, surviving offspring replace existing offspring and the evolution continues until either objective function is met or predetermined time/iteration numbers are exceeded. Initial population size (number of chromosomes) and mutation rates are programmatically set. Varying these values allows the developer to balance computation time with effective search space evaluation. Robustness of the EO is enhanced by random mutations introduced at the gene level (Michalewicz and Schmidt, 2002; Tylutki, 2002). Figure 1 shows a simple EO structure.

Constraint utilization within an EO differs from traditional linear optimization. In traditional linear systems, constraints result in simple true/false responses. That is, if a result fails to meet a constraint (i.e. a value of false returned), the solution is considered infeasible. In an EO, constraints can be multi-dimensional. This is accomplished by using 'soft' and 'hard' constraints coupled with penalty functions (Palisade, 1998; Tylutki, 2002). This methodology allows the modeler to allow acceptable ranges for parameters (soft constraints) while acknowledging wider bounds that are biologically possible and relevant. Each soft constraint can have different penalty styles ranging from linear through exponential, logarithmic, etc. Selection of the penalty function not only impacts the overall EO in terms of speed and robustness, it requires careful consideration regarding potential positive or negative effects on the animal being modeled. Figure 2 illustrates the behavior of different penalty

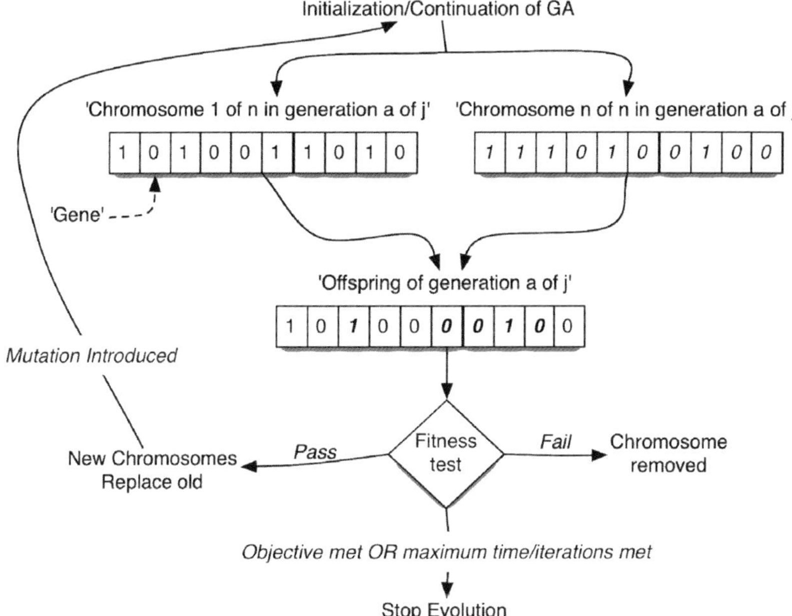

Figure 1. Generic evolutionary optimization structure highlighting chromosomes, genes, mutations, and crossing.

functions. In this example, the y-axis represents an adjusted solution (in this case, the difference from a desired solution of 5) and the x-axis is the number of generations. The simple residual curve would be used when it was desirable to be close to the desired (soft constraint) yet have a small negative impact as residuals increased. The logarithmic penalty exhibits interesting behavior in this example, where early solutions are heavily penalized followed by a period of small change. An example of this is meeting the amino acid requirements for an animal where performance is heavily penalized prior to meeting requirements. The linear penalty follows somewhat similar behavior as the logarithmic in that early solutions are penalized more heavily. Yet the overall penalty is much greater. A prime example of this behavior would be a traditional linear optimization where adding more energy to a dairy diet resulted in increased milk production.

The exponential penalty is very sensitive as clearly observed in Figure 2. This behavior offers high levels of flexibility in the optimization as it can dynamically expand bounds for a given situation. An example of this is dietary starch for lactating dairy cows. As starch levels are low, rumen health can be maintained (thus a very low penalty cost). As starch begins to increase, the potential for acidosis increases and once this increase begins, it can occur very rapidly. If physically effective NDF were also fitted with an exponential penalty function, rumen health can easily be preserved, as any solution outside of acceptable ranges would be quickly discarded.

Solutions (generations) that meet constraints are evaluated against one or more fitness tests. A fitness test is an evaluation to determine if this generation is equal to or better than the last generation. They are typically compared in relation to the objective function. An example would then be: does this solution result in a lower cost recipe then the last generation (if the objective was cost minimization). As the evolution progresses and solutions are discovered, the solutions are stored

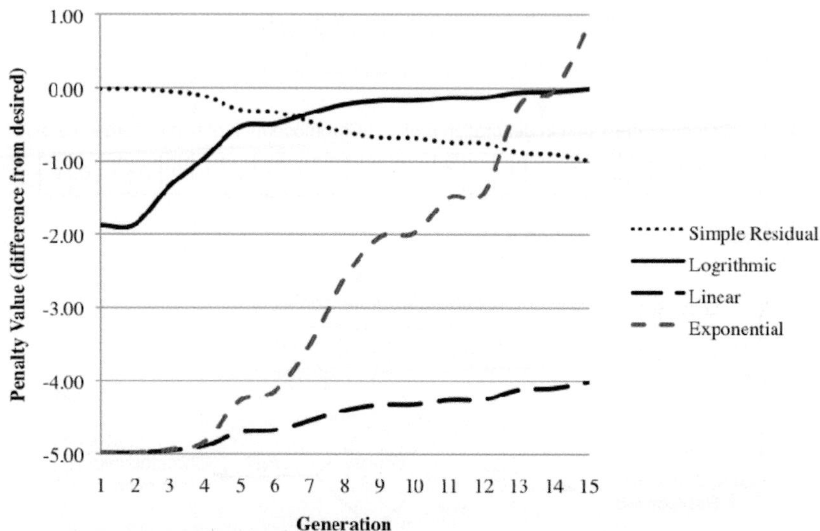

Figure 2. Example behavior of different penalty functions typically used by evolutionary optimizers.

and sorted by degree of fitness. Integrating penalty functions with fitness testing helps guide the EO for future generations as these functions further refine the search space. Fitness tests can be multi-variable; meaning that a calculated value can be used in evaluations that may rely on other components. An example of this would be optimizing a generic utility function where the utility was an integrated parameter with different weights assigned to each variable within it.

The search space of a problem grows in complexity as the number of non-linear functions increase within models. Simple linear problems can easily be evaluated two dimensionally since they are being evaluated for a single intersecting point. Non-linear systems result in three or more dimensions being simultaneously evaluated and can have multiple local optima while never achieving the global optimum. Figure 3 is a two dimensional representation of local versus global optima. The individual shapes within the figure represent local search spaces with feasible solutions. The grey rectangle itself represents the global search space and the areas in grey represent areas of infeasibility. Each o represents a solution and the x represents the local optima. The local search space with the G symbol represents the Global optima. Viewing this three dimensionally would be comparable to a topographical map.

The fitness tests and the constraints must be able to determine if the current evolutionary path is within a local or global optimum region. The ability of an EO to differentiate between local and global is typically where proprietary intellectual property resides for EO developers. This local/ global differentiation also impacts model design. A well-designed optimization may discover that the evolutionary path it is creating is very stable, thus it will move to completion quickly. However, simply forcing the optimization to begin with a non-optimal seed will force it to evaluate a larger search space.

Given the power, flexibility, and robustness of EOs, one must be aware of their limitations. As with any model, EOs can be over-parameterized. In an over-parameterized EO, either no solutions will be found or the search space has been so narrowed that no 'discovery' can be made. Furthermore, as with any optimizer, double (conflicting) bounds will cause failure. An example of this would be

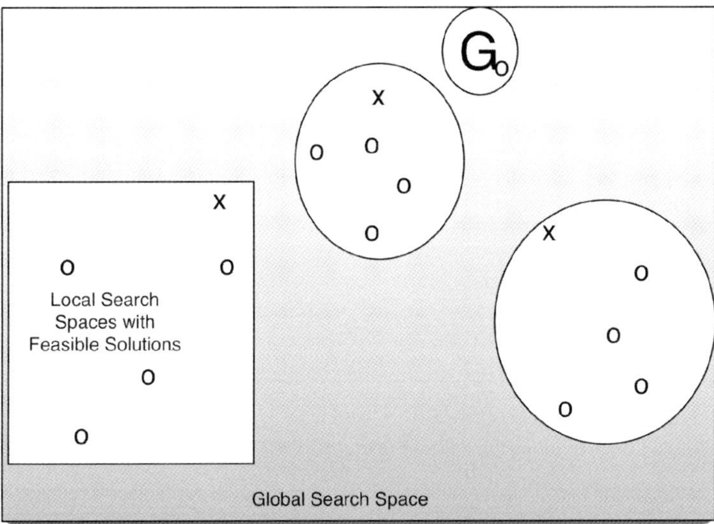

Figure 3. Two dimensional representation of local and global optima.

constraining roughage level in a diet (e.g. a minimum of 10 kg) while also placing a maximum on a roughage source (e.g. maximum inclusion of hay of 9 kg). When invoked with these constraints, the EO will run for an infinite period of time trying to meet them. Equally critical to EOs, and GAs in general, are time requirements. Given that each generation results in offspring that must be evaluated, tested for fitness, attempt to locate global optima, and meet all constraints, computational time can be significantly greater than that of linear optimization. However, if the optimization problem and the underlying model are defined appropriately, a single evolutionary process should result in biologically relevant solutions. While the solution set will be biologically relevant, it is nearly impossible for an EO to return the same solution set if the problem is optimized again. This is strictly due to different beginning chromosomes and evolutionary pathways. Theoretically, allowing the evolution to continue, a global optima will be discovered but the computational time requirements make this prohibitive.

A commercially available EO was integrated with AMTS.Cattle.Pro. AMTS.Cattle.Pro is a commercial implementation of the Cornell Net Carbohydrate and Protein System ver. 6.1. Prior optimization attempts by the modeling group at Cornell and AMTS were made focusing on linear and curve peeling approaches. The linear approach was limited to cost minimization and hard constraints. Additionally, the linear approach forced the user to optimize in terms of percent of requirement versus terms such as energy allowable milk production. This distinction, while small, allows potential negative feedback within the model to be included within optimization. Evolver EDK, developed by Palisades Corp. Ithaca NY (Palisade, 1998), is primarily used as an Excel add-in for financial modeling. As a commercial dynamic linked library, linkages can be established for optimization and Monte Carlo sampling of populations.

Optimization evaluation

A standard dairy cow weighing 703 kg, producing 36.3 liters of 3.7% fat, 3.1% true protein milk was modeled with the CNCPS version 6.1 (using AMTS.Cattle.Pro version 2). Table 1 and 2 list the nutrient feed constraints.

Table 1. Nutrient constraints implemented for evolutionary optimization evaluations.

Constraint	Safe minimum	Minimum	Maximum	Safe maximum
Dry matter intake, kg/d	24.40	24.45	24.49	24.72
ME allowable milk, l/d	36.24	36.28	36.64	37.19
MP allowable milk, l/d	36.24	36.28	36.64	38.10
Rumen NH3, % required	100	100	250	250
NFC, % DM	0	0	40	41
peNDF, % DM	21	22	35	35
Diet Fat, % DM	0	0	6.5	6.5
Lysine, % MP	6.3	6.4	7.6	7.6

Table 2. Ingredients available, their costs ($/mt), and their constraints (kg DM), for optimization evaluation.

Ingredient	Cost, $/mt	Minimum	Maximum
Maize silage	42.45	0	10.0
Alfalfa hay, 20% CP	198.46	0	7.7
Alfalfa hay, 17% CP	170.89	0	7.7
Maize gluten feed	148.84	0	2.72
Maize, ground	170.89	0	7.71
Dried beet pulp	385.89	0	2.27
Wheat middlings	154.35	0	1.36
Maize distillers grains	159.87	0	5.89
Soy oil cake, 47% CP	463.07	0	3.17
Blood meal	1,102.54	0	0.91
Expelled soy oil cake	534.73	0	3.17
Tallow	441.01	0	0.23
Urea	578.83	0	0.18
Smartamine M	13,230.43	0	0.07
Minerals/vitamins	826.90	0.45	0.46

Two simulations were evaluated. These simulations differed by the Lagrange objective function. The first simulation had diet cost minimization as the Lagrange. The second simulation had income over feed cost maximization as the Lagrange. Given the ability to use multi-variable parameters for the fitness test, income over feed cost is calculated for each trial. It is calculated as: Net Milk Price ($/l) × MINIMUM(ME Allowable Milk, MP Allowable Milk) – Ration Feed Cost ($/c). The evolutionary optimization was allowed to iterate until 7 feasible solutions were found for each simulation. In these simulations, the EO found 7 solutions within 90 seconds computational time; within expected time parameters.

The first five feasible solutions when the Lagrange was set to cost minimization are shown in Table 3. Depending upon sensitivity settings within the different EOs, notice that ingredient amounts typically vary in the second or greater decimal. Further program refinement may reduce this small step size. This could reduce computational time or allow a larger search area to be explored. Nutritionally these five solutions are nearly identical when evaluated with the CNCPS yet there is a five-cent

Table 3. First five solutions returned from evolutionary optimization for cost minimization.

Ingredient	Result 1	Result 2	Result 3	Result 4	Result 5
Maize silage	10.0	9.9	9.8	9.9	9.8
Alfalfa hay, 20% CP	2.2	2.2	2.2	2.2	2.2
Alfalfa hay, 17% CP	7.7	7.6	7.6	7.6	7.6
Maize gluten feed	0.986	0.991	0.987	0.992	0.992
Maize, ground	0.021	0.000	0.067	0.004	0.004
Dried beet pulp	1.027	1.015	1.015	1.015	1.017
Wheat middlings	0.811	0.813	0.813	0.810	0.810
Maize distillers grains	0.000	0.000	0.048	0.000	0.000
Soy oil cake, 47% CP	0.781	0.774	0.774	0.774	0.775
Blood meal	0.000	0.001	0.001	0.002	0.003
Expelled soy oil Cake	0.595	0.582	0.587	0.589	0.604
Tallow	0.062	0.062	0.062	0.062	0.062
Urea	0.011	0.010	0.010	0.010	0.010
Smartamine M	0.036	0.036	0.036	0.036	0.036
Minerals/vitamins	0.453	0.453	0.453	0.453	0.453
Total intake	24.71	24.51	24.56	24.46	24.41
Cost/cow	5.79	5.75	5.76	5.74	5.74
Income over feed cost/cow	5.09	5.01	5.06	4.99	4.96

per cow difference in cost. Unfortunately, this five-cent cost variance is accompanied by a ten-cent per cow loss in income over feed cost.

Allowing this simulation to return multiple solutions highlights the critical fact that no single, best solution is typically available. This level of variability does not include true model or feed chemical composition variance either. Unfortunately, this level of imperfect information permeates all ration formulation and model development and is seldom discussed.

Changing the Lagrange in order to maximize income over feed cost resulted in a diet that was lower in maize silage and slightly lower in diet roughage level (Table 4 and 5). The profit maximization simulation resulted in an 8.6% higher income over feed cost given the same diet cost. As seen in Table 5, ME and MP allowable milk were higher in the profit maximization simulations. This is expected behavior as it relates to increased marginal income. At this level of production, negative feedback via model behavior given optimized starch and peNDF levels does not exist. Thus, marginal returns to increased milk production will be high.

Discussion

The behavior of the EO suggests that non-linear animal production models can be successfully optimized. Based on these two simulations and previous work by Tylutki (2002), it appears that dairy farm profitability may be improved at least 5-10% by simply changing the Lagrange function. Traditional formulation programs, especially those utilizing linear optimization, cannot do this. Casual conversations with practicing nutritionists around the world support this idea that current formulation systems have greatly limited cow and farm profitability. Nutritionists have attempted to overcome this by adding new constraints or adjusting formulations post-optimization. While these practices may appear correct, the resulting rations may be far from global, or even local, optima.

Table 4. Model selected optima for cost minimization and profit maximization.

Ingredient (DM kg/d)	Cost minimization	Profit maximization
Maize silage	9.9	8.3
Alfalfa hay, 20% CP	2.2	5.8
Alfalfa hay, 17% CP	7.6	4.6
Maize gluten feed	0.992	0.839
Maize, ground	0.004	0.156
Dried beet pulp	1.015	1.242
Wheat middlings	0.810	1.143
Maize distillers grains	0.000	0.745
Soy oil cake, 47% CP	0.774	0.181
Blood meal	0.002	0.026
Expelled soy oil cake	0.589	0.740
Tallow	0.062	0.122
Urea	0.010	0.000
Smartamine M	0.036	0.027
Minerals/vitamins	0.453	0.453
Total intake	24.46	24.34

Table 5. Selected model output given selected optima for cost minimization and profit maximization.

Ingredient	Cost minimization	Profit maximization
Income over feed cost/cow/d	$4.99	5.42
Cost/cow/d	$5.74	5.74
Dry matter intake, kg/d	24.47	24.34
ME allowable milk, l/d	36.36	37.79
MP allowable milk, l/d	36.45	38.06
Rumen NH_3, % required	152	154
peNDF % DM	29.2	27.6
Roughage % DM	80.6	76.7
NFC % DM	34.3	33.8
Sugar % DM	5.59	6.04
Starch % DM	16.45	15.29
Soluble Fiber % DM	8.78	9.50
Diet Fat % DM	3.23	3.91
Lysine % MP	6.43	6.38
Methionine % MP	2.70	2.46
Crude Protein, % DM	15.6	16.2

Further refinements to the EO implementation should focus on improving the penalty functions and the integration of Monte Carlo sampling to include different variance sources (e.g. model variance, ingredient chemical analysis, animal description, etc.). Current penalty functions assume independence between constraints. In ruminant nutrition, integrating nutrients with penalty functions may be useful. For example, integrating dietary starch within a penalty function for peNDF could guide the evolution towards paths that increase peNDF levels with higher dietary starch. This may be accomplished with additional constraints; however, the penalty functions would remain independent. Other simulations (not shown) have suggested that speed can be enhanced via varying original population size and mutation rate based upon the number of constraints and ingredients. It may be possible to automate this process allowing mutation rates to vary through the simulation and 'remember' these evolutionary paths for future optimization. These stored paths may be further utilized when multi-variable Lagrange functions are used. This approach could be used to optimize a group, or multiple cattle groups, integrated with nitrogen and phosphorus excretion. Tylutki (2002) prototyped this utility function optimization and reported the ability to stabilize income over feed costs at different levels of environmental regulation. Multiple simulations could then be conducted to determine potential policy and farm profitability interactions.

References

Boston, R.C., Fox, D.G., Sniffen, C.J., Janczewski, R., Munsen, R. and Chalupa, W., 2000. The conversion of a scientific model describing dairy cow nutrition and production to an industry tool: the CPM Dairy project. In: McNamara, J.P., France, J. and Beever, D. (eds.) Modelling Nutrient Utilization in Farm Animals, Oxford: CABI Publishing. pp.361-377.

Michalewicz, Z. and Schmidt, M., 2002. Evolutionary Algorithms and Constrained Optimization. In: Sarker, R., Mohammadian, M. and Yao, X. (eds.) Evolutionary Optimization. Kluwer Academic Publishers, pp.57-86.

Palisade Corporation, 1998. Guide to Evolver. Palisade Corporation, Ithaca, NY, USA, 203 pp.

Tylutki, T.P., 2002. Improving herd nutrient management on dairy farms: (1) Daily milk production variance in high producing cows as an indicator of diet nutrient balance. (2) On-farm six sigma quality management of diet nutrient variance. (3) Feedstuff variance on a commercial dairy and the predicted associated milk production variance. (4) A model to predict cattle nitrogen and phosphorus excretion with alternative herd feed programs. (5) Accounting for uncertainty in ration formulation. Ph.D. Dissertation, Cornell University. Ithaca, NY, USA.

Tylutki, T.P., Fox, D.G., Durbal, V.M., Tedeschi, L.O., Russell, J.B., Van Amburgh, M.E., Overton, T.R., Chase, L.E. and Pell, A.N., 2008. Cornell Net Carbohydrate and Protein System: A model for precision feeding of dairy cattle. Animal Feed Science and Technology 143:174-202.

Assessing and optimizing the performance of a mechanistic mathematical model of the sheep mammary gland

C. Dimauro, A.S. Atzori and G. Pulina
Dipartimento di Scienze Zootecniche, Università di Sassari, Via Enrico De Nicola 9, 07100 Sassari, Italy; dimauro@uniss.it

Abstract

A mechanistic mathematical model of the mammary gland was constructed by using a computer-aided simulation via object-oriented modelling. The model was able to simulate the milk production process in dairy sheep taking into account the alveolar dynamics and the energy available for lactation. The overall modelling procedure was developed thorough three steps named understanding, assessment, and optimization. In the first step, physiological and biochemical processes regulating the phenomenon of milk secretion were translated into a differential equation system and theoretical solutions were found. In the assessment step, the ability of the model to reproduce realistic results was tested using real data. Finally, in the optimization step different scenarios were tested using data from an independent experiment. The main outputs of the model were: the actual and potential milk production, the alveolar dynamics and the animal energy balance during lactation Moreover, the model was able to simulate the effects on milk production of external stresses represented by the variation in milking frequency and a reduction in the energy supplied to the animal with the diet.

Keywords: mechanistic model, udder, lactation, dairy ewes

The udder as an 'animal parasite'

In lactating animals, the mammary gland is the largest energy user in the body, consuming more than 2/3 of the energy intake necessary for milk synthesis. During lactation, the whole animal is involved in fuelling the mammary gland (which could be considered a 'privileged' parasite), where the secretory cells ask for energy for their own maintenance and for milk production. In the mammary gland, the secretory cells are organized in alveolar structures that are considered the functional unit of milk production and are characterized by a milk secretion rate which can change during lactation according to the physiological status of the animal.

Most mechanistic mathematical models of the mammary gland are based on alveoli population dynamics: progenitor alveoli that become active, active alveoli that become quiescent, and quiescent alveoli that reactivate or become senescent. The milk yielded at a certain time t depends on the number of active alveoli, on the milk secretion capacity of each alveolus and on the energy available for lactation (from digestion and body fat mobilization). The number of active alveoli times the maximum alveolus secretion rate gives the potential milk production (in terms of energy), i.e. the milk that the animal could produce if the energy at disposal of the mammary gland was maximum. Therefore, the actual milk production depends on both the potential milk production and the net energy available for lactation. For example, it was demonstrated that high-yielding cows (Moe *et al.*, 1971) and ewes (Vernon, 1981) fed *ad libitum* during early lactation did not reach the potential milk production, even when utilizing body reserves. Moreover, in Chios dairy sheep, the nutrition level during pregnancy heavily affects the number of secretory alveoli at lambing (Bizelis *et al.*, 2000; Charismiadou *et al.*, 2000). For these reasons, a deep knowledge of the alveoli population dynamics and milk secretion rate per alveolus could be useful to maximize milk production and minimize the costs of nutrition by formulating the optimum energy allowance that should be administrated to a lactating animal in different periods of lactation. Currently, no experimental result regarding the number of active alveoli at any stage of lactation is available. Mathematical models are, therefore, the only tool to make an indirect estimation of the number and dynamics of secretory alveoli during lactation.

This chapter deals with a comprehensive mechanistic mathematical model of the mammary gland, including its relationship with milking frequency and energy available from metabolism and digestion in dairy sheep. The overall modelling procedure is developed into three steps: understanding, assessment, and optimization. The first step aims to gain a conceptual picture of how the mammary gland system might work. So, physiological and biochemical processes regulating the milk synthesis and ejection are translated into a differential equation system and theoretical solutions are found. In the assessment step the ability of the model to reproduce real results is tested using empirical measurements taken previously. Finally, in the optimization step different scenarios are tested using data from an independent experiment. The model used in this study is the mechanistic mathematical model of the sheep mammary gland that we proposed in 2007 (Dimauro *et al.*, 2007), which consisted of an adaptation of a model of the cow mammary gland (Vetharaniam *et al.*, 2003) to dairy sheep by using a computer-aided simulation via Stella® software (High Performance Systems Inc., Lebanon, NH, USA). Stella® is an object-oriented computer environment which was preferred to others for its capacity to create and present models using a mathematical underlying framework through stock-flow feedback relationships. In this way, model communicability and accessibility was assured being the model designed to be used and understood by researchers in the animal field and by decision makers.

The model

The mechanistic mathematical model of the mammary gland for dairy sheep proposed by Dimauro *et al.* (2007) is structured into three sub-models: alveolar, energy and milk.

Alveolar sub-model

The alveolar-and energy sub-models developed with Stella® software are shown in Figure 1. The secretory alveoli in the mammary gland are divided in three compartments that represent the state variables of this sub-model: P = progenitor alveoli, A = active alveoli, and Q = quiescent alveoli. At parturition, compartment A contains the active alveoli which had been formed in the last part of pregnancy. In the first stage of lactation, there is a rapid proliferation of new alveoli which pass

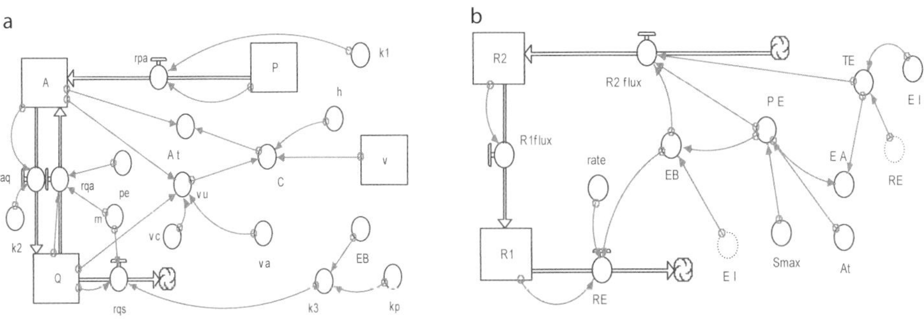

Figure 1. Alveolar (a) and energy (b) sub-models developed with Stella®.

from compartment P to compartment A at a specific rate k_1. Therefore, the number of alveoli r_{pa} that pass from compartment P to A can be expressed as follows:

$$r_{pa} = P \cdot k_1$$

Between milkings, the number of active alveoli which pass from compartment A to compartment Q (i.e. become quiescent) (r_{aq}) is calculated as:

$$r_{aq} = A \cdot k_2$$

where A is the number of alveoli present in compartment A and k_2 is the specific mean quiescence rate.

When all active alveoli become quiescent between two milkings, r_{aq} becomes equal to zero. Emptying the alveoli by milking or suckling reactivates milk secretion. Even if it is plausible that not all alveoli reactivate at exactly the same moment, Shorten *et al.* (2002) showed that the delay between emptying and reactivation could be ignored. Thus, the appearance in compartment A of reactivated alveoli can be modelled as an impulse which begins at the same instant that they are emptied. The number of reactivated alveoli (r_{qa}) at time t is thus influenced by the frequency of milking or suckling (m) and can be expressed as:

$$r_{qa} = p_e \times \text{PULSE}\,(Q, 24/m, 24/m)$$

where p_e is the fraction of emptied quiescent alveoli at each milking and represents the probability that an alveolus is emptied during milking; Q is the number of quiescent alveoli present in compartment Q at time t, and m is the daily frequency of milking. PULSE(*,*,*) is a function of Stella software with the following syntax: the first term represents the intensity of the impulse, the second the time t in which the impulse must begin, and the third the lag time between each impulse.

However, it is also important to consider that the mammary gland is not a closed system: alveoli are lost because of natural senescence which can be accelerated by a prolongation of the quiescent period. So, taking into account the daily milking frequency m, the number of alveoli that become senescent at time t (r_{qs}) can be represented simply as follows:

$$r_{qs} = \frac{k_3 Q}{m}$$

where Q is the number of quiescent alveoli present in compartment Q at time t and k_3 is the senescence rate. We assume that k_3 is not uniform during lactation, being a function of a potential quiescence rate k_p lower than or equal to k_3 (i.e. $k_p \leq k_3$) and of the energy status of the animal at any moment, thus:

$$k_3(t) = \frac{k_p}{E_B}$$

where E_B is a function which represents the energy balance – as a ratio – that is calculated in the energy sub-model.

The number of effectively active alveoli at instant t of lactation (A_t) depends not only on the combined kinetics of activation, quiescence and reactivation, but also on the state of current refilling of the mammary gland. This can be expressed as:

$$A_t = A \times C$$

where A is the number of active alveoli in compartment A and C is expressed by:

$$C = 1 - \left(\frac{v}{v_u}\right)^h$$

where v is the quantity of milk (l) present in the mammary gland at time t, v_u is the total milk capacity (l) of the mammary gland, and h (dimensionless) is a parameter that governs the rate at which C declines as v increases. The total capacity of the mammary gland (v_u) was calculated taking into consideration that in dairy sheep about 65% of the secreted milk accumulates in the cistern and in the large ducts (v_c), whereas the rest remains in the alveoli and small ducts (v_a) (Nudda et *al.*, 2000). The functional form adopted for v_u is the same as that used by Vetharaniam *et al.* (2003) as follows:

$$v_u = v_a(A + Q) + v_c$$

where A and Q are, respectively, the number of active and quiescent alveoli at time t, and v_a and v_c are as defined above.

Energy sub-model

The Stella® diagram developed to calculate energy is in Figure 1. Total net energy for milk production (E_T) is expressed as follows:

$$E_T = E_I + E_R$$

where E_I is the net energy available from feed for lactation, and E_R is the net energy that the animal can mobilize from its body reserves.

The energy necessary for the potential milk production (E_P) is calculated by the following formula:

$$E_P = S_{max} \times A_t$$

where S_{max} is the maximum secretion rate per alveolus and A_t is as defined previously.

So the energy balance (E_B) can be expressed as follows:

$$E_B = \frac{E_T}{E_p}$$

where E_B values may be higher or lower than 1.

The input values for E_I are inserted in the model using a graphical function (E_I) which allows the software Stella® to load external data. The energy that the animal mobilizes at time t from body reserves at lambing (E_R) is represented by the exit flow from stock *R1*. For this reason, using the *if-then-else* syntax implemented in Stella software, *if* $E_I < E_P$ *then* the flow *RE* is closed *else* it is opened, being the energy flux modulated by the converter *rate*. In late lactation, when E_T is greater than E_P (i.e. E_B >1), the energy surplus is deposited as body reserves as represented by stock *R2* through the *R2 flux* flow. The energy accumulated in *R2* will then be used in the next lactation using the *if-then-else* syntax in *R1 flux*.

Milk sub-model

Figure 2 shows the milk sub-model developed with Stella®. The quantity of milk (l) secreted by the mammary gland per unit of time (r_s), depends on the available energy for milk production and can be expressed as follows:

$$r_s = \frac{E_A}{R_{milk}}$$

where R_{milk} is the energy density of milk (MJ/l) (Pulina and Nudda, 2004), and E_A is a converter that if $E_T < E_P$ is equal to E_P else to E_T.

At every milking or suckling, milk is taken out of the mammary gland (r_h). This milk is assumed to be conserved in a *tank* from which the daily quantity (r_d) is then extracted. The quantity r_h can then be modelled using the Stella function *pulse*:

$$r_h = \text{PULSE}(v, 24/m, 24/m)$$

where v is the quantity of milk present in the mammary gland at time t and m is the daily frequency of milking.

Similarly, the milk is extracted (r_d) from the *tank* every 24 hours, that represents the daily milk production, can be expressed as follows:

$$r_d = \text{PULSE}(tank, 24, 24)$$

If the animals are milked once daily ($m = 1$), then obviously $r_d = r_h$. Total cumulative milk production at every instant of lactation is stocked in the final milk container (*total*).

The understanding step

At this step, no experimental data were directly used. The values of the model parameters were from literature or assumed as well as the initial value of progenitor, active and quiescent alveoli. Sensitivity analysis was used to refine the parameters and initial alveoli values to obtain a typical milk production pattern for Sarda dairy ewes with a body weight of about 50 kg at lambing, milked twice daily throughout a 240-day lactation period with a milk production at peak of 1.5-1.7 litres per day reached about 21-30 DIM, and a production of 0.5-0.6 litres at the end of lactation. For external energy inputs (net energy intake and energy from body reserves) simulated data were used. The complete list of values used in the model equations is reported in Table 1. The model was able to foresee the milk production (Figure 3a) and the number of active alveoli (Figure 3b) at different milking frequencies (m = 1, 2, 3).

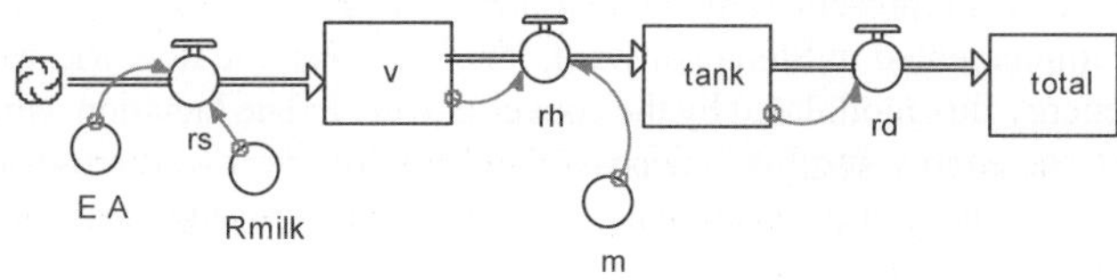

Figure 2. Milk sub-model developed with Stella®.

Table 1. Parameter values used for simulations.

Variables	Description	Value	Units	Source
A[1]	Active alveoli	10^{10}		Estimated
P[1]	Progenitor alveoli	3.5×10^{9}		Estimated
Q[1]	Quiescent alveoli	0		Estimated
S_{max}	Maximum secretion rate	7×10^{-11}	MJ/h	Estimated
R_{milk}	Energy density of milk	4.4	MJ/l	Pulina and Nudda (2004)
k_1	Alveoli proliferation rate	10^{-2}	h^{-1}	Calculated
k_2	Quiescence alveoli rate	10^{-1}	h^{-1}	Calculated
k_p	Senescence alveoli rate	1.6×10^{-3}	h^{-1}	Vetharaniam *et al.* (2003)
p_e	Fraction of emptied alveoli	9×10^{-1}		Vetharaniam *et al.* (2003)
v_a	Alveolus volume	10^{-10}	l	Nudda *et al.* (2000)
v_c	Cistern volume	0.65	l	Nudda *et al.* (2000)
NEI	Net energy intake	1,093	MJ	Simulated
BER	Body energy reserves	197	MJ	Simulated

[1]Initial value.
l = litres; h = hours; MJ = Mega Joule.

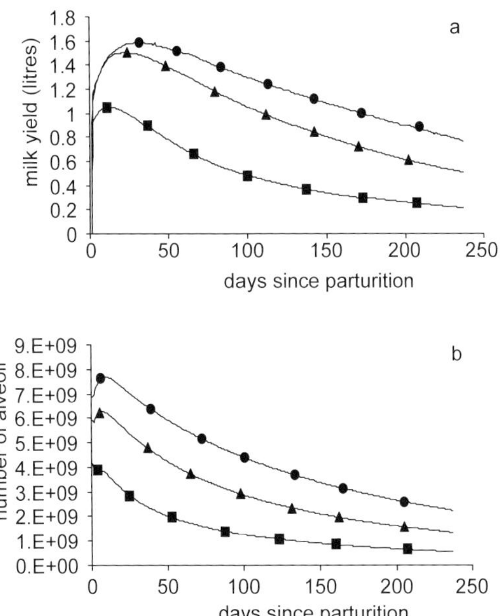

Figure 3. Output of simulation. Milk yield (a) and number of active alveoli (b) evolution in Sarda dairy ewes depending on daily milk frequency (■ one, ▲ two and ● three milkings per day).

The model was then used to simulate two different scenarios characterized by temporary stress factors: reduction in milking frequency (Figure 4a) or in energy from diet (Figure 4b). To simulate milking stress, the basis for simulation was m = 2 to which a once-daily milking stress was imposed for 10 days or 40 days. In the 10-day stress, the decrease in milk production was only temporary, whereas in the 40-day stress, although production increased after twice-daily milking was resumed, the quantity of milk produced was lower than that produced by sheep milked twice daily throughout lactation. This production decrease is clearly connected to a drastic senescence of the alveoli when the animals are milked once daily for a long time. To simulate energy stress, we assumed a 50% reduction in E_I for either 10 days or 30 days, starting at the 42nd day of lactation. The effects of simulated energy stress on milk production were similar to those of reduced milking frequency, with the difference that after regular feeding was resumed, milk production returned to almost normal levels for both cases of energy stress. All these results are in agreement with experimental results reported by Pulina *et al.* (2005) and Velez and Donkin (2005).

The assessment step

The assessment step was developed using data from several experiments conducted at the Department of Animal Science of the University of Sassari (Sassari, Italy). 120 Sarda dairy ewes with similar milk production, BCS, parity and feeding were selected. A standard lactation curve was fixed by fitting a Wood function to the production data, whereas body reserves over lactation were estimated using the mean BCS scores of the animals in four moments of lactation: at lambing, at 60 DIM, at 120 DIM and at the end of lactation (240 DIM). Using this data, the SRNS® model (http://nutritionmodels.tamu.edu/srns.htm) was applied to evaluate the daily net energy given to

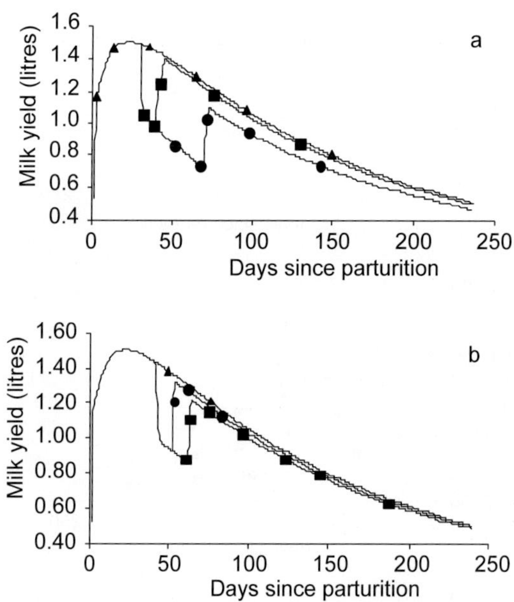

Figure 4. Output of simulation. Temporary fall and recovery of milk production of Sarda dairy ewes subjected to (a) milking stress (▲ continuous m = 2; ■ 10-day period at m = 1; ● 40-day period at m = 1) and (b) food restriction (▲100% energy intake; ■ 10 days at 50% energy intake; ● 30 days at 50% energy intake.

the sheep with a diet consisting of a mixture of forages (grass and legume hay) and concentrate (mainly corn grain).

The parameters of the Wood equation obtained by fitting, with a non-linear procedure (SAS-STAT software), the function $y = at^b\exp(ct)$ to actual data, were a = 1.02, b = 0.16 and c = -0.007, with an $R^2 = 0.86$. The estimated total production was 230 l, with a production of 1.44 l/day at peak (reached at 23 DIM) and a production of 0.46 l/day at the end of lactation (240 DIM). Values of the estimated milk production were uploaded in the Stella® model to obtain the reference lactation curve. Body energy reserves at lambing were fixed at 197 MJ with BCS of: 3.0 at lambing, 2.5 at 60 DIM, 2.8 at 150 DIM and 3.0 at the end of lactation. Body energy reserves at lambing and net energy intake during lactation were the external inputs of the model. As expected, milk production after the first run of the model did not follow the fixed one (the Wood's lactation curve), because the values of body energy reserves and energy intake had been changed to respect the values used in the understanding step. Therefore, the initial values of active and progenitor alveoli were recalibrated via sensitivity analysis. The optimal combination of the initial values of the number of active and progenitor alveoli that gave the desired lactation curve was 9×10^9 and 4.2×10^9, respectively.

One important feature of the model is that it is able to estimate both actual and potential milk production. The potential milk production is the milk that the animal is able to produce if the energy at disposal of mammary gland is maximum. Figure 5a displays the estimated milk yielded throughout lactation and the milk that could potentially be produced. The largest difference between the actual and the potential milk production, which occurred from lambing to around 50 DIM, gives the energy that should be supplied to the animal in order to produce the maximum quantity of milk allowed by the mammary gland. Moreover, the model was able to simulate the dynamics of body

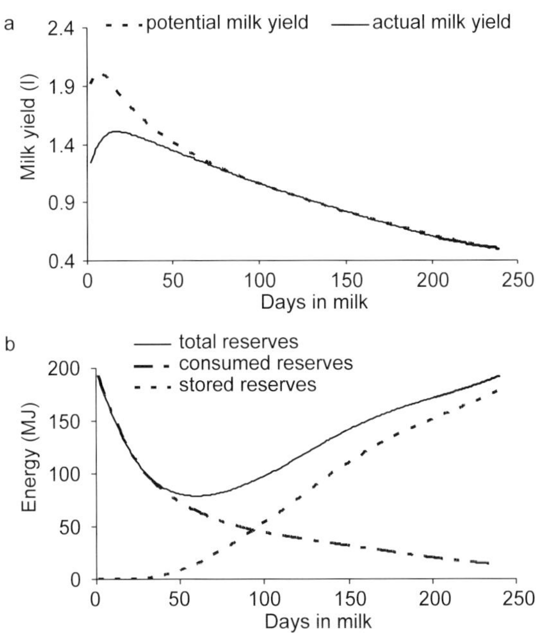

Figure 5. Output of simulation: (a) potential and actual milk yield of Sarda dairy ewes; (b) evolution of body energy available for lactation in Sarda dairy ewes.

energy reserves during lactation (Figure 5b). It can be seen that by the end of lactation the initial amount of reserves was set-up again.

The optimization step

Data used at this step were from a complex experiment reported by Bizelis *et al.* (2000) and Charismiadou *et al.* (2000). We focused our attention only on 13 Chios pregnant ewes, which were randomly assigned to high (HL) and low (LL) levels of nutrition, receiving 110% of their energy requirements for maintenance plus pregnancy for two fetuses and 90% of their maintenance energy requirements, respectively. After lambing, ewes of both groups were fed *ad libitum* and feed intake was recorded individually and daily for the first 3 weeks after lambing and then weekly up to the 12th week of lactation. Nutritional effects were monitored by recording body weight before feeding, daily for the first 3 weeks of lactation and then twice a week up to the end of the experiment (120 days after lambing). In the experiment, the authors found significant and large differences between the two groups of Chios ewes regarding mammary DNA content (a measure of its number of cells) at lambing and milk production during the first 120 days of lactation (for more details see Bizelis *et al.* (2000) and Charismiadou *et al.* (2000)). These results were used to optimize the model performance.

With this aim, two separate analysis were conducted: one for the HL group and one for the LL group. Technical results of the fitted Wood equation for both groups are listed in Table 2.

The values of the estimated milk production were uploaded in the Stella® model. The resulting lactation curves were the reference curves in the sensitivity analysis. Body reserves at lambing were 273 MJ for HL and 101 MJ for LL, whereas total net energy intake during the first 120 DIM were 1,012 MJ and 932 for HL and LL, respectively. Body energy reserves at lambing and net energy intake were the new external inputs to the model. Separated sensitivity analyses were conducted for the two groups (HL and LL) by varying the initial values of active and progenitor alveoli. The match of milk estimated by the model and that of the previously uploaded Wood lactation curves was reached only after the insertion of the maximum secretion rate for alveolus (S_{max}) in the sensitivity analysis. Table 3 shows the results of sensitivity analysis by fixing the initial values for active and progenitor alveoli and S_{max}.

Table 2. Production traits of lactation curves of Chios ewes fed two different energy regimes (High level (HL) = 110% of maintenance plus pregnancy for two fetuses, Low level (LL) = 90% of maintenance) during pregnancy.

	HL group	LL group
a	1.232	0.986
b	0.172	0.184
c	-0.008	-0.007
R^2	0.85	0.94
Time at peak (days)	21	25
Milk at peak (l/day)	1.76	1.48
Total milk (l)	180	155

Table 3. Results of sensitivity analysis by fixing the initial values for active and progenitor alveoli and S_{max}. obtained from data collected by Bizelis et al. *(2000) and Charismiadou* et al. *(2000) on Chios ewes fed two different energy regimes (High level (HL) = 110% of maintenance plus pregnancy for two fetuses, Low level (LL) = 90% of maintenance) during pregnancy.*

	HL group	LL group
Active alveoli	10^9	10^8
Progenitor alveoli	5×10^9	4×10^9
Total alveoli	6.3×10^9	4.1×10^9
S_{max} (MJk^{-1})	14.5×10^{-11}	9×10^{-11}

Conclusions

The mammary gland model constructed with a computer-aided design via object-oriented technique is very flexible and easily expandable. This model is able to interface with nutritional sub-models. The way in which it is constructed facilitates the task of starting from a simple model that can utilize the available quantitative and qualitative data. When additional data becomes available, the model can then be expanded by adding additional building objects and connections to it. Similarly, the whole structure of the model could be easily modified to represent a different or an updated understanding of the studied system. In those cases, model parameters should be re-optimised to obtain the best combination able to simulate the real milk production pattern.

The model was built to provide a plausible description of the evolution of milk production in Sarda dairy sheep as a function of milking frequency, milking stress and feed stress; then it was evaluated on real data from experiments conducted on Sarda sheep and, finally, its parameters were optimized by using experimental results of Chios sheep fed rations with different energy levels.

The model is well suited for simulating stress situations commonly found when rearing dairy sheep in Mediterranean countries. For its great flexibility, it can also be easily adapted for other sub models (e.g. hormonal) and can also deal with the evolution of other productive parameters (e.g. sheep milk fat content and protein content) over time.

Acknowledgments

We acknowledge prof. J.A. Bizelis who has kindly provided the original data of his experiments. We are grateful to dr. Ana H.D. Francesconi for revising the manuscript.

References

Bizelis, J.A., Charismiadou, M.A. and Rogdakis, E., 2000. Metabolic changes during the perinatal period in dairy sheep in relation to level of nutrition and breed. II. Early lactation. Journal of Animal Physiology and Animal Nutrition 84:73-84.

Charismiadou, M.A., Bizelis, J.A. and Rogdakis, E., 2000. Metabolic changes during the perinatal period in dairy sheep in relation to level of nutrition and breed. I. Late pregnancy. Journal Of Animal Physiology and Animal Nutrition 84:61-72.

Dimauro, C., Cappio-Borlino, A., Macciotta, N.P.P. and Pulina, G., 2007. Use of a computer-aided design to develop a stress simulation model for lactating dairy sheep. Livestock Science 106:200-109.

Moe, P.W., Tyrell, H.F. and Flat, W.P., 1971. Energetic of body tissue mobilization. Journal of Dairy Science 54:548-553.

Nudda, A., Pulina, G., Vallebella, R., Bencini, R. and Enne, G., 2000. Ultrasound technique for measuring mammary cistern size of dairy ewes. Journal of Dairy Research 67:101-106.

Pulina, G. and Nudda, A., 2004. Milk production. In: Pulina, G. (eds.) Dairy sheep nutrition. CABI Publishing, Wallingford, Oxfordshire, UK, pp 1-12.

Pulina, G., Nudda, A., Macciotta, N.P.P., Battacone, G., Fancellu, S. and Patta, C., 2005. Non-nutritional strategies to improve lactation persistency in dairy ewes. Proceedings of the 11th Great Lakes Dairy Sheep Symposium. 2005. Vermont, USA, pp.38-68.

Shorten, P.R., Vetharaniam, I., Soboleva, T.K, Wake, G.C. and Davis, S.R., 2002. Influence of milking frequency on mammary gland dynamics. Journal of Theoretical Biology 218:521-530.

Velez, J.C. and Donkin, S.S., 2005. Feed restriction induces pyruvate carboxilase but not phosphoenolpyruvate carboxychinase in dairy cows. Journal of Dairy Science 88:2938-2948.

Vernon, R.G., 1981. Lipid metabolism in the adipose tissue of ruminant animals. In: Christie, W.W. (ed.) Lipid Metabolism in Ruminant Animals. Pergamon Press, Oxford, England, pp.280-353.

Vetharaniam, I., Davis, S.R., Soboleva, T.K., Shorten, P.R. and Wake, G.C., 2003. Modelling the interaction of milking frequency end nutrition on mammary gland growth and lactation. Journal of Dairy Science 86:1987-1996.

A new development in pig growth modelling

P.C.H. Morel[1], D.L.J. Alexander[2], R.L. Sherriff[1], D. Sirisatien[3] and G.R. Wood[3]
[1]Animal Nutrition Division, IFNHH, Massey University, Palmerston North, New Zealand; p.c.morel@massey.ac.nz
[2]Mathematics of Materials, CSIRO, Clayton South, VIC 3169, Australia
[3]Department of Statistics, Macquarie University, NSW 2109, Australia

Abstract

Pig growth simulation models are used to determine feeding strategies that improve profitability on commercial farms. For a given farm, the number of diets fed, their energy, amino acid content, the quantity fed and the diet period can vary, thus giving a very large number of possible feeding strategies (F, as many as 10^{50}). Adding nonlinear optimisation methods to a growth model allows us to find an F yielding the maximum for a given objective function, usually the gross margin per pig or per pig place and year. Our simulation program links a linear program for a least-cost diet formulation, a stochastic pig growth model and a genetic algorithm (GA) to find the F giving a best solution. When finding F for maximum profitability, the gross margin obtained by the GA is higher than that found by random search or by feeding pigs to their maximal lean growth. In the growth model, pig genotypes are characterised by the maximal protein deposition potential (Pdmax), minimum lipid to protein ratio (MinLP) and the energy intake potential (p). In the model, variances and covariances of these quantities are used to grow a population of pigs instead of a single pig. A simulation study was conducted to investigate how different pig genotypes and different relative economic weightings for gross margin and nitrogen excretion affect the nitrogen retention and profitability. It was found that a large increase in nitrogen retention can be achieved through diet optimisation before profitability is compromised and that a lean genotype will have better nitrogen retention. Adding stochasticity to the model for a given population size showed that as the variances increase the variability in gross margin increases and with unchanging variances, the variability in gross margin decreases as the population size increases. Overall, using a feeding schedule which maximises gross margin for a single pig within a population of pigs results in a lower gross margin.

Keywords: stochasticity, optimisation, genetic algorithm

Introduction

This paper presents a summary of the work carried out by the authors in the area of pig growth modelling over the last decade. Pig growth simulation models are used to determine feeding strategies that improve profitability on commercial farms (de Lange *et al.*, 2001, Parsons *et al.*, 2007). For a given farm, the number of diets fed, their energy (d), amino acid content (r), the quantity fed (p) and the diet period (t) can vary, giving a very large number of possible feeding strategies (F, as many as 10^{50}, with useful grid choice) and only one or a few will result in maximal profitability. Non-linear optimisation will cope with this large number of possible feeding schedules when searching for a best solution. Usually pig growth models simulate growth for an average pig and the best solution found for such a single pig is unlikely to be the same as the one found for a population of pigs. This paper describes a stochastic pig growth model using non-linear optimisation and presents some results of a simulation study investigating the impact of population size and variance and covariance of pig genotype parameters on feeding strategies which maximise profitability.

Simulation program description

A flow chart representing how the different sub-programs and variables are linked together is presented in Figure 1.

The market situation is represented by the costs of the weaner pig, the ingredients and the return per kg carcass in New Zealand dollars. The return per kg carcass is based on back fat thickness and hot carcass weight.

For a given farm, the number of diets fed, their energy, their amino acid content, the quantity fed and the diet period can vary, thus giving a very large number of possible feeding strategies. Several nonlinear optimisation methods have been used to solve this problem (Alexander *et al.*, 2006) and a genetic algorithm (GA; Reeves and Rowe, 2002) has been found to be the most suitable technique. In the GA, the 'genome' is the feeding strategy *F* and crossovers and mutations act in a natural way. The GA has been extensively tested and generally a population of 20 genomes and a stopping criterion of 10 to 20 iterations with no changes have been found to be satisfactory.

A feeding strategy (*F*) that is generated by the GA is defined as a finite set of diets:

$$F = (d_1, r_1, p_1, t_1; d_2, r_2, p_2, t_2; \ldots; d_n, r_n, p_n, t_n),$$

where each diet consists of a quadruple (*d, r, p, t*) with:
d = Digestible energy density in the diet, in MJ/kg;
r = Minimum ileal digestible lysine to digestible energy ratio in the diet, in g/MJ;
p = Proportion of the standard NRC (1988) voluntary daily digestible energy intake;
t = Feeding period in days.

Parameters *d* and *r* are used for the least cost diet formulation, and *p* and *t* are input for the stochastic growth model.

The cost of the diets is minimized using a linear program (LP) which includes the prices and the nutrient composition of the different ingredients. In the LP the constraints *d* and the ileal digestible lysine level ($r \times d$) are fixed, while the constraints for the other essential amino acids (AA) are in

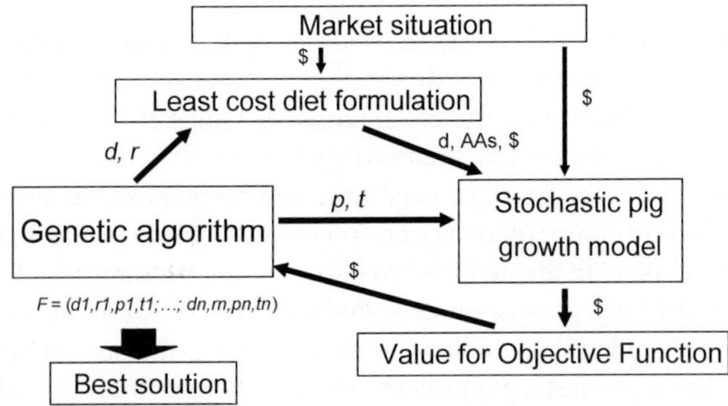

Figure 1. Overview of the simulation program linking a linear program for least-cost diet formulation, a stochastic pig growth model, a genetic algorithm (GA) to find a feeding strategy F giving a best solution for a given objective function in a specific market situation (see text for the abbreviations).

excess of the ideal balance. The constraints on mineral levels and ingredient inclusion levels are in line with commercial diet formulation. The diet composition (digestible energy, amino acid) and cost generated by the LP are inputs to the stochastic growth model.

The pig growth model used in the program is that described in detail by de Lange (1995). In this model, pig genotypes are characterised by the following three quantities: maximal protein deposition potential (Pdmax), minimum whole body lipid to protein ratio (MinLP) and the proportion of the standard NRC (1988) voluntary daily digestible energy intake (p). Stochasticity is applied on these parameters using the following correlation matrix. (R. Sherriff, personal communication):

	MinLP	Pdmax	p
MinLP	1		
Pdmax	-0.55	1	
p	0.3	0.25	1

The following steps were used to generate a pig population of size N:

1. Convert the correlation matrix to a covariance matrix using the set of variances determined by the coefficient of variation (CV).
2. Generate independent standard normal random variables, A ~ N(0, 1).
3. Calculate a multivariate random normal variable B from a distribution with the required correlation matrix, using $B = AL^T + \mu$, where L is the matrix square root (or a lower triangular matrix from a Cholesky factorization) of the covariance matrix and μ is the mean vector of pig parameters.
4. Repeat Steps 2 and 3 until the parameters for a pig population of size N have been generated.

For each F the growth model calculates the value ($) for a given objective function (OF). The OF to be maximised by the GA is usually the gross margin per pig (GMP) or per pig place and year (GMPPY),

OF = GMP = (Gross Return – Feed Costs – Weaner pig cost) per animal
OF = GMPPY = GMP × number of pigs grown per pig place per year

Results of simulation studies

Genetic algorithm performance from Alexander et al. *(2006)*

When finding F for maximum profitability, the gross margin obtained by the GA in 30 seconds is higher than that found by 100 hours of random search (Figure 2). To find out if a global optimum has been found by the GA, the nature of the objective function was investigated. The value of the objective function was examined along random cross-sections of the domain passing through the putative maximum. Figure 3 shows the Euclidean distance between a feeding schedule on the cross-section and the putative optimal feeding schedule, for the case of ten diets ($n=10$) and all feeding periods (t) of seven days. It was found that the objective function was shaped like a high dimensional 'craggy mountain' with one peak (Figure 3).

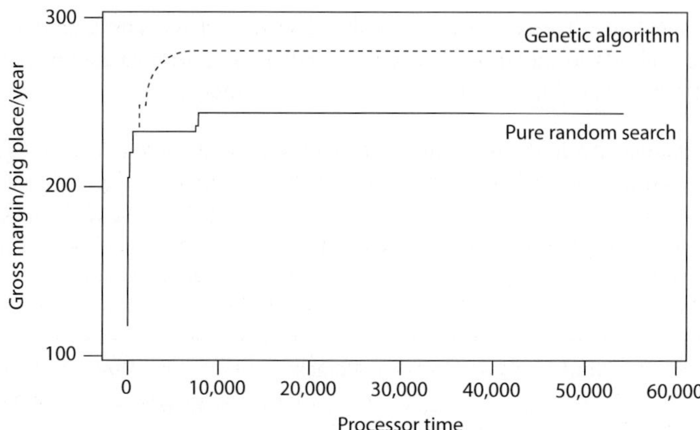

Figure 2. Changes in gross margin per pig place and year against processor time for pure random search and genetic algorithm optimisation.

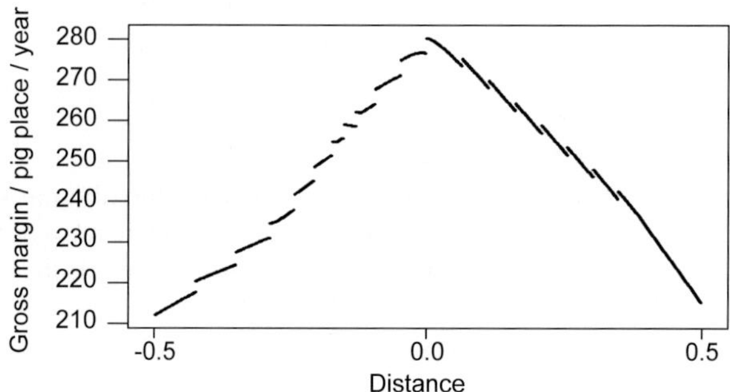

Figure 3. The craggy volcano shape of a section through the objective function at the best solution.

Increasing nitrogen retention (adapted from Morel and Wood, 2005)

A simulation study was conducted to investigate how different pig genotypes and different relative economic weightings for gross margin and nitrogen excretion affect the nitrogen retention and under different feeding strategies in Switzerland. A modified objective function used in the work was:

$$OF = a \times GMP + b \times NeP$$

where:
a,b are weighting factors;
GMP (CHF) is the gross margin per pig;
NeP (kg) is the total nitrogen excreted.

In changing the value for the factor a and b more or less importance can be given to a reduction in nitrogen excretion. For the simulations, the value of *a* was set to 1 and the values for *b* varied between 0 and 120. Growth was simulated for three genotypes having different Pdmax and MinLP (Fat genotype: 120 g/d and 0.9; Normal genotype: 160 g/d and 0.75; Lean genotype: 200 g/d and 0.6). The simulation used two feeding alternatives, one based on commercially available dietary specifications (practical) and one optimised with the GA. Each feeding alternative had 3 diets and the possible values of *d, r, p* and *t* for the two feeding alternatives and diets investigated are presented in Table 1.

With both the practical and optimised feeding schedules the Lean genotype had higher GMP and higher percentage of nitrogen retention (%NRet) than the Normal and Fat genotypes (Figure 4). For all genotypes, higher GMP and %NRet were observed for the Optimized diets when compared with Practical diets. As the *b* values in the objective function increased, so did the %NRet. With the optimized diets, %NRet can be improved without a large reduction in GMP over a wide range of *b* values. For example, across genotype, a 10-12% increase in %NRet (*b*=0 vs. *b*=20) was associated with a reduction in GMP of CHF 11.5 to CHF 13.5. In the case of the Lean genotype, with optimised diets, nitrogen retention up to 55% can be reached. The data show that a more sustainable pork meat production system can be achieved by using better genotypes and optimizing diet composition to match their genetic potential for lean growth.

Table 1. Investigated diets per feeding alternative.

	Practical			Optimised		
Diet	I	II	III	I	II	III
d	13.2	13.6	13.6	12-17	12-17	12-17
r	0.89	0.75	0.75	0.3-1.2	0.3-1.2	0.3-1.2
p	0.8	0.8	0.65-0.85	0.8	0.8	0.65-0.85
t	14	35	Open	14	35	Open

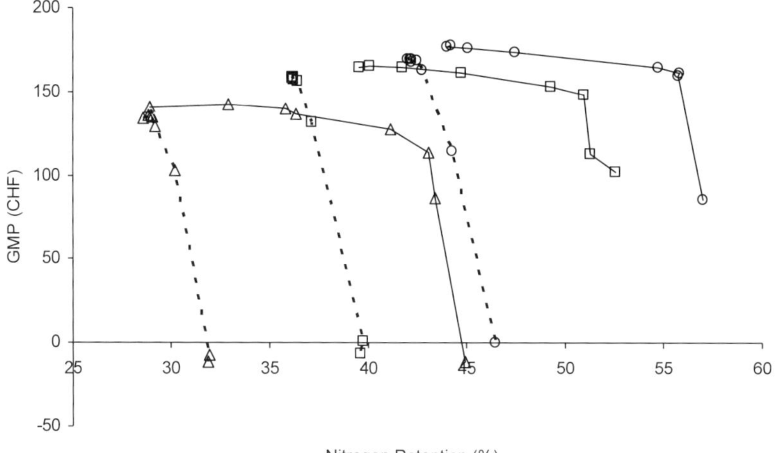

Figure 4. Gross margin per pig (GMP) and percentage nitrogen retained per pig for Fat (Δ), Normal (□) and Lean (○) pig genotypes fed either three practical diets (dotted line) or three optimized diets (solid line). For each line the first point on the left hand side represent the GMP value when b =0 and the last point on the right hand side the GMP value for b=120.

Effect of stochasticity on feeding strategies which maximise profitability (from Morel et al.*, 2008)*

In this study three genotypes varying in their Pdmax and MinLP were used: Fat genotype (120 g/d and 0.9), Normal genotype (160 g/d and 0.75), Lean genotype (200 g/d and 0.6) were considered. The population sizes (*N*) were set to either 1, 5, 25, 125 or 625. The stochasticity is applied to PdMax, MinLP and *p* via a variances and covariances matrix as described earlier in the text. The CV for the three parameters were set to 0 for *N*=1, and to 5%, 10% or 20% for *N*>1. The objective function was the gross margin per pig place and year (GMPPY). The market situation is represented by the price schedule and feed cost current in 2006 in New Zealand. Diets are changed on a weekly basis and growth is simulated between 20 and 85 kg live weight. Each combination: Genotype × *N* × CV was simulated 10 times, the GA had 20 genomes (in this case feeding strategies *F*) and the search was stopped when GMPPY had not improved for 10 iterations. The effect of including covariance was investigated only for *N*=125 and CV=10%. The mean GMPPY with standard deviations for all the Genotype × *N* × CV combination are given in Table 2.

Overall, higher GMPPY were achieved with the Lean genotype, thus highlighting the importance of using an improved pig genotype to maximise profitability.

Independently of the genotype, for a given level of variation (CV) between pigs, the variability (standard deviation for the 10 runs) in the maximal GMPPY found by the GA decreases as the population size (N) increases and for a given population size (N) the variability in GMPPY increases as the CV increases. The inclusion of covariances in the model has only a small effect on the GMPPY means and standard deviations.

Table 2. Mean and standard deviation (in brackets) for gross margin per pig place and year for three pig genotypes, differing population sizes and differing coefficients of variation in the pig genotype parameters. Covariance between pig parameters is considered for population size N=125.

Genotype	Coefficient of variation for pig parameters				
	N	0%	5%	10%	20%
120/0.9	1	135.9 (0.7)	-	-	-
Fat	5	-	146.2 (15.8)	140.40 (15.8)	117.8 (14.5)
	25	-	139.9 (1.2)	127.7 (9.0)	120.00 (20.0)
	125	-	139.2 (1.3)	126.6 (3.8)	114.8 (10.7)
	125 Cov	-	-	127.9 (4.5)	-
	625	-	138.9 (0.7)	126.1 (1.7)	115.3 (3.0)
160/0.75	1	308.4 (1.1)	-	-	-
Normal	5	-	305.2 (4.8)	298.5 (12.2)	257.1 (39.0)
	25	-	303.3 (2.9)	296.5 (6.2)	260.1 (17.8)
	125	-	305.3 (0.4)	293.7 (2.0)	264.1 (3.6)
	125 Cov	-	-	292.8 (3.1)	-
	625	-	305.3 (0.2)	295.0 (1.2)	258.0 (3.1)
200/0.6	1	356.2 (0.5)	-	-	-
Lean	5	-	351.3 (6.7)	347.3 (10.0)	306.7 (32.5)
	25	-	350.3 (2.0)	340.7 (6.9)	315.1 (8.7)
	125	-	350.6 (0.5)	342.3 (2.3)	319.6 (3.2)
	125 Cov	-	-	344.7 (2.0)	-
	625	-	350.7 (0.3)	342.1 (0.5)	318.0 (2.6)

The lysine to energy ratio (r) in the optimal feeding strategy determined for a single pig was always lower than that determined for a population of 125 pigs with covariance, and this difference is smaller for Lean genotype than Fat genotype (Figure 5). Feeding the diet which maximises gross margin for a single pig to a population of pigs results in lower gross margins of -$14.1, -$11.5 and -$6.4 for fat, normal and lean genotypes, respectively.

Summary

The addition of both linear and non-linear optimisation methods to a biological dynamic mechanistic pig growth model allows the rapid identification of feeding strategies which maximised or minimised specific objective functions (e.g. gross margin, minimal nitrogen excretion). Solutions provided by the optimisation method differ when the simulations are conducted for a single average pig compared to a population of pigs. This indicates the need to add stochasticity to growth models. Moreover, when growth models are used on commercial farms, it is important that the stochastic parameters (number of pigs, variances and covariances) match those observed on the farm.

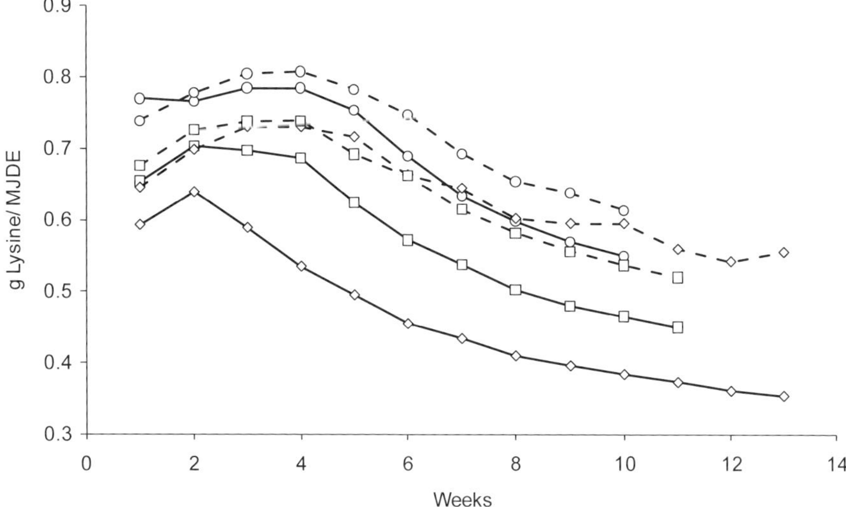

Figure 5. Lysine to digestible energy ratio (r; g/MJ) to maximise profitability for a single pig (solid lines) or a population of 125 pigs with a 10% coefficient of variation and covariance between pig parameters (dotted lines), for Fat (◊), Normal (□) and Lean (○) genotypes. In all cases, the optimal feeding schedule for the population demands a higher value of r.

References

Alexander, D.L.J., Morel, P.C.H. and Wood, G.R., 2006. Feeding strategies for maximizing gross margin in pig production. In: Pintér, J.D. (ed.) Global Optimization – Scientific and Engineering Case Studies, Springer Science+Business Media, *LLC,* 233 Spring Street, New York, NY 10013, USA, pp.33-43.

De Lange, C.F.M., 1995. Framework for a simplified model to demonstrate principles of nutrient partitioning for growth in the pig. In: Moughan, P.J., Verstegen, M.W.A. and Visser-Reyneveld, M.I. (eds.) Modelling growth in the pig, Wageningen Press, Wageningen, the Netherlands, pp.71-85.

De Lange, C.F.M., Marty, B.J., Birkett, S., Morel, P.C.H. and Szkotnicki, B., 2001. Application of pig growth models in commercial pork production. Canadian Journal of Animal Science 81:1-8.

Morel, P.C.H. and Wood, G.R. 2005. Optimisation of nutrient use to maximize profitability and minimise nitrogen excretion in pig meat production systems. Acta Horticulturae 674:269-275.

Morel, P.C.H., Wood, G.R. and Sirisatien, D. 2008. Effect of genotype, population size and genotype variation on optimal diet determination for growing pigs. Acta Horticulturae 802:287-292.

NRC (National Research Council), 1988. Nutrient requirements of swine. 9th edition, National Academic Press, Washington, DC, USA, 189 pp.

Parsons, D.J., Green, D.M., Schofield, C.P. and Whittemore, C.T., 2007. Real-time control of pig growth through an integrated management system. Biosystems Engineering 96:257-266.

Reeves, R.C. and Rowe, E.J., 2002. Genetic algorithms-principles and perspectives, a guide to GA theory. Kluwer Academic Publishers, Boston/Dordrecht/London, 332pp.

Identifiability and accuracy: a closer look at contemporary contributions and changes in these vital areas of mathematical modelling

L.O. Tedeschi[1] *and R. Boston*[2]
[1]*Department of Animal Science, Texas A&M University, College Station, TX 77843-2471, USA; luis.tedeschi@tamu.edu*
[2]*School of Veterinary Medicine, University of Pennsylvania, Kennett Square, PA 19348, USA*

Abstract

In the development of mathematical models we are plagued with two key concerns: is the model unique and accurate enough? Under the umbrella of the first question we will show whether the extent of the model structure in our data will shed light on all of the model's parameters. Identifiability in model advancement helps us with the question 'will a proposed experiment on a system enable us to determine values for the parameters of a model of that system?' The model is assumed to be known and to reflect the response of the system to the experiment. Identifiability is not concerned with the precision with which the parameters can be estimated. A comprehensive review and demonstrations of the practical considerations regarding identifiability are presented. We further introduced new concepts which when omitted from consideration in the identifiability process can lead to serious misjudgement about the resolution of important aspect of the model under investigation, and its identifiability classification per se. On the second question we discuss current methods and their pitfalls in evaluating model adequacy. The concordance correlation coefficient (CCC) has been commonly used to assess agreement of continuous data, such as agreement of a new assay and a gold-standard assay, observed versus model predicted values, different methods, raters, and reproducibility. The main advantage of CCC is that it incorporates precision and accuracy simultaneously. There are four methods to compute CCC; two of them are extremely dependent on normality whereas the other two are more robust to non-normality. Therefore, datasets that departures from normality or are not independent may pose a significant problem when using CCC with the squared distance payoff function. It has been shown that CCC may indicate an increasing agreement as the marginal distribution becomes different, suggesting the agreement should be compared over a similar range.

Keywords: assessment, adequacy, computer, development, methods, simulation

Introduction

Mathematical models are often plagued by unclear development methodology that is based on data sources that are undefined, and relations to the real world problem often are questionable. Models are developed based on quantitative and qualitative data, and complex interactions among variables (including time) to allow for a better understanding of the dynamics of key variables and their impact in the overall ability of the model to predict the problem being studied and the formulation of multiple hypothetical interventions. Therefore, mathematical models have to be tested for identifiability to reflect the responses of the system to experiments and for accuracy to yield predictions that are unbiased and trustworthy.

Identifiability in model advancement helps us with the question 'will a proposed experiment on a system enable us to determine values for the parameters of a model of that system?' The model is assumed to be known and to reflect the response of the system to the experiment. Identifiability is not concerned with the precision with which the parameters can be estimated. We further introduced new concepts which when omitted from consideration in the identifiability process can lead to serious misjudgement about the resolution of important aspect of the model under investigation, and its identifiability classification per se. The question then becomes 'if identifiability is so powerful why is it only occasionally used, and only for models of modest size and complexity?' Indeed, identifiability analysis is very complex and this situation has not been improved by the array of definitions and demonstrations that appear in the literature. Godfrey (1983) whose account of identifiability is by far the clearest, and uncompromising has said 'a number of definitions of terms associated with identifiability have appeared over the years, but these have led to some criticism'. Others (Cobelli and DiStefano, 1980) have proposed some rather formal definitions but several aspects of their definitions have been criticized (LeCourtier and Walter, 1981).

Model adequacy is the ability of a mathematical model to satisfactorily predict new information given the model structure. Precision and accuracy are two important measurements of model adequacy (Tedeschi, 2006). Precision is related to how closely the model predicts similar values consistently and accuracy is related to how closely the model predicts the 'true' values. Because the estimates will differ somewhat from the 'true' values, the problem becomes what level of difference is acceptable and how to measure it. There are several techniques to access accuracy of predictions (Tedeschi, 2006), including mean bias, mean square error of predictions, linear regression, kappa statistics, etc. Freese (1960) suggested that three elements are required when determining model accuracy: (a) desired accuracy required (or tolerated inaccuracy), (b) measurement of obtained accuracy using real experiments, and (c) a method to assess statistical acceptability; the author provided an assessment technique based on the χ^2. Later, Krippendorff (1970) introduced the concepts of agreement between two continuous variables and several modifications have been proposed (King and Chinchilli, 2001a; Liao, 2003; Lin, 1989), including for discrete data (Carrasco and Jover, 2005) and repeated measures (Carrasco *et al.*, 2009). These modifications resulted in the development of the concordance correlation coefficient (CCC) and its variations. Then, subsequently, total deviation index (TDI) and the coverage probability (CP) to measure agreement for continuous data were discussed (Lin *et al.*, 2002).

The objective of this paper is to highlight the benefits and pitfalls of identifiability and accuracy related to mathematical modelling. This paper provides a revised and updated appraisal to that presented by Boston *et al.* (2007).

Identifiability

The concept was likely discovered around 1956 (Berman and Schonfeld, 1956). Even thought exponential models provided an excellent fit for kinetic data (e.g. ratio iodine turnover in thyroid studies), there were limitations to the extent and structure of the model that could be developed from the experiments. Identifiability analysis embodies the efforts to assess whether a proposed experiment on a presumed model of a system will yield enough 'additional' information about the system to enable determination of all the unknown parameters of the model (Boston *et al.*, 2007). There is the need to use simple terms and definitions when discussing identifiability (Godfrey, 1983). Jacquez (1996, ch. 15) who takes several 'passes' to convey the meaning of identifiability said 'if one can check global identifiability, that is the way to go. But at times that process can be very difficult to carry out'. Then, he added 'in many applications in the biological sciences there are prior estimates of parameters or information to constrain the parameters … In such cases much can be learned by checking the ... identifiability … by numerical methods'.

Definitions

Figure 1 portrays an experiment on a two compartmental system. The system is 'represented' as a two-compartment model with exchange between the two compartments and irreversible loss from the first compartment (number 1). The experiment comprises a pulse input to the first compartment and subsequent sampling of that compartment, capturing the response of the system to the pulse. The equations to the response, presuming linear kinetics are as follows:

$$\underline{\dot{x}} = A\underline{x} + B\underline{u} \tag{1}$$

and the observations are given by:

$$\underline{y} = C\underline{x} \tag{2}$$

where u = system challenges (δ or k_i, for example), B = input (or application) matrix, C = observation (or output) matrix, x = predicted response and y = observed response, and A = condensed, or well-formed matrix representation of k_{ij}, viz.:

$$A = \begin{bmatrix} -a_{11} & a_{12} \\ a_{21} & -a_{22} \end{bmatrix},\ a_{11} = k_{01} + k_{21},\ a_{12} = k_{12},\ a_{21} = k_{21},\ \text{and } a_{22} = k_{02} + k_{12}\ (\text{if } k_{02} > 0).$$

By taking the Laplace transform of Equation 1 and Equation 2, we obtain: $Y = CX = C \times (sI - A)^{-1} \times$ BU, or, solving for Y we have $Y = g_n(s,\phi_n)/g_d(s,\phi_d)$, where k_{ij} = basic or model parameters, ϕ_i = observational parameters, V_j = volume of distribution (compartment j), basic parameter, and $g_{n/d}$ = numerator or denominator polynomial in s.

An unidentifiable system

Using Figure 1 to portray a system which is perturbed, or pulsed, in the first compartment and sampled in the first compartment; thus exposing the pattern of damping of the pulse, and propagation of the residue of the pulse throughout the system, assuming $u_2 = 0$; we have: $u = u_1 = [t = 0^+, \delta, 0]$ and $C = [1,0]$, which represent the input (compartment 1 pulse, δ) and output (compartment 1 sampled). The condensed form of the model parameters is as shown above. The state equations for this system can be expressed as $\dot{x}_1 = -a_{11}{\cdot}x_1 + a_{12}{\cdot}x_2 + u_1\ (= \delta)$ and $\dot{x}_2 = a_{21}{\cdot}x_1 - a_{22}{\cdot}x_2 + u_2\ (= 0)$

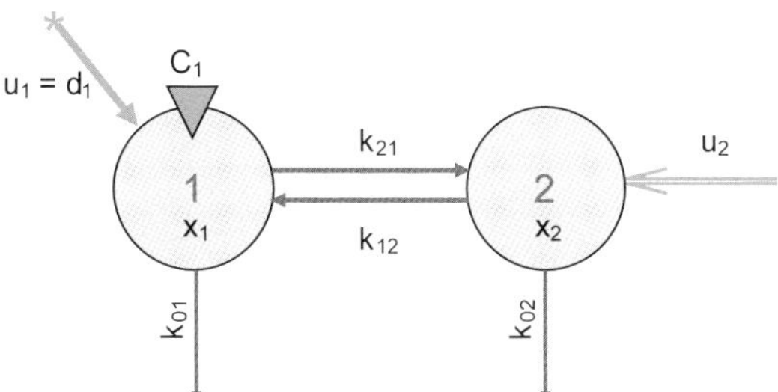

Figure 1. Experiment on a two compartmental system. A bolus injection ($u_1 = d_1$) into compartment 1, a steady state input (u_2) into compartment 2, and sampling C_1 from compartment 1. Note the three linear exchanges with basic parameters k_{21}, k_{12}, k_{01}, and k_{02}. The responses of compartments 1 and 2 are respectively x_1 and x_2.

Taking the Laplace transform of the state equations, assuming that uppercase X or Y represents the Laplace transform of the, lowercase, response functions x or y, we have:

$X_1(s + a_{11}) - X_2\, a_{12} = U_1$ and $-X_1\, a_{21} + X_2\,(s + a_{22}) = 0$

Solving this for X_1 and assuming that $u_1 = \delta = 1$, yields:

$$X_1 = \frac{\begin{vmatrix} U_1 & -a_{12} \\ 0 & (s+a_{22}) \end{vmatrix}}{\begin{vmatrix} (s+a_{11}) & -a_{12} \\ -a_{21} & (s+a_{22}) \end{vmatrix}} = \frac{U_1(s+a_{22})}{(s+a_{11})(s+a_{22}) - a_{12}a_{21}} = \frac{s+a_{22}}{s^2 + s(a_{11} - a_{22}) + (a_{11}a_{22} - a_{12}a_{21})}$$

$$\therefore X_1 = \frac{s+\phi_1}{s^2 + \phi_2 s + \phi_3}$$

We observe three parameters, $\phi 1$, $\phi 2$, and $\phi 3$ but we need four parameters. Therefore, this system is unidentifiable based on the current 'proposed' experiment. Figure 2 illustrates how such an experiment fails to produce adequate information for identifiability using WinSAAM (Stefanovski *et al.*, 2003).

An identifiable system

When an experiment on a system can be shown to render a model unidentifiable and the boundaries of parameter values afforded from such an experiment are excessively wide then we can modify our experiment in such a way that will expose enough information about the system as to allow all the model parameters to be identifiable. We found that the experiment, as designed above, provided three unique pieces of information about the model but, for identifiability purposes, the model called for four pieces of information. A suggested extension to the experiment is to admit a step infusion (u_2) into compartment two; thus we examine the output from compartment one using

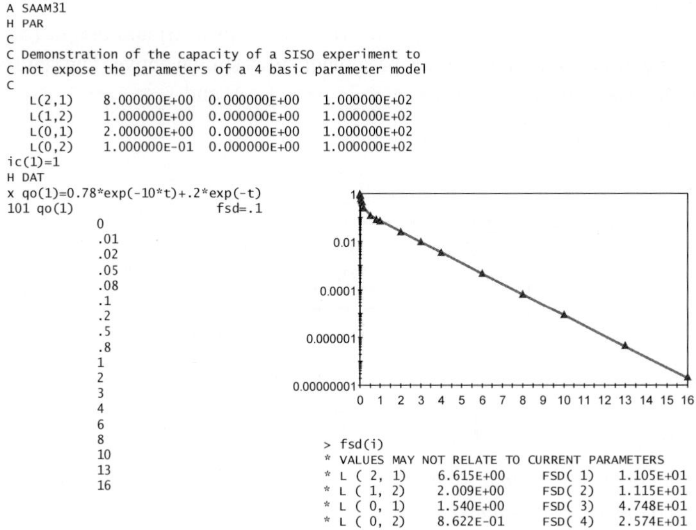

Figure 2. A WinSAAM model to estimate the basic parameters of the model of Figure 1 when V_1 is ignored.

a DISO (Double Input; bolus to compartment one, infusion to compartment two, Single Output; sample only compartment one) design.

Using our standard notation the state equations giving the anticipated response of the system, the state equations are $\dot{x}_1 = -a_{11}{\cdot}x_1 + a_{12}{\cdot}x_2 + \delta\ (t = 0^+, \mathrm{d})$ and $\dot{x}_2 = a_{21}{\cdot}x_1 - a_{22}{\cdot}x_2 + k$. The Laplace transforms yields $X_1(s + a_{11}) - X_2\, a_{12} = \delta$ and $-X_1\, a_{21} + X_2\,(s + a_{22}) = k_i / s$, and our output, y, is given by $y = x_1/C$. Solving these equations for X_1 we obtain:

$$X_1 = \frac{(d.s^2 + d.a_{22}s + k_i.a_{12})}{s(s^2 + \beta_1 s + \beta_2)}$$

We now see that we have gone from three to four, or five, observational parameters whereas before we reported just three observational parameters yielded from the experiment. Whether we need four or five parameters depends on our observation units.

Figure 3 depicts an identifiable system using WinSAAM (Stefanovski *et al.*, 2003). We changed (a) the observations 'qo(1)' and (b) the experiment by adding the infusion (u_2) into compartment two 'uf(2)', in addition to the pulse (u_1) into compartment one. We iteratively adjust the parameters and see that they converge (a) to the correct values and (b) to well-resolved estimates, thus supporting that the new experiment leads to identifiability of all model parameters, and thus the model is itself identifiable based on this new experiment.

Statistical Constraints. In our second example, we found that by eliminating one basic (model) parameter from the model of example one we moved from a situation where the model was unidentifiable to a situation with the model was globally identifiable. This is actually a case of conditional identifiability. There are virtually no situations now where investigations are advanced against the background where nothing is known. Either similar studies have already been undertaken, perhaps with different constraints or on different subjects, or *in vitro* investigations (versus *in vivo*) have yielded additional information to the study at hand. Using this previous, beforehand,

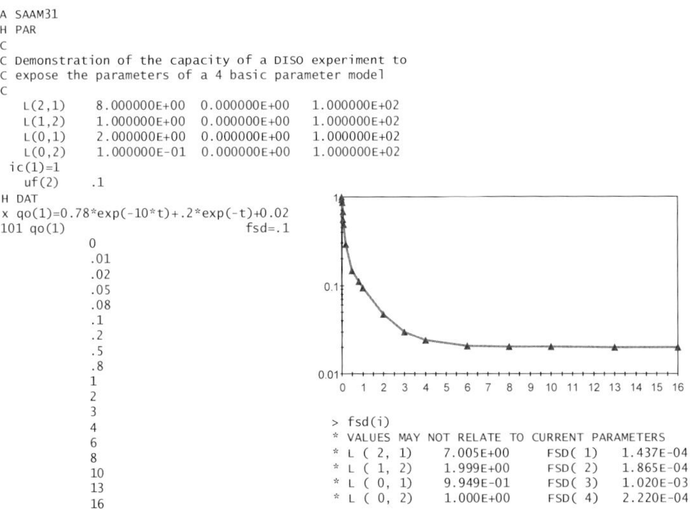

Figure 3. A WinSAAM model to estimate the basic parameters of the model of Figure 1 assuming an infusion to compartment two.

information is referred to as Bayesian Analysis and we advance our modelling objective by admitting key pertinent statistical knowledge available from prior related research. Further examples are provided by Boston *et al.* (2007).

Accuracy

As discussed in the accuracy section of the paper by Boston *et al.* (2007), the CCC index has been widely used to assess model adequacy. Because the CCC index measures the correlation between two variables assuming that their linear relationship is expected to have a slope of unity (45° line) and intercept zero, it combines precision and accuracy measurements at the same time. Equation (3) has the simplest arrangement to compute the CCC index between two variables. Assumptions and limitations of the CCC index have been discussed elsewhere (Atkinson and Nevill, 1997; Deyo *et al.*, 1991; Lin, 1989; Müller and Büttner, 1994; Nickerson, 1997).

$$\rho_C = \rho \times C_b \quad (3)$$

in which $C_b = 2/v + v^{-1} + u^2$, $v = \sigma_Y/\sigma_X$, and $u = \mu_Y - \mu_X/\sqrt{\sigma_Y/\sigma_X}$

where ρ_c is the concordance correlation coefficient estimate, ρ is the Person's correlation coefficient (precision estimate), C_b is the accuracy estimate, σ_Y and σ_X are the standard deviation of Y and X variables; respectively, and μ_Y and μ_X are the means of Y and X variables; respectively.

The concept of generalized estimating equations (GEE) has been introduced, including kappa for categorical data (Williamson *et al.*, 2000) and CCC for continuous data (Barnhart *et al.*, 2002; 2005). The GEE among k raters can be modelled with covariates adjustments, GEE do not require the full knowledge about the distribution of the data, and the estimates and the inferences for the estimates can be obtained simultaneously (Lin *et al.*, 2007).

Barnhart *et al.* (2005) considered the case of measuring CCC, among any two and among all k raters, with each rater measuring each of the n subjects multiple times (independent replications). A series of indices (intra-rater CCC, inter-rater CCC, and total CCC) have been proposed. The generalized CCC by King and Chinchilli (2001a) can be reduced to kappa statistics for categorical data and the original CCC (Lin, 1989) for continuous data.

Carrasco and Jover (2003) indicated the CCC calculations devised so far (Barnhart *et al.*, 2002; King and Chinchilli, 2001a; Lin, 1989) are special cases of the intra-class correlation (**ICC**) and it can be computed by variance components method with a two-way mixed, no interaction model, using maximum likelihood or restricted maximum likelihood methods. Lin *et al.* (2007) proposed an integrated approach based on those discussed by Barnhart *et al.* (2005) and Carrasco and Jover (2003).

Lin *et al.* (2007) indicated that when the number of replicates (m) tends to infinity (large number of replicates), the unified method to compute CCC is similar to that proposed by Barnhart *et al.* (2005); when m is equal to 1 the unified CCC is similar to the ICC proposed by Lin (1989), King and Chinchilli (2001a), Barnhart *et al.* (2002), and Carrasco and Jover (2003); and finally when m is equal to 1 and there are only 2 raters (k = 2 variables, such as Y and X), the unified CCC is similar to the original CCC proposed by Lin (1989). Lin *et al.* (2007) also described the inclusion of covariate when computing the unified CCC. However, Lin *et al.* (2007) suggested future work is still needed for non-normal continuous data (such as the inclusion of log and logistic functions) and the second problem is when there are missing data.

Carrasco *et al.* (2007) evaluated these four methods to compute CCC, namely moment method (Lin, 1989), variance components (Carrasco and Jover, 2003), U-statistics (King and Chinchilli, 2001b), and GEE (Barnhart *et al.*, 2002; Barnhart and Williamson, 2001). The two former approaches assume normal distribution whereas the two latter do not assume any distribution of the data. The authors concluded these four approaches underestimated the true CCC and its standard error when the distribution between-subjects was skewed. Only with small skewness and large sample size the coverage was satisfactory using the U-statistics and GGE. The U-statistics and the GEE approaches would be expected to be more tolerant to non-normal distributions.

We performed additional simulations (n = 1000), assuming k = 2 and 100 random numbers drawn from normal and log-normal distributions with two correlation coefficients (0.9 and 0.6). The results (Table 1) indicated that under normal distribution, the estimated precision (Pearson correlation) was very close to the theoretical value. The CCC, Cb, and TDI estimates were resistant to skewness whereas the Pearson correlation and CP were not. For log-normal distribution, Cb was resistant to skewness; all other statistics were not. The estimated Pearson correlation was considerably different from the theoretical value. Similarly, Lin *et al.* (2007) calculated CCC, Cb, TDI, and CP assuming normal distribution and different numbers of raters (k) and replicates (m). The estimates resembled the theoretical values. Except for CP-inter, the means of the estimated standard error are very close to the corresponding standard deviations of the estimates.

Boston *et al.* (2007) compared several methods used to compute CCC. They reported each method had different behaviours under different structures of error variance (non-normal distribution), depletion of data points, and sample size. They suggested the use of at least three techniques to calculate CCC when comparing different models or assessing calibration of equipments.

Further development and evaluations of the CCC were performed when repeated measurements are taken for each rater or method, such as longitudinal studies (Chinchilli *et al.*, 1996; King *et al.*, 2007). Chinchilli *et al.* (1996) initially proposed the use of CCC weighed by the amount of within-subject variability for each subject. While, King *et al.* (2007) used the population estimates rather than subject-specific estimates to compute CCC for repeated measures.

The accuracy of an approach to characterize an aspect of a system is the relative agreement between the results based on the approach compared with the results based on a 'gold' standard method. Concordance is a much more profound issue than merely correlation or regression because it needs to embrace the observation range over which the two series of measurements exist, it needs to

Table 1. Statistics of simulations (N=1000) of 100 randomly drawn values from normal and log-normal distributions assuming pre-established correlation coefficients and 2 raters.

Items	Normal distribution						Log-normal distribution					
	r = 0.9			r = 0.6			r = 0.9			r = 0.6		
	Mean	SD	Skew	Mean	SD	Skew	Mean	SD	Skew	Mean	SD	Skew
CCC	0.40	0.20	0.06	0.07	0.01	0.19	0.67	0.04	-2.20	0.07	0.01	0.31
r	0.89	0.04	-3.35	0.60	0.07	-0.39	0.96	0.05	-4.76	0.82	0.12	-1.81
Cb	0.45	0.21	0.03	0.11	0.01	0.17	0.70	0.04	-0.03	0.09	0.02	2.03
TDI	22.28	8.20	0.00	52.79	1.16	0.06	12.02	1.04	10.66	49.51	0.74	2.65
CP	0.01	0.01	0.51	0.00	0.00	10.88	0.01	0.01	2.67	0.00	0.00	13.49

penalize a quantification of agreement for not predicting the origin as a critical point, and also for not engaging a line of perfect agreement (slope unity) into its consideration (Boston *et al.*, 2007).

Conclusion

We have demonstrated the principles underlying identifiability and have illustrated them with examples. The mechanism leading to identifiability under Bayesian estimation is not hard to appreciate when we reflect on the reformulated structure of the extended estimation objective. Bayesian estimation using direct nonlinear methods (WinSAAM), or distribution (Gibbs) sampling and Monte Carlo Markov Chain methods are more common and will ultimately replace simple 'at hand' methods and the allied tools garnered to help them.

Accuracy can be determined using four different methods: moments, variance components, U-statistics, and GEE. These methods may yield biased estimates for CCC depending on the distribution and skewness of the data. Missing values might be a problem for some methods.

References

Atkinson, G. and Nevill, A., 1997. Comment on the use of concordance correlation to assess the agreement between two variables. Biometrics 53:775-777.

Barnhart, H.X., Haber, M. and Song, J., 2002. Overall Concordance Correlation Coefficient for Evaluating Agreement among Multiple Observers. Biometrics 58:1020-1027.

Barnhart, H.X., Song, J. and Haber, M.J., 2005. Assessing intra, inter and total agreement with replicated readings. Statistics in Medicine 24:1371-1384.

Barnhart, H.X. and Williamson, J.M., 2001. Modeling concordance correlation via GEE to evaluate reproducibility. Biometrics 57:931-940.

Berman, M. and Schonfeld, R.L., 1956. Invariants in experimental data on linear kinetics and the formulation of models. Journal of Applied Physics 27:1361-1370.

Boston, R.C., Wilkins, P.A., and Tedeschi, L.O., 2007. Identifiability and Accuracy: Two critical problems associated with the application of models in nutrition and the health sciences. In: Hanigan, M. (ed.) Mathematical Modeling for Nutrition and Health Sciences. Roanoke, VA, USA. pp.161-193.

Carrasco, J.L. and Jover, L., 2003. Estimating the generalized concordance correlation coefficient through variance components. Biometrics 59:849-858.

Carrasco, J.L. and Jover, L., 2005. Concordance correlation coefficient applied to discrete data. Statistics in Medicine 24:4021-4034.

Carrasco, J.L., Jover, L., King, T.S. and Chinchilli, V.M., 2007. Comparison of concordance correlation coefficient estimating approaches with skewed data. Journal of Biopharmaceutical Statistics, 17:673-684.

Carrasco, J.L., King, T.S. and Chinchilli, V.M., 2009. The concordance correlation coefficient for repeated measures estimated by variance components. Journal of Biopharmaceutical Statistics 19:90-105.

Chinchilli, V.M., Martel, J.K., Kuminyika, S. and Lloyd, T., 1996. A weighted concordance correlation coefficient for repeated measurement designs. Biometrics 52:341-353.

Cobelli, C. and DiStefano, III, J.J., 1980. Parameter and structural identifiability concepts and ambiguities: a critical review and analysis. American Journal of Physiology (Regulatory, Integrative and Comparative Physiology) 239:R7-24.

Deyo, R.A., Diehr, P. and Patrick, D.L., 1991. Reproducibility and responsiveness of health statis measures. Controlled Clinical Trials 12:142S-158S.

Freese, F., 1960. Testing accuracy. Forest Science 6:139-145.

Godfrey, K., 1983. Compartmental Models and their Application. Academic Press, London and New York.

Jacquez, J.A., 1996. Compartmental Analysis in Biology and Medicine (3rd ed.). BioMedware, Ann Arbor, Michigan.

King, T.S. and Chinchilli, V.M., 2001a. A generalized concordance correlation coefficient for continuous and categorical data. Statistics in Medicine 20:2131-2147.
King, T.S. and Chinchilli, V.M., 2001b. Robust estimator of the concordance correlation coefficient. Journal of Biopharmaceutical Statistics 11:83-105.
King, T.S., Chinchilli, V.M. and Carrasco, J.L., 2007. A repeated measures concordance correlation coefficient. Statistics in Medicine 26:3095-3113.
Krippendorff, K., 1970. Bivariate agreement coefficients for reliability of data. Sociological Methodology, 2:139-150.
LeCourtier, Y. and Walter, E., 1981. Comments on 'On parameter and structural identifiability: non unique observability/reconstructibility for identifiable systems, other ambiguities and new definitions'. IEEE Transactions and Automatic Control 26:800-801.
Liao, J.J.Z., 2003. An improved concordance correlation coefficient. Pharmaceutical Statistics 2:253-261.
Lin, L., Hedayat, A.S. and Wu, W., 2007. A unified approach for assessing agreement for continuous and categorical data. Journal of Biopharmaceutical Statistics 17:629-652.
Lin, L.I.-K., 1989. A concordance correlation coefficient to evaluate reproducibility. Biometrics, 45:255-268.
Lin, L.I.-K., Hedayat, A.S., Sinha, B. and Yang, M., 2002. Statistical methods in assessing agreement: Models, issues, and tools. Journal of the American Statistical Association 97:257-270.
Müller, R. and Büttner, P., 1994. A critical discussion of intraclass correlation coefficients. Statistics in Medicine 13:2465-2476.
Nickerson, C.A.E., 1997. A note on 'A concordance correlation coefficient to evaluate reproducibility'. Biometrics 53:1503-1507.
Stefanovski, D., Moate, P.J. and Boston, R.C., 2003. WinSAAM: A Windows-based compartmental modeling system. Metabolism: Clinical and Experimental 52:1153-1166.
Tedeschi, L.O., 2006. Assessment of the adequacy of mathematical models. Agricultural Systems, 89:225-247.
Williamson, J.M., Lipsitz, S.R. and Manatunga, A.K., 2000. Modeling kappa for measuring dependent categorical agreement data. Biostat 1:191-202.

King, T.S. and Chinchilli, V.M., 2001a. A generalized concordance correlation coefficient for continuous and categorical data. Statistics in Medicine 20: [illegible].

King, T.S. and Chinchilli, V.M., 2001b. Robust estimators of the concordance correlation coefficient. Journal of Biopharmaceutical Statistics 11: 83-105.

King, T.S., Chinchilli, V.M. and Carrasco, J.L., 2007. A repeated measures concordance correlation coefficient. Statistics in Medicine [illegible].

Krippendorff, K., 1970. Bivariate agreement coefficients for reliability of data. Sociological Methodology [illegible].

[illegible] and Wang, [illegible] [illegible] precision and [illegible] repeatability and reliability [illegible] definitions [illegible]

Liao, J.J.Z., 2003. An improved concordance correlation coefficient. Pharmaceutical Statistics [illegible].

Lin, L., Hedayat, A.S. and Wu, W., 2007. A unified approach for assessing agreement for continuous and categorical data. Journal of Biopharmaceutical Statistics [illegible].

Lin, L.I., 1989. A concordance correlation coefficient to evaluate reproducibility. Biometrics [illegible].

Lin, L.I., Hedayat, A.S., Sinha, B. and Yang, M., 2002. Statistical methods in assessing agreement: models, issues, and tools. Journal of the American Statistical Association [illegible].

[illegible], 1994. A critical [illegible] [illegible]

Part 2 – Modelling feeding behaviour and regulation of feed intake

Modelling short-term feeding behaviour

B.J. Tolkamp[1], J.A. Howie[1] and I. Kyriazakis[2]
[1]SAC Research Division, King's Buildings, West Mains Road, Edinburgh EH9 3JG, United Kingdom; bert.tolkamp@sac.ac.uk
[2]Veterinary Faculty, University of Thessaly, P.O. Box 199, 43100 Karditsa, Greece

Abstract

Previous analyses of feeding behaviour have either ignored the bouted structure of that behaviour or used log-survivorship analysis for the estimation of bout criteria. The latter is based on the null hypothesis that feeding behaviour is random within as well as between bouts. We falsify this null hypothesis by analysing how specific feeding events (such as visits to a feeder) by animals of a number species are distributed in time. They all show bouted, but non-random, feeding behaviour with a similar structure. This is caused by satiety at the end of feeding bouts, followed by an increase in the probability of animals starting to eat with time since feeding last. This results in a skewed normal distribution of the population of longer between-feeding, i.e. between-meal, intervals, which can be (almost) normalised by log-transformation of interval lengths. We show that either a Gaussian (for pooled data sets) or a Weibull (for individuals or data pooled across animals with similar feeding strategies) give excellent descriptions of the distribution of this population of log-transformed interval lengths. Depending on species, methodology and individual animal habits there may be one or more population(s) of shorter (i.e. within-meal) intervals. We developed methods that allow estimation of meal criteria even if the number of within-meal interval populations and their shape are not exactly known. Because the structure of feeding behaviour is the same across species, similar models to estimate bout criteria can be applied to data obtained with animals of species as diverse as cows, pigs, chickens, turkeys, ducks and dolphins. We demonstrate how analysis of data subsets, following disaggregation of pooled data, can elucidate underlying mechanisms that would otherwise remain obscure. The developed models can also be applied to other forms of animal behaviour that are affected by satiety-like principles.

Keywords: feeding structure, bout analysis, probability of behaviour

Introduction

Analysis of short-term feeding behaviour can be useful for a number of purposes. For farm animals, such analyses can test hypotheses on the biology of food intake regulation (e.g. Tolkamp *et al.*, 2002) and diet selection (e.g. Yeates *et al.*, 2003), to develop automated systems for the early detection of health and welfare problems (e.g. Gonzalez *et al.*, 2008) and to estimate relevant feeding behaviour traits that could be included in selection indices used in breeding programmes (e.g. Howie *et al.*, 2010). The methodology used to collect feeding behaviour data has a large effect on the unit in which it is recorded. Applied methodologies range from records of bites, registration of jaw movements, regular weighing of food troughs and observation of feeding behaviour to, and increasingly so, records of visits to electronic feeders (e.g. Tolkamp and Kyriazakis, 1997; Morgan *et al.*, 2000a; Bley and Bessei, 2008). Each of these methodologies results in its own definition of a 'feeding event' and feeding event data obtained with one methodology are difficult to compare with those obtained with another. Fortunately, feeding behaviour is structured because animals do not continually or randomly consume food. Instead, animals usually organize their feeding events in clusters, or bouts (Figure 1). Such bouts of feeding behaviour, if properly identified, can be called meals.

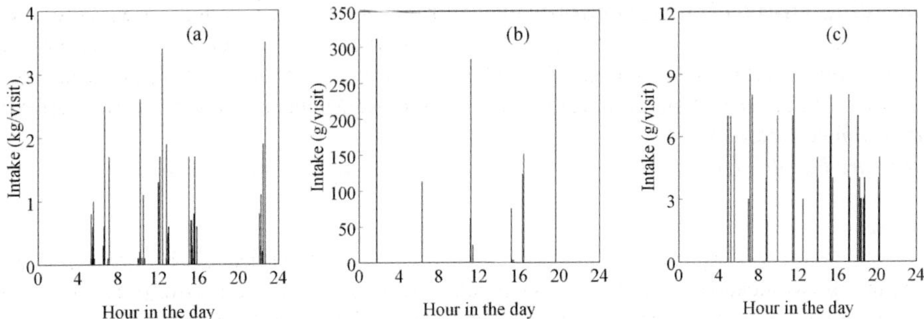

Figure 1. Typical examples of visits to feeders during a day by animals of three species. The graphs show intake per visit against visit start-time for: (a) a lactating dairy cow with access to feeders supplying a mixture of silage and concentrate (Tolkamp and Kyriazakis (1997), (b) a growing pig with access to a feeder supplying feed pellets (Morgan et al., *2000a) and (c) a growing Pekin duck with access to feeders supplying feed pellets (Bley and Bessei, 2008).*

For many decades, the preferred methods of identifying bouts of behaviour were log-survivorship (Slater and Lester, 1982) or log-frequency (Langton *et al.,* 1995) analyses. Such analyses are aimed at identifying the best meal criteria, i.e. the longest non-feeding interval that can be considered part of a meal. The null hypothesis underlying these methods is that animals behave randomly within as well as between feeding bouts. In this context this means that the probability of animals starting to feed is assumed to be independent of the time they fed last. Satisfactory meal criteria could not be obtained by these methods for housed (Tolkamp *et al.,* 1998) or grazing (Rook and Huckle, 1997) dairy cows. This necessitated the development of better models to group feeding behaviour into bouts.

It is the aim of this paper to, first, summarize the development of novel models to estimate meal criteria for dairy cows. Subsequently, we will provide examples of how such models can be applied to data obtained from other species. We argue that valuable information can be extracted from records of short-term feeding behaviour only if (1) appropriate methods are used to group feeding behaviour into bouts and (2) inappropriate pooling of data is avoided.

Dairy cow model development

Data collection

Data were collected with Hokofarm feeders as described in detail by Tolkamp and Kyriazakis (1997). In brief, lactating dairy cows wearing transponders were group-housed in a cubicle yard and had access to electronic feeders supplying a mixed ration consisting of silage and concentrates. Feeders were linked to a computer that recorded the start- and end-time and the feeder start- and end-weight at each visit. From these visit records, between-feeding intervals were calculated for each cow and these data formed the basis of our analysis.

Data analysis

Log-survivorship plots were calculated by log-transforming the number of intervals $\geq t$ (using natural logarithms, i.e. $\log_e$) for each between-feeding interval length t. A probability density function (pdf) consisting of two negative exponentials was fitted to the cumulative frequency distributions and drawn in a graph (after log-transformation) for comparison with observations. Meal criteria were estimated as described by Tolkamp *et al.* (1998). All curve fitting was done using the maximum likelihood procedure in GENSTAT (VSN 2008). Probabilities of animals starting to feed within x min at time t since feeding last were calculated as 1 minus the number of intervals $\geq (t + x)$ divided by the number of intervals $\geq t$.

Results

Figure 2a shows the log-survivorship curve and the fit of the 'broken-stick' model (i.e. a pdf consisting of two negative exponentials after log-transformation). The log-survivorship curve at longer intervals was not a straight line but a convex curve. This shows that the longer intervals did not have a negative exponential distribution, which implies that the probability of cows starting to feed is not independent of the time since feeding last (Tolkamp *et al.*, 1998). The meal criterion estimated by the parameters of the pdf consisting of two negative exponentials (8 min) does, therefore, have no biological meaning.

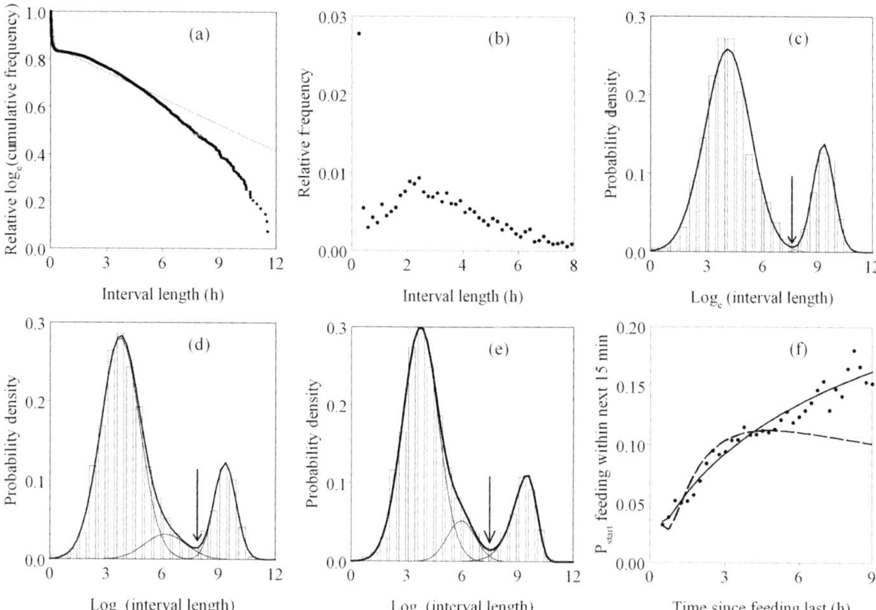

Figure 2. Selected feeding behaviour characteristics of lactating dairy cows (after data reported by Tolkamp et al. *(1998), Tolkamp and Kyriazakis (1999) and Yeates* et al. *(2001)). (a) The log-survivorship curve of between-feeding intervals (dots, appearing in parts as a thick line) and the fit of a double exponential model (after log-transformation, thin line). (b) Relative frequency distribution of between-feeding intervals (bin-width 10 min; the value for the first bin, i.e. 0.76, is not shown). Relative frequency distribution of log-transformed observed lengths of between-feeding intervals divided by bin-width, i.e. 0.5 $\log_e$-units (bars) and the fit of a probability density function consisting of two Gaussians (c), three Gaussians (d) or two Gaussians and a Weibull (e). The probability of cows starting to feed within the next 15 min in relation to time since feeding last (observations, dots; graph f) and the starting probabilities predicted by probability density functions that incorporated a Gaussian (broken line) or a Weibull (solid line) to describe the population of long log-transformed intervals. Arrows in graphs (c), (d) and (e) indicate meal criteria estimates (see Yeates* et al.*, 2001).*

A plot of the frequency distribution of between-feeding interval lengths (Figure 2b) confirmed that longer (i.e. between-meal) intervals were not distributed as a negative exponential but as a skewed normal. Log-transformation of between-feeding interval lengths, measured in sec, resulted in what appeared to be two clearly separated interval populations (one within- and one between-meal), each with an approximately normal distribution (Figure 2c).

To obtain meal criteria, at first two Gaussians were fitted to pooled and individual data sets (Tolkamp *et al.*, 1998; Figure 2c). The two Gaussians peaked at around 4.1 and 9.4 $\log_e$-units (i.e.interval lengths around 1 min and 3.5 h, respectively). Tolkamp *et al.* (1998) showed that the number of intervals assigned to the wrong population is minimal when meal criteria are chosen as the interval lengths were the two Gaussians cross. This can be calculated from the model parameters as around 7.5 log-units (i.e. around 30 min). Some cows proved to have an additional, again log-normally distributed, population of intervals (Tolkamp and Kyriazakis, 1999; Figure 2d). This population consisted of within-meal intervals during which cows went to drink water. This third population of intervals peaked at around 6 $\log_e$-units (i.e. around 7 min). Later analyses showed that the population(s) of within-meal intervals could indeed be described well by Gaussians but that the population of long (i.e. between-meal) intervals was slightly skewed. Statistically, the latter could be described significantly better by a Weibull than a Gaussian (Yeates *et al.*, 2001; Figure 2e). The parameters of the Weibull also predicted the change in the probability of cows starting to feed much better than the parameters of a Gaussian. As expected on the basis of the satiety concept, this probability was observed to increase with time since the last meal (Yeates *et al.*, 2001). This was described accurately from the Weibull parameters and not from those of the fitted Gaussian (Figure 2f). The same models could be applied to cows on mixed diets in which the concentrate to forage ratio was low (10:90) or high (30:70, both on fresh basis; see Tolkamp *et al.*, 2002) and have now successfully been applied for the estimation of meal criteria to data sets obtained with cows elsewhere (e.g. Salawu *et al.*, 2002; deVries *et al.*, 2003; Melin *et al.*, 2005; Abrahamse *et al.*, 2008). At the time we speculated that such models would also be suitable to analyse feeding behaviour of other species because the underlying biological principles that structure animal feeding behaviour are expected to be similar (e.g. Simpson, 1995).

Application to other species

Data description

Here we investigate whether the models we developed for cows would be suitable for other species. We provide a brief description of the data sets and refer the interested reader for more details to the following publications: on broilers (Howie *et al.*, 2009a,b), on pigs (Morgan *et al.*, 2000a; Wilkinson, 2007), on Pekin ducks (Bley and Bessei, 2008 and Howie *et al.*, 2010), on turkeys (Howie *et al.*, 2010) and on bottlenose dolphins (Jacobsen *et al.*, 2003). In brief, data from 1,058 female broilers were obtained between 14 and 35 d of age (mean start- and end-weight 0.47 and 2.28 kg) housed in pens of around 116 birds with access to 8 feeders supplying pelleted food (Howie *et al.*, 2009a). Data from 318 male turkeys were obtained between 18 and 22 wks of age (mean start- and end-weight 21.7 and 26.3 kg), housed in pens of around 207 birds with access to 16 feeders supplying pelleted food (Howie *et al.*, 2010). Data from 480 male and female Pekin ducks were obtained between 3 and 7 wks d of age (mean start- and end-weight 1.1 and 3.0 kg), housed in pens of around 160 birds with access to 16 feeders supplying pelleted food (Bley and Bessei, 2008; Howie *et al.*, 2010). Data from 60 pigs were collected between around 9 and 24 weeks (start- and end-weight around 24 and 114 kg), housed in pens of 10 animals with access to a single feeder supplying pelleted food (Wilkinson, 2007). Feeding behaviour of 11 bottlenose dolphin calves, housed in captivity with their dams, was observed up to 14 d post-partum (Jacobsen *et al.*, 2003). Dolphin calves feed by drinking after locking on to their dam's mammary gland. These lock-ons were recorded and from these records the lengths of intervals between lock-on were calculated.

Data analyses

Frequency distributions were calculated for all six species. Examples are provided of the effects of log-transformation of interval lengths on the frequency distribution and of the effects of disaggregation in sub-sets on the frequency distribution of log-transformed interval lengths and the probability of animals starting to feed. There were strong contrasts between species in diurnal patterns. Differences in feeding behaviour were least strong for cows and data pooled across day and night were, therefore used. Much lower feeding activity was recorded during the night (i.e. the dark phase) than during the day (i.e. the light phase) in broilers, ducks, turkeys and pigs. Pigs were housed in groups of 10 with access to a single feeder only. During the day, pigs were observed to queue frequently in front of the feeder that was occupied or blocked by a (dominant) pig. For that reason, pigs could not freely express their feeding behaviour during the day. They could, however, during the night when feeding activity was much reduced. For that reason, the night frequency distribution for pigs (but the day-time frequency distribution of the birds) was used for further analyses.

Results

Figure 3 shows the frequency distributions of between-feeding interval lengths for the six species. All graphs show that with an increase in interval length the frequency first declined sharply, then reached a nadir. Intervals longer than the length where the nadir occurred were all distributed as skewed normal distributions, as observed before in cows.

None of the graphs showed a distribution of longer between-feeding intervals in the shape of a negative exponential that would be consistent with a constant probability of animals starting to feed. This implies that conventional log-survivorship or log-frequency analyses, which are based on such a constant probability, are entirely inappropriate for any of these species. We tested whether or not log-transformation of interval lengths would lead to distributions as seen in dairy cows that would allow applications of cow-models to estimate meal criteria in other species. That was indeed directly the case for some species but for others disaggregation of the data was required before models could be fitted. Figure 4a shows the fit of a double lognormal model (as used by Tolkamp *et al.*, 1998 to analyze cow data) to the frequency distribution of log-transformed between-feeding interval lengths observed across 11 dolphin calves. The fit was excellent and resulted in meal-criteria estimates of 161 sec (compared with 104 sec estimated by the 'broken-stick' model; Jacobsen *et al.*, 2003).

Figure 4b and 4c show the effects of disaggregation of the pooled data obtained with ducks. From the pooled data it was not clear exactly how many populations of intervals existed and by which functions they could be described (Howie *et al.*, 2010). Inspection of individual distributions showed that for some of the birds there were only two populations of intervals (one within- and one between-meals; Figure 4b) while other birds clearly showed three separate distributions (Figure 4c). Ducks, like dairy cows, appear to be divided in sub-populations that do (Figure 4c) or do not (Figure 4b) drink during meals. After disaggregation, the models developed for cows can be fitted to data obtained from ducks and model parameters can subsequently be used to estimate meal-criteria in ducks. Other types of disaggregation (i.e. into sub-sets of intervals between (1) visits to the same feeder and (2) visits to different feeders) proved very successful for our understanding of the structure of feeding behaviour of species such as broilers and turkeys (Howie *et al.*, 2010).

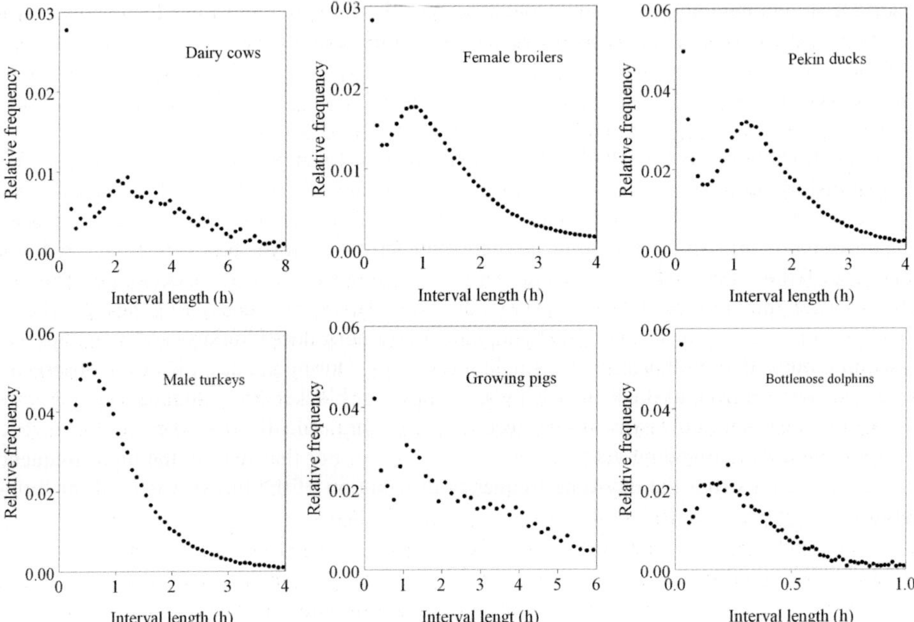

Figure 3. Frequency distributions of interval lengths between visits to feeders of 16 dairy cows (n = 13,258), 1,058 female broilers (n = 701,864), 480 Pekin ducks (n = 164,438), 318 turkeys (n = 84,574), 60 pigs during the night (n = 4,330) and of intervals between 'lock-ons' of 11 neonate bottlenose dolphin calves (n = 6,800). The distributions are for bin widths of 10 min (cows), 5 min (broilers, ducks, turkeys and pigs) and 1 min (dolphins). Not shown are the high values for the first bin, which were 0.766 (cows), 0.609 (broilers), 0.218 (ducks), 0.124 (turkeys), 0.640 (pigs) and 0.405 (dolphins). Plots are for data pooled across individuals (all species) and pooled across day and night (cows), or for the light period only (broilers, turkeys, ducks, dolphins) or for the dark period only (pigs).

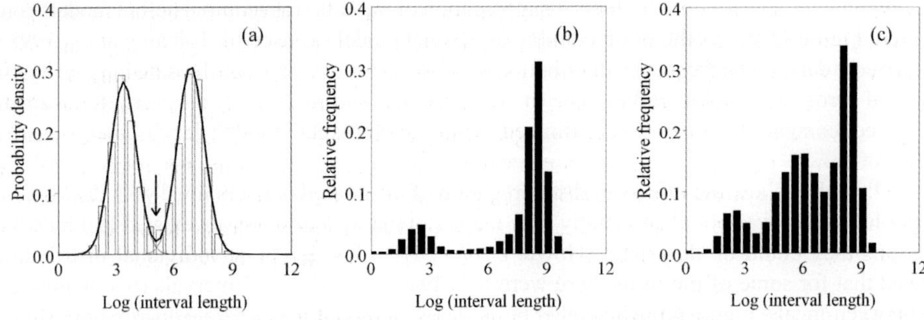

Figure 4. Frequency distributions of log-transformed interval lengths (all bin-width 0.5 $\log_e$-units) obtained from (a) bottlenose dolphin calves; the bars are the observed frequencies divided by bin-width, the lines show the fit of a double log-normal model and the arrow indicates the estimated meal criterion. (b) a subset of individual ducks that provided no evidence of a third population of intervals; (c) a subset of individual ducks that showed very clear evidence of a third population of intervals, presumably caused by within-meal drinking.

In our experience, frequency distribution(s) of log-transformed within-meal intervals can be described very well with one or two log-normals, for cows as well as other species. The frequency distribution of longer (i.e. between-meal) intervals is described well by either a log-normal (especially for pooled data) or by a log-Weibull (after disaggregation of the data into subsets of animals with similar feeding strategies; Howie *et al.*, 2009a,b, 2010).

General discussion

Animals in species ranging from insects (e.g. Simpson, 1995) to cattle (e.g. Metz, 1975) are expected to finish a meal when they are satiated. As a result of satiety, the probability of animals starting a meal immediately after finishing one is expected to be low. As the time since feeding last increases, satiety is expected to wear off and hunger/feeding motivation is likely to increase. The probability of animals starting to feed is, consequently, expected to increase with time since the last meal. This contrasts sharply with the null-hypothesis on which quantitative log-survivorship and log-frequency analyses are based. Meal criteria that are quantitatively estimated via such methods have, therefore, no biological basis.

Howie *et al.* (2009a,b, 2010) showed that the frequency distributions of day-time between-meal intervals pooled across age and feeding strategy is described better by a log-normal than by a log-Weibull in three poultry species. In agreement with this analysis, the probability of birds starting to feed first increased and then decreased with an increase in time since the last meal, which seems to be at odds with expectations based on the satiety concept. However, when sub-sets of birds with similar feeding strategies were analysed, the starting probability continued to increase with time since the last meal and log-Weibulls were more suitable to describe the frequency distribution of between-meal intervals (Howie *et al.*, 2009b, 2010). This shows how inappropriate pooling (in this case: across animals with different feeding strategies) can distort a basic underlying biological mechanism. The increase in starting probability and its consequences for the shape of log-survivorship curves can be masked entirely by inappropriate pooling for other reasons as well. For instance, Morgan *et al.* (2000a,b) found that the log-survivorship curve was in the typical 'broken-stick' shape for the pooled between-feeding intervals obtained from 16 single-housed pigs. However, pigs show most of their feeding activity during the day (i.e. the light period) and a much reduced activity during the night (i.e. the dark period). When observations were disaggregated into day and night sub-sets, the log-survivorship curves were not in the form of a straight line during the night nor during the day. In addition, the probability of animals starting to feed increased during the night as well as during the day, but this increase was not observed in the pooled data set. Pig feeding behaviour was, therefore, not random during the day and not random during the night although the pooled data gave the impression that it was (Morgan *et al.*, 2000b). The model calculations carried out by Yeates *et al.* (2003) have shown that the stronger the contrast in feeding behaviour is between sub-sets of data, the more bias is introduced by pooling of such data across sub-sets. Allcroft *et al.* (2004) showed that feeding behaviour cannot be modelled properly unless effects of non-feeding time on the probability of animals starting to feed are incorporated in the models.

Zorrilla *et al.* (2005) showed that unless an appropriate model is used to estimate meal criteria, interpretation of short-term feeding behaviour can be severely biased. The analyses of data obtained from the species presented here shows that the distribution of longer (i.e. between-meal) interval lengths is always in the shape of a skewed normal. This distribution can be (almost) normalised by log-transformation of interval lengths. The distribution of these log-transformed interval lengths can be described accurately for individuals or for pooled data of individuals with a similar feeding strategy by a Weibull function. The parameters of this function predict an increase in the probability of animals starting to feed with time since the last meal. This is consistent with expectations based on the satiety concept as well as with the observed increase in starting probabilities. If data are pooled across animals with different feeding strategies, however, log-normals were frequently better descriptors of such distributions. Pdf's that include a log-normal to describe the distribution of between-meal intervals can thus be used to estimate meal criteria in pooled data sets (Howie *et al.,* 2009a,b).

Conclusions

Analysis of short-term feeding behaviour produces the most biologically relevant information if it is based on the very unit in which animals organise their feeding behaviour (Tolkamp *et al.,* 2000). Animals of most species cluster their feeding behaviour in bouts, which can be called meals if they are properly identified. This requires estimation of meal criteria. Existing methods to do that, such as quantitative log-survivorship analyses, are based on the (implicit) assumption that feeding behaviour occurs randomly. We have shown here for six species that this is not the case. Better models are, therefore, required for the correct identification of meals.

Analyses of the distribution of log-transformed between-feeding intervals seem very promising for the identification of meal criteria. If inappropriate pooling is avoided, the population of longer (i.e. between-meal intervals) can be described accurately with a log-Weibull for all investigated species. The population(s) of shorter (i.e. within-meal) intervals can be described accurately with one or more log-normals, although disaggregation of pooled data sets is sometimes required before this can be observed. A methodology to accurately group feeding behaviour into meals is, therefore, now available, which can also be applied to other forms of behaviour that are affected by satiety-like principles (see Yeates *et al.*, 2001).

Acknowledgements

The authors gratefully acknowledge the contribution of T.B. Jacobsen and T.A.G. Bley who made data obtained with dolphins and Pekin ducks, respectively, available for our analyses.

References

Abrahamse, P.A., Vlaeminck, B., Tamminga, S. and Dijkstra, J., 2008. The effect of silage and concentrate type on intake behavior, rumen function, and milk production in dairy cows in early and late lactation. Journal of Dairy Science 91:4778-4792.

Allcroft, D.J., Tolkamp, B.J., Glasbey, C.A. and Kyriazakis, I., 2004. The importance of memory in statistical models for the analysis of animal behaviour. Behavioural Processes 67:99-109.

Bley, T.A.G. and Bessei, W., 2008. Recording of individual feed intake and feeding behavior of Pekin ducks kept in groups. Poultry Science, 87:215-221.

DeVries, T.J., von Keyserlingk, M.A.G., Weary, D.M. and Beauchemin, K.A., 2003. Measuring the feeding behavior of lactating dairy cows in early to peak lactation. Journal of Dairy Science 86:3354-3361.

Gonzalez, L.A., Tolkamp, B.J., Coffey, M.P., Ferret A. and Kyriazakis, I., 2008. Changes in feeding behavior as possible indicators for the automatic monitoring of health disorders in dairy cows. Journal of Dairy Science 91:1017-1028.

Howie, J.A., Tolkamp, B.J., Avendaño, S., and Kyriazakis, I,. 2009a. A novel flexible method to split feeding behaviour into bouts. Applied Animal Behaviour Science 116:101-109.

Howie, J.A., Tolkamp, B.J., Avendaño, S. and Kyriazakis, I., 2009b. The structure of feeding behavior in commercial broiler lines selected for different growth rates. Poultry Science 88:1143-1150.

Howie, J.A., Tolkamp, B.J., Bley, T.A.G. and Kyriazakis, I. 2010. Short-term feeding behaviour has a similar structure in broilers, ducks and turkeys. British Poultry Science, in press.

Jacobsen, T.B., Mayntz, M. and Amundin, M., 2003. Splitting suckling data of Bottlenose dolphin (*Tursiops truncatus*) Neonates in human care into suckling bouts. Zoo Biology 5:477-488.

Langton, S.D., Collett, D. and Sibly, R.M., 1995. Splitting behavior into bouts – A maximum-likelihood approach. Behaviour 132:781-799.

Melin, M., Wiktorsson, H. and Norell, L., 2005. Analysis of feeding and drinking patterns of dairy cows in two cow traffic situations in automatic milking systems. Journal of Dairy Science 88:71-85.

Metz, J.H.M., 1975. Time patterns of feeding and rumination in domestic cattle. Communications Agricultural University 75-12. Wageningen, the Netherlands, 71 pp.

Morgan, C.A., Emmans, G.C., Tolkamp, B.J. and Kyriazakis, I., 2000a. Analysis of the feeding behavior of pigs using different models. Physiology and Behavior 68:395-403.

Morgan, C.A., Tolkamp, B.J., Emmans, G.C. and Kyriazakis, I., 2000b. The way in which the data are combined affects the interpretation of short term feeding behaviour. Physiology and Behavior 70:391-396.

Rook, A.J. and Huckle, C.A., 1997. Activity bout criteria for grazing cows. Applied Animal Behaviour Science 54:89-96.

Salawu, M.B., Adesogan, A.T. and Dewhurst, R.J., 2002. Forage intake, meal patterns, and milk production of lactating dairy cows fed grass silage or pea-wheat bi-crop silages. Journal of Dairy Science 85:3035-3044.

Simpson, S.J., 1995. Regulation of a meal: chewing insects. In: Chapman R.F and de Boer, G. (eds.). Regulatory Mechanisms in Insect Feeding. Chapman and Hall, New York, NY, USA, pp.73-103.

Slater, P.J.B and Lester, N.P., 1982 Minimizing errors in splitting behavior into bouts. Behaviour, 79:153-161.

Tolkamp, B.J. and Kyriazakis, I., 1997. Measuring diet selection in dairy cows: effects of training on choice of dietary protein level. Animal Science 64:197-207.

Tolkamp, B.J., Allcroft, D.J., Austin, E.J., Nielsen, B.L. and Kyriazakis, I., 1998. Satiety splits feeding behaviour into bouts. Journal of Theoretical Biology 194:235-250.

Tolkamp, B.J. and Kyriazakis, I., 1999. To split behaviour into bouts, log-transform the intervals. Animal Behaviour 57:807-817.

Tolkamp, B.J., Schweitzer, D.P.N. and Kyriazakis, I., 2000. The biologically relevant unit for the analysis of short-term feeding behavior of dairy cows. Journal of Dairy Science 83:2057-2068.

Tolkamp, B.J., Friggens, N.C., Emmans, G.C., Kyriazakis, I. and Oldham, J.D., 2002. Meal patterns of dairy cows consuming diets with a high or a low ratio of concentrate to grass silage. Animal Science 74:369-382.

VSN International (2008) GENSTAT Version 11.

Wilkinson, S., 2007. The short-term feeding behaviour of group-housed commercially reared growing-finishing pigs. MRes Thesis, University of Edinburgh, UK.

Yeates, M.P., Tolkamp, B.J., Allcroft, D.J. and Kyriazakis, I., 2001. The use of mixed distribution models to determine bout criteria for analysis of animal behaviour. Journal of Theoretical Biology 213:413-425.

Yeates, M.P., Tolkamp, B.J. and Kyriazakis, I., 2003. Consequences of variation in feeding behaviour for the probability of animals starting a meal as estimated from pooled data. Animal Science 77:471-484.

Zorrilla, E.P., Inoue, K., Fekete, E.M., Tabarin, A., Valdez, G.R. and Koob, G.F., 2005. Measuring meals: structure of prandial food and water intake of rats. American Journal of Physiology-Regulatory Integrative and Comparative Physiology, 288:R1450-R1467.

A new Nordic structure evaluation system for diets fed to dairy cows: a meta analysis

P. Nørgaard[1], E. Nadeau[2] and Å.T. Randby[3]
[1]*Department of Basic Animal and Veterinary Sciences, Faculty of Life Sciences, University of Copenhagen, Groennegaardsvej 3, 1870 Frederiksberg, Denmark; pen@life.ku.dk*
[2]*Swedish University of Agricultural Sciences, Department of Animal Environment and Health, P.O. Box 234, 532 23 Skara, Sweden*
[3]*Norwegian University of Life Sciences, Department of Animal and Aquacultural Sciences, P.O. Box 5003, 1432 Ås, Norway*

Abstract

The overall aim was to establish a model for predicting chewing index (CI) values for ranking the fibrousnesses of feeds fed to dairy cows within the Nordic Chewing index system. The CI values are predicted as the sum of the eating (EI) and ruminating time index (RI) values. The EI values are assumed to be proportional with the NDF content and a particle size factor through the proportionality factor k_{EI}. The RI values are assumed to be proportional with the NDF content, a particle size factor and a hardness factor through the proportionality factor k_{RI}. The k_{EI}, k_{RI} values and the k_{EI}/k_{RI} ratios were parameterized as the mean eating (mET_f), mean ruminating time (mRT_f) per intake of forage NDF ($NDFI_f$) and the mET_f /mRT_f ratio by a Meta analysis of 75 published values from cattle fed three types of unchopped forages with or without supplementation with concentrates. The intake of NDF from ground concentrate ($NDFI_c$) and rolled barley ($NDFI_{RB}$) was related to $NDFI_f$ in the models, which included effects of BW, $NDFI_f$/BW, $NDFI_{RB}$/BW, $NDFI_{RB}$/$NDFI_f$, $NDFI_c$/$NDFI_f$, DM content of silage, interaction between forage type and physiological state of the cattle, method for recording chewing, and with studies as random effect for ruminating time. The mRT_f value per kg forage NDF decreased at increased BW and $NDFI_f$/BW. The mET_f value increased at increasing BW and at decreasing DM contents of grass silage. Intake of NDF from rolled barley stimulated ruminating time by ¾ of the stimuli from $NDFI_f$. The mET_f/mRT_f ratio, the mET_f and mRT_f values of grass silage fed to a standard cow, BW=625 kg, 0.7% $NDFI_f$ per kg of BW were predicted to 0.41 (min/min), and 41 and 109 (min/kg NDF), respectively.

Keywords: chewing index, mean eating time, mean ruminating time, intake forage NDF

Introduction

The intake of physically effective fibre (peNDF) (Mertens, 1997) stimulates eating time (ET), ruminating time (RT) (Allen, 1997), saliva secretion, rumen motility, formation of a ruminal mat and the establishment of an acceptable rumen pH. The dietary sources of peNDF fibre include forages, by-products and coarsely processed grain and oil cakes (Mertens, 1997). However, the content of neutral detergent fibre (NDF) of these feeds ranges from 5 to 80% in DM. In addition, the physical form of forages varies with maturity stage at harvest and particle size that can range from less than 2 mm to 500 mm.

Ranking fibrousness of feeds

The fibrousness of the individual feeds has been quantitatively ranked by a chewing index value (CI, min/kg DM) (CI) (Balch, 1971; Sudweeks *et al.*, 1981; Nørgaard, 1986). The CI value of the individual feeds within the Nordic Structure Evaluation System (Nørgaard *et al.*, 2008) is defined as the sum of an eating index value (EI) plus a ruminating index value (RI). The EI values are proportional with the NDF content in the forage and a particle size factor (Size_E) according to EI= k_{EI} × NDF × Size_E, where the k_{EI} is the proportionality factor. The RI values are proportional with the NDF content, a particle size factor (Size_R) and a Hardness factor according to RI= k_{RI} × NDF × Size_R × Hardness factor, where k_{RI} is the proportionality factor. The Size_E and Size_R values range from 0 for finely ground feed to 1 for unchopped forages. The Hardness factor value is estimated as: 0.75 + INDF/NDF, where the INDF/NDF ratio is the proportion of rumen indigestible NDF of NDF. The Hardness factor has a value of 1.0 for feeds with an INDF/NDF ratio of 0.25, which corresponds to an ADL/NDF ratio of about 0.085. The Hardness factor values for the individual feeds range from 0.8 for immature spring grass to 1.2 for mature alfalfa crops. The k_{EI} and k_{RI} values represent the mean eating (mET) and mean ruminating times (mRT) in lactating dairy cows with a BW of 625 kg fed a standard diet, which is formulated to balance the requirement for a high energy intake with a sufficient intake of peNDF. Consequently, the k_{EI} and k_{RI} values represent the mET and mRT (min/kg NDF) values of unchopped forage fed to a standard multiparous (MP) lactating dairy cow with a BW of 625 kg, an intake of 20 kg DM, a forage to concentrate ratio of 40:60 and an intake of forage NDF ($NDFI_f$) of 0.7% of BW.

Factors affecting chewing time

Aikman *et al.* (2008) found 30% higher mRT and 22% higher mET/mRT ratio in lactating Jersey compared with Holstein Frisian cows fed a complete diet. Beauchemin and Rode (1994) reported 15% higher mET and 20% higher mRT values in primiparous (PP) compared with multiparous (MP) lactating cows. Bae *et al.* (1983) documented that mET and mRT decreased curvilinearly with increased metabolic body weight of cattle of different breeds. The daily times spent eating and ruminating depend mainly on the $NDFI_f$ (Beauchemin, 1991). Chopping, grinding or pelleting of forage decrease mET and mRT per kg $NDFI_f$ (Woodford *et al.*, 1986; Nørgaard, 1989). Chopping of forage strongly reduces the mET values and also the mRT values depending on the reduction in particle size due to chopping (Soita *et al.*, 2000; Schwab *et al.*, 2002). Consequently, chopping of forage decreases the mET/mRT ratio, which requires different models for prediction of the Size_E and Size_R factors depending on feed particle length. Whole grain (Beauchemin *et al.*, 1994b) and coarsely processed concentrates (Grant and Weidner, 1992; Clark and Armentano, 1993; Weidner and Grant, 1994) also stimulate ruminating activity. De Boever *et al.* (1990) and Beauchemin et *al.* (2008) considered mET values of concentrates to be 4 min/kg DM. Published mET values per kg of $NDFI_f$ increased at decreasing DM content in grass silage fed to dairy cows (De Boever *et al.*, 1993). Increased mET values due to low DM content are not considered to increase the structure value within the Nordic Chewing system. Several experiments have documented that the mRT value increases and the mET value decreases due to restriction of forage intake (De Boever *et al.*, 1990).

Recording methods

The eating and ruminating activities in cattle have been measured by use of different recording methods, such as observation of chewing behaviour (Woodford *et al.*, 1986), visual identification of eating and ruminating time (RT) from pattern of recorded jaw movement oscillations (JMO) by use of different JMO sensors by Freer and Campling (1965), Castle *et al.* (1979), Bae *et al.* (1981), Van Bruchem *et al.* (1991) and De Boever *et al.* (1993), and electromyogram of the cheek muscles (Oshio, 2001) or by computerized identification of eating and ruminating times from JMO (Beauchemin *et al.*, 1991).

Parameterization

The parameterization of the k_{EI} and k_{RI} values is complicated by the fact that the published mET and mRT values originate from experiments with cattle of different size and in different physiological state fed different forage types, in different amounts with or without supplementation of concentrate (Table 1), and by using different methods for recording eating and ruminating activities. The objective of this meta-analysis was to study effects of animal characteristics, forage NDF intake, forage characteristics and effects of different processed concentrates on the mET/mRT ratio, mET, mRT and RT values in cattle fed unchopped forages to parameterize the k_{EI} and k_{RI} values.

Materials and methods

A meta-analysis of the mET mRT, mET/mRT and RT values from cattle fed unchopped forage with or without supplementation with ground or rolled barley including 79 different dietary treatments from 17 published studies (Table 1) was conducted. The mean number of experimental cattle per treatment was 4.9 ranging from two to nine. The experimental cattle were categorized into four physiological states (P): growing heifers (Beauchemin and Iwaasa, 1993) or steers (Luginbuhl *et al*., 1989), mature steers (Bae *et al*., 1981) or dry cows, lactating PP and MP cows. The mean body size (BW) ranged from 371 kg in heifers (Beauchemin and Iwaasa, 1993) to 728 kg in mature steers (Bae *et al*., 1981). The forages included 26 treatments with alfalfa hay (AH), 32 with grass hay (GH), 19 with grass silage (GS) and two with orchard grass hay (OGH). The forages were grouped in three types of forages: AH, GH or GS and OGH. The NDF concentrations of grass hay and silage in the studies by (Freer and Campling, 1965; Campling, 1966a,b,c; Castle *et al*., 1979) were estimated from the crude fibre content (CF) as $3.1+1.70 \times CF$, n=22, R^2=0.91, RMSE=1.7 based on values from Oshio (2001) and De Boever *et al*. (1993). The mET of forages per kg $NDFI_f$ (mET_f) was predicted as total recorded daily eating time minus 4 minutes per kg intake of concentrate DM. The mRT_f values (min/kg $NDFI_f$) of the individual treatments within studies were estimated as the daily recorded ruminating time (RT) divided by the $NDFI_f$. The intake of NDF from ground concentrate ($NDFI_c$) was estimated as $3 \times CF$ (NRC, 2001) in the studies by Campling, (1966b,c). The intake of NDF from steam rolled barley grain (Beauchemin and Rode, 1994; Beauchemin *et al*., 1991) or coarsely processed oil cakes (Campling, 1996b) was termed $NDFI_{RB}$ and included 22 treatments. The $(NDFI_c+NDFI_{RB})/NDFI_f$ ratios ranged between 0 and 1.5 with a mean value of 0.27±0.4. The effect of silage DM content on the mET_f values were studied by including the factor logDM=ln(min(DM %/40,1)). The effect of methods for recording eating and ruminating time was included in the models. The variation in the mET_f, mRT_f, RT, and mET_f/mRT_f values were analyzed by a meta analysis including P, BW, $NDFI_f$/BW, $NDFI_{RB}$/BW, $NDFI_c$/BW, $NDFI_{RB}/NDFI_f$, $NDFI_c/NDFI_f$, forage type, forage type × P, logDM, studies and recording method as potential effects. The analysis was done by use of the Mixed procedure in SAS (SAS system for Windows, release 9.1; SAS Institute Inc. Cary, NC, USA) with the number of animals per treatment as weight. The individual studies were considered as random effects in the model for RT. The selections of the final reduced models were based on minimizing the Bayesian information criterion (BIC) (Schwarz, 1978).

Final reduced models:

$$mET_f = \alpha + \beta_1(BW/100 - 6.25) + \beta_2 NDFI_f/BW + \beta_5 NDFI_{RB}/NDFI_f + \beta_6 NDFI_c/NDFI_f + \beta_7 logDM + \tau_k\pi_i + \theta_l + \varepsilon$$

$$mRT_f = \alpha + \beta_1(BW/100 - 6.25) + \beta_2 NDFI_f/BW + \beta_3 NDFI_{RB}/BW + \beta_5 NDFI_{RB}/NDFI_f + \beta_6 NDFI_c/NDFI_f + \tau_k\pi_i + \theta_l + \varepsilon$$

Table 1. Overview of experimental treatments included in the meta-analysis and mean recorded values±std for treatments within studies.

N[1]	Physiological state	BW (kg)	Forage[4]	Intake (kg DM/d)	Intake of forage NDF % of BW	Intake of concentrate (kg DM/d)		Eating (min/d)	Rumination (min/d)	Study
						Ground	Rolled barley			
4	Steers	728	GH	7.8±1.9	0.48±0.13	0.9	0	256±116	364±71	Bae *et al.* (1981)
2	Heifer	371	AH	5.4±0.01	0.61±0.1	0	0	373±18	319±57	Beauchemin and Iwaasa (1993)
2	Heifer	371	OGH	5.5±0.01	0.69±0.09	0	0	347±13	305±45	Beauchemin and Iwaasa (1993)
1	PP[2]	493	AH	10.5	0.62	2.7	6.0	397	417	Beauchemin and Rode (1994)
5	MP[3]	556	AH	12.7±1.4	0.67±0.15	3.3±0.8	7.0±1.5	404±25	452±21	Beauchemin and Rode (1994)
5	MP	670	AH	21±0.23	0.47±0.06	3.5±0.9	10±1.1	342±12	398±19	Beauchemin *et al.* (1991)
6	MP	637	AH	22±0.84	0.82±0.33	3.8±2.4	6±3.1	371±64	423±37	Beauchemin *et al.* (1994a)
2	MP	605	OGH	21±1.1	0.68±0.15	5.2±0.7	7.2±2.1	370±16	437±5	Beauchemin *et al.* (1994a)
4	MP	625	GS	14±0.6	0.71±0.05	2.6	0	282±16	551±27	De Boever *et al.* (1993)
2	Dry/MP	504	GH	8.2±1.2	0.72±0.12	0.45	0		514±54	Bruchem *et al.* (1991)
2	Dry/MP	504	GS	8.6±2.2	0.56±0.16	0.45	0		460±88	Bruchem *et al.* (1991)
5	Dry	525	GH	9.8±1.6	0.91±0.13	0.8±1.7	0	308±79	567±36	Campling (1966c)
5	Dry	525	GS	8.2±1.7	0.78±0.15	0.8±1.7	0	321±149	519±54	Campling (1966c)
1	Dry	500	GH	11.7	0.79	0	0	363	442	Campling (1966a)
6	Dry	620	GH	7.0±2.1	0.53±0.19	0.9	2.1±1.9	168±85	423±127	Campling (1966b)
1	Dry	460	GH	9.0	1.08	0	0	397	460	Castle *et al.* (1979)
1	Dry	460	GS	7.0	0.85	0	0	368	450	Castle *et al.* (1979)
3	Dry	540	GH	7.2±3.1	0.72±0.28	0	0	210±90	481±138	Freer and Campling (1965)
3	MP	650	AH	25.3±0.5	0.90±0.08	11.9±3.4	0	326±54	409±8	Kaiser and Combs (1989)
4	Steers	410	GH	4.7±1.2	0.88±0.22	0	0	305±79	441±61	Luginbuhl *et al.* (1989)
4	Heifer	455	GH	6.0±3.1	0.60±0.30	0	0	134±122	421±193	Oshio (2001)
2	PP	517	GS	10.5±1.2	0.80±0.02	2.6±3.7	0	506±44	421±51	Teller *et al.* (1993)
4	MP	650	AH	25±1.0	0.70±0.16	15±3.1	0	338±40	382±36	Woodford *et al.* (1986)

[1] Number of treatments.
[2] Primiparous cows.
[3] Multiparous cows.
[4] AH alfalfa hay, GH: grass hay, GS: grass silage, OGH: orchard grass hay.

$$RT = \alpha + \beta_0(NDFI_f/BW)^2 + \beta_2 NDFI_f/BW + \beta_3 NDFI_{RB}/BW + \beta_4 NDFI_c/BW + \tau_k\pi_i + \theta_l + \varphi_l + \varepsilon$$

$$mET_f/mRT_f = \alpha + \beta_2 NDFI_f/BW + \beta_7 \log DM + \tau_k\pi_i + \theta_l + \varepsilon$$

α = intercept
β_1 = effect of increased BW (kg)/100
β_2 = effect of increased $NDFI_f$ relative to BW (kg/kg)
β_3 = effect of increased $NDFI_{RB}$ relative to BW (kg/kg)
β_4 = effect of increased $NDFI_c$ relative to BW (kg/kg)
β_5 = effect of increased $NDFI_{RB}$ relative $NDFI_f$
β_6 = effect of increased $NDFI_c$ relative to $NDFI_f$
β_7 = effect of increased dry matter content in silage (logDM)
θ_l = fixed effect of recording method, l=1,2,..9
τ_k = fixed effect of type of forage, k=1,2 and 3
π_j = fixed effect of physiological state, j=1,2, 3 and 4
$\tau_k\pi_i$= interaction between forage type and physiological state
φ_l = random effect of study, l=1,2… 17
ε = residual error

Results

The method used for recording eating and ruminating times affected the RT, mET_f, mRT_f and the mET_f / mRT_f values (P<0.0001) (Table 2). Recording chewing from electromyogram in the cheek muscles (Oshio, 2001) appears to underestimate eating and ruminating time, whereas identification of recorded JMO by using different chewing halters (Freer and Campling, 1965; Bae *et al.*, 1981) appears to overestimate eating and ruminating time. Figure 1 shows a plot of observed vs. predicted mRT_f values.

The RT, mET_f and the mET_f /mRT_f values were affected by interactions between the physiological state of the cattle and the type of forage (P<0.002) and the mRT_f values tended to be affected by these interactions (P<0.07). The RT (P<0.0001) and mET_f / mRT_f values (P<0.0001) increased at increasing intake of forage NDF, whereas the mRT_f value decreased (P<0.0001). The mRT_f value decreased at increased BW of the cattle (P<0.0001), whereas the mET_f and the mET_f / mRT_f values were not significantly affected by BW. The mRT_f and mET_f values increased at increasing $NDFI_{RB}/NDFI_f$ from steam rolled barley (P<0.0001), whereas the intake of NDF from finely processed concentrate only tended to affect the mRT_f value (P<0.07). The mET_f values increased at decreasing DM content in silage (P<0.03). Increased intake of NDF from steam rolled barley (P<0.002) and from finely ground concentrate increased the mET_f value (P<0.03), which implicates that supplementation with both types of concentrate increased eating time with more than 4 minutes per kg concentrate DM. The daily time spent ruminating increased curvilinearly at increasing intake of forage NDF relative to BW (P<0.03), at increasing intake of NDF from rolled barley (P<0.0001) and increasing intake of NDF from finely processed concentrates (P<0.003).

Discussion

The ratios between the RMSE values and the mean RT, mRT_f, mET_f and the mET_f / mRT_f values were 0.10, 0.11, 0.30 and 0.30, respectively, which implicates better models for prediction of RT and mRT_f compared to mET_f values. Omitting the method of recording chewing or the interaction between forage type and physiological state in the models did not cause major changes in the P values of the other effects in the models for RT, mRT_f, mET_f and mET_f/ mRT_f values (Table 2).

Table 2. Effect of animal and feed characteristics on mean eating (mET_f) and mean ruminating time (mRT_f) in cattle fed unchopped or slightly chopped forage with or without supplementation with concentrate (see Table 1).

Chewing variable	Mean eating time (min/kg $NDFI_f$[b])			Mean ruminating time (min/kg $NDFI_f$[b])			Ratio between eating and ruminating time			Ruminating time (RT) (min)		
	mET_f	SE	P<	mRT_f	SE	P<	mET_f/mRT_f	SE	P<	RT	SE	P<
Number of studies	16			17			16			17		
Number of dietary treatments	75			79			75			79		
Degree of freedom	57			60			60			55		
Overall RMSE	26			12			0.21			45		
Intercept	60	21		115	10	0.0001	0.64	0.14		12	67	0.9
Effect of (BW, kg/100-6.25)	-21	4.8	0.0001	-20	2.3	0.0001						
Level of intake: $100*NDFI_f$[b]/BW	11	10	0.3	-52	4.8	0.0001	0.38	0.06	0.0001	555	87	0.0001
$(100*NDFI_f/BW)^2$										-120	54	0.03
$100*NDFI_{RB}$[c]/BW				-11	46	0.8				690	93	0.0001
$100*NDFI_c$[d]/BW										171	55	0.003
Effect of NDF_{RB}/NDF_f	75	16	0.0001	73	16	0.0001						
Effect of NDF_c/NDF_f	17	13	0.2	11	6.2	0.08						
Effect of logDM silage[f]	-21	9.6	0.03				-0.20	0.08	0.01			
Type of forage[f]* P[a]			0.0002			0.07			0.0001			0.0001
MP fed alfalfa hay	-13	17		6	8.5		-0.24	0.13		-32	44	
MP fed grass silage or hay	-23	18		30	9.2		-0.50	0.13		130	56	
PP fed alfalfa hay	-1	22		6	11		-0.11	0.16		40	50	
PP fed grass silage or hay	0	.		0	.		0	.		0	.	
Recording method			0.0001			0.0001			0.0001			0.0002
minimal, relative to observation	-109[g]	17		-12[g]	8.5		-0.72[g]	0.12		-87[g]	50	
maximal, relative to observation	38[i]	17		18[h]	4.2		0.08[i]	0.07		92[h]	26	

[a] Heifers, mature steers / dry cows, multi (MP) or primiparous (PP) cows.

[b] Intake of forage NDF (kg).

[c] Intake of NDF (kg) from steam rolled barley or coarsely processed concentrates.

[d] Intake of NDF (kg) from finely ground concentrates.

[e] Grass silage or hay, alfalfa hay and orchard grass hay/silage.

[f] ln(min(DM content in forage, %/40,1)) has a negative value for DM% <40% and a zero value for DM% ≥40%.

[g] Electromyogram cheek (Oshio, 2001), Jaw movements by different methods.

[h] Freer and Campling (1965).

[i] Bae *et al.* (1981).

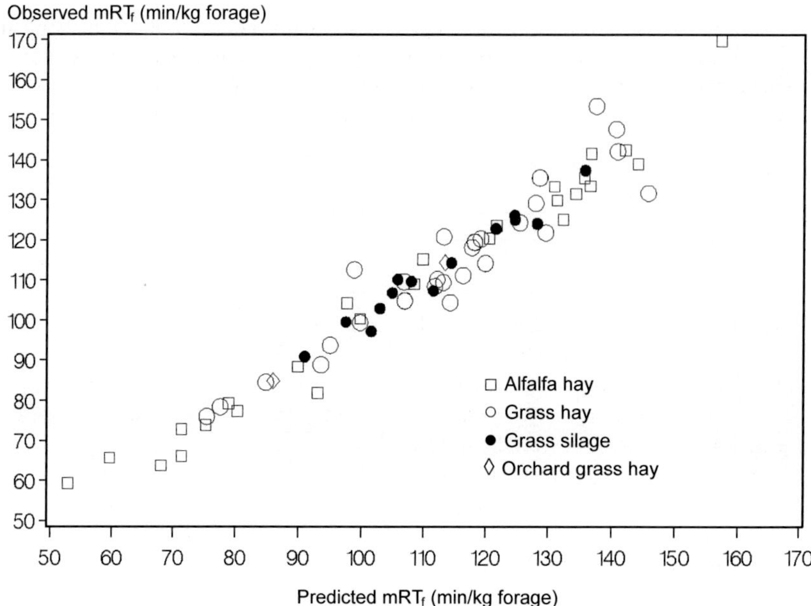

Figure 1. Observed vs. predicted mean ruminating time (mRT_f).

However, it had some minor effects on the numeric size of the estimated responses, β_1 to β_7, of BW, $NDFI_f$, $NDFI_{RB}$, $NDFI_c$ and logDM of silage on the RT, mRT_f, mET_f and the mET_f / mRT_f values. The mET_f and mRT_f values decreased by 0.21 and 0.20 min/kg $NDFI_f$ per kg increased BW, respectively (Table 2), whereas Bae *et al.* (1983) found that the mRT values decreased curvilinearly with increased $BW^{0.75}$ value. Substitution of BW for $BW^{0.75}$ in the present study increased the BIC values for the statistical models by more than 6 in this meta-analysis. The mRT_f decreased generally by 0.52±0.05 (min/kg $NDFI_f$) per kg increased $NDFI_f$/BW (kg/kg) (Table 2). A specific meta-analysis based on restricted forage intake (Bae *et al.*, 1981; Luginbuhl *et al.*, 1989; Van Bruchem *et al.*, 1991; Beauchemin and Iwaasa, 1993; Oshio, 2001) showed that the mRT_f values decreased by 0.42±0.07 (min/$NDFI_f$) per kg increased $NDFI_f$ /BW (kg/kg) and the mET_f values increased by 0.28±0.11 (min/kg $NDFI_f$) per kg increased $NDFI_f$ /BW (kg/kg).

Intake of finely processed concentrates appears to stimulate RT (P<0.003) and tended (P<0.08) to increase mRT_f by 11 min/kg $NDFI_c$ equal to about 3 min/kg concentrate. Intake of NDF from steam rolled barley grain increased mRT_f by 73 min/kg $NDFI_{RB}$, which is about 7 times higher compared with NDF from finely processed concentrates.

Lack of a sufficient number of ADL/NDF values in the included studies (Table 2) did not allow inclusion of the potential effects of Hardness of forage NDF fibre on the mRT_f value. Inclusion of NDF content in forage DM into the models with or without interaction with forage type did not decrease the BIC values and was not significant. The mET_f/mRT_f, mET_f and mRT_f values for a lactating MP cow with a BW of 625 kg, an intake of 20 kg DM/d and 0.7% $NDFI_f$ of BW can be predicted to 0.41, 45 and 109 (min/kg NDF from GS), respectively, and 0.67, 55 and 88 (min/kg NDF from AH), respectively. The mET_f/mRT_f ratio of 0.41 predicted from the model (Table 2) and the mET_f/mRT_f ratio of 45/109=0.41 estimated from the predicted mET_f and mRT_f values for a standard cow fed previlted GS or GH are similar. Consequently, the k_{EI}/k_{RI} ratio is approximately 0.5, and estimates of mET_f and mRI_f were fuzzed to a k_{EI} value of 50 min per kg NDF for unchopped forage, and a k_{RI} value of 100 min per kg NDF for unchopped forage by assuming a Hardness factor of one in the Nordic Chewing index system. Based on the present work, the chewing index system in use in the Nordic countries calculates CI (EI + RI) for forages and coarsely processed concentrates as:

EI = 0.5 × NDF,% × Size_E
RI = 1 × NDF,% × Size_R × Hardness factor (Nørgaard *et al.*, 2008).

Conclusions

The k_{EI} and the k_{RI} values for prediction of EI and RI values in the Nordic Structure Evaluation System were predicted from models including animal and feed characteristics for estimation of mET and mRT per kg of forage NDF. The mET values per kg forage NDF decreased at increasing body size, by supplementation with coarsely processed grain, and increased at decreasing DM contents below 40% in grass silage. The mET/mRT ratio increased at increasing intake of forage NDF/BW. The mRT values per kg of forage NDF decreased at increasing body size and forage NDF intake per BW. The present models appear to have potentials when comparing experimental results on RT, mRT and mET values with published literature values from cattle of different sizes and with different intakes of forage NDF.

References

Aikman, P.C., Reynolds, C.K. and Beever, D.E., 2008. Diet digestibility, rate of passage, and eating and rumination behavior of Jersey and Holstein cows. Journal of Dairy Science 91:1103-1114.
Allen, M.S., 1997. Relationship between fermentation acid production in the rumen and the requirement for physically effective fiber. Journal of Dairy Science 80:1447-1462.
Bae, D.H., Welch, J.G. and Gilman, B.E., 1983. Mastication and rumination in relation to body size of cattle. Journal of Dairy Science 66:2137-2141.
Bae, D.H., Welch, J.G. and Smith, A.M., 1981. Efficiency of mastication in relation to hay intake by cattle. Journal of Animal Science 52:1371-1375.
Balch, C.C., 1971. Proposal to use time spent chewing as an index to which diets for ruminants possess the physical property of fibrousness characteristic of roughages. British Journal of Nutrition 26:383-392.
Beauchemin, K.A. and Iwaasa, A.D., 1993. Eating and ruminating activities of cattle fed alfalfa or orchardgrass harvested at two stages of maturity. Canadian Journal of Animal Science 73:79-88.
Beauchemin, K.A., 1991. Ingestion and mastication of feed by dairy cattle. Veterinary Clinics of North America, Food Animal Practice 7:439-463.
Beauchemin, K.A., Eriksen, L., Norgaard, P. and Rode, L.M., 2008. Salivary secretion during meals in lactating dairy cattle. Journal of Dairy Science 91:2077-2081.
Beauchemin, K.A., Farr, B.I. and Rode, L.M., 1991. Enhancement of the effective fiber content of barley-based concentrates fed to dairy cows. Journal of Dairy Science 74:3128-3139.
Beauchemin, K.A., Farr, B.I., Rode, L.M. and Schaalje, G.B., 1994a. Optimal neutral detergent fiber concentration of barley-based diets for lactating dairy cows. Journal of Dairy Science 77:1013-1029.
Beauchemin, K., McAllister, T., Dong, Y., Farr, B. and Cheng, K., 1994b. Effects of mastication on digestion of whole cereal grains by cattle. Journal of Animal Science 72:236-246.
Beauchemin, K. and Rode, L., 1994. Compressed baled alfalfa hay for primiparous and multiparous dairy cows. Journal of Dairy Science 77:1003-1012.
Campling, R.C., 1966a. A preliminary study of the effect of pregnancy and lactation on th voluntary intake of food by cows. British Journal of Nutrition 20:25-39.
Campling, R.C., 1966b. The effect of concentrate on the rate of disappearance of digesta from the alimentary tract of cows given hay. Journal of Dairy Research 33:13-23.
Campling, R., 1966c. The intake of hay and silage by cows. Journal-of-the-British-Grassland-Society 21, 41-48.

Castle, M., Retter, W. and Watson, J., 1979. Silage and milk production: a comparison between grass silage of three different chop lengths. Grass and Forage Science 34:293-301.
Clark, P.W. and Armentano, L.E., 1993. Effectiveness of neutral detergent fiber in whole cottonseed and dried distillers grains compared with alfalfa haylage. Journal of Dairy Science 76:2644-2650.
De Boever, J., Andries, J., De Brabander, D.L., Cottyn, B. and Buysse, F., 1990. Chewing activity of ruminants as a measure of physical structure- a review of facors affecting it. Animal Feed Science and Technology 27:281-291.
De Boever, J.L., De Smet, A., De Brabander, D.L. and Boucque, C.V., 1993. Evaluation of physical structure. 1. Grass silage. Journal of Dairy Science 76:140-153.
Freer, M. and Campling, C., 1965. Factors affecting the voluntary intake of food by cows. 7. The behavior and reticular motility of cows given diets of hay, dried grass, concentrates and ground, pelleted hay. British Journal of Nutrition 19:195-207.
Grant, R.J. and Weidner, S.J., 1992. Effect of fat from whole soybeans on performance of dairy cows fed rations differing in fibre level and particle size. Journal of Dairy Science 75:2742-2751.
Kaiser, R. and Combs, D., 1989. Utilization of three maturities of alfalfa by dairy cows fed rations that contain similar concentrations of fiber. Journal of Dairy Science 72:2301-2307.
Luginbuhl, J.M., Pond, K.R., Burns, J.C. and Russ, J.C., 1989. Effects of ingestive mastication on particle dimensions and weight distribution of coastal bermudagrass hay fed to steers at four levels. Journal of Animal Science 67:538-546.
Mertens, D.R., 1997. Creating a system for meeting the fiber requirements of dairy cows. Journal of Dairy Science 80, 1463-1481.
Nørgaard, P., 1986. Physical structure of feeds for dairy cows. (A new system for evaluation of the physical structure in feedstuffs and rations for dairy cows). In: Neimann-Sørensen, A. (ed.) New developments and future perspectives in research of rumen function. Directorate-General for Agriculture, Coordination of Agricultural Reseach, Brussels, Belgium, pp.85-107.
Nørgaard, P., 1989. The influence of physical form of ration on chewing activity and rumen motility in lactating cows. Acta Agriculturae Scandinavica 39:187-202.
Nørgaard, P., Nadeau, E., Volden, H., Randby, A., Aaes, O. and Mehlqvist, M., 2008. A new Nordic structure evaluation system for diets fed to dairy cows. (Grassland Science in Europe, Volume 13). In: Hopkins, A., Gustafsson, T., Bertilsson, J., Dalin, G., Nilsdotter-Linde, N. and Sporndly, E. (eds.) Biodiversity and animal feed: future challenges for grassland production, pp.2008-2764.
NRC, 2001. Nutrient requirements of dairy cattle. National Academy Press, Washingon, D.C., USA.
Oshio, S., 2001. Effect of intake of Italian ryegrass hay on chewing behavior, rumen volume and fecal particle size in heifers. Animal Science Journal 72:410-415.
Schwab, E.C., Shaver, R.D., Shinners, K.J., Lauer, J.G. and Coors, J.G., 2002. Processing and chop length effects in brown-midrib corn silage on intake, digestion, and milk production by dairy cows. Journal of Dairy Science 19:47-48.
Schwarz, G.E., 1978. Estimating the Dimension of a Model. Annals of Statistics 6:461-464.
Soita, H.W., Christensen, D.A. and McKinnon, J.J., 2000. Influence of particle size on the effectiveness of the fiber in barley silage. Journal of Dairy Science 83:2295-2300.
Sudweeks, E.M., Ely, L.O., Mertens, D.R. and Sisk, L.R., 1981. Assessing minimum amounts and form of roughages in ruminant diets: roughage value index system. Journal of Animal Science 53:1406-1411.
Teller, E., Vanbelle, M. and Kamatali, P., 1993. Chewing behaviour and voluntary grass silage intake by cattle. Livestock Production Science 33:215-227.
Van Bruchem, J., Bosch, M.W., Lammers Wienhoven, S.C.W., Bangma, G.A. and Van, B.J., 1991. Intake, rumination, reticulo-rumen fluid and particle kinetics, and faecal particle size in heifers and cattle fed on grass hay and wilted grass silage. Livestock Production Science 27:297-308.
Weidner, S.J. and Grant, R.J., 1994. Altered ruminal mat consistency by high percentages of soybean hulls fed to lactating dairy cows. Journal of Dairy Science 77:522-532.
Woodford, J., Jorgensen, N. and Barrington, G., 1986. Impact of dietary fiber and physical form on performance of lactating dairy cows. Journal of Dairy Science 69:1035-1047.

Modelling within-day variability in feeding behaviour in relation to rumen pH: application to dairy goats receiving an acidogenic diet

S. Giger-Reverdin[1,2], M. Desnoyers[1], C. Duvaux-Ponter[1,2] and D. Sauvant[1,2]
[1]INRA, UMR791 Modélisation Systémique Appliquée aux Ruminants, 16 rue Claude Bernard, 75005 Paris, France; sylvie.giger-reverdin@agroparistech.fr
[2]AgroParisTech, 16 rue Claude Bernard, 75005 Paris, France

Abstract

As feeding behaviour and rumen pH are closely related, the aim of this work was to mechanistically model the post-prandial kinetics of feed intake, chewing behaviour and rumen pH using data from twelve mid-lactating goats receiving a high concentrate diet. The model comprised two subparts. The digestion submodel included the saliva secretion due to chewing, dry matter fermentation to volatile fatty acids, dry matter (DM) and liquid outflows. The behaviour submodel acted as a regulating one. It calculated the DM intake flow from DM intake rate, and from the possibilities of shifting between the 3 basic behaviours (eating, ruminating and resting). The model comprised 11 compartments, 8 auxiliary variables and 18 parameters and was developed on ModelMaker® software with a 1 min time step. A data set of 86 animal-days comprising simultaneous measurements of DM intake, feeding behaviour and pH during 13h after the afternoon feeding was used. A principal component analysis was performed to cluster 2 extreme groups of patterns on the first component. One group exhibited a high intake rate after feeding (group F as fast eaters) and the other group showed a lower rate of intake (group S as slow eaters). The variations in rumen pH were quite important for group F, but not for group S. The different steps of modelling were: adjustment of the patterns of feeding behaviour, integration of kinetics of DM intake and intake rates, determination of the parameters of the digestion submodel. Finally, pH kinetics were included in the fitting process. For group F, the model was able to simulate a realistic pH drop approximately 5 h after feeding. For the other group, the model tended to present a steady state situation. It was possible to build a fairly simple model relating digestion processes and post-prandial kinetics of chewing activities which simulated closely pH changes. It is the first step in modelling acidosis in a short term approach.

Keywords: model, ruminant, buffer, pH kinetics, Principal Component Analysis

Introduction

Post-prandial kinetics of rumen pH are frequently presented as the major index of the temporal pattern of digestion in this organ. This is particularly true with diets rich in concentrate because of the risk of acidosis. The meta-analysis developed by Dragomir *et al.* (2008) on rumen pH post-prandial kinetics obtained from the literature pointed out that these kinetics can vary a lot. Further, this approach did not include feeding behaviour while it is known to influence rumen pH (Desnoyers *et al.*, 2008a). The mechanistic models of intake published by Sauvant *et al.* (1996) or Imamidoost and Cant (2005) involved a dynamic and short term approach including chewing and digestion patterns. However, these models were based on data from several publications including only diets rich in forage. The aim of this work was to mechanistically model the post-prandial kinetics of feed intake, rumination behaviour and rumen pH variations in mid-lactating dairy goats using data obtained with a classical acidogenic diet rich in concentrates.

Material and methods

Experimental data

Twelve rumen cannulated goats (four Saanen and eight Alpine) were fed a total mixed ration *ad libitum* (35% grass hay, 15% pressed sugar-beet pulp and 50% concentrate on a dry matter basis). The diet had a dry matter content of 63%, and, on a dry matter basis, a NDF content of 34.7% and a crude protein content of 12.7%. Its energy value, calculated according to the INRA tables (Baumont *et al.*, 2007) was of 6.48 MJ NE/kg DM.

Feed intake was recorded every 2 min for several days by automatic weighing devices fitted under the feed trough. As animals were weighed weekly, feed intake was expressed in g of dry matter intake per kg of body weight using the value of the last weighing.

Feeding behaviour was estimated by a portable automatic system measuring jaw movements (APEC or *Appareil Portatif Pour l'Etude du Comportement*) developed at the INRA Theix France on sheep (Béchet *et al.*, 1995) and adapted to goats (Desnoyers *et al.*, 2009a). The data files obtained at the end of each recording period were analysed by a software package which classified chewing activity bouts as 'ruminating' or 'eating' every 2 min, taking into account the periodicity of jaw movements. An eating or a rumination period was a period exceeding seven min, and during which the animal maintained the same activity including short pauses in jaw movements of less than seven min. Jaw resting duration was estimated by difference.

Rumen pH was measured continuously by a self-cleaning pH probe (accuracy = 0.01 pH unit, Fisher Bioblock Scientific, Illkirch, France) fitted in the rumen through the cannula and linked to a portable device (Easy Log EL-2, Omega Engineering Inc., Stamford, CT, USA (Brossard *et al.*, 2003)). Rumen pH was recorded every min.

Every 20 min intervals after the afternoon feeding (41 20-min intervals altogether) eating, ruminating and resting durations, and feed intake were summed up and pH was averaged for each kinetic. Intake rate per 20-min interval was calculated as dry matter intake divided by eating duration and was expressed in g DM/min.

Model description

The model included two subparts: a digestion part taking into account rumen solid and liquid phases, and a behavioural regulation part controlling the probability of shifting between eating, ruminating and resting.

The digestion submodel took into account saliva secretion through chewing, DM fermentation to VFA, and DM and liquid flows (Figure 1).

Saliva production (SALIVA) was calculated from the duration of behavioural activities (eating, ruminating or resting) multiplied by the rate of saliva production which was considered to be higher for chewing activities than for the resting one (Méot *et al.*, 1997). Saliva flow was added to water inputs (i.e. food and non-food water) estimated to be proportional to dietary intake. Liquid outflow (Fliquout) depended on the volume of liquid in the rumen and its fractional rate of passage to the duodenum. Rumen liquid (LIQRUM) was the difference between the above mentioned flows (SALIVA and water inputs minus Fliquout). Dietary dry matter (DMDiet) was divided into degradable and non degradable fractions which constituted two compartments (dDMRUM and

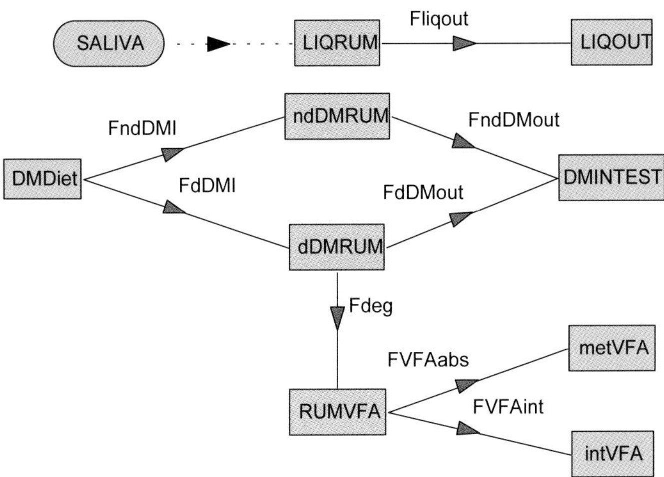

Figure 1. Diagram of the digestion submodel.

ndDMRUM). The dDMRUM fraction was a function of the input of digestible dry matter (FdDMI), the output of digestible dry matter (FdDMout) and the degradation of dry matter into volatile fatty acids (Fdeg). The ndDMRUM fraction followed the same principle excluding degradation. DM outflow to the intestine (DMINTEST) was considered as the sum of dDMRUM and ndDMRUM escaping the rumen according to a mass action law, both of them having the same fractional outflow rate. The VFA compartment (RUMVFA) was submitted to outflow and absorption processes according to mass action laws.

pH was calculated in two different ways:

- Firstly, pH was calculated empirically from VFA concentration according to the equation proposed by Tamminga and Van Vuuren (1988):
  pHVFAconc = 7.70 – 0.014 [VFA] mmoles/l
- Secondly, pH was calculated more mechanistically from the equilibrium between VFA and the buffering system HCO_3^-/CO_2 and partly inspired by the approach proposed by Imamidoost and Cant (2005):
  The dissociated [AA] and undissociated [HA] forms of volatile fatty acids were considered in equilibrium as described in Figure 2. The corresponding pH was then calculated according to the Henderson formula:
  pHacid = 4.8 + log ([AA]/[HA]).

The HA pool was increased by the input due to feed degradation (Fdeg) and decreased by the dissociation of HA to AA (forward flow, FAAHA; backward one, FHAAA), the removal from the rumen by the passage towards the intestine, and the absorption through the rumen wall (Nozière and Hoch, 2006). A similar process was followed for the AA pool.

Bicarbonates represented the major buffering system in the rumen (Figure 3).

The HCO_3^- pool was increased by the inflow of buffers entering the rumen via the saliva and decreased by the outflow from the rumen to the intestine and by the rate of equilibrium with the CO_2 pool. The aqueous form of CO_2 was considered as being in a permanent equilibrium with the gaseous form (Henry's Law), as it was assumed that CO_2 was eructed and that its partial pressure

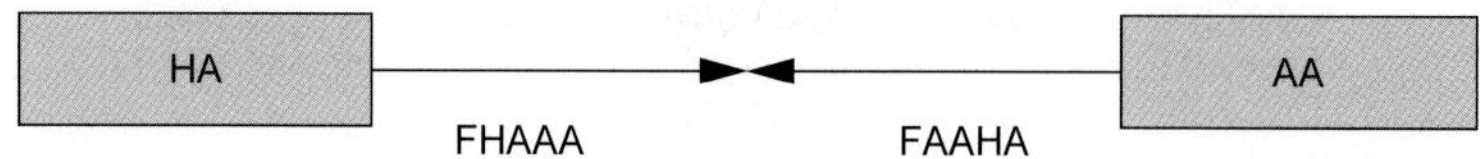

Figure 2. Diagram of the equilibrium between the two forms of volatile fatty acids.

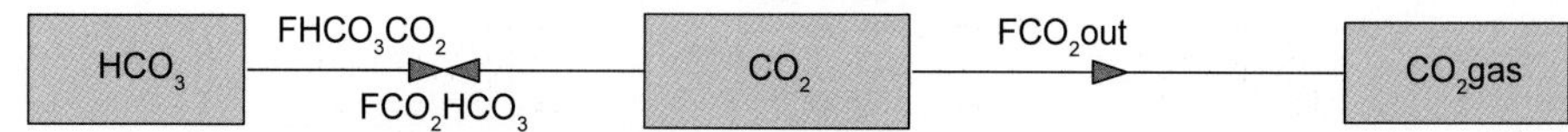

Figure 3. Diagram of the equilibrium between bicarbonates and carbon dioxide.

was always equal to 0.7 atm (70% of the eructed gas is constituted of carbon dioxide (Kohn and Dunlap, 1998)). From a chemistry point of view, the $pKCO_2$ is equal to 6.1. However, on one hand, there is an equilibrium between HCO_3- and CO_2 and, on the other hand, an equilibrium between the liquid and gaseous forms of CO_2. Therefore, assuming as Kohn and Dunlap (1998) that Henry's constant (k) for CO_2 was equal to 0.0229 mol/atm, the effective $pKCO_2$ became:

$$pKCO_2 = 6.1 + (-\log (0.229)) = 7.74.$$

The corresponding pH was estimated from the concentration of the buffering system HCO_3^-/CO_2, thus:

$$pHbuf = 7.74 + \log ([HCO_3^-]/pCO_2)$$

When the system is in equilibrium, pHacid and pHbuf are equal. Due to the integration step of 1 min compared to the longer delay of about 15 min necessary to achieve the equilibrium as shown in an *in vitro* experiment by Kohn and Dunlap (1998), the strict equality between the two calculated pH could not be achieved. Thus, the pH was calculated as the mean value of these two pH (pHacidbuf = (pHacid + pHbuf)/2) and the convergence between pHacid and pHbuf values to the mean value was obtained by using the principle of the 'ball placed in a groove' proposed by Sauvant and Lovatto (2000).

The behaviour submodel calculated the DM intake flow from the DM intake rate and from the possibilities of shifting between the 3 basic behaviours: eating, ruminating and resting. Therefore, there were six probabilities of shifting, two for each of the three activities. For example, the probability that an animal moved from eating to ruminating was called FEatRum, and the reverse one, FRumEat. At any given time, as the 3 behaviours were exclusive, the sum of the probabilities that the animal followed one activity or another was equal to 1. Such a principle was already applied by Sauvant *et al.* (1996):

$$\text{eating} + \text{ruminating} + \text{resting} = 1$$

The diagram flow of this submodel is presented in Figure 4.

The model comprised 11 compartments, 8 auxiliary variables and 18 parameters. It was developed on ModelMaker® software with a 1 min time step.

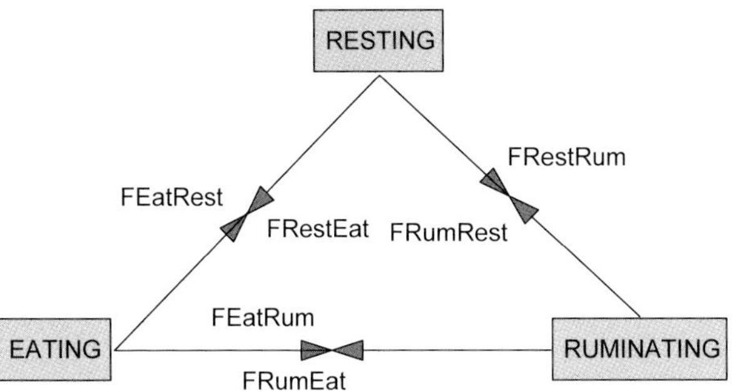

Figure 4. Diagram of the behaviour submodel.

Data set

A data set of 86 animal-days comprising simultaneous measurements of DM intake, eating and ruminating behaviour and pH during 13h and 40 min after the afternoon feeding, which meant 41 20-min intervals × 4 parameters = 164 variables, was used. A principal component analysis was performed with the 86 animal-days on these 164 variables to cluster 2 extreme groups of patterns on the 1st component as already described by Desnoyers *et al.* (2008b). The first component explained about 14.5% of the variance and exhibited the opposition between, on one hand, pH, and on the other hand, DM intake and intake duration during the first two hours after feeding. Therefore, this component discriminated the goats according to their rate of intake during the first two hours after feeding.

The projection of the data set on the first axis highlighted wide differences between animals. Two groups of 12 animal-days, each of them corresponding to 15% of the total data set, were formed. One of the groups exhibited an intensive eating pattern rate after feeding which corresponded to a high rate of intake (Figure 5). It was named group F (F for fast eaters).

After 6 hours, eating durations of group F were always shorter than ruminating ones for each 20 min interval. The difference in duration between the two types of chewing behaviour increased with time after feeding. It looked as if some secondary meals appeared after 7 hours.

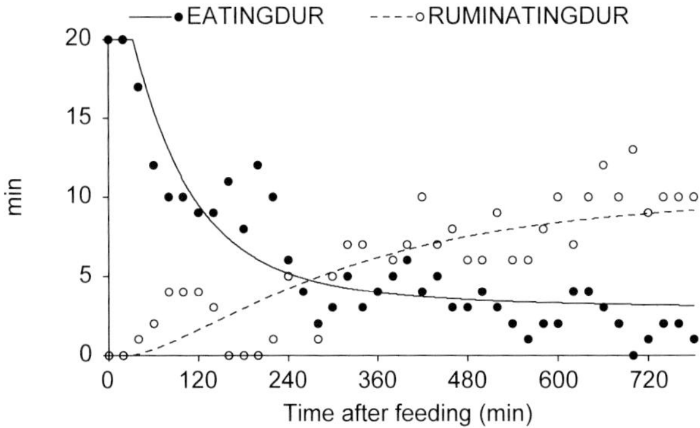

Figure 5. Evolution of chewing durations after feeding in 20 min intervals for the fast (F) eaters group clustered from the principal component analysis.

The other group showed a lower rate of intake and was named group S (S for slow eaters) as shown in Figure 6.

After 6 hours, eating durations for group S were also shorter than the ruminating ones, but the difference was not as large as that observed for group F.

The variations in rumen pH were quite important for group F. On the contrary, only small variations occurred in rumen pH for group S (Figure 7).

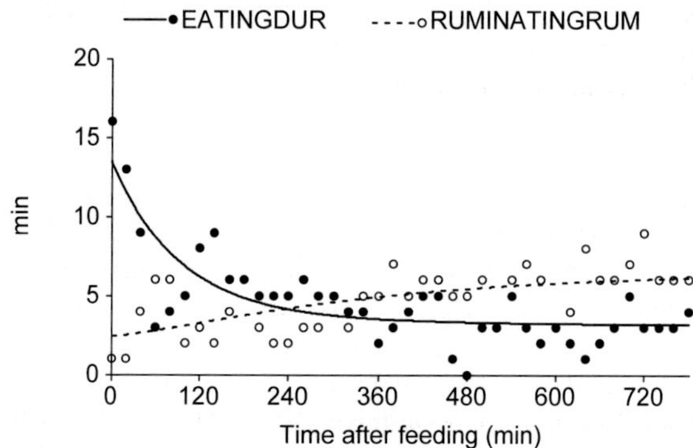

Figure 6. Evolution of chewing durations after feeding in 20 min intervals for the slow (S) eaters group clustered from the principal component analysis.

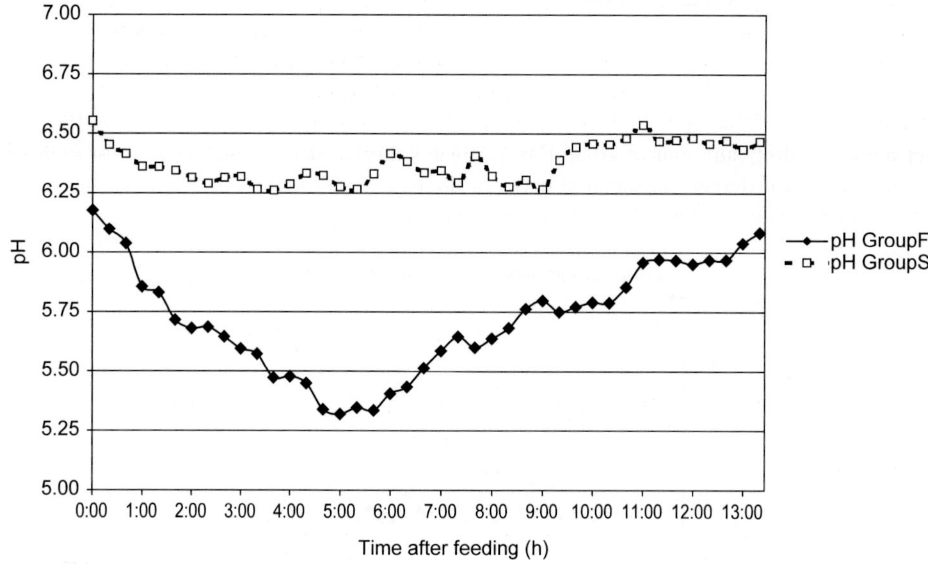

Figure 7. Evolution of pH after feeding for the fast (F) and slow (S) eaters groups clustered from the principal component analysis.

Even if the animals of group S received the same acidogenic diet as the animals of the other group, their pH was maintained within narrow limits: 6.25 to 6.50.

The modelling process was divided into four steps:

- In a first step, the patterns of feeding behaviour (eating, ruminating and resting) were adjusted on the 2 types of extreme patterns F and S (Figures 5 and 6). Secondary meals were neglected and only global trends were considered (Figures 5 and 6).
- In a second step, the kinetics of DM intake and intake rates were also included into the fitting process.
- In a third step, the parameters of the digestion submodel were determined assuming that some of them presented similar values for the two groups.
- At the end, the pH kinetics were included in the fitting process.

Results

For the group showing a marked behaviour pattern (group F), the model was able to simulate a globally realistic pH drop of approximately 0.70 unit with a nadir around 5 h after feeding (Figure 8). The model was therefore not adapted to the other group as it was built to simulate significant pH drops.

For the fast eaters group, the empirical modelling of pH based on the VFA concentration estimated from the dry matter degradation gave a quite good estimation of the observed pH. However, it must be kept in mind that measured pH kinetics were included in the fitting process.

The two calculated values of pH from acid and buffer systems were close. Nevertheless, even if the mechanistic modelling showed some agreement with the measured pH during the first two hours after feeding, it did not simulate a sufficient pH drop 5 hours after feeding. Moreover, at the end of the kinetic, the pH recovery was insufficient. Thus, the simulated kinetic appeared more smoothed than reality. Therefore, even if this last approach was more mechanistic and satisfactory from a scientific point of view it was not able to model a realistic kinetic of pH. This was likely due to the delay of about 15 min necessary to achieve the equilibrium (Kohn and Dunlap, 1998) in comparison to the one min step of the model. Some *in vitro* work is needed to determine the exact

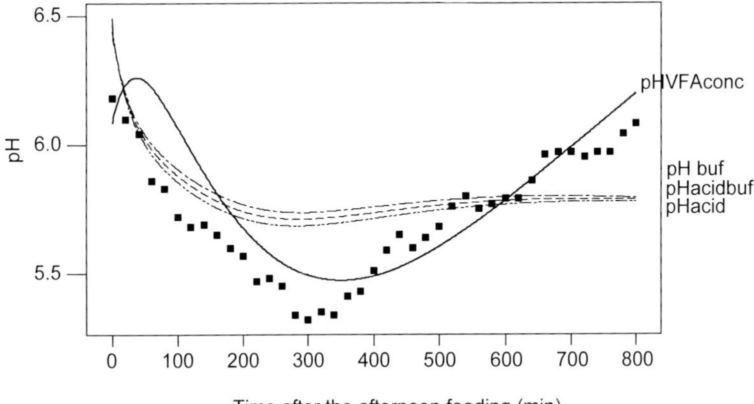

Figure 8. Comparison between observed pH and pH estimated from the VFA concentration or from the mechanistic model for the fast eaters group.

delay between the arrival of acid and buffer in such a system and the pH response. According to the *in vitro* results, changes in the current model will be envisaged.

For the other group (the slow eaters one), the model presented a *quasi* steady state situation, with a slight decrease of 0.15 pH unit 7 h after feeding.

Even if this model is not perfect to predict mechanistically pH kinetics, these preliminary results are quite encouraging. They show the interest of a short term approach after feeding which complements the approach proposed on a daily basis by Desnoyers *et al.* (2009b) with the same diet and animals.

It must also be stressed that with the same acidogenic diet, it was possible to cluster two groups of animals with quite different chewing patterns, and subsequently two very different types of kinetics for rumen pH. This pointed out the large between-animal variability of the parameters considered in this paper, and thus the probably highly variable susceptibility of animals to acidosis with a given diet.

Conclusion

It was possible to build a fairly simple mechanistic model relating digestion processes and post-prandial kinetics of chewing activities which closely simulated pH changes. As this result was obtained with an acidogenic diet, this model is a first step in modelling acidosis in a short term approach to complete the daily approach already published with the same diet. The more mechanistic calculation of rumen pH was not as satisfactory as it should have been, in comparison to a more empirical one. More work is needed on this aspect.

References

Baumont, R., Dulphy, J.P., Sauvant, D., Tran, G., Meschy, F., Aufrère, J., Peyraud, J.-L. and Champciaux, P., 2007. Les tables de la valeur des aliments. In: (eds.). Alimentation des bovins, ovins et caprins. Besoins des animaux – Valeurs des aliments. Tables Inra 2007, Editions Quae, Versailles, France, pp.181-286.

Béchet, G., Brun, J.P. and Petit, M., 1995. Comparison between two methods of measuring the grazing time of ewes. Annales de Zootechnie 44, suppl. 1:236.

Brossard, L., Fabre, M., Martin, C. and Michalet-Doreau, B., 2003. Validation of continuous ruminal pH measurements by indweling probes. In: Proceedings of the 26th Conference on gastrointestinal function. 2003/03/10-12. Chicago, IL, USA, p. 25.

Desnoyers, M., Duvaux-Ponter, C., Rigalma, K., Roussel, S., Martin, O. and Giger-Reverdin, S., 2008a. Effect of concentrate percentage on ruminal pH and time-budget in dairy goats. Animal 2:1802-1808.

Desnoyers, M., Giger-Reverdin, S., Duvaux-Ponter, C., Lebarbier, E. and Sauvant, D., 2008b. Modélisation des épisodes d'acidose sub-clinique et du comportement alimentaire associé: application à la chèvre laitière. In: 15èmes Rencontres autour des Recherches sur les Ruminants. 2008/12/03-04. Paris, France, pp.339-342.

Desnoyers, M., Béchet, G., Duvaux-Ponter, C., Morand-Fehr, P. and Giger-Reverdin, S., 2009a. Comparison of video recording and a portable electronic device for measuring the feeding behaviour of individually housed dairy goats. Small Ruminant Research 83:58-63.

Desnoyers, M., Giger-Reverdin, S., Duvaux-Ponter, C. and Sauvant, D., 2009b. Modeling of off-feed periods caused by subacute acidosis in intensive lactating ruminants: Application to goats. Journal of Dairy Science 92:3894-3906.

Dragomir, C., Sauvant, D., Peyraud, J.-L., Giger-Reverdin, S. and Michalet-Doreau, B., 2008. Meta-Analysis of 0-8 hours post-prandial kinetics of ruminal pH. Animal 2:1437-1448.

Imamidoost, R. and Cant, J.P., 2005. Non-steady-state modeling of effects of timing and level of concentrate supplementation on ruminal pH and forage intake in high-producing, grazing ewes. Journal of Animal Science 83:1102-1115.

Kohn, R.A. and Dunlap, T.F., 1998. Calculation of the buffering capacity of bicarbonate in the rumen and *in vitro*. Journal of Animal Science 76:1702-1709.

Méot, F., Cirio, A. and Boivin, R., 1997. Parotid secretion daily patterns and measurement with ultrasonic flow probes in conscious sheep. Experimental Physiology 82:905-923.

Nozière, P. and Hoch, T., 2006. Modelling fluxes of volatile fatty acids from rumen to portal blood. In: Kebreab, E., Dijkstra, J., Bannink, A., Gerrits, W.J.J. and France, J. (eds.) Nutrient digestion and utilization in farm animals: modelling approaches, CABI Publishing, Wallingford UK, pp.40-47.

Sauvant, D., Baumont, R. and Faverdin, P., 1996. Development of a mechanistic model of intake and chewing activities of sheep. Journal of Animal Science 74:2785-2802.

Sauvant, D. and Lovatto, P.A., 2000. Modelling relationships between homoeorhetic and homoeostatic control of metabolism: application to growing pigs. In: McNamara, J.P., France, J. and Beever, D.E. (eds.). Modelling nutrient utilization in farm animals, Cabi, Wallingford UK, pp.317-328.

Tamminga, S. and Van Vuuren, A.M., 1988. Formation and utilization of end products of lignocellulose degradation in ruminants. Animal Feed Science and Technology 21:141-159.

Modelling of the effects of heat stress on some feeding behaviour and physiological parameters in cows

T. Najar, M. Rejeb and M. Ben M Rad
Institut National Agronomique de Tunisie, 43 avenue Charles Nicolle, 1082 Tunis, Tunisia; najar.taha@inat.agrinet.tn

Abstract

A data base has been generated by using results reported in refereed papers that have assessed the effect of heat stress on dairy cows. Studied parameters have been evaluated using variance-covariance analysis to separate between and within variations. Simple and multiple polynomial or quadratic regressions were carried out to quantify the heat stress effects on nutritional parameters. The main results indicated that heat stress effects vary according to the physiological and dietary parameters. Heat stress effects on animal behaviour, physiological parameters and production appeared to occur at thermo hygrometric index (THI) ranging from 80 to 90. Respiratory rate and rectal temperature have the highest correlations with maximal ambient temperature and THI. Pronounced effect of heat stress was observed on dry matter intake (DMI). An average DMI reduction of 4.88 g/kg and a decrease of 11.5 g/kg MBW were noted when THI and NDF increase. These reductions were found to be moderate in diets with high level of concentrate and low fiber amount. The limited results with digestibility measurements show that digestibility of dry and organic matter based diets decreases during thermal stress conditions.

Keywords: heat stress, dairy cows, meta analysis

Introduction

Heat stress is a serious problem in dairy and beef production; it affects the animal behaviour, feed intake, digestibility and feed efficiency (Roman-Ponce *et al.*, 1977; Collier *et al.*, 1981; McGuire *et al.*, 1989). Controversial results from several studies were reported since a long time (Rhoad, 1936; McDowell *et al.*, 1969; West, 1999; Bouraoui *et al.*, 2002). These studies were carried out on a limited number of physiological and nutritional parameters, and more often using homogeneous environmental conditions with a limited variability. To overcome these limits, the Meta-analysis could be a successful statistical method that may comprehensively describe the main nutritional and physiological parameters using various environmental related parameters that were assessed over a wider range (Sauvant *et al.*, 2005; Schulin-Zeuthen *et al.*, 2007). This paper aimed to study by meta-analysis method effects of heat stress on some animal feeding behaviour parameters.

Material and methods

A meta-analysis was conducted using results reported in refereed papers that investigated potential effects of heat stress on animal behaviour. Data base was assembled from 96 papers corresponding to 370 treatments. Only papers with complete, precise and reliable information were included in the database. The major topics of the experiments is the study of the effects of heat stress on the physiological behaviour of cows, as well as the effect of the mode of housing, feed additives and the feed fibers in heat stress conditions. Main considered parameters were level of production, thermo hygrometric index (THI), heat stress period, intake time, forage intake, level of concentrate feed, feed and ration composition (crude protein, NDF, ADF, ADL), ration digestibility, external temperature, rectal temperature, and respiratory and heart rates. Different models to determine the

THI were used in the compiled papers. These models are based on measurements of temperature and relative humidity (Kibler 1964; Thom, 1959) or on dry wet bulb temperature measurements (Buffington *et al.*, 1981). When the ambient temperature (Ta) and the relative humidity (RH) are available, we calculated the THI by Kibler formulae: THI = 1.8 × Ta – (1 –RH) × (Ta – 14.3) + 32.

The studied parameters have been submitted to descriptive statistics that include sample size, means, minimum an maximum values and standard error; as well as variance and covariance analyses with general linear models procedure. Simple and multiple polynomial or quadratic regressions were carried out to quantify the heat stress effects on nutritional parameters. In each analysis, the best fit which provided the highest determination coefficient and the lower residual mean square error (RMSE) were retained.

Results

Descriptive statistics for the overall data file are in Table 1. Minimum and maximum values for THI indicate large variations of the environmental conditions between the experiments. It's the same for the dietary parameters indicating a wide variety of the fed rations, particularly as regards the protein content and the NDF who vary from 6.65% to 20.70% and from 24% to 68% respectively.

Variance heterogeneity tests were used to discard ouliers observations of physiological parameters. Confidence interval for variance was tested by the following formulae: $Pr\{a < \sigma^2 < b\} = 1 - \alpha$, where a and b are the endpoints of the interval. We used the $\chi2$ (chi-squared) distribution for the confidence interval with confidence level (α) of 95%.

No obvious relationship is observed between the respiratory rate (RR) and the THI (Figure 1) as well for the global than the intra experiment observations. A significant correlation is observed between the respiratory rate and the ambient temperature. This relation stipulates an increase of the respiratory rate by 2.40 point per °C. Considering intra experiment data improves the fitting between these two parameters (n=51, R2 = 0.26, RMSE = 2.92). Besides, adjustment of the residual of the regression between the RR and the THI with the milk yield (Figure 2) shows that the most important residual variations of this relation are related to low levels of milk. This observation could mean a less sensitivity of animals at moderated level of production to the effects of heat tress.

Table 1. Statistical description of the environmental, physiological and dietary parameters.

Parameter	n	Mean	Minimum	Maximum	RMSE
THI	242	80.76	59.5	109.4	6.8
Relative Humidity (%)	209	66.32	19	100	14.16
Temperature (°C)	305	31.5	16.6	49	4.43
Milk Yield (kg/day)	176	24.86	13	45	5.9
Respiratory rate/mn	218	81.18	30.5	143	20.7
Rectal temperature(°C)	273	39.86	32	46	2.27
Heart rate (/mn)	72	78.87	31	132	17.14
DMI (kg/d)	194	18.19	7.18	26.20	5.11
DMI (g/kg BW.75)	85	149	59	198	20.5
CP (%)	142	16.024	6.65	20.7	2.98
NDF (%)	154	32.66	24	68	9.77
ADF (%)	86	23.64	16	54	9.39

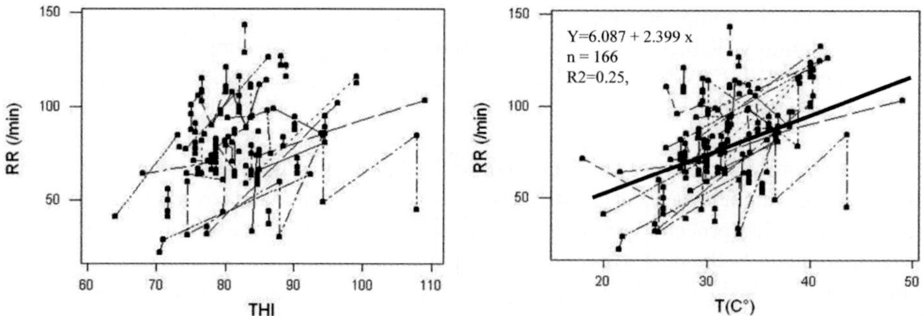

Figure 1. Relationship between respiratory rate (RR), thermo hygrometric index (THI) and ambient temperature (T(°C)).

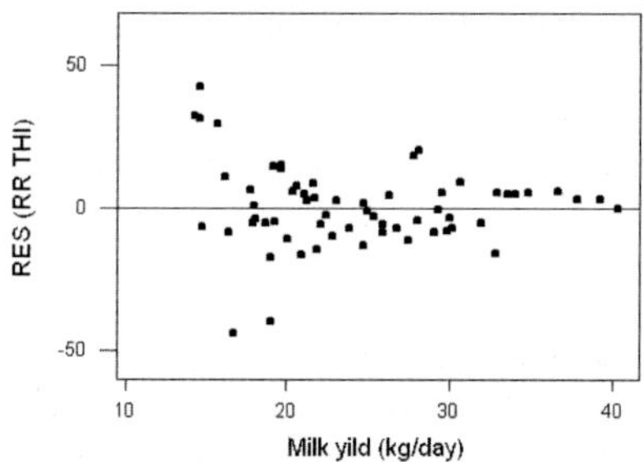

Figure 2. Relationship between the residual of the respiratory rate – thermo hygrometric index (THI) regression (RES (RR THI)), and the level of milk yield.

These results about effect of milk level is in agreement with observations of Tapki and Sahin (2006) who reported more sensitive behaviour of high producing dairy cows in a hot environment., but contrasting with some others in which the milk yield variations are related to differences in the stage of lactation of cows (Maust *et al.*, 1972). In this case, cows in early lactation use catabolise body fat and would be less influenced by the environmental stress. Kurihara and Shioya (2002) reported a decrease in milk production of cows which consumed more than 20 kg dry matter of feed per day at 25 °C. But non significant differences were observed with a higher temperature in the milk production of cows which consumed less than 20 kg dry matter of feed per day. Indeed, heat production for rumen activities and metabolic functions are proportional to level of feed intake (West, 1986). Decreased milk yield would be due to the cumulative effects of heat stress on feed intake, metabolism, and physiology of dairy cattle (Rhoads *et al.*, 2009). In some environmental conditions, the THI are not correlated with physiological responses of dairy cattle (Gomes da Silva *et al.*, 2007), and other indexes must be used.

Dairy cows succeed to maintain their body temperature indicated by the rectal temperature stable as long as the THI does not exceed 90 units (Figure 3). Beyond this value, which corresponds generally to an ambient temperature exceeding 37 °C, rectal temperature start to increases dramatically when THI increases. A significant quadratic relationship was observed when considering intra experiment observations (Equation 1).

$$\text{REC T} = 47.69 - 0.231\ \text{THI} - 0.00159\ \text{THI}^2\ (R^2=0.17, n=53, \text{RMSE}=0.0603) \quad (1)$$

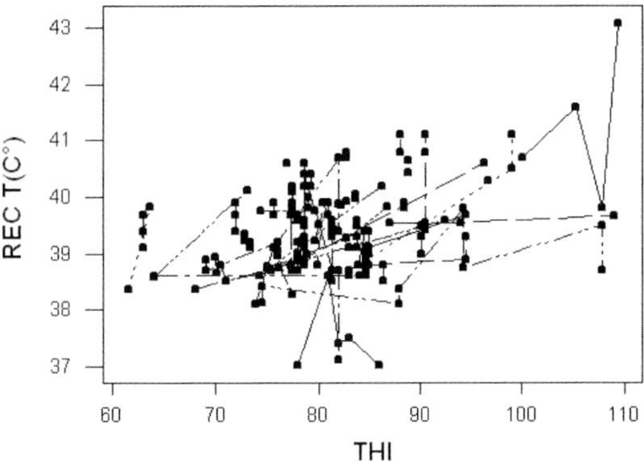

Figure 3. Variation of the rectal temperature (REC T) with the thermo hygrometric index (THI).

The mechanisms of heat losses by evaporation (sweating, sprinklers), convection, radiation, respiration and milk heat are not able to compensate the increases of heat provoked by the ambient temperature and humidity. Similar results were reported by Gomes Da Silva *et al.*, (2007) in Holstein cows.

Others have reported similar increases in respiration rate and rectal temperature when lactating cows were subjected to temperatures above their thermoneutral zone (Mohamed and Johnson, 1985; Stanley *et al.*, 1975). Results of Kabuga and Sarpong (1991) showed lack of influence of maximum temperature on rectal temperature when THI does not exceeded 80 units. This can be explained by a possible dissipation at night of excess heat gained in the day to sustain high production. Also, continuous access of animals to feed, increases intake during the cooler night hours and limited effects of heat stress. Slightly different result were found by Perissinotto *et al.* (2008), they reported that when the internal THI is less than or equal to 77, the thermal comfort is high, and when THI is higher than 80, thermal comfort is low.

The DMI is weakly correlated with the THI for the global data (Figure 4). A better fitting can be observed for the dry matter intake per kg metabolic body weight intra experiments (Equation 2 and 3).

DMI (g/kg MBW) = 485 – 4.41 THI (n=45; R^2=0.34, RMSE=7.17) (2)

DMI (g/kg MBW) = 538 – 5.15 THI (n=29; R^2=0.49, RMSE=6.33) (3)

This observation can be explained by the wide variation of the diet characteristics between the experiments, particularly concerning forage type and level and composition of concentrate feed. However, a decrease tendency can be noticed between these two parameters.

However, simple linear regression between the residual (Res) of the relationship (DMI-THI) revealed a high significant correlation with the NDF level in the diet for the global observations (Equation 4), and for the intra experiment observations (Equation 5).

Res (DMI-THI) = 42.0 – 1.23 NDF (R^2=35.5%, n=134, RMSE=9.45) (4)

Res (DMI-THI) = 105 – 2.45 NDF (R^2=38.5%, n=36, RMSE=7.23) (5)

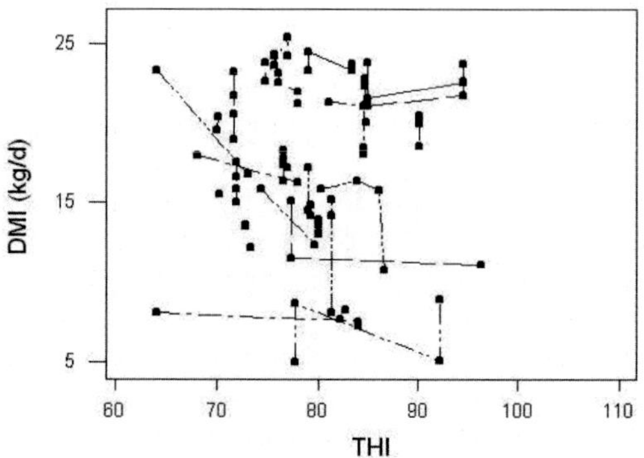

Figure 4. DMI variations with the thermo hygrometric index (THI).

Taking into account of the previous effects makes possible the fitting of the DMI according to the THI and to the NDF of the diet with a highly significant determination coefficient (Equation 6).

DMI (g/kg MBW) = 230 – 0.488 THI – 1.15 NDF, R^2=30.6%, n=36, RMSE=5.32) (6)

According to these results, DMI decreases by 4.88 g/kg MBW and by 11.5 g/kg MBW when THI and NDF increase by 10%. It was known for a long time, that cell wall content, assessed by NDF, have important role on the digestive tract distension, the digestibility and the level of intake (Tsai *et al.*, 1967; Van Soest, 1994; Reid *et al.*, 1988). Interaction between dietary fiber content and environmental conditions was observed by Magdub *et al.* (1982) and West *et al.*, (1999). They reported that increasing dietary NDF reduced DMI. However this reduction was comparable during both hot and cool weather. This result suggests that high-fiber diets were not affected by to a greater extend by heat stress. On the other hand, Cummins (1992) reported that the rate of decline in DMI with increasing environmental temperature was higher in diets with high fiber level, and possibly, this could be due to the greater heat generated from ruminal fermentation in cows fed diets containing more fibre. Likewise, Fuquay (1981) cited several reports from the sub-tropics which showed higher milk yields but low physiological responses of animals fed low fibre diets in comparison to those fed high fibre diets of equal energy levels.

Conclusions

This meta-analytical approach to evaluating the relationship between the heat stress parameters and some physiological and nutritional parameters revealed that the respiratory rate increases significantly with THI, this effect is related to level of milk production in cows. Effects of THI on rectal temperature start at THI close to 90 units. DMI decrease in heat stress conditions, the decline of DMI is related to dietary characteristics and is higher in high fibre diets.

Acknowledgements

The authors acknowledge Dr Salah Rezgui for the revision of the English writing of the document.

References

Bouraoui, R., Lahmar, M., Majdoub, A., Djemmali, M. and Belyea, R., 2002. The relationship of temperature-humidity index with milk production of dairy cows in a Mediterranean climate. Animal Research 51:479-491.

Buffington, D.E., Collazo Arocho, A., Canton, G.H. and Pitt, D., 1981. Black globe humidity index (BGHI) as a confort equation for dairy cows. Trans ASAE, St Josep, MI, 24:711-714.

Collier, R.J., Eley, R.M., Sharma, A.K., Pereira, R.M. and Buffington, D.E., 1981. Shade Management in Subtropical Environment for Milk Yield. Journal of Dairy Science 64:844-849.

Cummins, K.A., 1992. Effect of Dietary Acid Detergent Fiber on Responses to High Environmental Temperature. Journal of Dairy Science 75:1465-1471.

Fuquay, J.W., 1981. Heat stress as it affects animal production. Journal of Animal Science 52:164-174.

Gomes da Silva, R., Débora, A.E. Façanha Morais, M. and Guilhermino, M. 2007. Evaluation of thermal stress indexes for dairy cows in tropical regions. Revista Brasileira de Zootecnia 36:1192-1198.

Kabuga, J.D. and Sarpong K., 1991. Influence of weather conditions on milk production andrectal temperature of Holsteins fed two levels of concentrate. International Journal Biometeorology 34:226-230.

Kibler, H.H., 1964. Environmental physiology and shelter engineering. LXVII. Thermal effects of various temperature-humidity combinations on Holstein cattle as measured by eight physiological responses, Research. Bulletin of Missouri Agricultural Experiment. Station, p. 862.

Kurihara, M. and Shioya, S., 2002. Dairy cattle management in a hot environment. Available at: www.agnet.org/library/eb/529/eb529.pdf. Accessed June 2009.

Magdub, A., Johnson, H.D. and Belyea, R.L., 1982. Effect of environmental heat and dietary fiber on thyroid physiology of lactating cows. Journal of Dairy Science 65:2323-2331.

Maust, L.E., Mc Dowell, R.E. and Hooven, N.W., 1972. Effect of Summer weather on performance of Holstein cows in Three sieges of lactation. Journal of Dairy Science 55:1133-1139.

McDowell, R.E., Moody, E.G., Van Soest, P.J., Lehmann, R.P. and Ford, G.L., 1969. Effect of heat stress on energy and water utilization of lactating cows. Journal of Dairy Science 52:188-194.

McGuire, M., Beede, D.K., DeLorenzo, M.A., Wilcox, C.J., Huntington, G.B., Reynolds, C.K. and Collier, R.J., 1989. Effect of Growth Hormone on Milk Yields and Related Physiological Functions of Holstein Cows Exposed to Heat Stress. Journal of Animal Science 67:1050-1060.

Mohamed, M.E. and Johnson, H.D., 1985. Effect of growth hormone on milk yields and related physiological functions of Holstein cows exposed to heat stress. Journal of Dairy Science 68:1123-1129.

Perissinotto, M., Daniella, J.M. and Vasco Fitas da Cruz, 2008. Applied intelligent system for environmental control in dairy housing. Livestock Environment VIII, 31 August - 4 September 2008, Iguassu Falls, Brazil.

Reid, R.L., Jung, G.A. and Thayne, W.V., 1988. Relationship between nutritive quality and fiber components of cool season and warm season forages: a retrospective study. Journal of Animal Science 66:1275-1291.

Rhoad, A.O., 1936. The Dairy Cow in the Tropics. Journal of Animal Science 1936:212-214.

Rhoads, M.L., Rhoads, R.P., VanBaale, M.J., Collier, R.J., Sanders, S.R., Weber, W.J., Crooker, B.A. and Baumgard, L.H., 2009. Effects of heat stress and plane of nutrition on lactating Holstein cows: I. Production, metabolism, and aspects of circulating somatotropin. Journal of Dairy Science 92:1986-1997.

Roman-Ponce, H., Thatcher, W.W., Buffington, D.E., Wilcox, C.J. and Van Horn, H.H., 1977. Physiological and Production Responses o f Dairy Cattle t o a Shade Structure in a Subtropical Environment. Journal of Dairy Science 60:424-430.

Sauvant, D., Schmidely, P. and Daudin, J.J., 2005. Les méta-analyses des données expérimentales: applications en nutrition animale. INRA Production Animale 18:63-73.

Schulin-Zeuthen, M., Kebreab, E., Gerrits, W.J.J., Lopez, S., Fan, M.Z., Dias, R.S. and France, J., 2007. Meta-analysis of phosphorus balance data from growing pigs. Journal of Animal Science 85:1953-1961.

Stanley, R.W., Olbrich, S.E., Mark, F.A., Johnson, H.D. and Hilderbrand, E.S., 1975. Effect of roughage level and ambient temperature on milk production, milk composition, and ruminal volatile fatty acids. Tropical Agriculture 52:213.

Tapkı, B. and Sahin, A., 2006. Comparison of the thermoregulatory behaviours of low and high producing dairy cows in a hot environment. Applied Animal Behaviour Science 99:1-11.

Thom, E.C., 1959. The discomfort index. Weatherwise 12:57-60.

Tsai, Y.C., Castillo, L.S., Hardison, W.A. and Payne, W.J.A., 1967. Effect of dietary fiber level on lactating dairy cows in the humid tropics. Journal of Dairy Science 50:1126-1129.

Van Soest, J.P., 1994. Nutrition Ecology of the Ruminant. Second Edition. Cornell University Press. Ithaca, New York, NY, USA.

West, J.W., 1986, Effects of Heat-Stress on Production in Dairy Cattle. Journal of Dairy Science 86:2131-2144.

West, J.W., 1999. Nutritional strategies for managing the heat – stressed dairy cow. Journal of Animal Science 77(Suppl. 2):21-35.

West, J.W., Hill, G.M., Fernandez, J.M., Mandebvu, P. and Mullinix, B.G., 1999. Effects of dietary fiber on intake, milk yield, and digestion by lactating dairy cows during cool or hot, humid weather. Journal of Dairy Science 82:2455-2465.

Part 3 – Modelling fermentation, digestion and microbial interactions in the gut

A generic multi-stage compartmental model for interpreting gas production profiles

S. López[1], J. Dijkstra[2], M.S. Dhanoa[3], A. Bannink[4], E. Kebreab[5] and J. France[6]
[1]Instituto de Ganadería de Montaña (IGM), ULE – CSIC, Departamento de Producción Animal, Universidad de León, 24007 León, Spain; s.lopez@unileon.es
[2]Animal Nutrition Group, Wageningen University, Marijkeweg 40, 6709 PG Wageningen, the Netherlands
[3]North Wyke Research, Okehampton, Devon, EX20 2SB, United Kingdom
[4]Livestock Research, Animal Sciences Group, Wageningen UR, P.O. Box 65, 8200 AB Lelystad, the Netherlands
[5]Department of Animal Science, University of California, Davis 95616, USA
[6]Centre for Nutrition Modelling, Department of Animal and Poultry Science, University of Guelph, Guelph, Ontario, Canada

Abstract

The gas production technique has become a key tool in feed evaluation and rumen fermentation studies. The value of the technique relies on modelling experimental data to obtain estimates of rumen degradation parameters. One of the first models used to describe gas production profiles was the simple exponential equation, although it has some important limitations when applied to gas production curves: (1) the intercept is positive, (2) only fits diminishing returns profiles, and (3) models gas *per se* (i.e. fails to link gas production to substrate degradation). The first limitation is overcome mathematically by re-parameterisation, making the intercept zero. The second limitation can be resolved by introducing a discrete lag to mimic sigmoidicity or by using sigmoidal functions. The third limitation is overcome by modelling substrate degradation from gas production profiles, so that equations are derived from mechanistic principles, and all parameters have biological meaning. The link between substrate degradation and gas production allows for extent of substrate degradation in the rumen to be determined for a given passage rate. Several multi-phase models have been proposed, but these were originally derived empirically and assumptions made *a posteriori*. Based on the conceptual difference between stage and phase, a multi-stage approach is proposed, a generic model presented and the accompanying equations derived. A two-stage model with four pools (substrate, intermediate products, fermentation end-products and by-products such as fermentation gas) is illustrated. An interpretation of the breakdown of polysaccharides to monosaccharides (first stage) and the fermentation of these monosaccharides to yield gas and other products (VFA and microbial matter) (second stage) is presented. Gas production profiles were used to demonstrate fitting the two-stage model and to consider its ability to describe gas production curves.

Keywords: rumen fermentation kinetics, gas production curves, non-linear models, *in vitro*

Introduction

A number of *in vitro* methods have been developed to estimate digestibility and extent of microbial fermentation of feeds in the rumen, and to study their variation in response to different factors (López, 2005). The *in vitro* gas production technique has gained general recognition over the last decades, being used increasingly for feed evaluation, to investigate mechanisms of microbial fermentation, and for studying the modes of action of anti-nutritive factors, additives and feed supplements (Krishnamoorthy *et al.*, 2005). The technique is based on the assumption that the gas produced in batch cultures is just the consequence of fermentation of a given amount of substrate

(France *et al.*, 2000). Cumulative gas produced at different incubation times can be measured on a single and small sample, providing kinetic data from which extent and rate of feed degradation can be estimated. Within this context, mathematical modelling is a useful tool to link the data obtained *in vitro* with the processes occurring *in vivo*, and to estimate total-tract digestibility or rumen degradability from *in vitro* kinetic data (France *et al.*, 2000; López, 2005).

Fermentation kinetic parameters can be derived using non-linear equations able to describe diminishing returns or sigmoidal profiles (France *et al.*, 2000, 2005; Thornley and France, 2007). In general, such equations have provided good fits to observations and describe cumulative gas production curves accurately (López, 2008). To measure gas production from batch cultures containing a buffered medium and a microbial inoculum at several time intervals, different devices and apparatus have been designed (López, 2005). With automated systems it is possible to record gas released from substrate fermentation at short time intervals, so that the curves observed show some degree of irregularity with noticeable local bumps as a result of sudden slope changes (Schofield *et al.*, 1994; Groot *et al.*, 1996). This phenomenon has been interpreted as the result of autonomous fermentation of each feed component, so that total gas produced is the summation of fermented gas released from each component. Conventional non-linear equations are unsatisfactory for fitting these curves, and multi-phase models have been derived empirically to represent such a phenomenon.

The objectives of this publication were (1) to review briefly some aspects of modelling gas production kinetics and (2) to propose a new multi-stage compartmental model to resolve such kinetics.

Modelling gas production curves

One of the first models used to describe gas production profiles was the simple exponential equation proposed for *in situ* disappearance curves (Ørskov and McDonald, 1979). The equation used was $p = a + b(1 - e^{-ct})$, where p (g/g incubated) is substrate disappearance at an incubation time t (h), and a, b, c are degradation parameters. Appearance of *in vitro* gas production profiles with few data points is that of a smooth monotonically-increasing curve, and given the similarity in the shape between disappearance curves and gas production profiles, and the popularity of Ørskov & McDonald's equation, most researchers using the gas production technique opted to use the same exponential equation to fit gas production profiles, just changing the variables to gas produced (ml) from substrate fermented at a given incubation time t. However, this simple function has some significant limitations when applied to gas production curves which will be discussed.

Positive intercept

In the case of *in situ* disappearance curves, a positive intercept on the y-axis (constant a) is interpretable as the fraction that disappears instantly from the bag. However, in gas production profiles such a parameter (y-intercept) is unrealistic, because at the start of inoculation there is no gas derived from substrate fermentation in the vial where incubation is conducted. Therefore, direct application of the Ørskov & McDonald's equation to gas production profiles is inappropriate because inclusion of a positive y-intercept is inexplicable. This limitation is easily overcome mathematically by re-parameterisation so that the intercept is zero, resulting in the equation $G = A(1 - e^{-ct})$, where G (ml/g incubated) is gas production at an incubation time t (h), A (ml/g incubated) is asymptotic gas production (represents the value of G at the upper asymptote) and c (per h) is a fractional rate. The inconsistency of including a positive intercept in gas production models has occurred in other models (apart from the exponential equation) reported in the literature (e.g. Schofield *et al.*, 1994). In some cases, the curve shows a trend to cross the x-axis at an early incubation time resulting in a negative y-intercept when regressing against data. However, biologically the domain of the function should be restricted to the first quadrant (G and t should be both positive). In this case, a discrete lag parameter can be included in the equation, resulting in $G = A(1 - e^{-c(t-L)})$, where L (h) is a lag time before any gas production is observed.

Exponential vs. sigmoidal models

The exponential equation only fits diminishing returns profiles, but with some substrates (especially roughages and substrates which degrade slowly) some degree of sigmoidicity is observed in the curves (Robinson *et al.*, 1986). Therefore, the simple exponential equation is inadequate to fit such profiles. This limitation can be resolved by (1) introducing a discrete lag (see aforementioned model) to mimic sigmoidicity or (2) using sigmoidal functions. After the initial supremacy of the simple exponential model, some sigmoidal functions were proposed and subsequently used more or less extensively (Beuvink and Kogut, 1993; France *et al.*, 2000). One of the first sigmoidal functions proposed and widely used in the gas production technique is that derived by France *et al.* (1993), which is a generalized Mitscherlich model. Sigmoidal functions can be classified in two groups: (1) functions with a fixed inflexion point, such as the classical logistic (inflexion point at $A/2$) and Gompertz (inflexion point at A/e) growth functions, and (2) functions with a flexible (variable) inflexion point that can occur at any fraction of A and/or incubation time, and thus are able to represent curves with (sigmoidal) and without (diminishing returns) an inflexion point. Functions with a fixed inflexion point are conditioned by the position of this critical point, and in some functional forms they may contain an implicit y-intercept if equations are not properly re-parameterized to avoid such contradiction. France *et al.* (2000) derived logistic and Gompertz functions specially for fitting to gas production profiles. Functions with a variable inflexion point are generalized functions that usually include an additional shape parameter, and for specific values of that parameter encompass other simpler functions as special cases. The France equation (which encompasses the simple exponential equation) is an example of models belonging to this category, along with other models such as the Hill (Groot *et al.*, 1996; France *et al.*, 2000) and the Richards, which encompasses the simple exponential, Gompertz and logistic equations. Some other sigmoidal models have been used occasionally and rather empirically (e.g. as non-autonomous functions) as reviewed by Thornley and France (2007) and López (2008).

Linking gas production to substrate degradation

Most functions used, in particular the simple exponential equation, have been applied empirically to model gas *per se*, failing to link gas production to substrate degradation. To overcome this limitation, models can be derived to represent substrate degradation from gas profiles based on compartmental schemes, so that equations are derived from mechanistic principles, and all parameters have biological meaning (France *et al.*, 1990, 2000, 2005). Assuming that the feed comprises two fractions: a potentially degradable fraction S (degraded at a fractional rate μ (per h), after a discrete lag time L (h)) and an undegradable fraction U, the dynamic behaviour of these fractions in an *in vitro* system is described by the differential equations:

$$
\begin{aligned}
dS/dt &= 0, && 0 \leq t < L, \\
&= -\mu S, && t \geq L, \\
dU/dt &= 0, && t \geq 0.
\end{aligned}
$$

An essential aspect of modelling this process concerns the kinetics assumed for the process itself. The models may represent a variety of possible kinetic processes: zero-, first-, or second-order degradation kinetics (France *et al.*, 1990; López *et al.*, 2007), first-order degradation kinetics combined with a first-order lag process (Van Milgen *et al.*, 1991), and fractional degradation rates varying with time (France *et al.*, 1993, 2000, 2005; López *et al.*, 1999) or related to microbial activity (France *et al.*, 1990). Different candidate functions have been proposed for μ, in which this fractional degradation rate is a function of time. Once the functional form for μ is established, by substituting it in the differential equation for dS/dt above and integrating, an equation for S as a function of time can be derived. In its simple form, μ is constant resulting in a simple exponential equation for S. A suitable alternative is that μ varies with time resulting in a sigmoidal function for S. A third pool represented in the model is the accumulated fermentation gas. Following France *et al.* (2000, 2005), assuming that the rate at which gas is produced is directly proportional to the rate at which substrate is degraded, gas production (G, ml) is related to substrate degradation by:

$$\frac{dG}{dt} = -Y\frac{dS}{dt}, \quad \text{and} \quad G = Y(S_0 - S),$$

where Y (ml/g) is a constant yield factor and S_0 is the initial size of fraction S.

Rates of degradation and passage can be combined to calculate the extent of degradation (E, g degraded/g ingested) of the substrate in the rumen (France *et al.*, 1990, 1993) as:

$$E = \frac{\int_L^{\infty} \mu S dt}{S_0 + U},$$

where S which is subjected to both passage and degradation is the amount of potentially degradable substrate remaining in the rumen (López *et al.*, 1999; France *et al.*, 2000, 2005). This equation provides a general expression for calculating extent of degradation.

Multi-phase models

When gas production profiles are built up from a large number of data points (as with automated systems), curves can exhibit deviations from a smooth trend giving shapes with local bumps or irregularities. Such bumps may reflect rapid fermentation of a given feed component in that particular period, so that fermentation end-products (in particular gas) are formed as the result of the fermentation of each feed component. Aiming to represent this 'additive' process, a number of multi-phase models have been proposed to describe gas production profiles (Schofield *et al.*, 1994; Groot *et al.*, 1996; Cone *et al.*, 1997; Huhtanen *et al.*, 2008). These models result from the addition of two or more non-linear functions and thus contain a large number of parameters that can be estimated only when lots of data points are available for each gas profile, which is only possible with automated recording systems. These so-called multi-phasic models assume the existence of two or more 'phases' from which gas is produced (Figure 1).

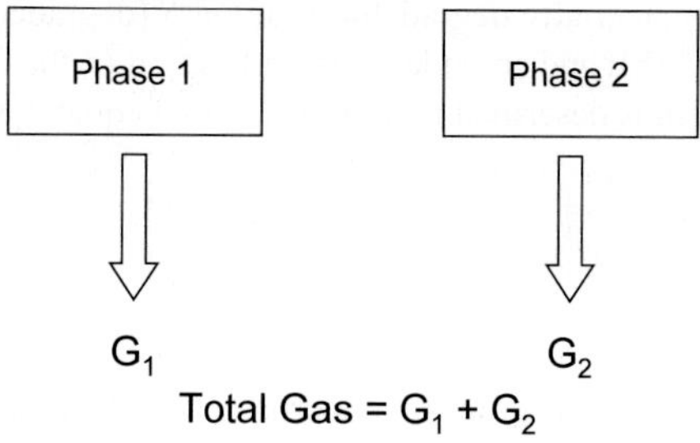

Figure 1. Multi-phase approach comprising two phases.

Each phase represents a fraction (chemical or nutritional) of the feed incubated. For instance, phase 1 could represent sugars or a soluble readily-fermentable fraction, whereas phase 2 could represent structural carbohydrates or an insoluble but potentially fermentable fraction (Stefanon *et al.*, 1996; Cone *et al.*, 1997). The most important concept is the assumption regarding fermentation kinetics of phases. Each phase behaves independently as a separate entity, and there is no interaction at all between phases, so that fermentation of each phase is not affected by and does not have any effect on other phases. Therefore, gas is released as a fermentation end-product from each phase, and total gas production from the feed is simply the sum of gas produced from all phases. In all multi-phase models reported in the literature, kinetics assumed for gas production are the same for all phases or feed fractions. Therefore, it can be stated that these models were originally derived empirically and assumptions made *a posteriori*. The best known multi-phase models are those proposed by Schofield *et al.* (1994) and Groot *et al.* (1996). Schofield *et al.* (1994) proposed both dual logistic and dual Gompertz models. They found that a dual exponential model was unsuitable to fit gas production profiles, whereas a modified dual logistic, with a single lag, was selected to model *in vitro* digestion of mixed substrates. Huhtanen *et al.* (2008) opted for a dual Gompertz model. Groot's model results from the summation of two or more (up to n) sigmoidal Hill (Hill, 1910) or C^{th}-order Michaelis-Menten equations:

$$G = \sum_{i=1}^{n} \frac{A_i}{1 + \frac{B_i^{C_i}}{t^{C_i}}}.$$

In this model, G (ml) denotes the amount of gas produced at time t (h) after incubation, A_i (ml) represents asymptotic gas production for phase i, B_i (h) is the time after incubation at which half of the asymptotic amount of gas has been formed for phase i, and C_i is a constant determining the sharpness of the switching characteristic of the phase i profile. The value of i indicates the number of phases in the profile ($i = 1, ..., n$).

Multi-stage models

In contrast with the concept of phase defined above, the term stage stands for a more mechanistic concept. A stage would be the dynamic transition between two compartments, and when there are two or more stages, they occur successively, so stage 1 is followed by stage 2, and this precedes stage 3 (Figure 2). This sequence determines that whatever happens in stage 1 will inevitably affect what will happen in stage 2. Compared with a multi-phase system, the final output of a multi-stage system is not just the sum of the outputs from the different stages constituting the system, for the interactions between stages will also have a significant contribution to the whole system.

In its simplest form the system would be composed of two compartments (substrate to be fermented and final products) with a single stage between them. The conversion of substrate into products

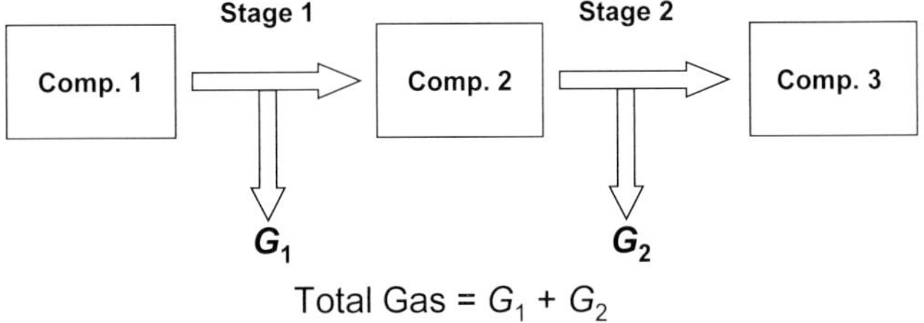

Figure 2. Multi-stage approach with two stages and three compartments.

gives rise to the production of gas. The kinetics of substrate (S) disappearance and product (P) and gas (G) formation can be represented by the conservation equations:

$$dS/dt = -kS \qquad S(0) = S_0 \tag{1}$$

$$dP/dt = kYS \qquad P(0) = P_0 \tag{2}$$

$$dG/dt = k(1 - Y)S \qquad G(0) = 0 \tag{3}$$

where k is the fractional fermentation rate (per h) and Y is a yield factor (g product/g substrate). The solution of Equation (3) for gas production gives:

$$(3) \Rightarrow dG/dt = k(1 - Y)S_0\, e^{-kt} \tag{4}$$

and after integration $G = (1 - Y)S_0(1 - e^{-kt})$, which is the simple exponential equation.

In a two stage model (such as that represented in Figure 2), there are three compartments and two stages. The compartments are substrate (S), intermediary (I) and product (P), with stage 1 representing the conversion of substrate to intermediary and stage 2 the fermentation of intermediary to final product. At each stage there is gas production (G_1 and G_2), and total gas production is the sum of both. The dynamics of the system can be represented by the conservation equations:

$$dS/dt = -k_1S \qquad S(0) = S_0 \tag{5}$$

$$dI/dt = k_1Y_1S - k_2I \qquad I(0) = I_0 \tag{6}$$

$$dG_1/dt = k_1(1 - Y_1)S \qquad G_1(0) = 0 \tag{7}$$

$$dP/dt = k_2Y_2I \qquad P(0) = P_0 \tag{8}$$

$$dG_2/dt = k_2(1 - Y_2)I \qquad G_2(0) = 0 \tag{9}$$

where k_1 and k_2 are the fractional fermentation rates (per h) for stage 1 and 2, respectively, and Y_1 and Y_2 are yield factors (g/g). The solution for equations describing gas production is:

$$(7) \Rightarrow dG_1/dt = k_1(1 - Y_1)S_0\, e^{-k_1t}$$

and after integration $G_1 = (1 - Y_1)S_0(1 - e^{-k_1t})$

$$(9) \Rightarrow \frac{dG_2}{dt} = k_2\left(1-Y_2\right)\left[\frac{k_1Y_1S_0}{k_2-k_1}e^{-k_1t} + \left(I_0 + \frac{k_1Y_1S_0}{k_1-k_2}\right)e^{-k_2t}\right]$$

and after integration

$$G_2 = \left(1-Y_2\right)I_0 + Y_1\left(1-Y_2\right)S_0 - \frac{k_2Y_1\left(1-Y_2\right)S_0}{k_2-k_1}e^{-k_1t} - \left(1-Y_2\right)\left(I_0 + \frac{k_1Y_1S_0}{k_1-k_2}\right)e^{-k_2t}$$

Total gas production, $G = G_1 + G_2 =$

$$= \left(1-Y_1\right)S_0\left(1-e^{-k_1t}\right) + \left(1-Y_2\right)I_0 + Y_1\left(1-Y_2\right)S_0 - \frac{k_2Y_1\left(1-Y_2\right)S_0}{k_2-k_1}e^{-k_1t} - \left(1-Y_2\right)\left(I_0 + \frac{k_1Y_1S_0}{k_1-k_2}\right)e^{-k_2t}$$

$$\text{or } G = \left(1-Y_1\right)S_0\left(1-e^{-k_1t}\right) + \left(1-Y_2\right)I_0\left(1-e^{-k_2t}\right) + Y_1\left(1-Y_2\right)S_0\left[1 - \frac{k_2e^{-k_1t}}{k_2-k_1} - \frac{k_1e^{-k_2t}}{k_1-k_2}\right] \tag{10}$$

G comprises three terms (two exponentials and a double exponential), each with its own upper asymptote $(1 - Y_1)S_0$, $(1 - Y_2)I_0$ and $Y_1(1 - Y_2)S_0$, respectively. The exponential terms have a diminishing returns profile, whereas the double exponential is sigmoidal with an inflexion point at

$$t^* = \frac{\ln(k_2 / k_1)}{k_2 - k_1},$$

providing $k_2 > k_1$. As derived, the model contains six parameters, namely S_0, I_0, Y_1, Y_2, k_1 and k_2. The model has no solution if $k_2 = k_1$. Furthermore, curve fitting does not allow separate estimates for S_0, I_0, Y_1, Y_2 to be obtained because they are not independent (but coupled) parameters. If asymptotes were identified as A_S, A_I and A_{IS}, respectively, the model would be:

$$G = A_S\left(1 - e^{-k_1 t}\right) + A_I\left(1 - e^{-k_2 t}\right) + A_{IS}\left[1 - \frac{k_2 e^{-k_1 t}}{k_2 - k_1} - \frac{k_1 e^{-k_2 t}}{k_1 - k_2}\right]. \qquad (11)$$

A plausible biological interpretation for the two-stage model, applied to the digestion of carbohydrates in the rumen, would be that the first stage represents the extracellular hydrolysis of polysaccharides (initial substrate) yielding monosaccharides (as intermediary compounds). In the second stage, monosaccharides are fermented by microbes (an intracellular process) yielding fermentation end-products (short-chain fatty acids and ATP as final products) and some by-products such as fermentation gas. In this case, the first exponential term of the model represents gas production derived from the hydrolysis of polysaccharides (starch, cellulose), the second exponential represents gas production derived from the fermentation of sugars initially contained in the feed before any polysaccharides are digested (represented in the model by I_0), and the double exponential term represents gas production derived from the fermentation of monosaccharides that are yielded as a result of hydrolysis of polysaccharides. From theoretical stoichiometry, gas is produced from fermentation of sugars, but no gas would be expected to be produced from the conversion of polysaccharides to monosaccharides. Thus, assuming this biological interpretation, the model could be simplified assuming $G_1 = 0$, resulting in:

$$G = A_I\left(1 - e^{-k_2 t}\right) + A_{IS}\left[1 - \frac{k_2 e^{-k_1 t}}{k_2 - k_1} - \frac{k_1 e^{-k_2 t}}{k_1 - k_2}\right].$$

As for curve fitting by non-linear regression, the model could be further re-parameterized to give $G = A - p_1 e^{-k_1 t} - p_2 e^{-k_2 t}$ (providing $k_2 \neq k_1$), where $A = A_S + A_I + A_{IS}$ is the upper asymptote, and p_1 and p_2 are constants.

This equation was fitted to gas production profiles obtained from the incubation of some pure substrates (glucose, maltose, cellobiose, starch). The simple exponential, France (generalized Mitscherlich) and two-phase (Groot *et al.*, 1996) models were fitted for comparison. Results are shown in Table 1.

Curve fitting with multi-phase or multi-stage models was slow and problematical, suggesting over-parameterization. Success rate was lower with the two-stage model (greater Akaike information criterion (AIC) than France and Groot models), confirming the results of Schofield *et al.* (1994) who found that a dual exponential model was unsuitable to fit gas production profiles. In those cases where fitting converged to an acceptable solution, goodness-of-fit was superior with Groot's model (smallest AIC).

Upon analysis of the problems arising when fitting the two-stage model, it was concluded that in some cases an acceptable solution could not be attained because k_2 tended to k_1, or that a solution could be met only if one of these parameters was zero.

Table 1. Goodness-of-fit of different models fitted to gas production curves of four pure substrates (values are for Akaike information criterion).

Substrate incubated	Curve	Exponential	France	Two-phase (Groot)	Two-stage
Glucose	1	468	442	423	448
	2	455	407	508	411
Maltose	1	427	382	320	408
	2	444	397	354	414
Cellobiose	1	480	461	432	478
	2	490	463	441	483
Starch	1	471	451	434	457
	2	462	432	411	435

Aiming to overcome these limitations, the model could be refined or new models derived to improve fitting performance, and some alternatives to reach this goal are:

- Change the mathematical functional form of the equations, using a truncated series expansion for e^x or expressing it in terms of hyperbolic sines and hyperbolic cosines.
- Assume different kinetics for the stages. In the model derived herein, first-order kinetics were assumed for both stages, but different kinetics (first-order but with a time-dependent fractional fermentation rate) could be assumed for one or both stages. For instance, first-order kinetics could easily be assumed for the fermentation of sugars, whereas kinetics resulting in a sigmoidal function could be assumed for the hydrolysis of structural polysaccharides. However, it is possible that this approach might result in differential equations with no analytical solution.
- Include a lag component (either a discrete lag or a lag function), as suggested by Palmer *et al.* (2005). This approach might offer another alternative to make the two-stage model easier to fit gas production profiles. However, a lag is essentially a mathematical artefact with little or no biological basis. In reality, a lag phase would correspond to the first stage(s) in any multi-stage representation.
- Combine multi-stage and multi-phase approaches, assuming that some feed fractions would behave as independent phases, and the fermentation of other fractions would follow a multi-stage process.

In conclusion, a multi-stage model for describing gas production profiles with meaningful biological interpretation has been derived based on compartmental modelling principles. Fitting performance of multi-stage and multi-phase (Groot) models seems to indicate over-parameterization, so the model developed needs to be refined or new models derived from the crucial concept of stage to obtain equations more suitable for fitting gas production data.

References

Beuvink, J.M.W. and Kogut, J., 1993. Modeling gas-production kinetics of grass silages incubated with buffered ruminal fluid. Journal of Animal Science 71:1041-1046.

Cone, J.W., Van Gelder, A.H. and Driehuis, F., 1997. Description of gas production profiles with a three-phasic model. Animal Feed Science and Technology 66:31-45.

France, J., Thornley, J.H.M., López, S., Siddons, R.C., Dhanoa, M.S., Van Soest, P.J. and Gill, M., 1990. On the two compartment model for estimating the rate and extent of feed degradation in the rumen. Journal of Theoretical Biology 146:269-287.

France, J., Dhanoa, M.S., Theodorou, M.K., Lister, S.J., Davies, D.R. and Isac, D., 1993. A model to interpret gas accumulation profiles associated with *in vitro* degradation of ruminant feeds. Journal of Theoretical Biology 163:99-111.

France, J., Dijkstra, J., Dhanoa, M.S., López, S. and Bannink, A., 2000. Estimating the extent of degradation of ruminant feeds from a description of their gas production profiles observed *in vitro*: Derivation of models and other mathematical considerations. British Journal of Nutrition 83:143-150.

France, J., López, S., Kebreab, E., Bannink, A., Dhanoa, M.S. and Dijkstra, J., 2005. A general compartmental model for interpreting gas production profiles. Animal Feed Science and Technology 123-124:473-485.

Groot, J.C.J., Cone, J.W., Williams, B.A., Debersaques, F.M.A. and Lantinga, E.A., 1996. Multiphasic analysis of gas production kinetics for *in vitro* fermentation of ruminant feeds. Animal Feed Science and Technology 64:77-89.

Hill, A.V., 1910. The possible effects of the aggregation of the molecules of haemoglobin on its dissociation curves. The Journal of Physiology 40:iv-vii.

Huhtanen, P., Seppaelae, A., Ots, M., Ahvenjaervi, S. and Rinne, M., 2008. *In vitro* gas production profiles to estimate extent and effective first-order rate of neutral detergent fiber digestion in the rumen. Journal of Animal Science 86:651-659.

Krishnamoorthy, U., Rymer, C. and Robinson, P.H. (eds.), 2005.The *in vitro* Gas Production Technique: Limitations and Opportunities. Animal Feed Science and Technology Volumes 123-124, Part 1. Elsevier B.V., Amsterdam. 578 pp.

López, S., 2005. *In vitro* and *in situ* techniques for estimating digestibility. In: Dijkstra, J., Forbes, J.M. and France, J. (eds.) Quantitative Aspects of Ruminant Digestion and Metabolism, 2nd edn. CAB International, Wallingford, UK, pp.87-121.

López, S., 2008. Non-linear functions in animal nutrition. In: France, J. and Kebreab, E. (eds.) Mathematical Modelling in Animal Nutrition. CAB International, Wallingford, UK, pp.47-88.

López, S., France, J., Dhanoa, M.S., Mould, F. and Dijkstra, J., 1999. Comparison of mathematical models to describe disappearance curves obtained using the polyester bag technique for incubating feeds in the rumen. Journal of Animal Science 77:1875-1888.

López, S., Dhanoa, M.S., Dijkstra, J., Bannink, A., Kebreab, E. and France, J., 2007. Some methodological and analytical considerations regarding application of the gas production technique. Animal Feed Science and Technology 135:139-156.

Ørskov, E.R. and McDonald, I., 1979. The estimation of protein degradability in the rumen from incubation measurements weighted according to rate of passage. Journal of Agricultural Science, Cambridge 92:499-503.

Palmer, M.J.A., Jessop, N.S., Fawcett, R. and Illius, A.W., 2005. Interference of indirect gas produced by grass silage fermentation acids in an *in vitro* gas production technique. Animal Feed Science and Technology 123:185-196.

Robinson, P.H., Fadel, J.G. and Tamminga, S., 1986. Evaluation of mathematical-models to describe neutral detergent residue in terms of its susceptibility to degradation in the rumen. Animal Feed Science and Technology 15:249-271.

Schofield, P., Pitt, R.E. and Pell, A.N., 1994. Kinetics of fiber digestion from *in vitro* gas production. Journal of Animal Science 72:2980-2991.

Stefanon, B., Pell, A.N. and Schofield, 1996. Effect of maturity on digestion kinetics of water-soluble and water-insoluble fractions of alfalfa and brome hay. Journal of Animal Science 74:1104-1115.

Thornley, J.H.M., and France, J., 2007. Mathematical Models in Agriculture. 2nd edn. CABI Publishing, Wallingford, UK, 923 pp.

Van Milgen, J., Murphy, M.R. and Berger, L.L., 1991. A compartmental model to analyze ruminal digestion. Journal of Dairy Science 74:2515-2529.

A mechanistic model of pH and gas exchanges in the rumen and its *in vitro* application

A. Serment[1,2] and D. Sauvant[1,2]
[1]INRA, UMR791 Modélisation Systémique Appliquée aux Ruminants, 16 rue Claude Bernard, 75005 Paris, France; amelie.serment@agroparistech.fr
[2]AgroParisTech, UMR791 Modélisation Systémique Appliquée aux Ruminants, 16 rue Claude Bernard, 75005 Paris, France

Abstract

Although rumen functioning is governed by both kinetic and thermodynamic laws, few mechanistic models have developed both aspects. However, variables such as pH and gas production are largely controlled by thermodynamics. This study proposes a mechanistic model, which focuses specifically on the mechanistic description of relations between pH, bicarbonate and gas production based on thermodynamic laws. The first version was developed in the context of an *in vitro* device. The model comprised 14 compartments. It simply described the processes of microbial degradation of proteins to peptides and ammonia (NH_3) and of fibre and non-fibre carbohydrates to soluble carbohydrates and then to volatile fatty acids (VFA), carbon dioxide (CO_2) and methane (CH_4). A specific sub-part of this model, composed of seven compartments, focuses on describing the dynamics of pH and gas production. The pH was determined from the chemical equilibrium between VFA represented by a weak acid and its associated base, and the bicarbonate and CO_2 couple. Gas produced in the system originated from CO_2 and CH_4 directly produced by the fermentation of soluble carbohydrates and from CO_2 produced by the acid-base reaction. As CO_2 and CH_4 were produced in the aqueous phase, there was *a priori* a delay between their production rate and their appearance in the gaseous phase. This process was described by Henry's law. The model was calibrated using experimental data of *in vitro* fermentation of pea seed fragments. The predicted pattern of concentrations of NH_3 and total gas production were realistic with regression coefficients of 1.06±0.020 and 0.93±0.016 between observed and predicted values, respectively (R^2=0.97, RMSE=11.12 mg/l, n=7 and R^2=0.99, RMSE=1.2 ml, n=7). The model provides the physicochemical basis which can be used in other fermentation models.

Keywords: modelling, gas test system, fermentation, buffer

Introduction

Reactions in the rumen are mainly controlled by kinetic and thermodynamic laws. Most of the existing mechanistic models of the rumen are based on a kinetic control (Kohn and Boston, 2000), while the role of thermodynamics has rarely been taken into account (Imamidoost and Cant, 2006; Kohn and Boston, 2000; Offner and Sauvant, 2006). However, variables such as the pH of ruminal fluid and gas production clearly depend on thermodynamics. In fact, the pH is the result of the equilibrium between the production of volatile fatty acids (VFA) and buffers through provided by the saliva. Gas production is partly a consequence of this equilibrium, depending also on the equilibrium between aqueous and gas phases in the rumen.

Most of the published mechanistic models use an empirical approach to calculate rumen pH, which is a rather unsatisfactory compromise. Attempts to develop a mechanistic representation of pH based on thermodynamic laws have not been entirely successful to date (Offner and Sauvant, 2006) or the link with gas production has not been developed (Imamidoost and Cant, 2005).

This study proposes a simple mechanistic model, using a minimum of kinetic and thermodynamic laws. It focuses specifically on the mechanistic description of the relations between pH, different forms of bicarbonate, and gas. The model represented a system of an *in vitro* device where a feedstuff was incubated with an inoculum of rumen fluid and a buffer solution. Because of its high level of standardization and its use by several research teams, this study was based on the syringe gas test method (Menke *et al.*, 1979).

Model description

System

The syringe was considered as a closed batch system within only an initial substrate supply and without outflow. The model describes the processes of degradation and fermentation of the feedstuff by the microbes of the inoculum. It comprises 14 compartments representing the main components of the feedstuff, a single population of microbes, fermentation products and bicarbonate in the buffer solution. The diagram of the model is summarized in Figure 1. Each compartment is defined by a dynamic differential equation. The flow from compartment 1 (C1) to compartment 2 (C2) is denoted by $F_{C1 \rightarrow C2}$. The diagram and the basic parameters of the model were defined from general knowledge about rumen digestion and from the literature.

Carbohydrate and nitrogen parts

The feedstuff part of the model was kept as simple as possible and the carbohydrate compartments (*CHO*, in mg) represented cell wall materiel (*NDF*) and non-fibre carbohydrates (*NFC*). They were degraded into soluble carbohydrates (*SC*), which were then fermented to produce VFA (*PVFA*, mmol/min), carbon dioxide (PCO_2, mmol/min), methane (PCH_4, mmol/min) and ATP (*PATP*, μmolATP/

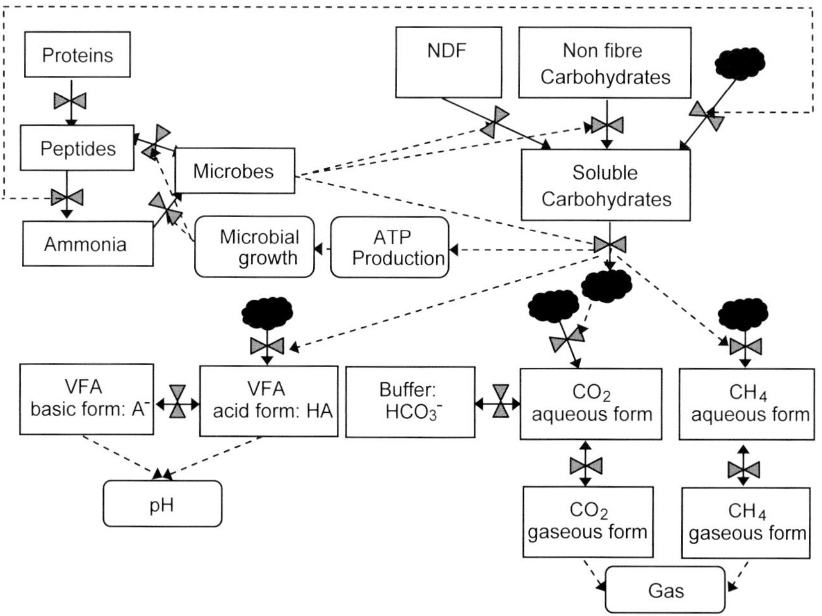

Figure 1. Simplified Forrester diagram of the physico-chemical model of in vitro *gas test model.*

min). The *CHO* degradation process was described by mass action laws depending on the mass of microbes (*MICR*) as suggested by Van Milgen *et al.* (1991):

$$dCHO/dt = -k_{CHO} \times CHO \times k_{MICR} \times MICR \quad (1)$$

where *CHO* is either *NDF*, *NFC* or *SC*, k_{CHO} the fractional rates of digestion or fermentation of *CHO* and k_{MICR} the coefficient of the influence of microbes.

Fermentation production rates were defined by Equation 2:

$$PATP = k_{ATP} \times F_{SC}$$

$$PVFA = k_{VFA} \times F_{SC}$$

$$PCO_2 = (2/3) \times k_{VFA} \times F_{SC} \quad (2)$$

$$PCH_4 = (1/3) \times k_{VFA} \times F_{SC}$$

where F_{sc} is the degradation flow of SC, k_{ATP} the rate of ATP production equal to 25 μmol ATP/mg of degraded *SC*, and k_{VFA} the rate of VFA production equal to 8.5 μmol total VFA/mg of degraded *SC*.

Four compartments represented the nitrogen fractions (in mg N): proteins (*PRO*), peptides (*PEP*), ammonia (NH_3) and microbes (*MICR*). Flows of degradation of *PRO* to *PEP* and *PEP* to NH_3 followed mass action law. Degradation of *PEP* to NH_3 also yielded carbon chains entering the *SC* compartment.

The equilibrium between microbial growth and catabolism controlled the N flows between *MICR* and *PEP* and NH_3. The *PATP* provided energy for microbial maintenance (*m*; μmolATP/min) and the remaining energy was used for microbial growth *GMICR* (mg of microbial N/min) (Lescoat and Sauvant, 1995):

$$m = k_m \times MICR \quad (3)$$

$$GMICR = YATP \times (PATP - m) \quad (4)$$

where k_m is the coefficient of energy requirement for maintenance (μmol ATP/mg of microbial N/min) and *YATP* the microbial growth efficiency (mg of microbial N/μmolATP).

The *GMICR* controlled $F_{PEP \to MICR}$ and $F_{NH3 \to MICR}$ according Equation 5:

$$F_{PEP \to MICR} = (1 - P_{NH}) \times GMICR$$

$$F_{NH3 \to MICR} = P_{NH} \times GMICR \quad (5)$$

$$P_{NH} = \frac{NH_3}{PEP + NH_3}$$

The lysis of microbes into *PEP* was accounted for by using a fractional rate of microbial lysis k_{lys}, which varied inversely with the fractional microbial growth rate (France *et al.*, 1982).

Determination of pH

In this model, two acid-base couples were involved in describing the change in pH: the bicarbonate of the buffer solution and the aqueous form of carbon dioxide (CO_{2aq}/HCO_3^-) and the acid form of total VFA and its associated base (HA/A^-). Only the bicarbonate system was taken into account because it was often considered as the prevalent ruminal buffer (Counotte *et al.*, 1979). To not omit a non-negligible part of the buffer capacity, we hypothesized that phosphate system could be assimilated to bicarbonate system in considering an identical pKa. The weak acid (HA and CO_{2aq}) and its basis (A^- and HCO_3^-) of each couple are in equilibrium according the following two reactions:

$$HCO_3^- + H_3O^+ \Leftrightarrow CO_{2aq} + 2\,H_2O \qquad \text{Reaction 1}$$

$$HA + H_2O \Leftrightarrow A^- + H_3O^+ \qquad \text{Reaction 2}$$

The equilibrium constants of these reactions are equal to the inverse of the dissociation constant $Ka_{bicarbonate}$ for the couple CO_{2aq}/HCO_3^- and to Ka_{VFA} for the couple HA/A^-:

$$\frac{1}{Ka_{bicarbonate}} = \frac{[CO_{2aq}]}{[HCO_3^-]*[H_3O^+]} \qquad (6)$$

$$Ka_{VFA} = \frac{[A^-]*[H_3O^+]}{[HA]} \qquad (7)$$

where $pKa = -log\ Ka$.

The $pKa_{bicarbonate}$ was equal to 6.25 (Counotte *et al.*, 1979) and pKa_{VFA} was estimated at 4.8, the mean pKa values for acetate, propionate and butyrate equal respectively to 4.76, 4.87 and 4.82 (Kohn and Dunlap, 1998).

The ion H_3O^+ is the intermediate element between Reaction 1 and Reaction 2. When this ion was considered as a compartment, the model became instable because of the very small size of this compartment and the fractional production and usage rates. In fact, within the pH range in the rumen (between 5.5 and 7.5), the concentration of H_3O^+ is comprised between $10^{-7.5}$ and $10^{-5.5}$mol/l. For simplification, this compartment was not maintained and its effect was represented by combining Reaction 1 and Reaction 2:

$$HCO_3^- + HA \Leftrightarrow CO_{2aq} + A^- + H_2O \qquad \text{Reaction 3}$$

Basically, the dissociation of x mol of HA was buffered by x mol of HCO_3^- to give x mol of CO_{2aq}, x mol of A and x mol of H_2O (Kohn and Dunlap, 1998). The equilibrium state of this reaction was calculated by the ratio of $Ka_{VFA}/Ka_{bicarbonate}$ and was equal to $10^{1.45}$. The hypothesis in the model was that at each incubation time step, bicarbonate and VFA systems were at equilibrium state, thus:

$$\frac{Ka_{VFA}}{Ka_{bicarbonate}} = \frac{[CO_{2aq}]_{t+dt}*[A^-]_{t+dt}}{[HCO_3^-]_{t+dt}*[HA]_{t+dt}}$$

$$= \frac{([CO_{2aq}]_t + \frac{x}{Vol} + \frac{PCO_2}{Vol} - \frac{F_{CO_{2aq} \to CO_{2gas}}}{Vol} + \frac{F_{CO_{2gas} \to CO_{2aq}}}{Vol})*([A^-]_t + \frac{x}{Vol})}{([HCO_3^-]_t - \frac{x}{Vol})*([HA]_t - \frac{x}{Vol})} \qquad (8)$$

where [species] is the concentration of the chemical species, *Vol* the volume of fermentation juice in a syringe (30 ml), and $F_{CO2aq \rightarrow CO2gas}$ and $F_{CO2gas \rightarrow CO2aq}$ the flows of CO_2 between the aqueous and gaseous phases, which are described later.

Equation 8 is a quadratic equation with x as unknown. This equation was evaluated and solved at each incubation time. Differential equations of the chemical species, which were involved in the acid-base equilibrium, are described in Table 1. Using the Henderson-Hasselbach equation made it possible to calculate *pH*:

$$pH = pKa_{VFA} + log ([A^-] / [HA]) \quad (9)$$

Gas

Gas produced in the syringe originated from CO_2 and CH_4 directly produced by the fermentation of *SC* (PCO_2 and PCH_4) and from CO_2 produced by the acid-base reaction between VFA and bicarbonate systems (Reaction 3; Blümmel *et al.*, 1997). As CO_2 and CH_4 are produced in the aqueous phase, there is *a priori* a delay between their production rate and the actual appearance in the gaseous phase. Therefore, it was necessary to describe gas diffusion between these phases. Four compartments (in mmol) were used to represent gases: CO_{2aq} and CH_{4aq} in the aqueous phase, and CO_{2gas} and CH_{4gas} in the gaseous phase. Cumulative gas (in ml) was calculated from the sum of CO_{2gas} and CH_{4gas}. The solubility of a gas in a liquid is proportional to the partial pressure of this gas in the gaseous phase. This phenomenon is called 'Henry's law':

$$[gas_{aq}] = k_H \times P(gas) \quad (10)$$

Table 1. Compartments of the model describing the change in pH and gas exchanges.

Symbol[1]	Unit	Differential equation[2]
HA Acid form of VFA	mmol	$dHA/dt = PVFA - x$
A^- Basic form of VFA	mmol	$dA^-/dt = +x$
HCO_3^- Bicarbonate	mmol	$dHCO_3^-/dt = -x$
CO_{2aq} Aqueous carbon dioxide	mmol	$dCO_{2aq}/dt = +x + PCO_2 + F_{CO2gas \rightarrow CO2aq} - F_{CO2aq \rightarrow CO2gas}$
CO_{2gas} Gaseous carbon dioxide	mmol	$dCO_{2gas}/dt = -F_{CO2gas \rightarrow CO2aq} + F_{CO2aq \rightarrow CO2gas}$
CH_{4aq} Aqueous methane	mmol	$dCH_{4aq}/dt = PCH_4 + F_{CH4gas \rightarrow CH4aq} - F_{CH4aq \rightarrow CH4gas}$
CH_{4gas} Gaseous methane	mmol	$dCH_{4gas}/dt = -F_{CH4gas \rightarrow CH4aq} + F_{CH4aq \rightarrow CH4gas}$

[1]VFA: volatile fatty acids.
[2]$F_{C1 \rightarrow C2}$ is the flow from compartment1 (C1) to compartment2 (C2). x is the flow exchanged at each time step in the acid-base reaction between VFA and bicarbonate systems.

where $[gas_{aq}]$ is the concentration of a chemical species in the aqueous phase (in mmol/l) and $P(gas)$ the partial pressure of the same species in the gaseous phase (in Pa). The constant, k_H (in mmol/l. Pa^{-1}) is called the Henry constant. It depends on the solvent, the species and the temperature.

According to Henry's law, molecules of CO_{2aq} or CH_{4aq} are in a permanent state of agitation. Some of them, which are near the exchange surface, can have an energy status sufficient to leave the aqueous phase and enter the gaseous phase. This event takes place at a constant rate. The 'escape' flow is proportional to the concentration of CO_{2aq} and CH_{4aq} (Equation 11). In the same way, molecules of gas in the gaseous phase fall on the exchange surface. Part of this CO_{2gas} and CH_{4gas} bounces while a constant quantity enters the liquid phase. This 'entrance' flow is proportional to the partial pressure of the gas in the gaseous phase (Equation 12):

$$F_{aq \rightarrow gas} = k_{aq \rightarrow gas} \times [gas_{aq}] \tag{11}$$

$$F_{gas \rightarrow aq} = k_{gas \rightarrow aq} \times P(gas) \tag{12}$$

where $F_{aq \rightarrow gas}$ and $F_{gas \rightarrow aq}$ are expressed in mmol/min, $k_{aq \rightarrow gas}$ and $k_{gas \rightarrow aq}$ in l/min are the coefficients of transfer between phases. At equilibrium, $F_{aq \rightarrow gas}$ and $F_{gas \rightarrow aq}$ are equal and the Henry constant is equal to the ratio of $k_{gas \rightarrow aq}$ and $k_{aq \rightarrow gas}$. Equation 12 then becomes:

$$F_{gas \rightarrow aq} = k_{aq \rightarrow gas} \times k_H \times P(gas) \tag{13}$$

Differential equations of the chemical species involved in the acid-base equilibrium and the aqueous-gaseous phase equilibrium are described in Table 1.

Numerical techniques

The model was developed and adjusted with ModelMaker 3 software. The Euler integration method was used with an integration time step of 1 minute. The simulation time was 48 hours.

Parameter optimisation was carried out with the Simplex method, repeated until the residual sum of squares reached a minimum value.

This model was calibrated with experimental data of fermentation in a syringe (fractional degradation rates and initial concentrations). In the first phase of model development, the tested feedstuffs were a whole pea seed and nine granulometric fractions of this pea seed. Those fractions were obtained by grinding pea seeds in a hammer mill and by fractioning the flour with a continuous dry sieving classifier. The dimensions of the mesh sieves were considered as identifiers of the fractions. The fractions F2.5 and F0.76 were the fractions retained on 2.5 and 0.76 mesh sieves. The pan fraction was the fraction retained on the pan of the sieve (Maaroufi *et al.*, 2009).

Results and discussion

Ammonia concentration

The model reflected fairly well the kinetics of ammonia concentration in the syringe (Figure 2). Output values of the model were compared with the mean of the experimental data that had been used for calibrating model (7 incubation times). The regression coefficient between observed and predicted values was 1.06±0.020, (R^2=0.97, RMSE=11.12 mg/l, n=7). Ammonia concentration increased with time. Moreover, between 6 and 12 hours of incubation, a transitory decrease was

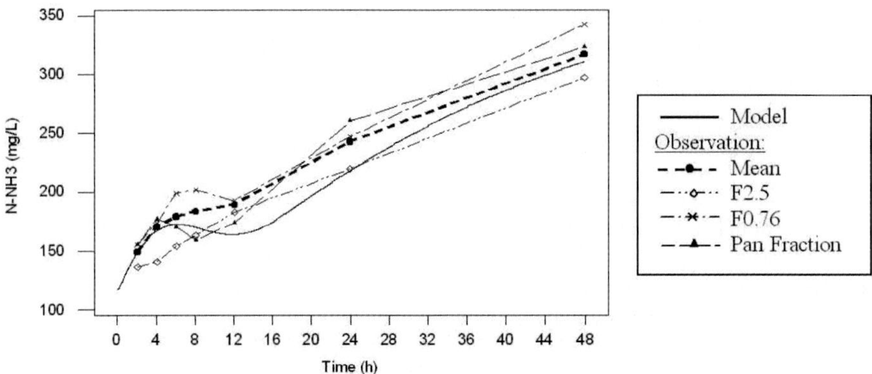

Figure 2. Dynamics of concentration of ammonia in a syringe: values predicted by the model and experimental observations (Maaroufi et al.*, 2009). The experimental observations corresponded to the fermentations of a whole pea seed and its granulometric fractions. 'Mean' was the mean the observations. F2.5 and F0.76 were the fractions retained on 2.5 and 0.76 mesh sieves. The pan fraction was the fraction retained on the pan of the sieve.*

observed, which occurred when the relative degradation of substrate energy was higher than that of proteins and amino-acids. As a consequence, NH_3 was recaptured by the microbes. It was experimentally shown that this transient drop was more marked for the finest fractions of pea seed (F0.76 and pan fraction vs. F2.5) (Maaroufi *et al.*, 2009). The model simulated this drop well even though its amplitude may require further investigation. After about 20 hours of incubation, NH_3 accumulation was due to the microbial lysis rate, which was greater than their growth rate. Cone *et al.* (1997) observed the same type of dynamics for ammonia.

Gas production and pH determination

The prediction of gas production was satisfactory compared with the experimental data, with a regression coefficient of 0.93±0.016 between the values of the means of the observations and the predicted values (R^2=0.99, RMSE=1.2 ml, n=7). Simulated gas followed the behaviour of the *in vitro* gas profile (Figure 3) with a lag phase at the beginning of the incubation, an increase of production of gas and lastly saturation. During the lag time, a certain degree of instability was observed between compartments CO_{2gas} and CO_{2aq}, and CH_{4gas} and CH_{4aq}, which was due to rapid exchange of matter between these elements. The structure of the model simulated the delay between the production of gases and their appearance in the gaseous phase fairly well: 46 min for CO_2 and 15 min for CH_4. The delay was longer for CO_2 because of the existence of the buffer system and a greater Henry constant, which increased $F_{CO2gas \rightarrow CO2aq}$. A sensitivity analysis indicated that an increased initial quantity of buffer increased the production of gas.

The predicted values of pH were fairly realistic compared with the observed values (Figure 4). The pH values ranged from 6.2 to 7.2. At the beginning of the incubation, the production of VFA led to a rapid drop of pH followed by a slow decrease due to the reduction of available soluble carbohydrate. It was a progress compared to the model of Offner and Sauvant (2006) where the simulated pH did not follow a right trajectory. However, the change in pH was not completely satisfactory with a coefficient regression between observed and predicted values of only 0.46±0.049 (R^2=0.94; RMSE=0.023, n=7). The predicted pH decrease was more important than observed.

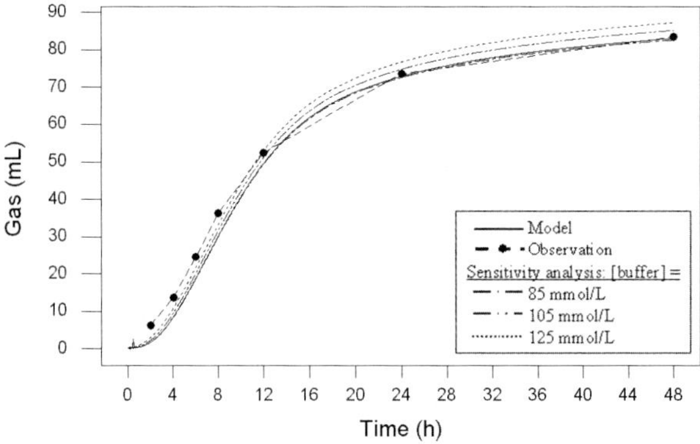

Figure 3. Dynamics of production of gas in a syringe: sensitivity analysis of the initial concentration of buffer. Comparison of values predicted by the model and experimental observations (Maaroufi et al., *2009). 'Observation' corresponded to the mean values of fermentations of a whole pea seed and of its granulometric fractions.*

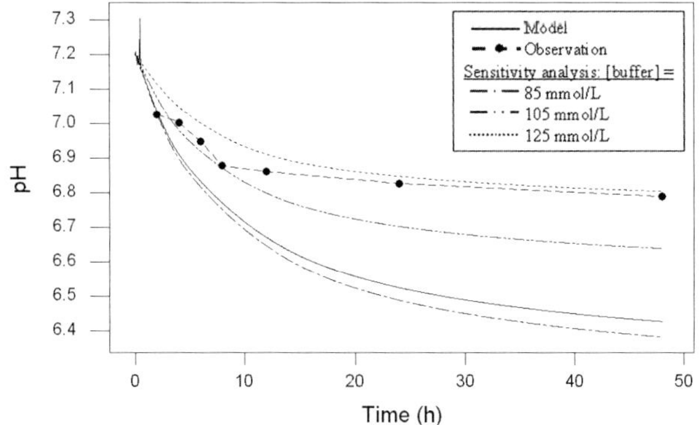

Figure 4. Dynamics of pH in a syringe: sensitivity analysis of the initial concentration of buffer. Comparison of values predicted by the model and experimental observations (Maaroufi et al., *2009). 'Observation' corresponded to the mean values of fermentations of a whole pea seed and of its granulometric fractions.*

At the beginning of the incubation, equilibrium between VFA and the bicarbonate system was disturbed for one hour by the instability previously described between the compartments CO_{2gas} and CO_{2aq}. Further work should be carried out to better understand and model this phenomenon.

A sensitivity analysis of the initial quantity of buffer (Figure 4) showed better prediction results when the quantity of buffer increased (125 mmol/l instead of 87.8 mmol/l). Thus, buffer capacity of the model may have been under-estimated. Rumen juice of the inoculum certainly contained HCO_3^- and other buffer substances, which were not taken into account and not measured experimentally by Maaroufi *et al.* (2009).

In this model, as a first approach, we chose to add the phosphate ions to the bicarbonate system in considering an identical pKa. The insertion of the phosphate system into the model to calculate pH could be envisaged to increase the buffer capacity of the system because of its superior value of pKa (7.2).Chosen pKa values were those used most in the literature for rumen pH. It however consists of values at 273 K. Kohn and Dunlap (1998) proposed a value of 6.1 for $pKa_{bicarbonate}$ at 310 K. By using Equation 14, new values of pKa can be calculated for the temperature in the syringe (312 K): 4.6 and 6.1 for respectively pKa_{VFA} and $pKa_{bicarbonate}$.

$$\Delta G = -RT \ln Ka. \quad (14)$$

where ΔG is the Gibbs free energy of reaction in $m.kg.s^{-2}.mol^{-1}$, T the temperature of the reaction in K and R the gas constant ($8.314\ m^2.kg.s^{-2}.K^{-1}.mol^{-1}$). The impact on the model would be limited because new values of pKa would not change the equilibrium constant of Reaction 3. Moreover, in an empirical evaluation of pH determination, Kohn and Dunlap (1998) waited 15 min to obtain stable equilibrium pH. In our model, we considered that equilibrium is reached at each minute. Further *in vitro* experiments are necessary to determine the time to reach equilibrium.

Conclusion and further developments of the model

At this stage of development, this model shows promising results. The pH is determined in a mechanistic way and with a total description contrasting with the corresponding part of the model of Immamidost and Cant (2005) and it is linked to the production of gas.

More external evaluations are expected on data of kinetics of VFA, NH_3 and N microbial concentrations, pH and gas from the *in vitro* literature. After that, an *in vivo* version can be foreseen. This approach of a mechanistic physicochemical model could be easily included as a submodel in a larger, mechanistic model of the rumen.

References

Blümmel, M., Makkar, H.P.S. and Becker, K., 1997. *In vitro* gas production: a technique revisited. Journal of Animal Physiology and Animal Nutrition 77:24-34.

Cone, J.W., Van Gelder, A.H. and Driehuis, F., 1997. Description of gas production profiles with a three-phasic model. Animal Feed Science and Technology 66:31-45.

Counotte, G.H.M., Van't Klooster, A.T., Van der Kuilen, J. and Prins, R.A., 1979. An analysis of the buffer system in the rumen of dairy cattle. Journal of Animal Science 49:1536-1544.

France, J., Thornley, J.H.M. and Beever, D.E., 1982. A mathematical model of the rumen. Journal of Agricultural Science 99:343-353.

Imamidoost, R. and Cant, J.P., 2005. Non-steady-state modeling of effects of timing and level of concentrate supplementation on ruminal pH and forage intake in high-producing, grazing ewes. Journal of Animal Science 83:1102-1115.

Kohn, R.A. and Boston, R.C., 2000. The role of thermodynamics in controlling rumen metabolism. In: McNamara, J.P., France, J. and Beever, D.E. (eds.) Modelling nutrient utilization in farm animals, Cabi, Wallingford, UK, pp.11-24.

Kohn, R.A. and Dunlap, T.F., 1998. Calculation of the buffering capacity of bicarbonate in the rumen and *in vitro*. Journal of Animal Science 76:1702-1709.

Lescoat, P. and Sauvant, D., 1995. Development of a mechanistic model for rumen digestion validated using the duodenal flux of amino acids. Reproduction, Nutrition, Development 35:45-70.

Maaroufi, C., Chapoutot, P., Sauvant, D. and Giger-Reverdin, S., 2009. Fractionation of pea flour with pilot scale sieving. II. *In vitro* fermentation of pea seed fractions of different particle sizes. Animal Feed Science and Technology 154:135-150.

Menke, K.H., Raab, L., Salewski, A., Steingass, H., Fritz, D. and Schneider, W., 1979. The estimation of the digestibility and metabolizable energy content of ruminant feedingstuffs from the gas production when they are incubated with rumen liquor *in vitro*. Journal of Agricultural Science 93:217-222.

Offner, A. and Sauvant, D., 2006. Thermodynamic modeling of ruminal fermentations. Animal Research 55:343-365.

Van Milgen, J., Murphy, M.R. and Berger, L.L., 1991. A compartmental model to analyze ruminal digestion. Journal of Dairy Science 74:2515-2529.

Modelling rumen volatile fatty acids and its evaluation on net portal fluxes in ruminants

P. Nozière[1], F. Glasser[1], C. Loncke[1], I. Ortigues Marty[1], J. Vernet[1] and D. Sauvant[2]
[1]INRA, UR1213 Herbivores, Site de Theix, 63122 Saint Genès Champanelle, France; noziere@clermont.inra.fr
[2]INRA-AgroParisTech, UMR 0791, Modélisation Systémique de la Nutrition des Ruminants, 75231 Paris, France

Abstract

The production of volatile fatty acids (VFA) is still poorly predicted by rumen models despite its major contribution to ruminant energy supply. We aimed at building an empirical model describing production rates (PR) of individual VFA in the rumen. Three distinct bibliographic data bases were used. They gathered published *in vivo* measurements on ruminal VFA-PR, digestive fluxes and molar proportion of VFA in the rumen, and net portal fluxes of VFA. These data bases cover a wide range of intake levels and dietary composition. A meta-analysis was performed, using within-experiment models. Models were derived, describing total VFA-PR and individual VFA molar proportions in the rumen from variation in dry matter intake level and dietary composition. The VFA-PR was explained by intake of rumen fermentable organic matter (RfOM), calculated from detailed knowledge of the chemical composition of diets according to INRA Feed Tables, with the effect of other factors remaining non significant. The VFA molar proportions were explained by dry matter intake, digestible organic matter, digestible neutral detergent fibre and rumen starch digestibility. Based on these results, we proposed an empirical model to predict individual VFA-PR from simple predictors which represent the nature of digested substrates, as well as their site and extent of digestion. The model can be used under a wide range of levels of feed intake and dietary composition. Evaluation of predicted VFA-PR indicated close relationships with observed net portal appearance of VFA. The relationships were consistent with quantitative knowledge on VFA metabolism by the digestive tissues.

Keywords: volatile fatty acids, rumen, portal vein, meta-analysis

Introduction

The volatile fatty acids (VFA) produced in the rumen are the main source of energy absorbed by ruminants. The amount and profile of VFA (proportion of acetate [C2], propionate [C3] and butyrate [C4]) have consequences on animal responses to dietary changes (i.e. efficiency of energy utilization, production of methane, and composition of animal products). Thus, a prediction of VFA production rate (VFA-PR) is necessary to progress towards new feed evaluation systems. The VFA profile depends on the type of fermented substrate, the microbial population, and intraruminal environment affecting microbial metabolism. The amount and profile of VFA is governed by enzymatic and thermodynamic laws that several mechanistic models attempt to represent (Murphy *et al.*, 1982; Bannink *et al.*, 2006; Kohn and Boston, 2000; Offner and Sauvant, 2006). So far, the prediction of VFA-PR by mechanistic models is still unsatisfactory (Bannink *et al.*, 1997; Offner and Sauvant, 2004). One of the major problems is that models were not derived from observed VFA-PR but from observed VFA concentrations, which are also far from being described accurately. An empirical model, based on measured data of total VFA-PR and VFA molar proportions, may be a helpful tool to know the « most probable values » of individual VFA-PR. The approach presented in the present paper is a meta-analysis of published results which allows to distinguish between- and within-experiment variation in VFA. The aim of this study was (1) to develop an empirical model predicting PR of individual VFA; (2) to evaluate the model with independent observations of net portal appearance (NPA) of VFA.

Material and methods

Development of the VFA-PR model

The model was developed in several steps: (1) representation of total VFA (tVFA) PR in the rumen (tVFA-PR, mmol/d/kg BW); (2) representation of the individual VFA (iVFA) molar proportions in the rumen fluid ([iVFA] / [tVFA], mol/mol); (3) estimation of a correction factor (Cor-iVFA) for the difference between concentration and production in VFA molar proportions. Lastly, the individual VFA production rate in the rumen (iVFA-PR, mmol/d/kg BW) was calculated as follows:

$$\text{Predicted iVFA-PR} = \text{Predicted tVFA-PR} \times \text{Predicted [iVFA] / [tVFA]} \times \text{Cor-iVFA}.$$

For prediction of tVFA-PR, and quantification of Cor-iVFA, the *VFA-Prod* database (Nozière *et al.*, 2007) was used with published *in vivo* measurements on rumen VFA-PR using dilution techniques in ovine, growing buffalo or cattle, and cows. A consistent and homogeneous description of chemical composition and nutritional values of feeds and diets included in the *VFA-Prod* database was made using INRA Feed Tables (2007), assuming additivity of tabulated values, as described by Loncke *et al.* (2009).

For prediction of [iVFA] / [tVFA], the *BoviDig* database (Sauvant *et al.*, 2000) was used with reported simultaneous measurements of digestive fluxes (total tract apparent digestibility of organic matter (OM) and neutral detergent fibre (NDF), and/or duodenal fluxes of OM, NDF, starch) and molar proportion of VFA in the rumen of bovines, including lactating, growing, and non-producing animals.

Both *VFA-Prod* and *BoviDig* databases included a large range of dry matter (DM) intake (4 to 45 g DM/d/kg BW) and diet composition (12 to 79% NDF). The VFA molar proportions in the rumen fluid presented a wide variability, 44 to 79% for acetate (C2), 13 to 44% for propionate (C3), 5 to 20% for butyrate (C4), 0.3 to 10.5% for the minor VFA. For each step during model development, a first covariable was identified as the main predictor, from a physiological perspective and based on previous results on prediction of NPA of VFA from INRA feed tables, as described by Loncke *et al.* (2009). The intake of rumen fermentable OM (RfOM intake, g/d/kg BW) was assessed as the main predictor of tVFA-PR, with RfOM = digestible OM in total tract – fat – non degradable CP in sacco – non degradable starch in sacco – fermentation products of silages (INRA, 2007). The dNDF/dOM ratio, calculated from measured digestible NDF (dNDF) and OM (dOM) in total tract, was assessed as predictor of molar proportions of individual VFA in the rumen fluid. Within-experiment models were adjusted using the GLM procedure, with experiment as a fixed effect (Sauvant *et al.*, 2008). Significant additional predictors, including intake level, dietary composition, site and extent of digestion, were detected using analysis of interfering factors on residuals, LSMeans and slopes, and subsequently tested as additional covariables. For variables retained in the final models the criterion was that they should be available from INRA Feed Tables.

Evaluation of the VFA-PR model

For the evaluation of the model against independent data, the *Flora* database (Vernet and Ortigues-Marty, 2006) was used with published *in vivo* measurements on NPA of VFA on multicatheterized ruminants. The database contained a consistent and homogeneous description of feeds and diets according to INRA Feed Tables (Loncke *et al.*, 2009). Model prediction of iVFA-PR was compared for each experimental treatment with measured iVFA-NPA. The treatments covered a wide range of levels of feed intake (6 to 42 g DM/d/kg BW) and diet composition (14 to 71% NDF). For diets containing silages, predicted tVFA-PR was incremented, assuming 11 mol of absorbable VFA/kg fermentation product of silages (INRA, 2007). The relationships between predicted iVFA-PR and measured iVFA-NPA was analysed using the GLM procedure with the experiment as a fixed effect, and using analysis of interfering factors on residuals, LSMeans and slopes.

Data were analyzed using Minitab statistical software. For each model, the intercept and slopes (±SE) are reported, as well as number of experiments (Nexp) and experimental treatments (Nt) used for model adjustment, R^2 and residual standard deviation (Syx).

Results

Prediction of total VFA production rate in the rumen

The tVFA-PR was linearly and closely related to RfOM intake (Figure 1). Changes in RfOM intake (mean SD = 1.8 g/d/kg BW) were related to changes in DM intake (25% of experiments), in dietary content of RfOM (27% of experiments) or in both DM intake and dietary content of RfOM (48% of experiments). The average slope and intercept did not significantly depend on the origin of changes in RfOM intake. Also, no factors were detected that significantly interfered with slope. Factors related to dietary N level (crude protein [CP], fermentable CP) were positively related to LSMeans and residuals; however, in experiments where dietary N level was the experimental factor (Nexp = 28, Nt = 74) no significant effect of CP or fermentable CP on the total VFA-PR was shown, and their introduction as an additional covariable did not improve the explanation of observed variation by the model.

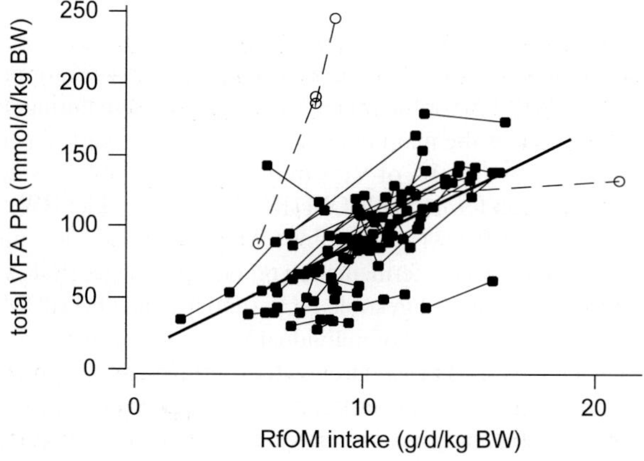

Figure 1. Within-experiment relationship between the intake of rumen fermentable organic matter (RfOM) and the total volatile fatty acid production rate (tVFA-PR). Statistical outliers are presented on dotted lines and excluded from the model. The bold line represents the adjusted model; Nexp and Nt are respectively the number of experiments and treatments; ***P<0.001*; NS: not significant (P<0.10): tVFA-PR = 8.9*^{NS} *(±6.6) + 8.03*** (±0.64) RfOM intake (Nexp=35; Nt=107;* R^2*=0.89; Syx=11.8).*

Prediction of the individual VFA molar proportions in the rumen fluid

The molar proportions of individual VFAs (% total VFA) were closely related to the dNDF/dOM ratio (% dOM), and the relationships were far from linear, even after logarithmic transformation (Nexp=133; Nt=375; $P<0.001$). With quadratic models based on log (dNDF/dOM) less than 4% of the data set appeared to be outliers statistically. The residuals of the model were significantly related to ruminal starch digestibility (RSD, % starch intake, $P<0.05$), pH ($P<0.001$) and DM intake (DMi, kg/d/100 kg BW, $P<0.01$). When RSD and DMi were included as covariables (Nexp=44, Nt=124 including 96 on dairy cows, mean dNDF/dOM = 0.294, mean RSD = 68.4, mean DMi = 2.88), the quadratic term of log (dNDF/dOM) became non significant, and models were significantly improved for C2, C3, C4, and C2/C3 (Table 1). No other interfering factor, including pH, was detected on these final models, based on log dNDF/dOM, RSD, and DMi.

Quantification of a correction for VFA molar proportions between concentration and production

In the *VFA-Prod* database, a combination of observed proportions of individual VFA in the tVFA-PR (i.e. the infused labelled VFA was quantified in both the individual and the total VFA pools) and observed proportions in rumen VFA concentration was available in 63, 44 and 42 treatments for C2, C3 and C4, respectively. The experiments focused on the effect of level of feed intake and/or dietary composition (60%), or on effects of other factors, i.e. additives or animal species (40%). The experiments covered a wide range of feed intake level (4 to 39 g DM/d/kg BW), dietary composition (15 to 79% NDF) and VFA profiles (range 39 to 77% for C2, 14 to 47% for C3, and 4 to 17% for C4). Within this data set, molar proportions of individual VFA (mol iVFA/100 mol tVFA) in the production and in the concentration were significantly related. Within-experiment relationships were linear for C2 and C3, and quadratic for C4 (Figure 2). No significant interfering factors were detected for C3. A significant effect of the methodology of tracer infusion (continuous infusion versus pulse dose) was observed on the LSMeans for C2 and C4 (+2 and -2 mol/100 mol, respectively, with continuous infusion). Given that these within-experiment relationships were close to the first bisector (Y=X) over a wide range of VFA molar proportion, and that between-

Table 1. Responses of individual volatile fatty acid molar proportions to variations in the proportion of digested neutral detergent fibre (dNDF, g/kg DM) in the digested organic matter (dOM, g/kg DM) expressed in log 100 dNDF/dOM, ruminal starch digestibility (RSD, %starch intake), and dry matter intake (DMi, g/100 g BW).

	Nexp	Nt	intercept	slopes			R^2	Syx
				log 100 dNDF/dOM	RSD	DMi		
Acetate (mol/100 mol VFA)	44	124	54.2*** (±4.7)	12.0*** (±2.0)	-0.052* (±0.021)	-1.99* (±0.88)	0.92	1.23
Propionate (mol/100 mol VFA)	44	124	19.7*** (±5.5)	-6.63** (±2.40)	0.070** (±0.025)	2.62* (±1.04)	0.92	1.45
Butyrate (mol/100 mol VFA)	44	124	19.0*** (±2.4)	-3.99** (±1.41)	-0.026† (±0.015)	NS	0.76	0.88
Acetate/Propionate (mol/mol)	44	124	2.82** (±0.87)	1.46*** (±0.38)	-0.011** (±0.004)	-0.42* (±0.16)	0.92	0.23

*** $P<0.001$; ** $P<0.01$; * $P<0.05$; † $P<0.10$; NS: not significant ($P<0.10$).

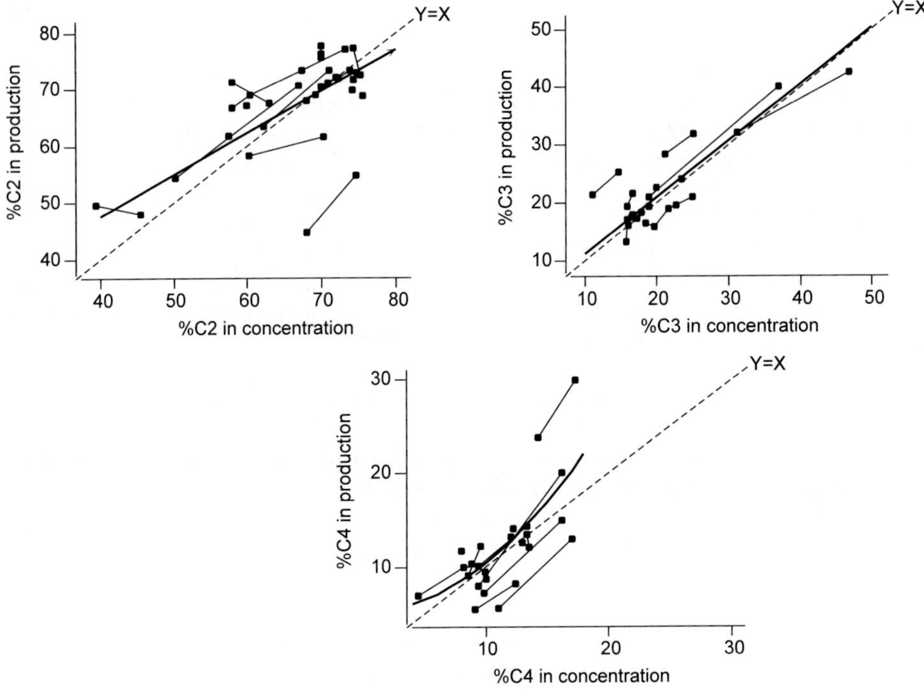

Figure 2. Within-experiment relationships between the relative proportions of individual volatile fatty acid in rumen concentrations (conc, on the X axis) and in the production rates (prod, on the Y axis). The bold lines represent the adjusted models for acetate (C2), propionate (C3) and butyrate (C4); Nexp and Nt are respectively the number of experiments and treatments; ***P<0.001; *P<0.05; *NS: not significant (*P<0.10*): %C2prod = 18* (±6) + 0.74*** (±0.09) %C2conc (Nexp = 22; Nt=56;* R^2 *= 0.96; Syx = 2.1); %C3prod =* 1.4^{NS} *(±1.5) + 0.98*** (±0.07) %C3conc (Nexp = 17; Nt= 41;* R^2*=0.98; Syx = 1.4); %C4prod = 5.2*** (±0.6) + 0.052*** (±0.004)* $\%C4conc^2$ *(Nexp=17; Nt=39;* R^2*=0.96; Syx = 1.1).*

experiment differences were mainly explained by methodological factors, the differences in VFA composition between production and concentration were not taken into account. CoriVFA was set to unity, and iVFA-PR was calculated as follows:

Predicted iVFA-PR = Predicted tVFA-PR × Predicted [iVFA] / [tVFA]

Evaluation of prediction of rumen production rate of individual VFA against independent observations of net portal appearance of VFA

Predicted iVFA-PR for all dietary treatments present in the *Flora* database were compared to measurements of NPA of VFA and β-hydroxybutyrate (β-OH). Measured NPA of individual VFA were closely and linearly related to their predicted ruminal PR (Figure 3). Following correction for uptake of arterial C2 by portal-drained viscera (PDV, according to the model proposed by Nozière and Hoch, 2006), the relationship between C2-corrected NPA and predicted C2-PR did not significantly differ from the 1st bisector (Y=X). The residuals were normality distributed. Interfering factors related to dNDF (negative) and digestible N (positive) contents were observed on slopes,

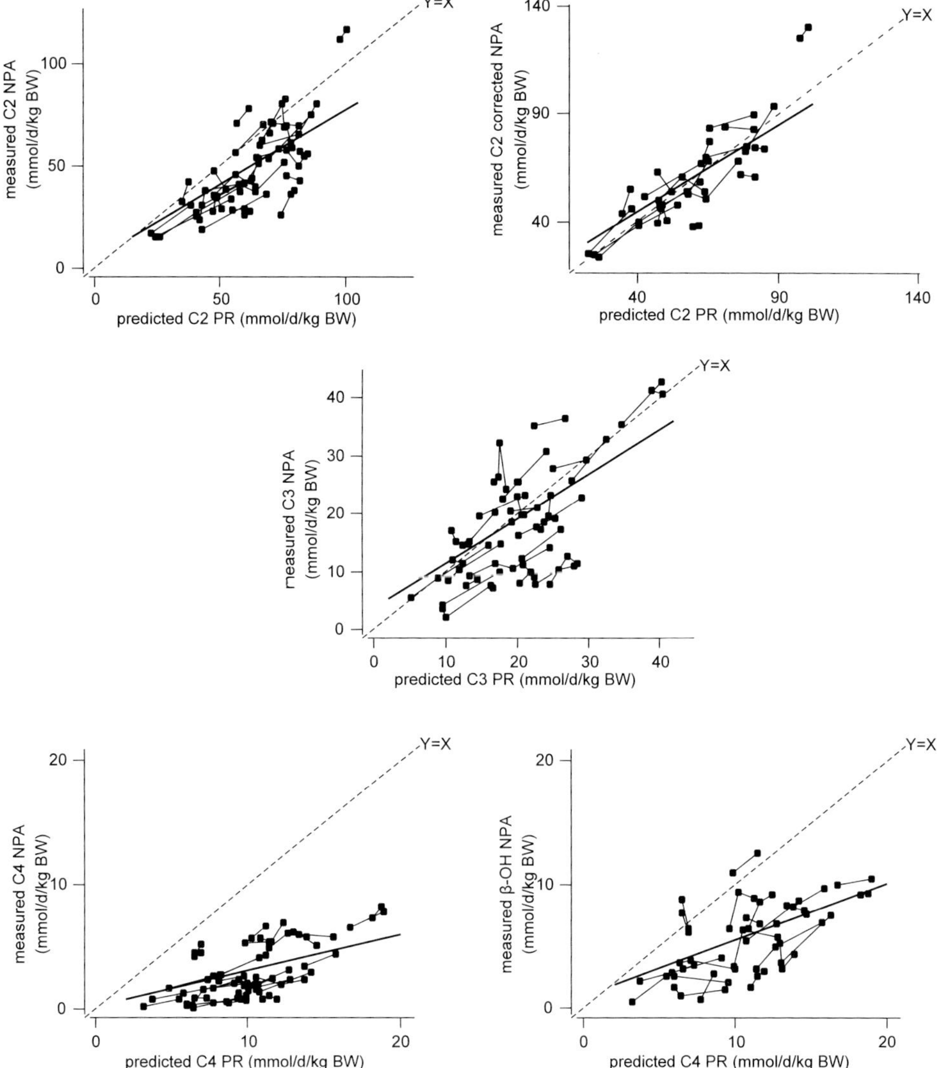

Figure 3. Within-experiment relationships between the predicted volatile fatty acid production rate (PR) and their measured net portal appearance (NPA). The bold lines represent the adjusted models for acetate (C2), propionate (C3), butyrate (C4) and β-hydroxybutyrate (β-OH); Nexp and Nt are respectively the number of experiments and treatments: ***P<0.001; *P<0.05; †P<0.10; *NS: not significant (P<0.10): C2-NPA = 4.0^{NS} (±5.6) + 0.73*** (±0.09) predicted C2-PR (Nexp = 30; Nt=72; R^2 = 0.94; Syx = 5.0); C2-corrected NPA = 11.8† (±6.5) + 0.82*** (±0.10) pred C2-PR (Nexp = 18; Nt=44; R^2 = 0.95; Syx = 5.2); C3-NPA = 3.8* (±1.6) + 0.72*** (±0.08) predicted C3-PR (Nexp = 30; Nt=73; R^2 = 0.97; Syx = 1.6); C4-NPA = 0.14^{NS} (±0.42) + 0.29*** (±0.04) predicted C4-PR (Nexp = 30; Nt=74; R^2 = 0.96; Syx = 0.43); β-OH-NPA = 0.86 ^{NS} (±1.02) + 0.46*** (±0.10) predicted C4-PR (Nexp = 28; Nt=69; R^2 = 0.90; Syx = 0.96).*

LSMeans and residuals of these models, but their introduction as an additional covariable did not improve the models.

Discussion

Total VFA production rate in the rumen

Despite the major contribution of VFA to ruminant energy supply, the VFA-PR has been quantified *in vivo* in less than 80 publications during the last 4 decades. This may largely be explained by the cost of performing experiments requiring isotopes. Moreover, simultaneous measurement of both VFA-PR and OM truly fermented in the rumen (i.e. OM apparently digested plus microbial synthesis) is dramatically scarce. Nevertheless, by gathering all available data of VFA-PR, and homogeneously characterizing feedstuffs and diets according to INRA feed tables, we showed in the present work a close relationship between VFA-PR and the calculated amount of RfOM, over a wide diversity of intake levels and dietary composition. This systematic characterisation of diets also allowed assessment of putative interfering factors reflecting intake level and dietary composition. This showed that the relationship between VFA-PR and RfOM intake presented no significant interfering factors, and did not depend on whether changes in RfOM intake are due to changes in DM intake and/or in RfOM dietary content. The additivity of RfOM of the different feeds of the diet was assumed. The lack of interfering factors likely reflects the low level of digestive interactions in the present data set (estimated as 3.1±1.2% for diet OM digestibility, according to the empirical model of Sauvant, 2003). The average slope of the model corresponds to 8 moles VFA/kg OM fermented, that is consistent with laws of stoichiometry, and with the efficiency of microbial synthesis assumed in the French PDI system. This empirical approach provides a helpful tool to evaluate VFA production from dietary characteristics.

Individual VFA molar proportions in the rumen fluid

In most mechanistic rumen models, the prediction of the VFA composition is based on stoichiometric coefficients describing partition of VFA for each substrate considered. These coefficients have been calculated empirically from *in vivo* data using stoichiometric models (Murphy *et al.*, 1982; Bannink *et al.*, 2006; Sveinbjörnsson *et al.*, 2006). Unfortunately, their validation remains scarce and the inaccuracy of such approaches for a generic application has been pointed out by several authors (Bannink *et al.*, 1997; Friggens *et al.*, 1998; Offner and Sauvant, 2004). The approach developed in the present work (which could be defined as a 'per VFA' approach) is based on empirical equations with dietary and digestive factors as explanatory variables for each VFA. It clearly differs from the classical stoichiometric approach describing partition of VFA for each substrate considered (which could be defined as a 'per substrate' approach).

The dNDF/dOM ratio, which reflects the nature of OM digested at the whole tract level, appeared as the main factor involved in changes in VFA molar proportions in the rumen through a wide diversity of experimental factors. As expected, RSD explained a substantial part of residual variability, reflecting the influence of the site of starch digestion (rumen vs. intestines) on VFA molar proportions. For each VFA, the sign of coefficients obtained for dNDF/dOM and RSD was similar to those obtained by Friggens *et al.* (1998) for cellulose and starch, using multiple regression techniques to relate VFA profile to dietary chemical composition. Even when both dNDF/dOM and RSD (reflecting type, site and extent of digested substrates) were considered, we established an effect of DM intake on molar proportions of C2 (negative) and C3 (positive), as also observed by Sveinbjörnsson *et al.* (2006) with a large data set on dairy cows restricted to Nordic diets. Other independent additional factors may affect VFA profiles (i.e. particle size, buffer supply) and contribute to reduce the residual variability. In the present study they could not be quantified however, due to the structure of the meta-design. Nevertheless, the final models obtained in the present study included dNDF/dOM, RDS and DMi, and had a low Syx compared to other works derived from stoichiometric models which did not correct for experiment effects and hence this variation remain included in Syx (Bannink *et al.*, 2006) or from within-experiment models based on a more aggregated covariable such as percentage of concentrate (Lescoat and Sauvant, 1995) or NDF dietary content (Sauvant, 2003). The rumen pH was not an interfering factor which suggests that its effect on VFA stoichiometry was largely confounded with RSD and/or DMi.

Equations of VFA molar proportions proposed in the present paper have been developed on measured digestive fluxes of OM, NDF and starch. Therefore, their use for predicting purposes (using tabulated data of dOM, dNDF or RSD) requires caution because these parameters are subjected to digestive interactions, depending on DM intake and diet composition. An empirical model of digestive interactions (Sauvant, 2003) demonstrated that the ratio dNDF/dOM is less sensitive to digestive interactions than dOM or dNDF taken separately, which is an advantage for the use of this ratio for predictive purposes. Also, Offner and Sauvant (2004b) proposed an empirical model for predicting RSD from tabulated in sacco starch degradability including effect of DM intake.

Individual VFA molar proportions: productions vs. concentrations

In all studies reporting production rate of individual VFA, a certain discrepancy is observed in relative molar proportion of individual VFA produced and present in rumen fluid. This likely originates from differences between absorption rates of VFA by the rumen wall, and/or from interconversions between VFAs. The within-experiment relationships obtained in the present work over a wide range of VFA profiles, suggest that changes in molar proportion of C2 and C4 production are respectively slightly overestimated by C2 concentration and underestimated by C4 concentration in the rumen. This may be explained by differences in fractional absorption rate between VFAs ($C2<C3<C4$, Bergman, 1990). To assess whether different absorption rates may explain the differences between production and concentration profiles across studies, we carried out dynamic simulations of rumen VFA pools, with VFA fractional absorption rates depending on rumen pH (estimated from VFA concentration if not reported in the publications), VFA concentration and rumen volume (Dijkstra *et al.*, 1993, Nozière and Hoch, 2006). The simulations indicated that differences in pH and VFA concentrations were insufficient to explain between-experiment differences between production and concentration profiles. However, the present work stresses that these between-experiment differences were mainly explained by methodological rather than by nutritional factors. Therefore, the differences in molar proportion of individual VFA produced and present in the rumen were not taken into account in the prediction of individual VFA production rate. We thus assumed that ruminal production of individual VFA can simply be derived from tVFA-PR and the molar proportion of individual VFA in rumen fluid.

Evaluation of the predicted ruminal VFA production rates against observed net portal appearance of VFA

The estimated production rate of individual VFA calculated with our model is closely related to their measured net portal appearance. Our results suggest that on average C2 is fully recovered at first pass in the portal vein, C3 is largely recovered, whereas 25% of C4 is recovered as C4 and 46% as β-OH. These results are highly consistent with the knowledge on the extent of PDV metabolism of VFA. Indeed, results obtained with the temporarily-isolated washed rumen method (Kristensen and Harmon, 2006) presented similar recovery rates. The results are also in agreement with portal recoveries of individual VFA predicted by steady-state simulations with a mechanistic model of Bannink *et al.* (2008).

Conclusion

This work quantified the effects of amount and nature of OM fermented on VFA production and VFA molar proportion in the rumen, using an approach alternative to the classical stoichiometric approach. It also showed that within-experiment variations in VFA profile in production are close to those measured from their concentration in the rumen fluid. This suggests that individual VFA fluxes may be estimated from their ruminal concentrations profiles. Based on this assumption, we proposed an empirical model of individual VFA-PR based on simple predictors available from the INRA Feed Tables representing the nature of digested substrates as well as their site and extent of digestion. Models were built using the most generic predictors. A systematic analysis of interfering factors was carried out in order to reduce as much as possible the experiment effects which led a model including RfOM, dOM, dNDF, RSD and DMi. Using this model for predictive purposes by using tabulated data requires that digestive interactions are taken into account. Predicted VFA production rates were compared with their net portal appearance. Percentages of VFA disappearing by PDV metabolism were consistent with the knowledge on VFA metabolism by PDV, which indicates that the representation of transport of VFA from rumen to portal blood may be simplified to follow principles of mass-action laws. This work will help to progress towards the development of feed evaluation systems based on nutrient fluxes.

References

Bannink, A., De Visser, H. and Van Vuuren, A.M., 1997. Comparison and evaluation of mechanistic rumen models. British Journal of Nutrition 78:563-581.

Bannink, A., Kogut, J., Dijkstra, J., Kebreab, E., France, J., Tamminga, A. and Van Vuuren, A.M., 2006. Estimation of the stoichiometry of volatile fatty acid production in the rumen of lactating cows. Journal of Theoretical Biology 238:36-51.

Bannink, A., France, J., Lopez, S., Gerrits, W.J.J., Kebreab, E., Tamminga, S. and Dijkstra, J., 2008. Modelling the implications of feeding strategy on rumen fermentation and functioning of the rumen wall. Animal Feed Science and Technology 143:3-26.

Bergman, E.N., 1990. Energy contribution of volatile fatty acids from the gastrointestinal tract in various species. Physiological Reviews 70:567-587.

Dijkstra, J., Boer, H., Van Bruchem, J., Bruining, M. and Tamminga, S., 1993. Absorption of volatile fatty acids from the rumen of lactating dairy cows as influenced by volatile fatty acids concentration, pH and rumen liquid volume. British Journal of Nutrition 69:385-396.

Friggens, N.C., Oldham, J.D., Dewhurst, R.J. and Horgan, G., 1998. Proportions of volatile fatty acids in relation to the chemical composition of feeds based on grass silage. Journal of Dairy Science 81:1331-1344.

INRA (ed.), 2007. Alimentation des bovins, ovins et caprins – Besoins des animaux – Valeurs des aliments – Tables INRA 2007. Quae (ed.), INRA, Versailles, France, 307 pp.

Kohn, R.A. and Boston, R.C. 2000. The role of thermodynamics in controlling rumen metabolism. In: McNamara, J.P., France, J. and Beever, D.E. (eds.) Modelling nutrient utilization in farm animals. CAB International, Wallingford, UK, pp.11-24.

Kristensen, N.B. and Harmon, D.L. 2006. Splanchnic metabolism of short chain fatty acids in ruminant. In: Sejrsen, K., Hvelplund, T. and Nielsen, M.O. (eds.) Digestion, metabolism and impact of nutrition on gene expression, immunology and stress. Wageningen Academic Publishers, Wageningen, the Netherlands, pp.249-268.

Lescoat, P. and Sauvant, D., 1995. Development of a mechanistic model for rumen digestion validated using the duodenal flux of amino acids. Reproduction Nutrition Development 35:45-70.

Loncke, C., Ortigues-Marty, I., Vernet, J., Lapierre, H., Sauvant, D. and Nozière, P., 2009. Empirical prediction of net portal appearance of volatile fatty acids, glucose and their secondary metabolites (β-hydroxybutyrate, lactate) from dietary characteristics in ruminants: a meta-analysis approach. Journal of Animal Science 87:253-268.

Murphy, M.R., Baldwin, R.L. and Koong, L.J., 1982. Estimation of stoichiometric parameters for rumen fermentation of roughage and concentrate diets. Journal of Animal Science 55:411-421.

Nozière, P. and Hoch, T., 2006. Modelling fluxes of volatile fatty acids from rumen to portal blood. In: Kebreab, E., Dijkstra, J., Bannink, A., Gerrits, W.J.J., and France, J. (eds.) Nutrient digestion and utilization in farm animals: modelling approaches. CAB International, Wallingford, UK., pp.40-47.

Nozière, P., Glasser, F., Martin, C. and Sauvant, D., 2007. Predicting *in vivo* production of volatile fatty acids in the rumen from dietary characteristics by meta-analysis: Description of available data. In: Ortigues-Marty, I., Miraux, N. and Brand-Williams, W. (eds.) Energy and protein metabolism and nutrition, EAAP Publication No 124, Wageningen Academic Publishers, Wageningen, the Netherlands, pp.585-586.

Offner, A. and Sauvant, D., 2004. Prediction of *in vivo* starch digestion in cattle from *in situ* data. Animal Feed Science and Technology 111:41-56.

Offner, A. and Sauvant, D., 2006. Thermodynamic modeling of ruminal fermentations. Animal Research 55:343-365.

Sauvant, D., Martin, O. and Mertens, D., 2000. Mise au point d'un modèle empirique de réponses multiples de la digestion du bovin aux régimes. Rencontres Recherches Ruminants 7:341.

Sauvant, D., 2003. Modélisation des effets des interactions entre aliments sur les flux digestifs et métaboliques chez les bovins. Rencontres Recherches Ruminants 10:151-158.

Sauvant, D., Schmidely, P., Daudin, J.J. and St-Pierre, N.R., 2008. Meta-analyses of experimental data in animal nutrition. Animal 2:1203-1214.

Sveinbjörnsson, J., Huhtanen, P., and Udén, P., 2006. The Nordic dairy cow model, Karoline – Development of volatile fatty acid sub-model. In: Kebreab, E., Dijkstra, J., Bannink, A., Gerrits, W.J.J., and France, J. (eds.) Nutrient digestion and utilization in farm animals: modelling approaches. CAB International, Wallingford, UK., pp.1-14.

Vernet, J. and Ortigues-Marty, I., 2006. Conception and development of a bibliographical database of blood nutrient fluxes across organs and tissues in ruminants: data gathering and management prior to meta-analysis. Reproduction Nutrition Development 5:527-546.

Representing tissue mass and morphology in mechanistic models of digestive function in ruminants

A Bannink[1], J Dijkstra[2] and J France[3]
[1]Livestock Research, Wageningen UR, P.O. Box 65, 8200 AB Lelystad, the Netherlands; andre.bannink@wur.nl
[2]Animal Nutrition Group, Wageningen University, P.O. Box 338, 6700 AH Wageningen, the Netherlands
[3]Centre for Nutrition Modelling, Department of Animal and Poultry Science, University of Guelph, Guelph, Ontario N1G 2W1, Canada

Abstract

Representing changes in morphological and histological characteristics of epithelial tissue in the rumen and intestine and to evaluate their implications for absorption and tissue mass in models of digestive function requires a quantitative approach. The aim of the present study was to quantify tissue mass (M) and absorptive area (AA) from parameters that are easily derived from morphological inspection and histology of tissue biopsies, and to compare this representation with approaches in current model of digestive function. Relatively small changes of 5% in some morphological and histological characteristics were calculated to affect absorptive area (AA) and epithelial tissue mass (M) strongly in the rumen and the intestine of cattle. The cumulative effect of changes in volume, height and width of primary protrusions and of secondary protrusions of the rumen wall was 18% for rumen mucosal AA and 29% for rumen serosal AA. It was 24% for intestinal mucosal AA. The cumulative effect on rumen and intestinal M was 20% and 22%, respectively. The simulations indicate that allometric functions that relate volume or weight to AA and M require exponentiation with an exponent higher than those based on geometric shape or volume.

Keywords: modelling, morphology, gastrointestinal epithelia, cattle

Introduction

Omics-techniques (genomics, transcriptomics, proteomics, metabolomics) are thought to result in a paradigm shift in mathematical modelling of physiological functions. New and more detailed information on patterns of expression of genes, on proteins, and on metabolite profiles is obtained from tissues taken from the animal which will offer new opportunities to science. Information on regulatory networks and signalling pathways may improve insight into physiological functioning and into the response of tissues to altered physiological and environmental conditions. There are two crucial notions, however, with respect to the application of omics-techniques to answer physiological questions. First, information is gathered from samples taken from the animal which does not represent the functioning of the whole tissue or organ. Second, if physiological state remains not well characterized the information from omics-techniques can not become exploited to its full extent. Physiological problems do not reside in samples, nor can they be solved satisfactorily by just marking physiological states. Instead, various sources of information must be integrated and a vision is needed of how biology functions at higher levels of organization than the molecular level of omics-data.

With the aim of modelling gastrointestinal tract functioning on the basis of concepts and data derived from omics-techniques, also information on prior physiological issues remains to receive a prominent role. Hence, a description of the mass and morphology of the gastrointestinal epithelia in animals remains necessary. It is well known that besides changes in the enzymatic activity per unit of tissue mass (Harmon *et al.*, 1991; McBride *et al.*, 1990), tissue mass and morphology in particular adapt to nutritional changes (Dirksen, 1984; Bannink *et al.*, 2008). High adaptive capacity reflects their principal role as the boundary between external and internal milieu of the animal, as well as the route of transport to retrieve nutrients from the external milieu (Bannink *et al.*, 2006). Hence, mechanistic models of gastrointestinal physiology need to consider tissue mass and morphology as explanatory variables when representing these principal roles. In most efforts to model physiological function, these aspects either have not been represented at all, or by simple equations relating nutrient absorption to volume (e.g. Dijkstra *et al.*, 1992) or metabolic weight.

The objective of the present study was to explore how changes in mass and morphology of gastrointestinal epithelia in the ruminant can be represented, and how these may relate to absorptive function. Using simple geometrical principles to represent observed morphology and plasticity of rumen and small intestinal epithelia, their mass and surface area were calculated. The aim was to quantify mass and absorptive area from measurable morphological and histological characteristics of rumen and intestinal epithelia. A further aim was to compare this representation with approaches in current models of digestive function.

Biology and model representation

Biology of rumen epithelia

Rumen epithelia adapt strongly not only to the pronounced increase in feed intake by dairy cows after parturition (Dirksen, 1984), but also to changes in feeding strategy and in dietary composition. For example, in periparturient cows, Bannink *et al.* (2008) observed a differential response by rumen epithelia to gradual increments in concentrate allowance within 10 or 20 days after parturition to a maximum of 12 kg dry matter per day. Temporal effects during the first three weeks after parturition included a doubling of papillae size, whereas epithelial thickness varied between 20% thinner and 20% thicker. The changes were estimated to result in nearly a doubling of papillae surface area and epithelial tissue mass. These large changes were associated with only a modest increase in rumen concentrations of volatile fatty acids despite a large increase in their production rate (Bannink *et al.*, 2008). These results confirm the strong adaptive capacity of the rumen wall. Nevertheless, abnormalities in rumen epithelia may occur, and (subacute) rumen acidosis receives much attention because it is considered a serious risk to cattle health and productivity (Enemark *et al.*, 2002). In this regard, inspection of the rumen epithelia (Figure 1) as an essential part of rumen wall functioning and controlling the conditions in the rumen deserves just as much attention as the processes taking place in the lumen (Bannink, 2007).

Modelling rumen epithelia

In this study several parameters were used to represent general geometrical characteristics of rumen epithelial tissue (Figure 1; Table 1) and to allow quantification of effects on epithelial mass (M) and absorptive area (AA). The AA and M are quantified using the following functions:

AA= function (volume, area covered by protrusions, shape of protrusions)
M = function (AA, histological parameters epithelia)

The lumen of the rumen can ideally be considered to be a sphere of radius r_r (cm). The AA of such a sphere, AA_r (cm^2), is calculated as

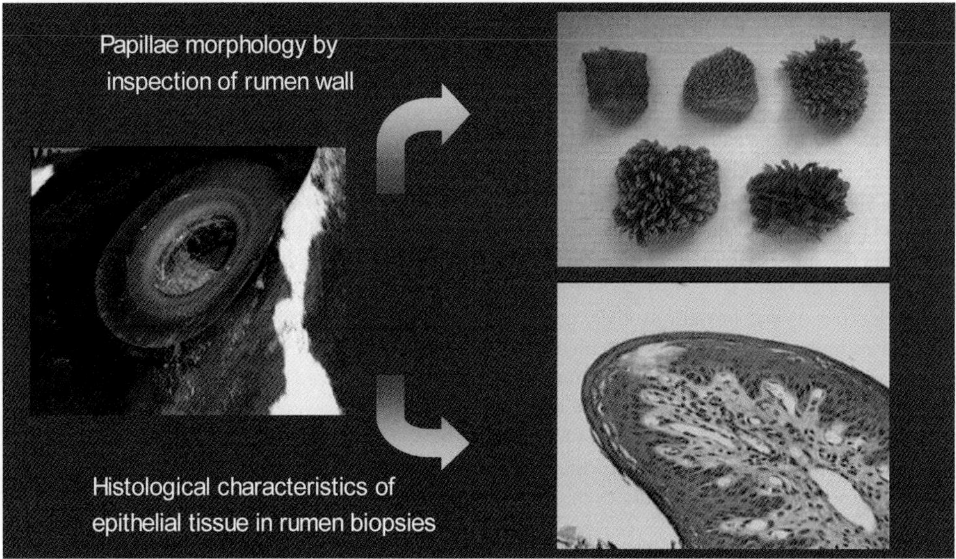

Figure 1. Morphological and histological inspection of rumen epithelia.

Table 1. Summary of parameters easily derived from morphological and histological inspection of rumen and intestinal tissue that can be used to quantify tissue mass and tissue surface area.

Tissue	Notation	Meaning	Reference value and units
Rumen epithelia			
	r_r	radius of rumen fluid volume	25 cm
	r_{p1}, r_{p2}	radius 1 & 2 of ellipsoid ground base of *papillae*	1.5 mm
	h_p	height *papillae*	5.0 mm
	f_r	fraction of rumen wall surface covered with *papillae*	0.8
	d_p	density of *papillae* on covered rumen wall surface	0.5/mm^2
	h_{ret}	height of *retepegs*	25 μm
	r_{ret}	radius of *retepegs*	25 μm
	x_b	thickness of epithelial basal sheet (excluding *retepegs*)	100 μm
Intestinal epithelia			
	r_{si}	radius of cylindrical tube of intestine	2.5 cm
	l_{si}	length of intestinal tube	m [1]
	f_{si}	fraction of intestinal tube surface covered with *villi*	0.95
	r_v	radius of *villi* (from width)	50 μm
	h_v	height of *villi*	500 μm
	r_{mv}	radius of *microvilli* (from width w_{mv})	0.05 μm
	h_{mv}	height of *microvilli* (brush-border)	2.0 μm
	h_c	crypt depth	250 μm

[1] Length is not relevant for calculations of the effect of morphological and histological parameters on absorptive area and tissue mass at a specific location in the intestine.

$$AA_r = 4 \times \Pi \times r_r^2$$

A fraction f_r of this surface is covered with protrusions of the rumen wall called *papillae* with density d_p (number of papillae per mm^2). Each *papillum* has ideally a cylindrical shape with an elliptic ground base with radii r_{p1} and r_{p2} (mm), and half a sphere as *papillum* tip (which was neglected here and simplified to an elliptic ground base). From these parameters a cylinder with a circular ground base and identical surface with radius r_p (mm) can be calculated. Then, at the mucosal side the AA of such a rumen *papillum*, AA_p (mm^2), with height h_p (mm; excluding the ground base because that does not contain the basal epithelial sheet) and with a cylindrical shape can be calculated as

$$AA_p = 2 \times \Pi \times r_p \times h_p + \Pi \times r_p^2$$

Because the epithelial tissue forms a basal sheet with thickness x_b (mm) along the mucosal side of the rumen wall (Figure 1) and has a specific weight *sw* (g/mm^3), which is either covered with *papillae* or not, the M of the basal epithelial sheet of the rumen sphere including the *papillae* protrusions, M_b (g), can be calculated as

$$M_b = (AA_r \times (1\text{-}f_r) \times 100 + AA_p \times f_r) \times x_b \times sw$$

At the serosal side of the basal epithelial sheet in rumen *papillae* the epithelial surface is well exposed to blood circulation and has secondary protrusions, when well-developed called *retepegs,* of the basal epithelial sheet which ideally have a cone shape with radius r_{ret} (mm) of the ground base (half of cone width w_{ret}) and height h_{ret} (mm; Figure 1). The AA_{ret} (mm^2; excluding ground base) and M_{ret} (g) of these *retepegs* are then calculated by

$$AA_{ret} = \Pi \times r_{ret} \times \sqrt{(r_{ret}^2 + h_{ret}^2)}$$
$$M_{ret} = (\Pi \times r_{ret}^2 \times h_{ret}) / 3 \times sw$$

Effects on total mucosal AA and serosal AA and on total M were calculated from changes in AA_r, AA_p and AA_{ret}, and in M_b and M_{ret}, taking into account their contribution to total AA and M which is determined by f_r, d_p and r_{ret}.

Finally, for comparison with extant models, rumen volume, Vol_r (cm^3), is calculated as

$$Vol_r = 4/3 \times \Pi \times r_r^3$$

Biology of intestinal epithelia

Similar to the rumen, changes in feed intake will affect the amount of digesta flowing into the intestine and affect functioning of the intestinal wall. Zitnan *et al.* (2003) concluded from observations on intensively reared cattle that with a higher concentrates allowance, the height of *villi* in duodenum and jejunum increased and that there was a positive correlation between morphometric parameters of the rumen and intestinal mucosa. These findings indicate an adaptive response of intestinal epithelia to nutrition. The intestinal wall acts as a barrier, as a source of digestive enzymes and as the site of nutrient absorption (Bannink *et al.*, 2006). In young growing ruminants in particular, external factors that induce immune responses in the ruminant as a host may seriously affect morphology and digestive function of the small intestine (Frink *et al.*, 2002). Morphological and histological characteristics of intestinal tissue can be used to obtain an indication of (suboptimal) intestinal functioning.

Modelling intestinal epithelia

The small intestine can assumed to be a cylindrical tube of length l_{si} with a circular base of radius r_{si} (cm; Figure 2). For a fraction f_{si} the intestinal wall is covered with *villi* which ideally can be considered cylindrical protrusions with radius r_v (mm) and height h_v (mm). In addition, the *villi* are covered with a brush-border that holds an unstirred water layer and includes a very dense layer of *microvilli*, which are considered secondary protrusions of cylindrical shape with radius r_{mv} (mm) and height h_{mv} (mm). The *villi* are assumed to be completely covered by *microvilli.*

Although it varies with nutritiona and type of ruminant, the value of l_{si} (m) is not relevant when effects of morphometric changes on AA and M are calculated per unit of intestinal length. The intestinal epithelia are considered to have a basal epithelial layer with *crypts* from which epithelial cells migrate to the *villi* protrusions (Figure 2). The contribution of *crypts* to the epithelial tissue is represented by *crypt* depth h_c (mm). These *crypts* can ideally be considered to have an inversed cylindrical shape compared to the cylindrical shapes of the *villi*. Hence, it was assumed that the epithelial sheet in *crypts* can be treated similarly to that in *villi*. However, in the present study it was assumed that *crypts* do not contribute to AA and a basal epithelial sheet without protrusions was assumed on the location of the *crypts*. A similar contribution to M was assumed as for *villi* with an ideally cylindrical shape. Further, it was assumed that *villi* and *crypts* are evenly distributed across the intestinal wall creating an identical density of both.

Calculations of AA and M for intestinal epithelia were in principal similar to those indicated above for rumen epithelia, but with some differences. The rumen is considered to be a sphere whereas the intestine is assumed to be a cylinder. Also, a basal epithelial tissue layer is evident that exactly follows the surface of the rumen wall, whereas this is less clear for the intestinal wall. Further, the primary protrusions which are themselves cylindrical (*papillae* and *villi*, respectively) have secondary protrusions themselves which are of a different shape in the rumen compared to those

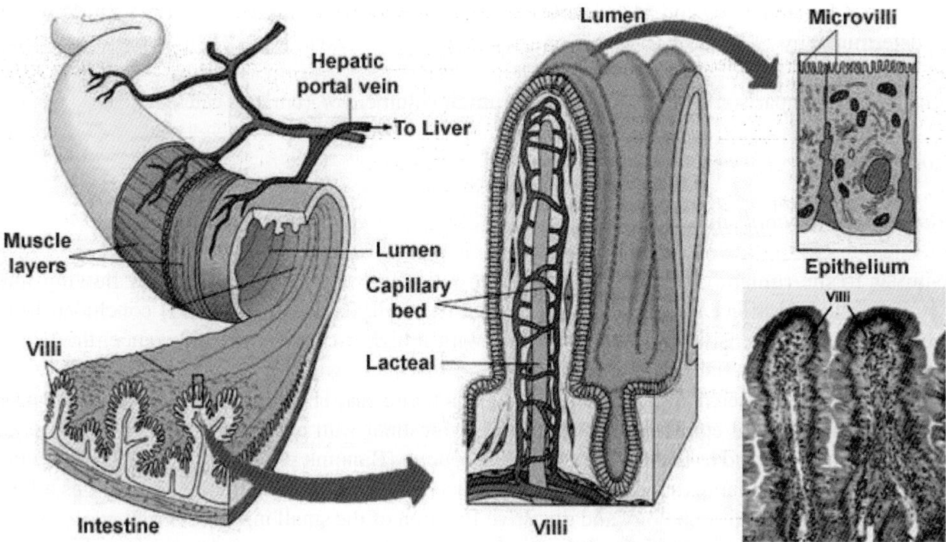

Figure 2. Morphological and histological inspection of intestinal epithelia (www.colorado.edu/intphys/Class/IPHY3430-200/image/villi.jpg).

in the intestine (cone and cylindrical, respectively) and at a different side (serosal and mucosal, respectively). Finally, the dimensions of both types of epithelial tissue strongly differed.

Extant models of digestive function

Morphological and histological changes in epithelia can affect AA. However, in extant models of digestive function (e.g. Dijkstra *et al.*, 1992, 1993) absorptive function and tissue mass are related to digesta volume or metabolic weight by

$$AA = \text{function}(\text{volume}^{2/3})$$
$$M = \text{function}(\text{volume}^{3/4} \text{ or weight}^{3/4})$$

These representations based on volume or metabolic weight are generally used to accommodate the effects of nutrition and tissue mass, and will be compared to the equations developed in the present study derived from geometrical principles.

Results and discussion

In the reference situation of rumen and intestinal epithelia (indicated by parameters values in Table 1), the presence of *papillae* on the rumen wall increased rumen mucosal AA by 10-fold and serosal AA by more than 20-fold. Papillae abundance and papillae shape hence must be expected to have a strong impact on mucosal and serosal AA as well as on M. The presence of *retepegs* seems to have a strong impact on rumen serosal AA but obviously does not affect mucosal AA as retepegs occur only on the serosal side. The calculations for intestinal epithelia indicate that the presence of *villi* increased intestinal AA by 20-fold and *microvilli* increased it by a factor of more than 1,200 at the mucosal side only. Thus, also the shape of *villi* and *microvilli* are expected to have a strong impact on intestinal AA, with the effect of *crypt* depth remaining uncertain.

The sensitivity of AA and M to changes in parameters related to morphological and histological characteristics of rumen and small intestine epithelia was explored (Table 2). In the case of rumen epithelia, a 5% increase in rumen fluid volume, achieved by changing r_r, was matched by an increase of AA to the exponent 2/3, according to geometrical principles. However, when this increase in volume is accompanied by minor changes of 5% in parameter values for epithelia characteristics (Table 2) the changes in AA cannot be represented anymore with an exponent of 2/3, but a much higher exponent is needed than this (up to 3.4; Table 2) to explain a similar increase of AA from a change of volume only. A change of 5% in parameter values is considered to be minor compared to the variability observed *in vivo* dairy cows (Bannink *et al.*, 2008). Suppose that an increased volume (e.g. result of increased feed intake) causes solely a 0.1 mm increase of *papillae* width (less than measurement error), then the change in AA is reproducible by exponentation of volume by 3/4. Adding changes in other parameter values (an additional fraction of 0.01 of area covered with *papillae* and an increase of *papillae* height by 0.1 mm) would require an exponent of 6/7.

Changes in parameter values for intestinal epithelia also affect intestinal AA and M (Table 2). Assuming an unchanged digesta volume, changes of *villi* height resulted in similar relative changes in calculated AA. In particular the *microvilli* led to a severe increase of AA at the mucosal side and a change in *microvilli* height (width assumed not to vary) led to a relative change of mucosal AA.

The sensitivity of calculated AA and M to small changes in morphological and histological parameters (which are known to vary substantially *in vivo*) illustrates the high adaptive nature of gastrointestinal epithelia. Omics-techniques may be used to obtain gene expression and enzyme

Table 2. Calculated change in absorptive area (AA) and epithelial mass (M).

Parameter changed	Change of parameter value (%[1])	Change of AA estimate mucosal side (%)	Change of AA estimate serosal side (%)	Exponent to describe same change in mucosal AA[2]	Change of M estimate (%)	Exponent to describe same relative change in M[3]
Rumen epithelia (*parameter involved)*						
volume (r_r)	+5	+3	+3	0.7	--- [3]	--- [3]
+ *papillae* width (r_p)	+5	+8	+8	1.6	+5	0.76
+ fraction covered (f_r)	+5	+13	+13	2.5	+9	0.77
+ *papillae* height (h_p)	+5	+18	+19	3.4	+15	0.78
+ thickness epithelial base (x_b)	+5	+18	+19	3.4	+19	0.79
+ *retepeg* height (h_{ret})	+5	+18	+23	3.4	+20	0.79
+ *retepeg* width (r_{ret})	-5	+18	+29	3.4	+20	0.79
Intestinal epithelia (*parameter involved)*						
Volume/diameter (r_g)	+5	+3	+3	--- [4]	+3	--- [4]
+ *villi* height (h_v)	+5	+7	+7		+7	
+ *villi* width (r_v)	+5	+13	+13		+13	
+ fraction covered (f_g)	+5	+18	+18		+18	
+ *microvilli* height (h_b)	+5	+24	+24		+18	
+ *crypt* depth (h_c)	+5	+24	+24 [5]		+22	

[1] The cumulative effect on AA and M of subsequent changes in various parameter values, added to an initial effect of a 5% increase of volume.
[2] The exponent value needed to obtain an identical increase in $Volume^{Exponent}$ with a 5% change in volume to the increase in AA calculated with adding a 5% change in various parameter values.
[3] The exponent value needed to obtain an identical increase in $Volume^{Exponent}$ from $Volume^{3/4}$ to the increase calculated for M as a result of a 5% change in various parameter values (volume remained unchanged because this does not apply to the intestine, in contrast to calculations for rumen AA; see note 2).
[4] Changes in AA in the intestine cannot be represented by $Volume^{Exponent}$ and hence calculations of exponent values not considered.
[5] Assuming changes in *crypt* depth contribute to M and not AA. A similar contribution to M was assumed as for *villi* with an ideally cylindrical shape.

activity data from rumen epithelia in response to dietary changes (e.g. Steele *et al.*, 2009) and, combined with observed or simulated AA and M changes, provide an integrated view. The approach presented in this study might seve as a basis for including a representation of gastrointestinal wall tissues in models of digestive function. For example, modelling rumen digestion requires that rumen acidity to be taken into account. Rumen acidity depends on a number of factors including rumen volume but in particular on absorptive capacity of rumen epithelia (Bannink *et al.*, 2008). The results of the present study should be considered preliminary because major simplifications were made to represent epithelia, calculations apply to the reference situation with relatively small perturbations, and non-linear effects on AA en M were not yet explored.

Conclusions

With major nutritional changes, the accompanying adaptations to gastrointestinal epithelia may be severe. Changes in morphological and histological characteristics of rumen and intestinal epithelia were shown to markedly affect AA and M. These changes appear more indicative than a representation of geometrical change of digesta volume. The allometric functions of digesta volume generally applied appear unsuitable to represent changes in epithelial characteristics. Much higher exponents would be needed to represent the effects on AA and M calculated in the present study. The present approach may be used to develop representations of changes in AA and M in mechanistic models that aim to predict nutrient absorption or nutrient requirement and functioning of gastrointestinal epithelia.

References

Bannink, A., Dijkstra, J., Koopmans, S.-J. and Mroz, Z., 2006. Physiology, regulation and multifunctional activity of the gut wall, a rationale for multicompartmental modelling. Nutrition Research Reviews 19:227-253.

Bannink, A. 2007. Modelling Volatile Fatty Acid Dynamics and Rumen Function in Lactating Cows. PhD Thesis Wageningen University, Wageningen, the Netherlands.

Bannink, A., France, J., López, S., Gerrits, W.J.J., Kebreab, E., Tamminga, S. and Dijkstra, J., 2008. Modelling the implications of feeding strategy on rumen fermentation and functioning of the rumen wall. Animal Feed Science and Technology 143:3-26.

Dijkstra, J., Neal, H.D.St.C., Beever, D.E. and France, J., 1992. Simulation of nutrient digestion, absorption and outflow in the rumen: model description. Journal of Nutrition 122:2239-2256.

Dijkstra, J., Boer, H., Van Bruchem, J., Bruining, M. and Tamminga, S., 1993. Absorption of volatile fatty acids from the rumen of lactating dairy cows as influenced by volatile fatty acid concentration, pH and rumen liquid volume. British Journal of Nutrition 69:385-396.

Dirksen, G., Liebich, H.G., Brosi, G., Hagemeister, H. and Mayer, E., 1984. Morphologie der pansenschleimhaut und fettsaureresorption beim Rind- bedeutende faktoren fur gesundheit und Leistung. Zentralblatt fur Veterinarmedizin A 31:414-430.

Enemark, J.M., Jorgensen, R.J. and Enemark, P.St., 2002. Rumen acidosis with special emphasis on diagnostic aspects of subclinical rumen acidosis: a review. Veterinarija ir Zootechnika 20:16-29.

Frink, S., Grummer, B., Pohlenz, J.F. and Liebler-Tenorio, E.M., 2002. Changes in distribution and numbers of CD4+ and CD8+ T-lymphocytes in lymphoid tissues and intestinal mucosa in the early phase of experimentally induced early onset mucosal disease in Cattle Journal of Veterinary Medicine B 49:476-483.

Harmon, D.L., Groos, K.L., Krehbiel, C.R., Kreikemeier, K.K., Bauer M.L. and Britton, R.A., 1991. Influence of dietary forage and energy intake on metabolism of acyl-CoA synthetase activity in bovine ruminal epithelial tissue. Journal of Animal Science 69:4117-4127.

McBride, B.W. and Kelly, J.M., 1990. Energy costs of absorption and metabolism in the ruminant gastrointestinal tract and liver: a review. Journal of Animal Science 8:291-340.

Steele, M.A., AlZahal, O., Hook., S.E., Greenwood, S. and McBride, B.W., 2009. A temporal characterization of the rumen epithelium response to dramatic shifts in dietary fermentable carbohydrates. Journal of Animal Science 87 (E-suppl 2)/Journal of Dairy Science 92 (E-suppl 1): 132.

Zitnan, R., Kuhla, S., Nurnberg, K., Schoenhusen, U., Ceresnakova, Z., Sommer, A., Baran, M., Greserova, G. and Voigt, J., 2003. Influence of the diet on the morphology of ruminal and intestinal mucosa and on intestinal carbohydrase levels in cattle. Veterinarni Medicina 48:177-182.

Fluctuations in methane emission in response to feeding pattern in lactating dairy cows

L.A. Crompton[1], J.A.N. Mills[1], C.K. Reynolds[1] and J. France[2]
[1]Animal Science Research Group, School of Agriculture, Policy and Development, University of Reading, Whiteknights, Reading RG6 6AR, United Kingdom; l.a.crompton@reading.ac.uk
[2]Centre for Nutrition Modelling, Department of Animal & Poultry Science, University of Guelph, Ontario N1G 2W1, Canada

Abstract

Methane from enteric fermentation of organic matter by ruminants is considered a key contributor to climate change. This study examined the effect of feeding a total mixed ration at different intervals, either once, twice or four times daily, on pattern of methane emission by lactating dairy cows and developed a response function based on exponentials to describe the observed patterns of methane emission. The function describes an asymmetrical shape exhibiting a continuous rise to a peak followed by a period of linear decline. There were differences between treatments in terms of total methane output and the pattern of emission, with peaks observed following feedings. The rate of decline in methane production post-prandially was linked to amount of dry matter consumed following each feeding. The simple model fitted the data satisfactorily and provides a biological description for fluctuations in methane release in response to changes in feeding pattern. The response function could be applied more widely as part of methane emission inventories following further work to examine the differences in eating behaviour and methane emission across different production systems.

Keywords: methane, diurnal pattern, feeding frequency, dairy cows

Introduction

Ruminant agriculture in the UK and Europe is responsible for the production of a broad range of high value food products and it represents a mainstay of many rural communities. However, the inefficient utilisation of dietary nutrients by ruminants results in the excretion of a variety of pollutants, with adverse environmental consequences. At the same time, nutrients lost to the environment do not appear in saleable product and therefore, represent a significant waste of money to producers. Nutrition encompasses not only the range of nutrients available, but also their form, availability and delivery. Practical constraints including feedstuff availability and labour and equipment requirements further impact on nutrition management at the farm level. Fortunately, the very factors that give rise to this complexity and confusion (feed availability, variation between feedstuffs, feeding frequency, etc.) provide opportunities to improve the existing nutritional management on farm.

Methane from enteric fermentation is one of the principal greenhouse gases from ruminant agriculture and it also represents an energy loss that reduces feed conversion efficiency. For a broad range of diets, the quantity of methane released is closely related to the amount of dry matter consumed with further variation according to nutrient composition (Blaxter and Clapperton, 1965; Mills *et al.*, 2003). However, previous research has not established whether factors such as feeding frequency affect methane output for a given level of intake. Therefore, the aim of this study was to assess fluctuations in methane emission in response to changes in diurnal feeding pattern and develop a mathematical model for use in predicting methane release in response to feeding.

Materials and methods

Four rumen-fistulated Holstein-Friesian dairy cows in mid-lactation were fed *ad libitum* a total mixed ration of forage and a cereal based concentrate in a 50:50 ratio on a dry matter basis. The forage was a mixture of maize and grass silage in the ratio of 75:25 on a dry matter basis. The three treatments were the addition of 2 kg of a rumen-protected protein supplement (PS) (Amino Green, SCA, NuTec) to the rumen in combination with different feeding frequency during the day. The daily feed allowance for each animal was fed as one, two or four equal feedings daily. The protein supplement was added directly into the rumen prior to the morning feed using the established fistulae to reduce variation caused by palatability and rate of eating. Cows were randomly assigned to treatments in a 4×4 balanced Latin Square design with 4 week periods. Measurements of respiratory exchange, including methane production were obtained over four days during the final week of each period when cows were housed in open-circuit respiration chambers.

Results

The daily mean results for feed intake, milk production and methane emission are shown in Table 1. Dry matter intake tended to increase as the number of daily feed provisions increased. Intakes ranged from 18.06 to 19.56 kg/d between one and four feed provisions daily. There was a trend for milk yield to increase with increased feeding frequency, ranging from 28.0 to 29.4 kg/day. The addition of the protein supplement caused an increase in milk yield compared to the control. There was no significant change in milk composition in response to either feeding frequency or rumen-protected protein (data not shown).

Daily methane emission increased from 524 to 632 l/d as a result of increased dry matter intake with greater feeding frequency, but this was not significant (NS). After correcting for differences in intake (assuming a constant rate of methane production per kg for this specific diet), a decrease in methane of 35 l/d was attributable to reducing feeding frequency from twice to once daily (NS). Compared with more frequent feeding patterns, once daily feeding was associated with a 6.5% decreases in methane per kg of dry matter intake and methane per kg milk yield (NS).

Response functions

The mean diurnal patterns of methane production for the four treatments are shown in Figure 1. The results demonstrate a clear relationship between methane emission and feed consumption and microbial fermentation within the rumen. Methane production increases after each feeding and

Table 1. Effect of feeding frequency on dry matter intake, milk yield and methane emission in lactating dairy cows.

Diet	Control	PS diet	PS diet	PS diet	SED
Feeding pattern	2 × daily	1 × daily	2 × daily	4 × daily	
Protein supplement (kg DM)	0	1.75	1.75	1.75	
DMI (incl. PS) (kg/d)	18.63	18.06	19.23	19.56	0.51
Milk yield (kg/d)	26.6	28.0	29.4	28.3	0.86
Methane (l/d)	571	524	594	633	20.6
Methane (l/kg DMI)	30.6	29.1	31.0	32.4	0.96
Methane (l/kg milk)	22.2	20.2	21.5	23.0	0.84

DM: dry matter, DMI: dry matter intake, SED: standard error of the difference, PS: protein supplement.

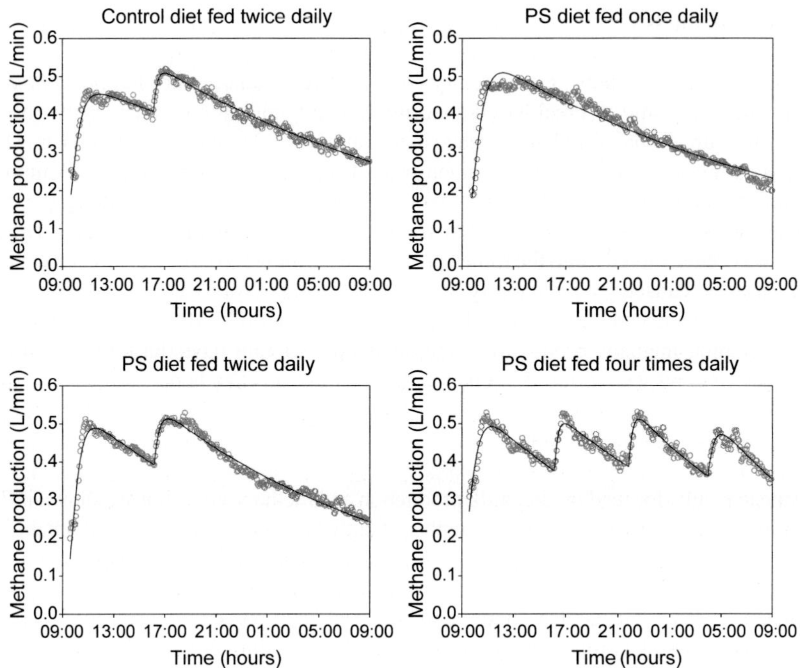

Figure 1. Effect of feeding frequency on diurnal patterns of methane emission in lactating dairy cows. (Observations are grey open circles; solid lines are the fitted response function). PS: protein supplement.

the rate of methane emission closely matched the number of feed provisions. Response functions based on exponentials and the gamma function were examined to describe the observed relationship between the daily pattern of feed intake (x axis, min) and the rate of methane production (y axis, l/min) (Thornley and France, 2007). The following equation based on exponentials was chosen to describe the methane emission profiles:

$$\text{Methane emission rate} = a \exp\left[b\,(1 - e^{-ct}) / c - dt\right]$$

where t is the time from feeding and a, b, c and d (all > 0) are parameters that define the scale and shape of the emission profile post feeding. The function describes an asymmetrical shape exhibiting a continuous rise to a peak followed by a period of fairly linear decline, and it provides a statistically robust model that is easy to fit.

The rate of methane emission increased after each feeding ranging from 0.17 to 0.60 l/min, with output rising quickly towards a peak 45 to 140 min following the commencement of feeding, depending on the amount of feed offered. The pattern of methane production closely matched the number of feeds provided. Smaller feeds were associated with a more rapid rise to peak methane emission rate, when compared to larger feeds. The rate of decline in methane emission rate post-prandially appears to be linked to the amount of dry matter consumed at each feeding. When animals were fed once daily, methane production declined slowly throughout the day due to the large amount of organic matter available to the rumen microbes compared to animals fed four times daily where the decrease in methane production is more rapid.

The response function was fitted to the mean data for each treatment and parameter estimates obtained for each feed provision across the treatments. Estimated parameters were then used to calculate specific attributes describing the methane emission profile as shown in Table 2. The response function fitted the data satisfactorily as evidenced by the R^2 values.

Table 2 highlights the differences in post feeding methane emission between the four treatments. Despite the daily feed allowance being divided equally for two and four times daily feed provision, there exist clear differences in the pattern of methane production related to each feeding.

Discussion

Methane emission is known to be linked to intakes of dry matter and more specifically digestible organic matter, with further variation according to nutrient composition. However, the decrease in methane production observed in this study when changing to once daily feeding from twice daily feeding was greater than could be accounted for by reduced intake alone given a consistent composition. Once daily feeding may have altered rumen VFA proportions through changes in the rate and quantity of organic matter fermented. The link between rumen VFA stoichiometry and methane output is well known and has formed the basis of mechanistic models of methanogensis (Baldwin *et al.*, 1987). Previous *in vivo* research has shown a tendency for the VFA profile to alter in response to feeding frequency (Sutton *et al.*, 1986), but the effects across several studies have been inconsistent. The proportion of propionic acid has been reported to increase (Knox and Ward, 1961), decrease (Kaufmann *et al.*, 1975; Jensen and Wolstrup, 1977; French *et al.*, 1990; Soto-Navarro *et al.*, 2000) or remain unchanged (Bath and Rook, 1963; Jorgensen *et al.*, 1965; Robles *et al.*, 2007) in response to more frequent feeding. Once daily feeding was associated with numerical reductions in methane per kg of dry matter intake and methane per litre of liquid milk. The simple response function model fitted the data satisfactorily and provides a biological description for fluctuations in methane emission in response to changes in feeding pattern. The response function supports the established relationship between daily dry matter intake and methane production reported previously (Mills *et al.*, 2003), whilst emphasising feeding regime as a potential source of variation in methane production (Beauchemin *et al.*, 2008; Martin *et al.*, 2010).

Table 2. Estimated methane curve characteristics after fitting the response function to each diurnal pattern.

Diet	Feeding time (h)	Initial CH_4 (l/min)†	Time to peak (min)†	Peak CH_4 (l/min)†	Rate of decline (/min)†	Total CH_4 (l/feed)†	R^2
Con × 2	10:00	0.19	133	0.45	-0.0005	157	0.889
	16:00	0.38	55	0.51	-0.0007	357	0.976
PS × 1	10:00	0.18	141	0.51	-0.0007	502	0.959
PS × 2	10:00	0.14	117	0.49	-0.0009	167	0.911
	16:00	0.40	65	0.51	-0.0008	367	0.976
PS × 4	10:00	0.27	102	0.49	-0.0011	168	0.853
	16:00	0.36	48	0.50	-0.0009	149	0.847
	22:00	0.39	47	0.51	-0.0011	159	0.937
	04:00	0.36	60	0.47	-0.0014	123	0.863

†The appropriate formula is given in Thornley and France (2007), section 16.2.3.
PS: protein supplement.

The function could be linked with models of intake to further refine assessment of methane release from livestock for national greenhouse gas inventories. The model also has potential to assess the contribution of intake behaviour relative to nutrient profile in the observed difference in methane emission, thereby shedding light on the as yet unanswered questions concerning the relative importance of animal and diet based factors in determining methane release. The model may also be useful in describing the underlying variation observed between methane output from grazing animals and those confined within housed production systems and fed conserved forages.

Acknowledgements

The financial support of UK Defra project LS3656 and the Canada Research Chairs Program are gratefully acknowledged.

References

Baldwin, R.L., Thornley, J.H.M. and Beever, D.E., 1987. Metabolism of the lactating cow. II. Digestive elements of a mechanistic model. Journal of Dairy Research 54:107-131.

Bath, I.H. and Rook, J.A.F., 1963. The evaluation of cattle foods and diets in terms of the ruminal concentration of volatile fatty acids. I. The effects of level of intake, frequency of feeding, the ratio of hay to concentrates in the diet, and of supplementary feeds. Journal of Agricultural Science, Cambridge 61:341-348.

Beauchemin, K.A., Kreuzer, M., O'Mara, F. and McAllister, T.A., 2008. Nutritional management for enteric methane abatement. Australian Journal of Experimental Agriculture 48:21-27.

Blaxter, K.L. and Clapperton, J.L., 1965. Prediction of the amount of methane produced by ruminants. British Journal of Nutrition 19:511-522.

French, N. and Kennelly, J.J., 1990. Effects of feeding frequency on ruminal parameters, plasma insulin, milk yield, and milk composition in Holstein heifers. Journal of Dairy Science 73:1857-1863.

Jensen, K. and Wolstrup, J., 1977. Effect of feeding frequency on fermentation pattern and microbial activity in the bovine rumen. Acta Veterinaria Scandinavica 18:108-121.

Jorgensen, N.A., Schultz, L.H. and Barr, G.R., 1965. Factors influencing milk fat depression on rations high in concentrates. Journal of Dairy Science 48:1031-1039.

Kaufmann, W., Rohr, K., Daenicke, R. and Hagemeister, H., 1975. Experiments on the influence of the frequency of feeding on rumen fermentation, food intake and milk yield Sonderheft der Berichte iiber Landwirtschaft 191:269-295.

Knox, K.L. and Ward, G.M., 1961. Rumen concentrations of volatile fatty acids as affected by feeding frequency. Journal of Dairy Science 44:1550-1553.

Martin, C., Morgavi, D.P. aand Doreau, M., 2010. Methane mitigation in ruminants: from microbe to the farm scale. Animal 4:351-365.

Mills, J.A.N., Kebreab, E., Yates, C.M., Crompton, L.A., Cammell, S.B., Dhanoa, M.S., Agnew, R.E. and France, J., 2003. Alternative approaches to predicting methane emissions from dairy cows. Journal of Animal Science 81:3141-3150.

Robles, V., González, L.A., Ferret, A., Manteca, X. and Calsamiglia, S., 2007. Effects of feeding frequency on intake, ruminal fermentation, and feeding behaviour in heifers fed high-concentrate diets. Journal of Animal Science 85:2538-2547.

Soto-Navarro, S.A., Krehbiel, C.R., Duff, G.C., Galyean, M.L., Brown, M.S. and Steiner, R.L., 2000. Influence of feed intake fluctuation and frequency of feeding on nutrient of digestion, digesta kinetics, and ruminal fermentation profiles in limit-fed steers. Journal of Animal Science 78:2215-2222.

Sutton, J.D., Hart, I.C., Broster, W.H., Elliott, R.J. and Schuller, E., 1986. Feeding frequency for lactating cows: effects on rumen fermentation and blood metabolites and hormones. British Journal of Nutrition 56:181-192.

Thornley, J.H.M. and France, J., 2007. Mathematical Models in Agriculture, revised 2nd Edition. CAB International, Wallingford, UK, xvii + 906pp.

Prediction of methane production in beef cattle within a mechanistic digestion model

J.L. Ellis[1], J. Dijkstra[2], E. Kebreab[3], S. Archibeque[4], J. France[1] and A. Bannink[5]
[1]Centre for Nutrition Modelling, Department of Animal and Poultry Science, University of Guelph, Guelph, ON, Canada; jellis@uoguelph.ca
[2]Animal Nutrition Group, Wageningen University, P.O. Box 338, 6700 AH Wageningen, the Netherlands
[3]Department of Animal Science, University of California, Davis, CA 95616, USA
[4]Animal Sciences, Colorado State University, Fort Collins, CO, USA
[5]Livestock Research, Wageningen UR, P.O. Box 65, 8200 AB Lelystad, the Netherlands

Abstract

Methane is produced by ruminants as the result of microbial digestion, it represents an energy loss to the animal, and it is also a potent greenhouse gas. Mechanistic modelling can lend insight into dietary strategies aimed at reducing methane emissions from cattle, but require proper representation of aspects of underlying rumen fermentation and digestion. Proper prediction of the production of volatile fatty acids (VFA) is central to accurate methane prediction. This study evaluated the newly updated and expanded model of VFA dynamics developed by Bannink *et al.* (2008), in comparison to a previous model version (Bannink *et al.*, 2006), within a rumen model (Dijkstra *et al.*, 1992; modified by Mills *et al.*, 2001) for its methane prediction ability in beef cattle fed high-grain diets. In an evaluation of the rumen model performed by Kebreab *et al.* (2008) using the Bannink *et al.* (2006) VFA stoichiometry, the model performed well on dairy cow data, but poorly on beef cattle data in predicting methane emission. Several modifications were therefore made to the model to adapt it for typical high-grain beef cattle diets and then evaluated for its accuracy to predict methane emissions from feedlot cattle. Passage rate of protozoa was increased proportionally with the grain percent of the diet, and the protozoal proportion of the amylolytic microbial pool was reduced accordingly. This allowed a small cellulolytic microbial pool to remain in the rumen, where it previously went extinct. Also, stoichiometry of rumen VFA production was adjusted for the use of monensin in the observed data. Preliminary results showed that while the representation of some central aspects of rumen fermentation probably improved, the model over-predicted methane production with a root mean square prediction error value of 2.86 MJ/d, with 56% of that error coming from bias and 44% from random sources. Concordance correlation coefficient value was 0.194. Results indicate that other areas of the model require improvement for predicting methane production accurately in high grain feedlot diets, such as hind gut fibre digestibility.

Keywords: beef cattle, high grain, monensin, methane, modelling

Introduction

Central to the accurate prediction of enteric methane (CH_4) production, within dynamic mechanistic models of rumen fermentation and digestion, is accurate representation of volatile fatty acid (VFA) stoichiometry (Bannink *et al.*, 2008; Ellis *et al.*, 2008). Accurate CH_4 prediction is closely tied to VFA stoichiometry, as hydrogen, the main substrate for methanogens, is produced with acetate and butyrate formation, and utilized with propionate and valerate production. To date, accurate representation of VFA dynamics represents a significant weakness in models of rumen function (Dijkstra *et al.*, 2008). Methane is of considerable interest because of its impact on the environment and large contribution to total farm greenhouse gas emissions (e.g. see Beauchemin *et al.*, 2009), and improved representation of VFA dynamics should improve prediction of CH_4 in these models across various diets.

In an evaluation of the Dijkstra *et al.* (1992) rumen model (as modified by Mills *et al.* (2001) and Bannink *et al.* (2006)) performed by Kebreab *et al.* (2008), the model performed well on dairy cow data, but poorly on beef cattle data in predicting CH_4 production. Results on the dairy data showed the model performed best in terms of mean square prediction error and concordance correlation coefficient values (14.5% and 0.713, respectively) when compared to the other models (MOLLY, 2007; Moe and Tyrrell, 1979; IPCC, 2006), but the model performed worst among the models when tested on the beef cattle data. Results showed a tendency for the model to over-predict CH_4 production on the beef cattle data (μ value of -0.35), with a CCC value of 0.160 and a RMSPE value of 53.6 percent of the observed mean (Kebreab *et al.*, 2008).

In an attempt to advance VFA representation in the model, Bannink *et al.* (2008) developed a mechanistic description of VFA handling in the rumen, including pH-dependent stoichiometric parameters for rapidly fermentable carbohydrate and starch. Since typical high grain beef cattle feedlot diets are high in fermentable carbohydrate and starch and tend to be associated with lower rumen pH, the hypothesis of this study was that CH_4 predictions by the Dijkstra model would improve on the Kebreab *et al.* (2008) beef data if the model was used with the updated Bannink *et al.* (2008) pH dependent VFA stoichiometry. Some preliminary results will be presented.

Materials and methods

Database

The database compiled for this study consisted of 64 individual animal observations previously used in model evaluation by Kebreab *et al.* (2008) and originally published by Archibeque *et al.* (2006; 2007a,b,c). The range of dietary variables is summarized in Table 1. Datasets included information on CH_4 production, bodyweight, chemical composition of the diet and supplementation. All animals were treated with monensin (Elanco Animal Health, Greenfield, IN, USA). In the original publications, observed CH_4 was measured using indirect calorimetry for 6 h and scaled up to a 24 h basis.

Table 1. Summary of database across studies.

Variable	n	Mean	Median	SD	Minimum value	Maximum value
DMI (kg/d)	64	6.88	6.86	0.982	4.32	9.07
Forage (% of diet)	64	10	10	0.0	10	10
BW (kg)	64	384	368	61.3	280	526
CH_4 (MJ/d)	64	4.56	4.65	2.168	0.87	10.22
CH_4 (% GEI)	64	3.42	3.44	1.484	0.81	5.92
NDF (g/kg DM)	64	133	133	4.4	128	141
Starch (g/kg DM)	64	614	615	20.8	578	642
Water soluble carbohydrate (g/kg DM)	64	42	46	12.8	24.7	61.4
Protein (g/kg DM)	64	133	135	16.3	97	155
Monensin (mg/kg DM)	64	15.6	17.6	3.36	9.90	17.6

Simulation model

The model version evaluated in this study consisted of the Dijkstra *et al.* (1992) rumen model as later modified by Mills *et al.* (2001) to include prediction of CH_4 production and representation of hindgut fermentation and digestion, and more recently by Bannink *et al.* (2008) to include pH dependent VFA stoichiometry coefficients for soluble carbohydrate and starch. This integrative model is based on a series of dynamic, deterministic differential equations developed on dairy cow data.

Model evaluation

The CH_4 predictions were evaluated using two methods. Firstly, mean square prediction error (MSPE), calculated as:

$$\text{MSPE} = \sum_{i=1}^{n} (O_i - P_i)^2 / n \qquad (1)$$

where n is the total number of observations, O_i is the observed value, and P_i is the predicted value. Square root of the MSPE (RMSPE), expressed as a percentage of the observed mean, gives an estimate of the overall prediction error. The RMSPE can be decomposed into error due to overall bias (ECT), error due to deviation of the regression slope from unity (ER) and error due to the disturbance (random error) (ED) (Bibby and Toutenburg, 1977).

Secondly, concordance correlation coefficient analysis (CCC) was performed (Lin, 1989), where CCC is calculated as:

$$\text{CCC} = R \times C_b \qquad (2)$$

where R is the Pearson correlation coefficient and C_b is a bias correction factor.

The R variable gives a measure of precision, while C_b is a measure of accuracy. The C_b variable is calculated as:

$$C_b = \frac{2}{[v + 1/v + \mu^2]} \qquad (3)$$

where v provides a measure of scale shift, and μ provides a measure of location shift.

The v value indicates the change in standard deviation, if any, between predicted and observed values. A positive μ value indicates under-prediction, while a negative μ indicates over-prediction.

Results and discussion

Fibrolytic microbial mass pool

Initial runs of the model revealed major problems maintaining the fibrolytic microbial mass (Mf) pool within the rumen on high grain low forage beef cattle diets. Examination of the model inputs showed that the Mf pool was zero because of a small dietary fermentable fibre pool (Ff), and a high fractional degradation rate (K_d) value associated with the dietary fibre. However, Mf bacteria in the large intestine survived this diet, which suggests the reason for their disappearance in the rumen may lie in differences between rumen and large intestine representation in the model. The rumen and large intestine sub-models are identical in degradation and growth equations, but differ in volume, passage rate, pH parameters and the presence of protozoa in the rumen.

Predicted fractional rumen solid and liquid passage rates by the model were 0.68 (±0.002) and 1.77 (±0.014) /d, respectively. Predicted rumen pH averaged 6.41 (±0.012), where the average minimum pH was 6.01 (±0.011), and the time spent below critical pH of 6.3 averaged 8.82 h/d (±0.180). The model predicts that hydrolysis of Ff by fibrolytic microbes is inhibited at pH less than 6.3. The average predicted hind gut pH was 6.99 (±0.011) and time below critical pH was 5.13 h/d (±0.227). Even when the model was hypothetically modified to have time below critical pH to be zero, the predicted Mf pool remained zero.

Within the model, the rumen protozoal mass is calculated as a fraction of the mass of amylolytic micro-organisms (Ma). Protozoa are represented to predate on Mf bacteria, and hence their predicted mass has a strong impact on the predicted survival of the Mf. Of interest was the fractional passage rate of protozoa from the rumen. Not all studies support the premise that defaunation on high concentrate diets is the result of low rumen pH (Slyter *et al.*, 1970; Lyle *et al.*, 1981; Towne *et al.*, 1990). Instead we propose that reduced protozoal numbers in the rumen with high concentrate diets may be the result of decreased bulk in the rumen and therefore decreased places to attach and be selectively retained in the rumen.

Based on this premise, an equation was developed to predict protozoal passage rate (K_{popa}; /d) based on solid passage rate and the roughage percent of the diet, if roughage was less than 50% of the diet:

$$K_{popa} = (1.0 - (0.5 + (1.0 \times (R_P - 50))/100)) \times K_{sopa} \quad (4)$$

where K_{sopa} (/d) is fractional solid passage rate according to Mills *et al.* (2001), and R_P is roughage percent. At roughage percentages greater than 50, K_{popa} was $0.5 \times K_{sopa}$. The fraction of the Ma bacterial pool that was protozoa then decreased in accordance with K_{popa}. This modification to the model allowed the rumen to maintain a small Mf pool.

Monensin

Not addressed in previous evaluations on these data was the feeding of monensin in all diets of the evaluation database. Monensin shifts the VFA profile of the rumen towards propionate production and away from acetate production (e.g. see Nagaraja *et al.*, 1997), and this shift is not accountable for in the structure of the Dijkstra model. Not accounting for monensin supplementation would lead to over-estimation of CH_4 production, as available hydrogen is utilized with the production of propionate, making less hydrogen available for methanogenesis. This may well be part of the reason CH_4 was over-estimated by the Dijkstra model in the study by Kebreab *et al.* (2008). Nagaraja *et al.* (1997) showed that the magnitude of the shift in VFA profile differed for high forage (70%) versus medium forage (50%) diets. Therefore, a meta-analysis of the literature for low forage diets was conducted here, to determine the magnitude in VFA profile shift that would be expected from treatment with monensin in our database. A database of 24 treatment means from 9 studies was compiled where roughage was less than 20% of the diet, and consisted of studies from Surber and Bowman (1998), Randall *et al.* (1978), Richardson *et al.* (1976), Zinn *et al.* (1994), Zinn and Borques (1993), Ralston and Davidson (1976), Fandino *et al.* (2008), Thorton and Owens (1981) and Meyer *et al.* (2009). The database was analysed in PROC MIXED of SAS (SAS, 2001), treating study as a random effect. The range of monensin feeding levels in this data ranged from 0.0 to 79.0 mg/kg DM. The following equations were produced in SAS and applied to the VFA yield equations within the model:

Acetate (% change) = Monensin (mg/kg DM) × (-0.163±0.0340) (P<0.0001, RMSE = 3.36) (5)
Propionate (% change) = Monensin (mg/kg DM) × (0.560±0.1199) (P=0.0001, RMSE = 38.8) (6)
Butyrate (% change) = Monensin (mg/kg DM) × (-0.320±0.1598) (P=0.057, RMSE = 49.1) (7)

where values are percent change in the original VFA stoichiometry in the model.

VFA prediction

Predicted VFA proportions by the model were 55.5(±0.66)% acetate, 31.8(±0.90)% propionate, 10.0(±0.25)% butyrate and 2.70(±0.170)% valerate, averaged across all datasets. No observed values were available to compare these values against. Beauchemin and McGinn (2005) fed a similar corn grain based diet, consisting of 81.4% dry rolled corn and a monensin feeding level of 33 mg/kg DM, and had VFA molar proportions of 43.6% acetate, 44.3% propionate, 7.7% butyrate and 2.25% valerate. However, observed rumen pH was lower in the Beauchemin and McGinn (2005) trial, at 5.69, and they fed a higher level of monensin than in the observed data used here. Devant *et al.* (2001) fed an 88% concentrate corn and barley grain based diet, with no monensin, and reported VFA molar proportions of 45.1% acetate, 46.2% propionate, 5.9% butyrate and 1.1% valerate. These limited data suggest that the model may still over-estimate acetate and butyrate %, and under-estimate propionate % in the rumen with these diets.

CH_4 prediction

Predicted versus observed CH_4, accounting for the above two modifications, is presented in Figure 1 and results of MSPE and CCC analysis are presented in Table 2. Results still show a tendency for CH_4 to be over-predicted by the model, particularly at very low observed CH_4 values. The observed database contains several extremely low observed CH_4 production values, of less than 1.5% GE intake per day, on the same diets that produced CH_4 values of 3-5% GE intake per day with other animals. While values as low as 2% GE intake per day are typically observed with high grain fed beef cattle, some of this low end variation may be an artefact of the short time period for which CH_4 was measured (6 hours), and how that time period related to timing of the last meal. A more reliable comparison could have been made had CH_4 been measured for 24 hours. As is, the model seems unable to predict such low CH_4 production levels for the diets specified. This may indicate

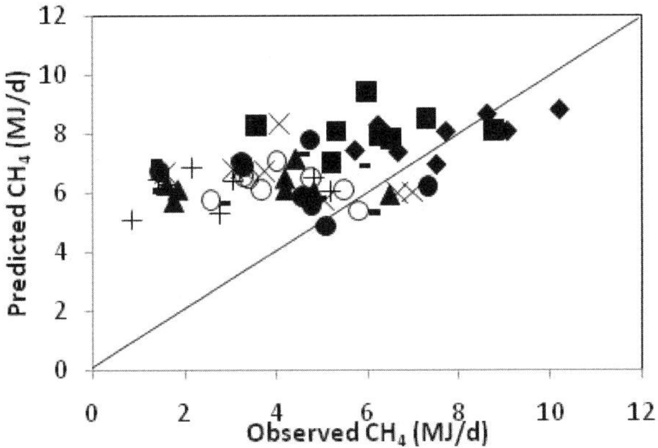

Figure 1. Predicted versus observed CH_4 (MJ/d) where the predicted values have been modified to include the effect of roughage percent on protozoal passage, and to include the effect of monensin. Points represent individual animals and different symbols represent treatments from different experiments.

Table 2. Summary of evaluation statistics for the revised model.

Model	New VFA and Protozoa[1]	New VFA, Protozoa and Monensin[2]
Predicted mean (MJ/d)[3]	7.20 (±1.01)	6.75 (±1.02)
RMSPE[4]	3.23	2.86
RMSPE %[5]	70.2	62.1
ECT %[6]	64.2	56.2
ER %[7]	0.11	0.01
ED %[8]	35.7	43.8
CCC[9]	0.149	0.194
R[10]	0.422	0.457
C_b[11]	0.353	0.425
v[12]	2.12	2.12
μ[13]	−1.75	−1.46

[1] Results when the updated VFA stoichiometry (Bannink *et al.*, 2008) and modification to protozoal fractional passage rate are included in the model.
[2] Results when the updated VFA stoichiometry (Bannink *et al.*, 2008), modification to protozoal fractional passage rate and monensin correction are included in the model.
[3] Where values are expressed as mean ± SD, and the observed CH_4 (MJ/d) mean is 4.56 (±2.17).
[4] Root mean square prediction error (MJ/d)
[5] Root mean square prediction error expressed as a percentage of the observed mean.
[6] Error due to bias, as a % of total MSPE.
[7] Error due to regression, as a % of total MSPE.
[8] Error due to disturbance, as a % of total MSPE.
[9] Condordance correlation coefficient, where CCC = $R \times C_b$.
[10] Pearson correlation coefficient.
[11] Bias correction factor.
[12] Scale shift.
[13] Location shift.

an inadequacy in the observed values, or that further modifications to the model are required to make it applicable to high concentrate beef cattle diets. As suggested above, VFA stoichiometry may need to be adjusted for these high grain beef cattle diets. As well, whole tract % digestibility of fibre in the model averages 65% while rumen fibre digestibility averages 35%, indicating an overestimation of fibre digestibility in the hind gut. This likely contributes to an overestimation of CH_4 production. The hind gut fermentation model developed by Mills *et al.* (2001) was based on the rumen model, but may need modifications to handle these high grain beef cattle diets. A lower hind gut fermentable fibre digestibility would lower predicted CH_4 values by the model, and may allow them to be more in line with observed values.

Results here represent an improvement over Kebreab *et al.* (2008) in terms of CCC analysis, but not MSPE analysis. In Kebreab *et al.* (2008), with no survival of rumen Mf, predicted CH_4 values were incorrectly lower than they should have been. Allowing survival of the Mf microbial pool increased predicted CH_4 in this study, and adjusting for monensin feeding lowered it slightly. In addition, while Kebreab *et al.* (2008) excluded some of the extremely low CH_4 observations from their analysis, all data points were retained in the current analysis. This is the cause of the lower observed CH_4 mean (4.56±2.17) in this study compared to Kebreab *et al.* (2008) (5.11±1.89), and contributes to the differing results.

Conclusions

The preliminary results indicate that model modifications were required to have the model running adequately with high concentrate beef cattle diets. Modification of protozoal passage rate and representation of the effects of monensin on VFA stoichiometry improved its functionality, but did not overly improve CH_4 predictions. Further modification of the model VFA stoichiometry may be required, and representation of hind gut fermentation may need to be improved.

References

Archibeque, S.L., Miller, D.N., Freetly, H.C. and Ferrell, C.L., 2006. Feeding high-moisture corn instead of dry-rolled corn reduces odorous compound production in manure of finishing beef cattle without decreasing performance. Journal of Animal Science 84:1767-1777.

Archibeque, S. L., Miller, D.N., Freetly, H.C., Berry, E.D. and Ferrell, C.L., 2007a. The influence of oscillating dietary protein concentrations on finishing cattle. I. Feedlot performance and odorous compound production. Journal of Animal Science 85:1487-1495.

Archibeque, S.L, Freetly, H.C., Cole, N.A. and Ferrell, C.L., 2007b. The influence of oscillating dietary protein concentrations on finishing cattle. II. Nutrient retention and ammonia emissions. Journal of Animal Science 85:1496-1503.

Archibeque, S.L, Miller, D.N., Parker, D.B., Freetly, H.C. and Ferrell, C.L., 2007c. Effects of feeding steam-rolled corn in lieu of dry-rolled corn on the production of odorous compounds in finishing beef steer manure. In: ASABE, Proceedings of International Symposium on Air Quality and Waste Management for Agriculture, June 17-20, 2007. Minneapolis, MN, USA,

Baldwin, R.L., 1995. Modeling Ruminant Digestion and Metabolism. Chapman & Hall, London, UK.

Bannink, A., Kogut, J., Dijkstra, J., France, J., Kebreab, E., Van Vuuren, A.M. and Tamminga, S., 2006. Estimation of the stoichiometry of volatile fatty acid production in the rumen of lactating cows. Journal of Theoretical Biology 238:36-51.

Bannink, A., France, J., Lopez, S., Gerrits, W.J.J., Kebreab, E., Tamminga, S. and Dijkstra, J., 2008. Modelling the implications of feeding strategy on rumen fermentation and functioning of the rumen wall. Animal Feed Science and Technology 143:3-26.

Beauchemin, K.A., McAllister, T.A. and McGinn, S.M., 2009. Dietary mitigation of enteric methane from cattle. CAB Reviews: Perspectives in Agriculture, Veterinary Science, Nutrition and Natural Resources 4, No. 035.

Beauchemin, K.A. and McGinn, S.M., 2005. Methane emissions from feedlot cattle fed barley or corn diets. Journal of Animal Science 83:653-661.

Bibby J. and Toutenburg, T., 1977. Prediction and Improved Estimation in Linear Models. John Wiley & Sons, Chichester, UK.

Devant, M., Ferret, A., Calsamiglia, S., Casals, R. and Gasa, J., 2001. Effect of nitrogen source in high-concentrate, low-protein beef cattle diets on microbial fermentation studied *in vivo* and *in vitro*. Journal of Animal Science 79:1944-1953.

Dijkstra, J., Kebreab, E., France, J. and Bannink, A., 2008. Modelling protozoal metabolism and VFA production in the rumen. In: France, J. and Kebreab, E. (eds.) Mathematical modelling in animal nutrition. CABI Publishing, Wallingford, UK, pp.170-188.

Dijkstra, J., Neal., D. St. C., Beever, D.E. and France. J., 1992. Simulation of nutrient digestion, absorption and outflow in the rumen: Model description. Journal of Nutrition 122:2239-2255.

Ellis, J. L., Dijkstra, J., Kebreab, E., Bannink, A., Odongo, N. E., McBride, B. W. and France, J., 2008. Aspects of rumen microbiology central to mechanistic modelling of methane production in cattle. Journal of Agricultural Science 146:213-233.

Fandio, I., Calsamiglia, S., Ferret, A. and Blanch, M., 2008. Anise and capsicum as alternatives to monensin to modify rumen fermentation in beef heifers fed a high concentrate diet. Animal Feed Science and Technology 145:409-417.

IPCC (Intergovernmental Panel on Climate Change), 2006. IPCC Guidelines for National Greenhouse gas Inventories. IGES, Hayama, Kanagawa, Japan. Available at: http://www.ipcc-nggip.iges.or.jp/ Accessed Nov. 3, 2009.

Kebreab, E., Johnson, K.A., Archibeque, S.L., Pape, D. and Wirth, T., 2008. Model for estimating enteric methane emissions from United States dairy and feedlot cattle. Journal of Animal Science 86:2738-2748.

Lin, L.I.K., 1989. A concordance correlation coefficient to evaluate reproducibility. Biometrics, 45:255-268.

Lyle, R. R., Johnson, R.R. and Backus, W.R., 1981. Ruminal characteristics as affected by monensin, type of protein supplement and proportions of whole wheat and corn in forage-free diets fed to finishing steers. Journal of Animal Science 53:1377-1382.

Meyer, N.F., Erickson, G.E., Klopfenstein, T.J., Greenquist, M.A., Luebbe, M.K., Williams, P. and Engstrom, M.A., 2009. Effect of essential oils, tylosin, and monensin on finishing steer performance, carcass characteristics, liver abscesses, ruminal fermentation, and digestibility. Journal of Animal Science 87:2346-2354.

Mills, J.A.N., Dijkstra, J., Bannink, A., Cammell, S.B., Kebreab. E. and France, J., 2001. A mechanistic model of whole-tract digestion and methanogenesis in the lactating dairy cow: Model development, evaluation and application. Journal of Animal Science 79:1584-1597.

Moe, P.W. and Tyrrell, H.F., 1979. Methane production in dairy cows. Journal of Dairy Science 62:1583-1586.

MOLLY. 2007. Version 3.0. Available at: http://animalscience.ucdavis.edu/research/molly Accessed Nov. 3, 2009.

Nagaraja, T.G., Newbold, C.J., Van Nevel, C.J. and Demeyer, D.I., 1997. Manipulation of ruminal fermentation. In: Hobson, P.N. and Stewart, C.S. (eds.). The Rumen Microbial Ecosystem. Chapman & Hall, New York, NY, USA, pp.523-632.

Richardson, L.F., Raun, A.P., Potter, E.L., Cooley, C.O. and Rathmacher, R.P., 1976. Effect of monensin on rumen fermentation *in vitro* and *in vivo*. Journal of Animal Science 43:657-664.

Ralston, A.T. and Davidson, T.P., 1976. The Effect of Varying Levels of Monensin in Finishing Rations for Beef Cattle. Special Report 452. Agriculture Experiment Station, Oregon State University, Corvallis, OR, USA.

SAS User's Guide: Statistics, 2000. SAS Inst. Incorporated., Cary, NC, USA.

Surber, L.M. and Bowman, J.G., 1998. Monensin effects on digestion of corn or barley high-concentrate diets. Journal of Animal Science 76:1945-1954.

Slyter, L.L, Oltjen, R.R., Kern, D.L. and Blank, F.C., 1970. Influence of type and level of grain and diethylstilbestrol on the rumen microbial population of steers fed all-concentrate diets. Journal of Animal Science 31:996-1002.

Thorton, J.H. and Owens, F.N., 1981. Monensin supplementation and *in vivo* methane production by steers. Journal of Animal Science 52:628-634.

Towne, G., Nagaraja, T.G., Brandt, R.T. and Kemp, K.E., 1990. Ruminal ciliated protozoa in cattle fed finishing diets with or without supplemental fat. Journal of Animal Science 68:2150-2155.

Van Maanen, R.W., Herbein, J.H., McGilliard, A.D. and Young, J.W., 1978. Effect of monensin on *in vivo* rumen propionate production and blood glucose kinetics in cattle. Journal of Nutrition 108:1002-1007.

Zinn, R.A., Plascencia, A. and Barajas, R., 1994. Interaction of forage level and monensin in diets for feedlot cattle on growth performance and digestive function. Journal of Animal Science 72:2209-2215.

Zinn, R.A. and Borques, J.L., 1993. Influence of sodium bicarbonate and monensin on utilization of a fat-supplemented, high-energy growing-finishing diet by feedlot steers. Journal of Animal Science 71:18-25.

Relationship between passage rate and extrinsic diet characteristics derived from rumen evacuation studies performed with dairy cows

S. Krizsan[1], S. Ahvenjärvi[2], H. Volden[1] and P. Huhtanen[3]
[1]Department of Animal and Aquacultural Sciences, Norwegian University of Life Sciences, P.O. Box 5003, 1432 Ås, Norway; sophie.krizsan@njv.slu.se
[2]MTT-Agrifood Research Finland, Animal Production Research, 31600 Jokioinen, Finland
[3]Dept. of Agricultural Research for Northern Sweden, Swedish University of Agricultural Sciences, 90183 Umeå, Sweden

Abstract

A meta-analysis based on experiments with dairy cows was conducted to study the effects of extrinsic diet characteristics on passage rate (Kp) of indigestible neutral detergent fibre (iNDF). A data set was collected that included 108 dietary treatment means from 29 studies. Dietary treatments consisted of different forages supplemented with varying amounts of concentrate feed. Minimum prerequisite for an experiment to be included in the analysis was that Kp was calculated using the flux/compartmental pool method, and that live weight (LW), total and forage dry matter intake (DMI), neutral detergent fibre (NDF) and iNDF diet concentrations were given or could be estimated. Total DMI and intake of NDF (NDFI), proportion of concentrate in diet (CProp), and milk yield in the data varied from: 8.71 to 39.0 g/kg LW, 3.76 to 16.6 g/kg LW, 0 to 0.693, and 0 to 35 kg/d, respectively. A mixed model regression was used to analyse responses of fixed factors on Kp. Root mean square error (RMSE) for the models are given. The 3 best equations were: (1) Kp = 0.0159 + 0.000790 × NDFI (g/kg LW) (RMSE = 0.00295/h); (2) Kp = 0.0129 + 0.000755 × NDFI (g/kg LW) + 0.0189 × CProp (NDF basis) (RMSE = 0.0031/h); and (3) Kp = 0.00661 + 0.000938 × NDFI (g/kg LW) + 0.0220 × CProp (NDF basis) + 0.0114 × iNDF/NDF (RMSE = 0.00328/h). Equation 3 expressed on DM basis was: Kp = 0.00816 + 0.000400 × DMI (g/kg LW) + 0.00401 × CProp + 0.00908 × iNDF/NDF (RMSE = 0.00379/h). Model performance suggested that Kp is more closely related to NDFI than DMI. The positive coefficients for CProp on NDF basis and iNDF/NDF is related to the faster Kp of concentrate particles compared with forage particles and that Kp increases when NDF potential digestibility decreases.

Keywords: dairy cow, indigestible neutral detergent fibre, passage rate, rumen evacuation

Introduction

Ruminant animals are unique because of their special digestive system based on microbial degradation in the fore-stomach. The utilization of fibrous plant material is made possible by a long retention time of feed particulate matter in the rumen. Rates of degradation and passage of fibre are the main determinants of the level of rumen contents. Further, rates of degradation and passage are considered to be competitive processes that set the limits for ruminal digestibility. Prediction equations of passage rate (Kp) is used in calculations of ruminal digestibility of carbohydrate and protein fractions in one-compartment models and included in predictions of microbial efficiency (NRC, 2001; Fox *et al.*, 2004; Danfær *et al.*, 2006). Generally, Kp equations have been developed separately for forage and concentrate feed, and for predictions of ruminal liquid outflow. Most of the equations depend on feed intake determined on dry matter (DM) basis (NRC, 2001; Seo *et al.*, 2006). The equations have been developed based on large sets of empirical data using rare-earths or Cr-mordanted fibre as Kp markers. However, compartmental mean retention time estimated from the descending phase of marker excretion curves has been markedly shorter than the proportion of forestomachs to total mean retention time determined from lignin recovery in slaughter studies (Huhtanen and Ahvenjärvi, 2008). Huhtanen *et al.* (2006) suggested from a limited number of studies (N = 41) where Kp had been calculated for indigestible neutral detergent fibre (iNDF) using rumen evacuation technique that intake of neutral detergent fibre (NDFI) was a better predictor of Kp of iNDF than dry matter intake (DMI). Indigestible NDF is an internal marker and an ideal nutritional entity i.e. rate of digestion is zero and Kp would be determined accurately using flux/compartmental pool method.

Meta-analysis approaches are suitable to summarise findings across published studies and develop empirical predictive equations of biological responses (St-Pierre, 2001; Sauvant *et al*., 2008). The objective with this study was to analyse experimental data to generate prediction equations of Kp of iNDF by conducting a meta-analysis of data from studies performed with dairy cows.

Materials and methods

Data material and calculations

A database was constructed from experiments with dairy cows where the experimental objective was to study dietary effects on digestion kinetics, or where Kp of iNDF was related to the primary objective, using the flux/compartmental pool method. A total of 29 studies, using ruminally cannulated cows, comprising 108 treatment means were pooled in a European database. The list of studies used in this meta-analysis is given in Appendix 1. The diets in the experiments consisted of different types of forages (e.g. fresh grass, grass hay or silage, legume-, whole crop barley-, and maize silage) and different levels of concentrate feeding. The concentrate feeds fed with the different forages differed both in the amount and composition but were offered at fixed levels throughout a study. Nineteen experiments were targeted on type of forage, 6 studies dealt with different sources or levels of concentrate supplementation, and 4 studies had factorial arrangement of treatments including both forage type and level of concentrate supplementation. The prerequisite for an experiment to be included in the analysis was that Kp was calculated for iNDF, preferably based on flow of iNDF from intake (N = 95), else faecal output of iNDF (N = 13), and rumen pool size of iNDF. The equation used to calculate fractional rate of passage assuming a steady-state situation is given below (Equation 1). In one study Kp was determined based on the indigestible component of acid detergent fibre (N = 2) instead of iNDF. Further prerequisite for inclusion of an experiment was that production parameters (forage and total DMI, milk production and live weight (LW)), and diet chemical composition (concentrations of crude protein, neutral detergent fibre (NDF), and iNDF) were determined or could be estimated. When forage iNDF concentrations were not reported, the estimates were back-calculated according to Equation 1 using given numbers of Kp of iNDF, concentrate concentration of iNDF, concentrate and forage DMI and rumen iNDF pool size. If concentrate chemical composition was not reported, default feed table values from the country where the experiment was performed were used. Concentrate iNDF concentrations were based on either determined values (N = 104) or values derived from unpublished data sets from MTT Agrifood Research Finland (N = 4). Despite this, estimation of values restricted the number of treatment means included in bivariate and multivariate mixed model regressions to 106. In all experiments, even when not reported, iNDF concentration in feed or feed ingredients, rumen contents and faecal samples was determined by long time rumen incubations of the samples in nylon bags of a minimum of 96 h.

$$\text{Kp (Rate of passage; 1/h)} = \frac{\text{Flux of indigestible component into the compartment (kg/h)/}}{\text{Rumen pool of indigestible component (kg)}} \quad (1)$$

Statistical analysis

The relationships between Kp of iNDF and independent variables were analyzed using the mixed model procedure of SAS (Littell *et al.*, 1996). The model was $Y = B_0 + B_1X_{1ij} + b_0 + b_1X_{1ij} + B_2X_{2ij} + ... + B_nX_{nij} + e_{ij}$, where B_0, B_1X_{1ij}, B_2X_{2ij}, ..., X_{nij} are the fixed effects and b_0, b_1, and e_{ij} are the random experiment effects (intercept, slope and error), where i = 1, ... n studies and j = 1, ..., n_i values. In multivariate models only the first independent variable was treated as a random factor. Variation in the experimental units, experimental designs, objectives of the experiments, different measurement methods, and laboratory assays would contribute to the random effect of study in these regressions.

In the tables, root mean squared error (RMSE) values after adjusting for the random study effect are presented. Rationale and further details of using mixed model analysis to integrate quantitative findings from multiple studies are described by St-Pierre (2001) and Sauvant *et al.* (2008).

Results

The experimental data are described in Table 1. There was considerable variation in both diet composition and animal parameters. Grass silages of different maturities were clearly the predominant forage component of the diets, but dried forages, mixes of grass and legumes or pure legume silages were also frequently represented. Cereal grains, mainly barley, oats and wheat, were the main energy components of the concentrates. The most frequently used protein supplement was rapeseed meal. Variation in DMI was associated with proportion of concentrate in the diet (CProp) on DM basis (CProp (DM)) [DMI (g/kg LW) = 23.4 + 18.4 × CProp (DM); $P<0.0001$, $R^2 = 0.23$, RMSE = 5.7 g/kg LW]. Likewise, NDFI was positively related to CProp on NDF basis (CProp

Table 1. Description of the experimental animal and diet characteristics in the data base for prediction of passage rate of indigestible neutral detergent fibre in dairy cows[1].

Item	N	Mean	SD[2]	Minimum	Maximum
Intake (g/kg LW)					
DM	108	29.6	6.38	8.7	39.0
NDF	108	11.3	2.75	3.8	16.6
Forage DM	108	19.1	4.91	5.8	32.5
Forage NDF	106	9.3	2.39	3.0	16.0
Milk production (kg/d)	108	23.8	8.51	0	35.1
Live weight (kg)	108	608	35.4	545	686
Diet composition (g/kg of DM)					
NDF	108	389	81.3	184	661
CP	108	107	26.6	61.7	200
iNDF	108	87	34.8	24	174
Concentrate proportion (DM basis)	108	0.34	0.165	0	0.69
Concentrate proportion (NDF basis)	106	0.18	0.131	0	0.52
Passage rate of iNDF (1/h)	108	0.0245	0.00631	0.0120	0.0425

[1] LW: live weight, DM: dry matter, NDF: neutral detergent fibre, CP: crude protein, and iNDF: indigestible NDF.
[2] SD = standard deviation.

(NDF)) [NDFI (g/kg LW) = 10.4 + 5.29 × CProp (NDF); P=0.01, R^2 = 0.06, RMSE = 2.7 g/kg LW], but this relationships was not very strong.

The effects of extrinsic diet characteristics on Kp in simple and multiple linear regressions estimated by mixed model regression analysis are presented in Table 2. Intake of DM and NDF were closest associated with Kp of iNDF among the single linear regressions ($P \leq 0.02$; Table 2). Both parameters were positively related to Kp of iNDF with only a slightly higher precision in explained variation by NDFI (Figure 1) than DMI (Figure 2) as indicated by a lower adjusted RMSE and higher R^2. The mean response in Kp of iNDF to increased intake of NDF was more than twice as high as for increased intake of DM (0.000790 versus 0.000362/h per g intake of NDF and DM/kg LW, respectively). Neither concentrate DMI (expressed as g/kg LW) nor CProp (DM) had a significant effect ($P \geq 0.64$; Table 2) when included in bivariate models with DMI. However, including CProp on NDF basis (CProp (NDF)) in a bivariate model with NDFI showed clear significant effects of both first and second independent variables on Kp of iNDF ($P \leq 0.02$; Table 2). Segregating total intake to forage and concentrate intake on DM or NDF basis in bivariate models did not yield any further improvements. Including the dietary ratio of iNDF to NDF (iNDF/NDF) as third variable rendered the effect of NDFI as first independent variable to become significant (P=0.04; Table 2) when combined with concentrate intake of NDF. However, the multivariate model with NDFI, CProp (NDF) and iNDF/NDF explained Kp of iNDF even more precisely as was seen from a decrease in RMSE from 0.00337 to 0.00328/h.

The 3 best equations for Kp of iNDF generated from this data material were:

$$Kp = 0.0159 + 0.000790 \times NDFI\ (g/kg\ LW) \quad (2)$$
(RMSE = 0.00295/h and R^2 = 0.35)

$$Kp = 0.0129 + 0.000755 \times NDFI\ (g/kg\ LW) + 0.0189 \times CProp\ (NDF) \quad (3)$$
(RMSE = 0.00310/h and R^2 = 0.58)

$$Kp = 0.00661 + 0.000938 \times NDFI\ (g/kg\ LW) + 0.0220 \times CProp\ (NDF) + 0.0114\ iNDF/NDF \quad (4)$$
(RMSE = 0.00328/h and R^2 = 0.64)

Equation 4 expressed on DM basis (i.e. using DM based intake and proportion of concentrate in the diet) reduced the precision of the explained variation (RMSE = 0.00379/h and R^2=0.37) and did not provide significant ($P \geq 0.14$) association of neither the second or third independent variable to Kp of iNDF.

Discussion

Methodology of passage kinetics

Procedures used to estimate feed particle passage rates in ruminants have differed with regard to marker type, compartmental model, sampling site, and applied amount and particle size of labelled feed. These factors have contributed to ambiguous biological interpretations i.e. relative differences due to animal and diets have not been clearly separated from methodology effects. Trivalent metal ions (most often Cr) and rare earths (most often Yb) have been the most commonly used passage kinetic markers. Ideally external markers should be indigestible, associated with feed residues that are undigested, resemble the same flow characteristics as the labelled fraction and not separate from the labelled fraction throughout the digestive tract (Ellis *et al.*, 1994). Further, various mathematical models exist for the estimation of passage kinetics. Marker excretion curves from

Table 2. Effects of extrinsic diet characteristics on passage rate of indigestible neutral detergent fibre (1/h) estimated by mixed model regression analysis ($Y = A + BX_1 + CX_2 + DX_3$)[1].

X_1	X_2	X_3	A[2]	B	P	C	P	D	P	RMSE[3]	R^2
DMI			0.0138	0.000362	<0.01					0.00349	0.31
NDFI			0.0159	0.000790	0.02					0.00295	0.35
FDMI			0.0211	0.000218	0.30					0.00293	0.12
FNDFI			0.0195	0.000551	0.13					0.00286	0.18
DMI	CDMI		0.0139	0.000330	0.02	0.0000800	0.64			0.00352	0.32
DMI	CProp (DM)		0.0138	0.000364	<0.01	-0.000260	0.96			0.00350	0.31
NDFI	CNDFI		0.0160	0.000461	0.15	0.00173	<0.01			0.00312	0.57
NDFI	CProp (NDF)		0.0129	0.000755	0.02	0.0189	<0.01			0.00310	0.58
FDMI	CDMI		0.0155	0.000299	0.15	0.000383	0.03			0.00297	0.37
FNDFI	CNDFI		0.0150	0.000592	0.08	0.00199	<0.01			0.00301	0.56
NDFI	CProp (NDF)	NDF	0.0124	0.000727	0.03	0.0194	<0.01	1.83×10^{-6}	0.82	0.00312	0.58
NDFI	CProp (NDF)	INDF/NDF	0.00661	0.000938	<0.01	0.0220	<0.01	0.0114	0.04	0.00328	0.64
NDFI	CNDFI	NDF	0.0157	0.000443	0.21	0.00175	<0.01	9.70×10^{-7}	0.91	0.00314	0.57
NDFI	CNDFI	INDF/NDF	0.00952	0.000612	0.04	0.00216	<0.01	0.0125	0.02	0.00337	0.64

[1] DMI: dry matter (DM) intake (g/kg live weight (LW)), NDFI: intake of neutral detergent fibre (g/kg LW), FDMI: forage DMI (g/kg LW), FNDFI: forage NDFI (g/kg LW), CDMI: concentrate DMI (g/kg LW), CProp (DM): concentrate proportion on DM basis, CNDFI: concentrate NDFI (g/kg LW), CProp (NDF): concentrate proportion on NDF basis, NDF: diet NDF concentration (g/kg of DM), INDF/NDF: ratio between diet iNDF and NDF concentrations, NDFD: NDF digestibility.

[2] A ($P<0.01$).

[3] RMSE = root mean squared error.

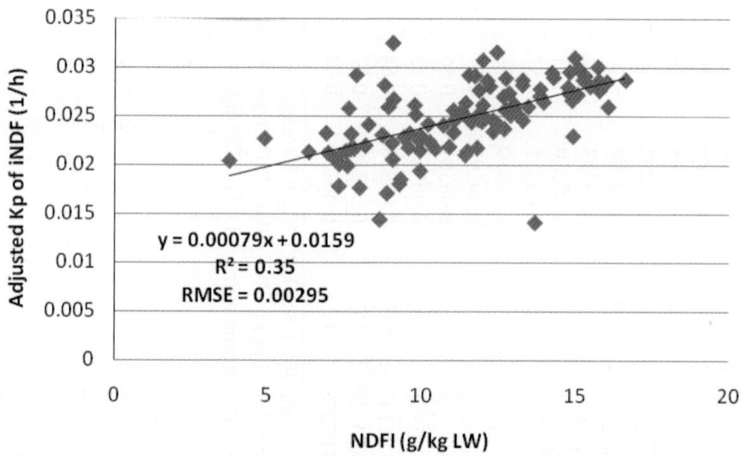

Figure 1. Intake of NDF (g/kg LW) vs. adjusted Kp of iNDF (1/h). RMSE = root mean squared error.

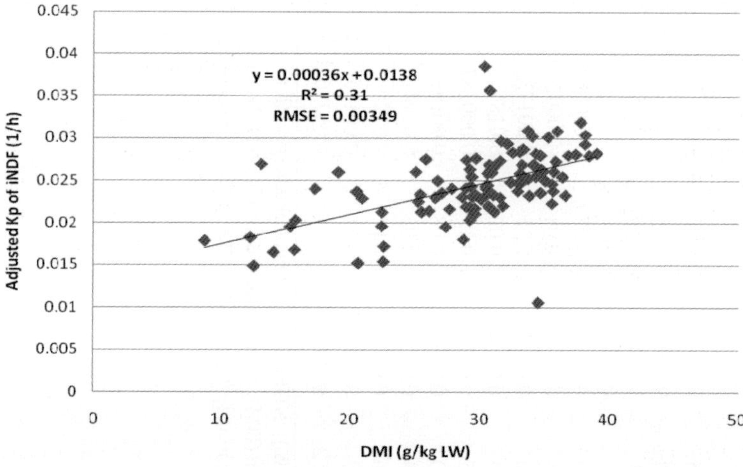

Figure 2. Intake of DM (g/kg LW) vs. adjusted Kp of iNDF (1/h). RMSE = root mean squared error.

duodenal and faecal samples have indicated that passage of feed particles in ruminants is at least a two-compartmental process. Gamma functions with differentiated age-dependency in the first compartment followed by an exponential distribution of the residence time in the escapable pool (see Ellis *et al.*, 1994) have often resulted in the best fit of the model to the data. External markers have been criticised to give different estimates of the pre-duodenal compartmental mean retention time because of migration, preferential binding, and/or density and digestibility changes. Moreover, the ascending phase of duodenal marker excretion curve has been interpreted as selective retention of feed particles in a non-escapable pool and the concomitant descending phase of the curve as a compartment with mass action dilution turnover. An alternative method to measure passage kinetics of fibre is the flux/compartmental pool method using rumen evacuation (Robinson *et al.*, 1987) or slaughter technique (Paloheimo and Mäkelä, 1959). The rumen evacuation technique demands a precise estimate of the rumen pool size and that the evacuations do not interfere with normal rumen functions (Huhtanen *et al.*, 2007). Further, the technique is based on first order kinetics and a one-compartment model. It has been claimed that the validity of marker systems may be tested by comparing the marker retention time to that estimated using the rumen evacuation technique for an internal marker naturally included in the feed (Huhtanen *et al.*, 2006). Huhtanen and Kukkonen (1995) obtained similar estimates of Kp with the rumen evacuation technique based on indigestible acid detergent fibre and iNDF. The largest disadvantages of the rumen evacuation technique are that passage kinetics is determined for the whole diet and that the omasal fibre pool is ignored (Huhtanen *et al.*, 2006).

Intake of DM versus intake of NDF

In most feed evaluation systems an increase in Kp of forage particles is predicted with increased intake of DM (NRC, 2001; Seo *et al.*, 2006). The relationships between passage rate and DMI have been developed from marker kinetic data. With this data set there was only a marginal response in precision with NDFI as independent variable in the simple regressions when compared to DMI. Huhtanen *et al.* (2006) observed a much higher improvement in precision when replacing DMI with NDFI in single regression models with Kp of iNDF (R^2 of 0.31 versus 0.68, respectively). Lund (2002) and Rinne *et al.* (2002) observed faster Kp of iNDF in dairy cows fed grass silages made from material harvested at later maturity stages. It is well recognized that intake of grass silage DM decrease with delayed harvest of the ensiled material. However, intake of NDF increases with greater maturity due to the proportionally higher increase in concentration of NDF in silage DM (Huhtanen *et al.*, 2006). Further, Rinne *et al.* (2002) observed increases in rumen pool sizes of fresh matter, DM, OM, NDF, digestible NDF and iNDF in dairy cows fed grass silages of progressing maturity. In the bivariate and multivariate models that additional accounted for the effect of CProp and dietary NDF characteristic Kp was much better predicted on NDF basis than DM basis. The coefficient for CProp on NDF basis was positive in all models. This is in agreement with a shorter retention time of concentrate particles than that of forage particles in marker kinetic studies (Colucci *et al.*, 1990; Mambrini and Peyraud, 1997).

Passage rate of forage and concentrate versus whole diet

One of the main drawbacks of the rumen evacuation technique is that digestion kinetics is determined for the whole diet and is not separated for individual feeds. Passage kinetics of forages can still be studied in animals fed unsupplemented forage diets without confounding extrinsic effects of the diet. Robinson *et al.* (1987), Huhtanen and Jaakkola (1993) and Stensig *et al.* (1998) observed decreased passage rate using the rumen evacuation technique when amount of concentrate in the diet increased. However, no effect of increased concentrate proportion in the diet to dairy cows was observed at higher levels of feeding (Colucci *et al.*, 1990). A proportionality factor of 1.6 was introduced in the Nordic dairy cow model Karoline to account for the higher Kp of concentrates in relation to Kp of forages (Danfær *et al.*, 2006). The proportionality factor in this material of 1.1 (when comparing the slopes of Kp of iNDF for NDFI with and without the inclusion of CProp (NDF)) was much lower. The differences between these proportionality factors suggest that regression analysis may not work very well to separate passage rates of concentrate and forage particles; the passage rates are non linear parameters and an aggregated Kp is not a simple weighted mean of forage and concentrate particle passage rates. There are two mechanisms that seem to affect the Kp estimate in different directions: (1) the observed rate has decreased as a response to an increased proportion of concentrate in the diet, but on the other hand (2) the observed rate is increasing with increased proportion of concentrate in the diet owing to faster Kp of concentrate particles than forage particles.

Conclusions

The equations suggested that Kp is more closely related to intake of NDF than DM. The positive coefficients for CProp on NDF basis and iNDF/NDF is related to the faster Kp of concentrate particles compared with forage particles and that Kp increases when NDF potential digestibility decreases. The Kp estimates were lower than estimates derived from marker excretion curves, probably because rumen evacuation derived Kp estimates include the retention time in the large particle pool, whereas Kp estimated from marker kinetics are derived from descending phase of marker curve that represents retention time in the small particle compartment.

Acknowledgements

M. R. Weisbjerg, T. Eriksson, E. Prestløkken, K. Kuoppala and J. Bertilsson that have complemented published data or contributed with unpublished data.

References

Colucci, P.E., Macleod, G.K., Grovum, W.L., McMillan, I. and Barney, D.J., 1990. Digesta kinetics in sheep and cattle fed diets with different forage to concentrate ratios at high and low intakes. Journal of Dairy Science 73:2143-2156.

Danfær, A., Huhtanen, P., Udén, P., Sveinbjörnsson, J. and Volden, H., 2006. The Nordic dairy cow model, Karoline – Description. In: Kebreab, E., Dijkstra, J., Bannink., A., Gerrits, J. and France, J. (eds.). Nutrient digestion and utilization in farm animals: Modelling approaches. CAB International, Wallingford, UK, pp.383-406.

Ellis, W.C., Matis, J.H., Hill, T.M. and Murphy, M.R. 1994. Methodology for estimating digestion and passage kinetics of forages. In: Fahey, G.C., Jr., Collins, M., Mertens, D.R. and Moser, L.E. (eds.) Forage quality, evaluation, and utilization. American Society of Agronomy, WI, US, pp.682-756.

Fox, D.G., Tedeschi, L.O., Tylutki, T.P., Russell, J.B., Van Amburgh, M.E., Chase, L.E., Pell, A.N. and Overton, T.R. 2004. The Cornell Net Carbohydrate and Protein System model for evaluating herd nutrition and nutrient excretion. Animal Feed Science and Technology 112:29-78.

Huhtanen, P. and Ahvenjärvi, S., 2008. Prediction of rumen residence time using markers or rumen evacuation and slaughter data. Canadian Journal of Animal Science 88:733-734.

Huhtanen, P., Ahvenjärvi, S., Weisbjerg, M.R. and Nørgaard, P., 2006. Digestion and passage of fibre in ruminants. In: Sejrsen, K., Hvelplund, T. and Nielsen, M.O. (eds.) Ruminant physiology: Digestion, metabolism and impact of nutrition in gene impression, immunology and stress. Wageningen Academic Publishers, Wageningen, the Netherlands, pp.87-135.

Huhtanen, P., Asikainen, U., Arkkila, M. and Jaakkola, S., 2007. Cell wall digestion and passage kinetics estimated by marker and in situ methods or by rumen evacuations in cattle fed hay 2 or 18 times daily. Animal Feed Science and Technology 133:206-227.

Huhtanen, P. and Jaakkola, S., 1993. The effects of forage preservation method and proportion of concentrate on digestion of cell wall carbohydrates and rumen digesta pool size in cattle. Grass and Forage Science 48:155-165.

Huhtanen, P. and Kukkonen, U., 1995. Comparison of methods, markers and sampling sites for estimating digesta passage kinetics in cattle fed at two levels of intake. Animal Feed Science and Technology 52:141-158.

Littell, R.C., Milliken, G.A., Stroup, W.W. and Wolfinger, R.D.,1996. SAS System for Mixed Models. SAS Inst. Incorporated, Cary, NC, USA.

Lund, P., 2002. The effect of forage type on passage kinetics and digestibility of fibre in dairy cows. PhD Thesis. The Royal Veterinary and Agricultural University, Copenhagen, Denmark.

Mambrini, M. and Peyraud, J.L., 1997. Retention time of feed particles and liquids in the stomach of ruminants and intestines of dairy cows. Direct measurements and calculations based on faecal collection. Reproduction Nutrition Development 37:427-442.
NRC (National Research Council), 2001. Nutrient requirements of dairy cattle. 7th revised edition, National Academy Press, Washington, DC, USA, 381 pp.
Paloheimo, L. and Mäkelä, A., 1959. Further studies on the retention time of food in the digestive tract of cows. Acta Agralia Fenniae 94:15-39.
Rinne, M., Huhtanen, P. and Jaakkola, S., 2002. Digestive processes of dairy cows fed silages harvested at four stages of grass maturity. Journal of Animal Science 80:1986-1998.
Robinson, P.H., Tamminga, S. and VanVuuren, A.M., 1987. Influence of declining level of feed intake and varying the proportion of starch in the concentrate on rumen ingesta quantity, composition and kinetics of ingesta turnover in dairy cows. Livestock Production Science 17:37-62.
Sauvant, D., Schmidely, P., Daudin, J.J. and St-Pierre, N.R., 2008. Meta-analyses of experimental data in animal nutrition. Animal 2:1203-1214.
Seo, S., Tedeschi, L.O., Lanzas, C., Schwab, C.G. and Fox, D.G., 2006. Development and evaluation of empirical equations to predict feed passage rate in cattle. Animal Feed Science and Technology 128:67-83.
Stensig, T., Weisbjerg, M.R. and Hvelplund, T., 1998. Digestion and passage kinetics of fibre in dairy cows as affected by the proportion of wheat starch or sucrose in the diet. Acta Agriculturae Scandinavica, Section A – Animal Sciences 48:129-140.
St-Pierre, N.R., 2001. Integrating quantitative findings from multiple studies using mixed model methodology. Journal of Dairy Science 84:741-755.

Appendix 1. References used for the meta-analysis

Ahvenjärvi, S., Joki-Tokola, E., Vanhatalo, A., Jaakkola, S., and Huhtanen, P., 2006. Effects of replacing grass silage with barley silage in dairy cow diets. Journal of Dairy Science 89:1678-1687.
Ahvenjärvi, S., Vanhatalo, A. and Huhtanen, P. Ruminal metabolism of silage soluble N fractions. Unpublished.
Ahvenjärvi, S., Korhonen, M., Vanhatalo, A. and Huhtanen, P. Digestion and passage kinetics of fiber in dairy cows fed primary and secondary growth grass silage supplemented with two levels of concentrate. Unpublished.
Ahvenjärvi S., Rinne M., Heikkilä T. and Huhtanen P. The effect of forage diet and intrinsic characteristics of forage particles on passage kinetics in dairy cows. Unpublished.
Bertilsson, J., 2005. Feeding mixed grass-clover silages with elevated sugar contents to dairy cows. Proceedings of the XIVth international silage conference, Belfast, Northern Ireland. In: R.S. Park and M.D. Stronge (eds.). Wageningen Academic Publishers, Wageningen, the Netherlands, 227 pp.
Bertilsson, J. and Murphy, M., 2003. Effects of feeding clover silages on feed intake, milk production and digestion in dairy cows. Grass and Forage Science 58:309-322.
Bosch, M.W. and Bruining, M., 1995. Passage rate and total clearance rate from the rumen of cows fed on grass silages differing in cell-wall content. British Journal of Nutrition 73:41-49.
Eriksson, T., Murphy, M., Ciszuk, P. and Burstedt, E., 2004. Nitrogen balance, microbial protein production, and milk production in dairy cows fed fodder beets and potatoes, or barley. Journal of Dairy Science 87:1057-1070.
Eriksson, T. and Nilsdotter-Linde, N. Milk production, nitrogen balance and digestion in dairy cows fed grass-legume silage with either birdsfoot trefoil or white clover. Unpublished.
Gasa, J., Holtenius, K., Suttons, J.D., Dhanoa, M.S. and Napper, D.J., 1991. Rumen fill and digesta kinetics in lactating Friesian cows given two levels of concentrates with two types of grass silage *ad lib.* British Journal of Nutrition 66:381-398.
Karp, V., 2005. The effect of whole crop pea-barley silage on ruminal fermentation, microbial synthesis, digestibility, and intake in dairy cows. MS Thesis. Helsinki University, Helsinki, Finland.

Khalili, H. and Huhtanen, P., 2002. Effect of casein infusion in the rumen, duodenum or both sites on factors affecting forage intake and performance of dairy cows fed red clover-grass silage. Journal of Dairy Science 85:909-918.

Kuoppala, K., Rinne, M., Ahvenjärvi, S., Nousiainen, J. and Huhtanen, P., 2004. Digestion kinetics of NDF in dairy cows fed silages from primary growth or regrowth of grass. Journal of Animal and Feed Sciences 13(Suppl.1):127-130.

Kuoppala K., Ahvenjärvi S., Rinne M. and Vanhatalo A., 2009. Effects of feeding grass or red clover silage cut at two maturity stages in dairy cows. 2. Dry matter intake and cell wall digestion kinetics. Journal of Dairy Science 92:5634-5644.

Kuusonen, U., 2001. Red clover and oriental goat's rue fed solely or as a mixture with meadow fescue silage to the dairy cows. MS Thesis. Helsinki University, Helsinki, Finland.

Lund, P., 2002. The effect of forage type on passage kinetics and digestibility of fibre in dairy cows. PhD Thesis. The Royal Veterinary and Agricultural University, Copenhagen, Denmark.

Minde, A. and Rygh, A.J., 1997. Metoder for å bestemme nedbrytningshastighet, passasjehastighet og vomfordøyelighet av NDF hos melkeku (Methods for determination of rates of degradation and passage, and ruminal digestibility of neutral detergent fibre in dairy cows). MS Thesis, Norwegian University of Life Sciences, Ås, Norway (in Norwegian).

Murphy, M., Åkerlind, M., and Holtenius, K., 2000. Rumen fermentation in lactating cows selected for milk fat content fed two forage to concentrate ratios with hay or silage. Journal of Dairy Science 83:756-764.

Prestløkken, E. Randby, Å. T. and Garmo, T., Effect of harvesting time and wilting on feed intake and production by dairy cows. Unpublished.

Rinne, M., Huhtanen, P., and Jaakkola, S., 2002. Digestive processes of dairy cows fed silages harvested at four stages of grass maturity. Journal of Animal Science 80:1986-1998.

Robinson, P.H., Tamminga, S. and VanVuuren, A.M., 1986. Influence of declining level of feed intake and varying the proportion of starch in the concentrate on rumen fermentation in dairy cows. Livestock Production Science 15:173-189.

Robinson, P.H., Tamminga, S. and VanVuuren, A.M., 1987. Influence of declining level of feed intake and varying the proportion of starch in the concentrate on milk production and whole tract digestibility in dairy cows. Livestock Production Science 17:19-35.

Robinson, P.H., Tamminga, S. and VanVuuren, A.M., 1987. Influence of declining level of feed intake and varying the proportion of starch in the concentrate on rumen ingesta quantity, composition and kinetics of ingesta turnover in dairy cows. Livestock Production Science 17:37-62.

Sairanen, A., Khalili, H., Nousiainen, J.I., Ahvenjärvi, S., and Huhtanen, P., 2005. The effect of concentrate supplementation on nutrient flow to the omasum in dairy cows receiving freshly cut grass. Journal of Dairy Science 88:1443-1453.

Stensig, T. and Robinson, P.H., 1997. Digestion and passage kinetics of forage fiber in dairy cows as affected by fiber-free concentrate in the diet. Journal of Dairy Science 80:1339-1352.

Stensig, T., Weisbjerg, M.R. and Hvelplund, T., 1998. Digestion and passage kinetics of fibre in dairy cows as affected by the proportion of wheat starch or sucrose in the diet. Acta Agriculturae Scandinavica, Section A – Animal Sciences 48:129-140.

Tamminga, S., Robinson, P.H., Vogt, M. and Boer, H., 1989. Rumen ingesta kinetics of cell wall components in dairy cows. Animal Feed Science and Technology 25:89-98.

Weisbjerg, M.R., Hvelplund, T., Kristensen, V.F. and Stensig, T., 1998. The requirement for rumen degradable protein and the potential for nitrogen recycling to the rumen in dairy cows. In: Kifaro, G.C., Ndemanisho, E.E. and Kakengi, A.M.V. (eds.). Approaches to increased livestock production in the 21st century: Proceedings of the 25th scientific conference AICC-Arusha. August 5-7, 1998. TSAP Conferences Series, Arusha, Tanzania, pp.110-118.

Ability of mathematical models to predict faecal output with a pulse dose of an external marker in sheep and goat

A. Moharrery
Animal Science Department, Agricultural College, Shahrekord University P.O. Box 115, Shahrekord, Iran; alimoh@mailcity.com

Abstract

Faecal output estimates derived from a one-compartment, Gamma-2, age-dependent model were compared with estimates derived algebraically by computing the area under the marker excretion curve for adult sheep and goats given a pulse dose of chromium sesquioxide (Cr_2O_3). Animals were fed one of three diets (as fed basis): 100% alfalfa hay, 64:36 alfalfa hay: barley grain, and 50:50 alfalfa hay: concentrate mixture. For the one-compartment model, faecal output was calculated as the dose of chromium sesquioxide divided by the initial concentration in the compartment (milligrams of Cr_2O_3/gram of DM) multiplied by the age-dependent rate constant (per hours). For the algebraic method, faecal output was calculated as the dose of Cr_2O_3 divided by the area under the marker excretion curve ([milligrams of %/gram of faecal DM].hours), both with the full complement of faecal samples and with faecal samples collected at 12-h intervals. Faecal output estimated by the three methods did not differ ($P<0.48$) from measured faecal output (total collection). Marker retention time calculated from the one-compartment, age-dependent model was greater ($P<0.05$) than retention time calculated algebraically (sum of concentration × time divided by sum of concentrations weighted for collection interval) for animals fed all three diets. No interaction was found for species × diet or species × method of prediction ($P<0.05$). These results suggest that the area under the marker excretion curve generated from a pulse dose of Cr_2O_3 will provide estimates of faecal output that do not differ from those calculated from a one-compartment, age-dependent model.

Keywords: sheep, goat, algebraic models, faecal output, retention time

Introduction

Indigestible markers or reference substances have been employed extensively in grazing research for determination of digestibility and intake. Indigestible markers can also be used to determine the rate of passage of nutrients through the gastrointestinal tract, site and extent of digestion, and microbial protein synthesis in ruminants. Faecal marker concentration curve, i.e. plots of concentration (mg marker/g dry matter (DM) faeces against time (hour)), are constructed from grab or bulked samples of faeces taken at different times following the single-dose infusion of an indigestible, non-absorbable digest-flow marker (France *et al.*, 1988).

Compartmental models frequently are used to describe marker excretion curves generated from a pulse dose of an external marker. These nonlinear models allow estimation of compartmental volume (mass of digesta in the compartment) and rate constants for flow from the compartment (Pond *et al.*, 1988). In many research applications with grazing ruminants, an estimate of faecal output for use in subsequent prediction of intake is the primary reason for using pulse-dose marker techniques.

Mathematic methods are available for calculating marker mean retention time from a pulse dose of an external marker, (Kaske and Engelhardt, 1990). The aim of the work described in the present paper was to compare the estimated faecal output calculated algebraically from the area under the marker excretion curve with that calculated from a one-compartment, age-dependent model in sheep and goat given a pulse dose of chromium sesquioxide (Cr_2O_3).

Materials and methods

Animals and diets

Four male sheep and four male goat (of a native breed, Lory-Bakhtiary and Iranian Saanen) with body weight of 58±5.4 and 38±4.4 kg (sheep and goat, respectively) were selected for the experiment. All animals were housed in individual metabolism cages and were fed one of three diets (as fed basis): 100% alfalfa hay, 64:36 alfalfa hay: barley grain, and 50:50 alfalfa hay: concentrate mixture. Sun-dried good quality alfalfa hay was chopped to about five to six centimeters in length. The ingredients used in the preparation of the total mixed ration (TMR) are shown in Table 1.

Daily feed intake was recorded by collection of any feed refusals before each morning feeding. The experiment was divided into three periods of 21 days (14 days for adaptation to each diet and 7 days for samplings). By the end of the each period, the animal groups were changed to the other diet in a cross-over design.

Faecal collection and chromium analysis

Markers were administered as a pulse dose to each animal at the rate of 160 mg Cr as Cr_2O_3 per each kg live weight. They were prepared as capsules and individually orally dosed. Faecal grab samples were collected after 0, 3, 6, 9, 12, 18, 24, 30, 36 and 48h after the dosing, oven-dried and stored for later analysis. Total feaces for each animal was collected under the metabolism cage and

Table 1. Ingredients and nutrient composition of experimental diets.

	Diet 1	Diet 2	Diet 3
Dietary components (%)			
Alfalfa hay (fine chopped)	100.00	74.00	50.00
Barley grain	-	26.00	40.00
Beet pulp	-	-	8.00
Cotton seed meal	-	-	2.00
Nutrient composition (% DM)			
Dry matter	90.00	89.48	89.42
DOM	71.14	77.10	79.07
ME (Mcal/kg)[1]	1.91	2.22	2.47
CP	15.00	14.62	14.62
RDP[1]	72.00	72.21	70.88
Fat	2.00	2.04	2.01
NDF	50.00	42.07	36.85
ADF	36.67	29.03	23.62
Ca[1]	1.26	0.95	0.70
P[1]	0.22	0.26	0.26

DM: dry matter, DOM: digestible organic matter, ME: metabolizable energy, CP: crude protein, RDP: rumen degradable protein as a percentage of CP, NDF: neutral detergent fibre, ADF: acid detergent fibre, Ca: calcium, P: phosphorus.

[1]Calculated according to NRC (2007) for diet component.

weighed during the last seven days of each period and 10% of that was taken and dried at 60 °C. Faeces samples were oven dried at 60 °C. All samples were ground on a 1-mm screen prior to determining chemical composition.

Chromium concentration in faeces samples were analysed according to the modified procedure of Dansky and Hill (1952). Samples for Cr analysis were ground as above mentioned and wet-ashed and the ash fused with sodium peroxide to convert chromic oxide to chromate. After appropriate dilution, the color density is measured spectrophotometery at 440 nm. A standard reference curve was established at this wavelength by determining the optical density of solutions prepared from known amounts of chromic oxide carried through the fusion process.

Calculations

All data of sheep and goats were used as a source of marker excretion curves. Chromium concentrations (milligrams of Cr_2O_3, Cr/gram of faecal DM) over time were fitted to a one-compartment, age-dependent model (Krysl *et al.*, 1988) to derive estimates of k_0 (scaling factor with units of [milligrams/gram].hours), k_l (age-dependent rate parameter for the Gamma-2 distribution), and τ (the time from dose until first appearance of marker in the feaces). Gastrointestinal mean retention time was calculated as $2/k_l + \tau$, and faecal output (grams/ hour) was calculated as dose of Cr (milligrams) divided by k_0 ([milligrams/gram].hours). Calculation of faecal output by this approach is equivalent to calculations described by Pond *et al.* (1988), in which faecal output equals dose divided by initial concentration in the compartment multiplied by the age-dependent passage rate for the one-compartment model.

For the algebraic approach, area under the Cr excretion curve (AUC) is equivalent to calculations described by Galyean (1993) as follows: area under the curve ([milligrams of Cr/gram of faecal DM].hours) = $\Sigma(c_i + c_p).(t_i - t_p)/2$, where c_i = the current concentration (milligrams/gram), c_p = the concentration at the previous collection time (milligrams of Cr/gram of faecal DM), t_i = the current collection time (hours), and t_p = the previous collection time (hours). Faecal output per hour was calculated as the dose of Cr (milligrams) divided by the area under the curve ([milligrams of Cr/gram of faecal DM].hours).

To examine effects of faecal sampling schedule on calculations with the algebraic approach, area under the curve also was calculated from samples taken at 12-h intervals during the 48-h faecal sampling period. Hence, samples taken at 3, 6, 9, 18, and 30h were deleted before calculation of area under the marker excretion curve. All other computations were as described previously for the full data set. Marker retention time was calculated algebraically according to the method described by Kaske and Engelhardt (1990) and extended by Galyean (1993) as follows: marker retention time (hours) = $\Sigma c_i \cdot t_i \cdot \varphi / \Sigma c_i \cdot \varphi$, where c_i = marker concentration at collection time = t_i, and φ= the interval between the current and previous collection time.

Criteria of goodness of the approach

For comparison between observed and predicted values a suitable method, which explained by Zhao *et al.* (2001) is used. In this method, bias and accuracy factors are used as a quantitative measure of the goodness of the prediction model. The bias factor indicates by how much, on average, a model over-predicts (bias factor >1) or under-predicts (bias factor <1) the observed data.

$$\text{Bias factor} = 10^{\frac{1}{n}\sum_{i=1}^{n}\log_{10}\left(\frac{\textit{predicted value}_i}{\textit{observed value}_i}\right)}$$

The accuracy factor indicates by how much, the predictions differ from the observed data.

$$\text{Accuracy factor} = 10^{\frac{1}{n}\sum_{i=1}^{n}\left|\log_{10}\left(\frac{\textit{predicted value}_i}{\textit{observed value}_i}\right)\right|}$$

In both equations, n is the number of observations used in the calculations. In a perfect model, both the bias and accuracy factors are equal to one. Additionally, statistical analysis was used to compare the predicted values with the results of actual measurement using SAS (2003).

Statistical analysis

Excretion curves for all animals were fit to one-compartment age-dependent model (Krysl *et al.*, 1988) using the nonlinear procedure of SAS (2003; PROC NLIN, iterative Marquardt method) as described by Krysl *et al.* (1988). Detailed explanations of the terminology were also presented by Pond *et al.* (1988). The complete randomised model was used to analyse data for faecal output, and related parameters (SAS, 2003). In this regard, in 4×2×3 factorial arrangement (four methods for measured, one-compartment, and area under the curve [full data set and 12-h interval data], two Species for sheep and goat and three diets). Retention time data for the algebraic and one-compartment methods (full data set) were analyzed by species and diets with the same statistical model (SAS, 2003). The data were analysed using the general linear model procedure of SAS (2003). Values are given as means, and the homogeneity of variance was tested (Bartlett's test).

Results and discussion

Daily mean faecal output measured and estimates from different methods are shown in Table 2. All three methods estimates same amounts of faecal output to compare measured values ($P<0.05$). Sheep species showed significant higher faecal production compared to goats ($P<0.05$). When faecal output was expressed per unit of body weight all differences disappeared. Sheep and goats defecated about 7.0 g of faecal DM/kg of body weight in all diets. In this regards, no significant difference was shown between different diets for daily faecal production ($P<0.05$). No interaction was found for species × diet or species × method of prediction ($P<0.05$).

The means of observed and predicted values along with bias and accuracy factors are reported in Table 2. Statistical analysis was used to compare the predicted values with the results of actual measurement. No significant difference was detected between measured values and predicted values for all parameters.

The faecal output calculated from the one-compartment model, the area under the curve with all faecal samples, and the area under the curve with samples collected at 12-h intervals did not show any significant difference ($P<0.05$). Error percentage for estimation of faecal production has shown highest value when samples collected at 12-h intervals (Table 2) and it has reduced to 3.7% with all faecal samples. This result would be expected because increased frequency of sampling should increase the accuracy of faecal output estimates with the algebraic area under the curve approach and more accurately reflect the true area under the marker excretion curve. However, for grazing ruminants, samples collected at 12-h intervals might decrease the time needed to estimate faecal output compared with the regular dosing chromic oxide technique, which requires steady state conditions (often 5 days), followed by faecal sampling for an additional extended period (often 5 days) but again problem exist, for example diurnal fluctuation in marker concentration in the faeces (Le Du and Penning, 1982).

Table 2. Comparison of results for observed and model predicted values.

	Methods[1]				Species		Diets		
	Measured	1CMPT	AUC-All	AUC-12	Sheep	Goat	1	2	3
Faecal (g/day)	353	354	335	326	401[a]	283[b]	328	349	349
SE[2]	19.67				13.91		17.04		
Error, %[3]		-1.588	3.727	6.848	7.082	-1.090	3.482	-0.226	5.732
SE		4.058			3.313		4.058		
Bias		0.007	-0.016	-0.030	-0.031	0.005	-0.015	0.001	-0.024
SE		0.018			0.015		0.018		
Accuracy		0.063	0.060	0.073	0.054	0.076	0.056[ab]	0.049[b]	0.091[a]
SE		0.013			0.010		0.013		

[1] 1CMPT: one-compartment model, AUC-All: algebraically method with all faecal samples collected, AUC-12: algebraically method with samples collected at 12-h intervals.
[2] SE: standard error.
[3] Error percentage = (((mean of observed – mean of predicted)/Mean)×100)
while Mean = (mean of observed + mean of predicted)/2.
[a,b] Means with the same letter in each row and for each section are not significantly different ($P<0.05$).

Additionally, we used bias and accuracy factors to compare the goodness of estimations. Bias and accuracy factors indicate a close degree of goodness of fit between observed and predicted values. The discrepancy between bias and accuracy factors is negligible for each character. However, for the reason that the observed value is in the denominator of both equations for bias and accuracy factors, and we do not have any zero in the observed or predicted values, using this method is logical for system evaluation. One of the most important advantages of the predicted methods is that they predict values accurately, because no values were closer to zero or one. This finding disagrees with Zhao *et al.* (2001), that they mentioned due to the limitation of their formulae, bias and accuracy factors cannot be used for evaluation data which is equal or close to zero or one. As a result, bias and accuracy factors show that prediction values by three methods are close to measured values. In this regard, bias and accuracy factors demonstrate no over-prediction or under-prediction for data set. In respect to the data set (without zero value), if the bias factor is one, a conclusion made for method fitness.

Divergence of estimates from actual values was smaller with the diet No 2 than with the two other alfalfa based diets (Table 2). In this manner, accuracy factor for diet No 2 showed smallest and significant difference ($P<0.05$) compared with other diets. Krysl *et al.* (1985) reported that faecal output estimated from the one-compartment model was 100±2% for lambs fed alfalfa hay and 103±3% for lambs fed prairie hay.

In all three diets of present experiment there was an agreement between actual (measured) and predicted (estimated) faecal output for sheep and goats (Table 3). Estimates by all three calculation methods did not differ ($P<0.05$) from measured faecal output, although numerical differences were noted. No interaction was found for species × diet or species × method of prediction ($P<0.05$).

Dry matter intake was very low in goats compared with intake for sheep (1.32 vs. 1.83 kg, respectively). But, dry matter intake per kg body weight$^{0.75}$ was same among two species (82 g/kg

Table 3. Faecal output (gram/day) measured and estimates from the different methods for all animals fed three diets.

	Methods							
	Measured		1CMPT		AUC-All		AUC-12	
Treat	Sheep	Goat	Sheep	Goat	Sheep	Goat	Sheep	Goat
Diet 1	386	284	399	287	387	254	388	233
Diet 2	449	264	431	288	415	277	397	271
Diet 3	448	288	393	324	369	303	353	312

1CMPT: one-compartment model, AUC-All: algebraically method with all faecal samples collected, AUC-12: algebraically method with samples collected at 12-h intervals.

$wt^{0.75}$). The time between the marker dose and the first appearance in the feces (Tau) for pooled data was found 3.33 hours. Marker retention time calculated from the one-compartment, age-dependent model was greater ($P<0.05$) than retention time calculated algebraically (sum of concentration × time divided by sum of concentrations weighted for collection interval) for animals fed all three diets (Table 4). In present study (Table 4), total tract mean retention time was 22% longer ($P<0.05$) for the one-compartment model to compare algebraic methods. For goats alimentary passage rate was 8.6% slower ($P<0.05$) than sheep. These finding is in contrary with Huston *et al.* (1986) how has reported that the retention time of feed particles is much shorter in goats to compare sheep. However, in their experiment Ytterbium chloride (Yr) was used as a marker in the forage. In Huston *et al.* (1986) retention time in total alimentary tract for goats is much longer than whatever find in the present study (28 vs. 16.8 hours). The authors offer no explanation for this shorter retention time. Such behavior by native sheep and goats when removed from natural surrounding may be due to excited animals. It has been shown that excited animals have higher alimentary movements.

Despite the species and method of calculation, all diets had showed same retention time. Moore *et al.* (1992) reported that the pelleted diet had faster passage and shorter retention times than did the hay diet. In their study, the hay diet exited the rumen at 83% of the passage rate of the pelleted diet when Cr was the marker.

Estimates of k_0, and area under the curve (both have units of [milligrams of Cr_2O_3/gram of faecal DM]xhours) have shown significant difference among the diets (Table 4; $P<0.05$). The diet No 2 shows about 13% less marker concentration per gram of DM faecal to compare two other diets. This result would be expected because pulse dose of Cr_2O_3 for diet 1 to 3 were 7,700, 7,300 and 8,200 milligram per animal, respectively. In this manner, faecal output was not different among the calculation methods; dose of marker divided by k_0 or the area under the excretion curve equals faecal output. Marker mean retention time generally was less for the algebraic method than for the one-compartment model ($P<0.05$). Lalles *et al.* (1991) also noted that estimates were generally lower for marker mean retention time with the mathematical method than with compartmental models in early-weaned calves fed concentrate diets. In agreement with the Galyean (1993), the present data suggest that the area under the marker excretion curve generated from a pulse dose of Cr_2O_3 can be used to estimate faecal output. In contrary, Moore *et al.* (1993) suggested that the algebraic approach would be less precise than a compartmental modeling approach to calculation of fecal output. They have recommended that experimenters consider both methods.

Table 4. Estimates of K_0 and marker mean retention time (hours), from the different methods for all animals fed three diets.

Treat	Parameters							
	Retention time (hours)				[mg/g DM faecal] × hours			
	1CMPT		AUC-All		K_0		Area under the curve	
	Sheep	Goat	Sheep	Goat	Sheep	Goat	Sheep	Goat
Diet 1	18.4	17.9	15.1	14.9	610	550	594	507
Diet 2	16.3	18.1	13.7	14.0	526	491	507	474
Diet 3	15.9	19.2	13.5	15.8	634	547	598	519
Statistical analysis								
Diets	1	2	3	SE	1	2	3	SE
	16.59	15.74	16.09	0.475	566^{ab}	500^{b}	575^{a}	23.78
Methods	1CMPT	AUC-All		SE	1CMPT	AUC-All		SE
	17.63^{a}	14.65^{b}		0.389	560	533		19.41
Species	Sheep	Goat		SE	Sheep	Goat		SE
	15.47^{b}	16.81^{a}		0.389	579	515		19.41

1CMPT: one-compartment model, AUC-All: algebraically method with all faecal samples collected, SE: standard error.
Means with the same letter in each row in statistical section are not significantly different ($P<0.05$).

Implications

Pulse-dosing typically is used to estimate faecal output and retention time in specific parts of the gut. Mean retention time also can be calculated mathematically from marker excretion data. Result from present study showed that the mathematical methods can be used with highest precision for estimation of faecal output. Estimates derived by this way did not differ significantly from those calculated from a one-compartment, age-dependent model. Compartmental models rely on assumptions about the kinetics of digesta flow in ruminants and their adequacy of fit should help us understand the biology of digesta flow; this end cannot be accomplished with algebraic methods.

Acknowledgements

I am grateful to all the staff from the Section of Research Department in the Shahrekord University.

References

Dansky, L.M. and Hill, F.W., 1952. Application of the chromic oxide indicator method to balance studies with growing chickens: Two figures. Journal of Nutrition 47:449-459.

France, J., Dhanoa, M.S., Siddons, R.C., Thornley, J.M. and Poppi, D.P., 1988. Estimating the production of faeces by ruminants from faecal marker concentration curves. Journal of Theoritical Biology 135:383-391.

Galyean, M.L., 1993. Technical Note: An algebraic method for calculating faecal output from a pulse dose of an external marker. Journal of Animal Sciences 71:3466-3469.

Huston, J.E., Rector, B.S., Ellis, W.C and Allen, M.L., 1986. Dynamics of digestion in cattle, sheep, goats and deer. Journal of Animal Sciences 62:208-215.

Lalles, J.P., Delval, E. and Poncet, D., 1991. Mean retention time of dietary residues within the gastrointestinal tract of the young ruminant: A comparison of noncompartmental (algebraic) and compartmental (modelling) estimation methods. Animal Feed Science and Technololgy 35:139-159.

Le Du, Y.L.P. and Pennining, P.D., 1982. Animal based techniques for estimating herbage intake. In: Leaver, J.D. (ed.). Herbage Intake Handbook, Berks: The British Grassland Society, Hurley, UK, pp.37-75.

Kaske, M. and Von Engelhardt, W., 1990. The effect of size and density on mean retention time of particles in the gastrointestinal tract of sheep. British Journal of Nutrition 63:457-465.

Krysl, L.J., Galyean, M.L., Estell, R.E. and Sowell, B.F., 1988. Estimating digestibility and faecal output in lambs using internal and external markers. Journal of Agricaltural Sciences, (Camb.) 111:19-25.

Krysl, L.J., McCollum, F.T., and Galyean, M.L., 1985. Estimation of faecal output and particulate passage rate with a pulse dose of ytterbium-labeled forage. The Journal of Range Management 38:180-182.

Moore, J.A., Pond, K.R.,. Poore, M.H. and Goodwin, T.G., 1992. Influence of Model and Marker on Digesta Kinetic Estimates for Sheep. Journal of Animal Sciences 70:3528-3540.

Moore, J.A., Fisher, D.S. and Matis, J.H., 1993. Letters to the Editor: Reply to the letter by G.R. Reese. Journal of Animal Sciences 71:1667.

NRC (National Research Council), 2007. Nutrient requirements of small ruminants: sheep, goats, cervids, and New World camelids. National Academy Press, Washington, DC, USA.

Pond, K.R., Ellis, W.C., Matis, J.H., Ferreiro, H.M. and Sutton, J.D., 1988. Compartment models for estimating attributes of digesta flow in cattle. British Journal of Nutrition 60:571-595.

SAS User's Guide: Statistics, Version 9.1 Edition. 2003. SAS Inst., Incorporated Cary, NC, USA.

Zhao L., Chen, Y. and Schaffner, D.W., 2001. Comparison of logistic regression and linear regression in modeling percentage data. Applied Environmental Microbiology 67:2129-2135.

Part 4 – Modelling interactions between nutrients and physiological functions: consequence on product quality and animal health

Dynamic modelling of contractile and metabolic properties of bovine muscle

N.M. Schreurs[1,2], *F. Garcia-Launay*[1], *T. Hoch*[1], *C. Jurie*[1], *J. Agabriel*[1], *D. Micol*[1] *and B. Picard*[2]
[1]*UR1213 Recherches sur les Herbivores, INRA Theix, 63122 Saint Genès Champanelle, France*
[2]*Institute of Food, Nutrition and Human Health, Massey University PN452, Private Bag 11222, Palmerston North 4442, New Zealand; N.M.Schreurs@massey.ac.nz*

Abstract

During bovine growth, a dynamic multifactor process influences the development of muscle characteristics that influence meat quality. This paper fits non-linear equations to data from growing cattle in order to provide a mathematical description of the development of contractile and metabolic muscle characteristics. The muscle characteristics considered were: the mean cross-sectional area of muscle fibres, fibre type proportions and the glycolytic and oxidative enzymatic activities represented by lactate dehydrogenase (LDH, glycolytic) and isocitrate dehydrogenase (ICDH, oxidative). The Gompertz equation was used to describe the increase in fibre cross-sectional area as cattle mature. The proportion of oxidative fibres in the muscle was modelled using a 3-parameter equation to denote an initially fast increase in the proportion of oxidative fibres which then slowed with increased degree of maturity. A 4-parameter equation represented the proportion of glycolytic fibres and LDH activity in the muscle which increased with increasing maturity and then decreased after puberty. An equation was fitted to the data of ICDH activity in the muscle which had an exponential decrease followed by a linear increase with increasing maturity. For each muscle considered, the rate of change in the contractile and metabolic muscle characteristics differed between beef breeds (e.g. Charolais) compared with non-beef breeds (e.g. dual-purpose or dairy breeds) and also between bulls compared to cows and steers. These differences in muscle development were accounted for by fitting different parameters for each breed and sex group. Assessment of the mean square deviation indicated that parameterisation was logical given the biological responses observed from animal experiments.

Keywords: cattle, beef, meat, fibre, enzyme

Introduction

As an animal grows there are many factors that can influence the meat quality characteristics. The factors can be intrinsic to the animal, such as the animal's breed or sex, or factors that result from on-farm management decisions, such as nutrition and timing of slaughter. Factors influencing meat quality have their effect through changes in the physical and chemical properties of the muscle (Purchas, 1989). Muscle fibre size, the mix of muscle fibre types and the activities of metabolic enzymes within the muscle of cattle change in response to growth and production requirements as the animal matures. The responses are different for cattle of different ages, sex, breed and between muscles within the animal (Wegner *et al.*, 2000).

The number of muscle fibres is fixed at birth and muscle fibre hypertrophy results in an increase in muscle size measured as the fibre cross-sectional area. Large fibre size has been negatively correlated with meat tenderness (Renand *et al.*, 2001; Dransfield *et al.*, 2003). Muscle fibres that make up the structure of the muscle are classified as slow or fast contracting. Muscle fibres are also classified on their metabolic mechanism for regenerating ATP. Adequate blood supply to the muscle fibre ensures that oxygen is readily available and muscles will utilise oxidative metabolism. If oxygen is limiting an alternative glycolytic pathway is used to create ATP whereby pyruvate is converted to lactate. Due to their more constant use, slow contracting fibres tend to utilise predominately oxidative metabolism to regenerate ATP while fast contracting fibres have a more glycolytic metabolism. Therefore, the main muscle fibre types are the slow-oxidative, fast-glycolytic and the intermediate fast-oxidative glycolytic. Utilisation of the substrates of glycolytic and oxidative metabolism is dependent on the activity of enzymes required to catalyse the biochemical steps towards ATP generation. Isocitrate dehydrogenase (ICDH) and lactate dehydrogenase (LDH) are two common measures of muscle oxidative and glycolytic activity, respectively. Oxidative metabolism is associated with increased redness, finer texture, pH and improved juiciness of the meat (Purchas, 1989).

The aim of this work was to model the changes in muscle characteristics as an animal matures. The main hypothesis is that at any particular age or level of maturity the sex and breed of the animal are likely to be strong driving forces in the development of muscle characteristics (Hoch *et al.*, 2002). The objective was to fit non-linear equations that will describe the changes in the fibre cross-sectional area, fibre type proportions, and the activity of metabolic enzymes during post-natal growth. Muscle characteristics differ between muscles within an animal (Schreurs *et al.*, 2008) so, in order to look at the effects of breed and sex, only one muscle has been considered here.

Data sets and data categorisation for non-linear fitting

Non-linear fitting of equations for the surface area and proportion of fibre types was done using data from 11 experiments with a total of 1,760 observations. For modelling enzyme activities, more experimental results were available and data from 31 experiments with a total of approximately 2,860 observations were used. To make comparisons between breeds, data were classified according to three breed categories: beef, dairy and dual-purpose. Beef breeds are those breeds where the sole purpose is to produce meat e.g. Charolais and Limousin. The dairy breeds are those breeds where the main purpose is to produce milk e.g. Holstein however, males are slaughtered for meat, as are the cows after a period of fattening at the end of their productive life. The dual-purpose breeds are noted as being intermediate to dairy and beef breeds in their capacity to produce milk and meat e.g. Salers and Aubrac. To investigate the effects of sex on the development of muscle characteristics, data were also classified according to the sex of the animal. The three sexes considered were: bulls, cows and steers.

Initial hypotheses and choice of equations

A meta-analysis of results from 36 experiments indicated that the muscle characteristics developed differently between the muscles, breeds and sexes of cattle (Schreurs *et al.*, 2008). To specifically look at the effect of breed and sex, only the *longissimus thoracis* muscle was considered. Animal maturity is a strong driving force for physiological development and the degree of maturity rather than age was chosen to represent this. This allowed the comparison of cattle of different breeds and sexes at the same level of maturity, whatever their age. The degree of maturity is defined as the animal's live weight at the time of the measurement divided by the industry defined mature weight. The objective was to describe post-natal development of muscle characteristics starting at birth when the degree of maturity is 0.05.

Although a meta-analysis of the data (Schreurs *et al.*, 2008) indicated differences between breeds and sexes in the rates of change and mature values obtained for a particular muscle characteristic, the general pattern and direction of development was similar between all animal types. This indicated that the same non-linear equation could be used to describe the development of a muscle characteristic for different animals and different values of the parameters would explain differences between cattle breeds and sexes.

Reviewing the literature and performing a meta-analysis provided indications of the changes in the metabolic and contractile characteristics of the muscle as cattle grow and mature. The meta-analysis was able to specify how the muscle characteristics were changing over time (e.g. if they were increasing or decreasing) and if there were likely to be differences between the breeds and sexes in the rate of that change. The meta-analysis and further investigation in the literature gave an initial indication of the inflections or changes in the rate of muscle characteristic development. From this information a subset of non-linear equations which were likely candidates to appropriately describe the development of each muscle characteristic was created. After sequential fitting of each equation and refining of the equations the best equation to describe the dynamic of the development of each muscle characteristic was elucidated.

Cross-sectional surface area of muscle fibres.

The Gompertz equation (Equation 1) was used to model the increase in fibre cross-sectional area in both slow-oxidative and fast-glycolytic fibres. This growth function was chosen as the development of the cross-sectional area (CSA) mimics the typical sigmoidal body-growth curves of cattle (Thornley and France, 2007).

$$CSA(t) = \alpha e^{\frac{\beta}{\gamma}(1-e^{-\gamma t})} \quad (1)$$

The α parameter represents fibre area at birth and $\alpha e^{\frac{\beta}{\gamma}}$ expresses the asymptotic plateau that is reached at infinite t and that the cross-sectional area of muscle fibres approaches in mature cattle. The parameter g describes the velocity with which this limit is reached. The t indicates the degree of maturity adjusted by 0.05 to have the model starting at birth. The degree of maturity corresponds here to the proportion of the live weight that the animal has achieved in relation to the industry recognised mature live weights for the breeds and sexes in the data base (Schreurs *et al.*, 2008). A value of t greater than 1 is possible when an animal grows to be heavier than the industry standard mature live weight.

Percentage of fibre types

The proportion of slow oxidative fibres (%SO) in the muscle was modelled using a three parameter equation (Equation 2) to denote an initially fast increase in the proportion of oxidative fibres which then slowed with increased degree of maturity.

$$\%SO(t) = \alpha - \beta e^{-\gamma t^2} \quad (2)$$

In this case, α and $\alpha - \beta$ represent the respective maximum and minimum values for the percentage of slow-oxidative fibre in the muscle. The γ parameter expresses the velocity with which the maximum value is reached.

The proportion of fast-glycolytic fibres (%FG) in the muscle increases before puberty followed by a gradual decline, which can be represented by a 4 parameter model with a positive asymptote (Equation 3).

$$\%FG(t) = (\alpha - \delta + \beta t)e^{-\gamma t} + \delta \quad (3)$$

The α and δ parameters represent the percentage of fast-glycolytic fibres in the muscle at birth and approached at maturity, respectively. The β parameter describes the rate of increase prior to puberty and γ the rate of decrease after puberty.

Metabolic enzyme activities

The equation used to describe the changes in LDH activity as cattle grow is the same as that used to model the percentage of fast-glycolytic fibres in the muscle (Equation 3). This is logical due to both being involved in the glycolytic metabolism of the muscle. The ICDH activity in muscle develops in opposite direction to that of the LDH activity. Therefore, ICDH activity in the muscle was considered to fit an equation with an exponential decrease followed by a linear increase with increasing maturity (Equation 4).

$$ICDH(t) = \alpha e^{-\beta t} + \gamma t \qquad (4)$$

where α is the ICDH activity at birth. The rate of the exponential decrease is represented by the β parameter and the rate of the linear increase represented by the γ parameter.

Model fitting

Model parameters were estimated using the NLMIXED procedure of SAS (SAS 9.1, 2003). The experiments from which data were obtained were incorporated as a random effect. The predictions based on the parameter values obtained were evaluated using the Mean Squared Deviation (MSD) in preference to other approaches as it works well with biological models (Gauch *et al.*, 2003). The MSD considers the average squared difference between values predicted by the model compared to observed values. Due to the relationship of the MSD components with regression parameters we simplistically denoted if there was any translation, rotation or a lack of correlation between the predicted and observed values. Graphical assessment by plotting observed versus predicted values was also used.

An initial attempt at fitting the models indicated that the lack of data for newborn cattle (degree of maturity less than 0.2) resulted in poor fitting of the α parameter in the equations for LDH and ICDH activity and also for the percentage of fast-glycolytic fibres in the muscle. It was decided to fix the values of α in this case based on intercept values obtained in the meta-analysis and from published data (Gagnière *et al.*, 1997).

Parameters were initially fitted to compare between breeds by fitting the full model to each breed type (beef, dual-purpose and dairy) and then separately fitting to compare between sexes (bulls, steers and cows). By comparing if the parameter values were significantly different ($P<0.05$) and in conjunction with MSD evaluations and graphical analysis of the non-linear curve with the data we identified that there was a clear effect of sex, while there was no effect of breed on the development of muscle fibre cross-sectional areas and the percentage of the fibre types in the muscle. Both breed and sex had an influence on the parameter values of the development of ICDH and LDH activities in the muscle as the cattle grew.

Further consideration of the parameter values obtained for the different breeds and sexes showed that the values obtained for cows and steers were not different so these two sexes were grouped in subsequent iterations (Figures 1 and 2; Table 1). For the models of ICDH and LDH activity the dual-purpose and the dairy breeds also had parameter values that were similar and these two breed groups were also considered as one group in further iterations.

The parameterisation of the non-linear equations for ICDH and LDH activity then went on to consider breed and sex together, rather than separately as in previous iterations. This revealed that for both the equations describing ICDH and LDH activity in the muscle, the sex of the cattle gave different values of the g parameter and the breed gave different values of the β parameter. Therefore, in the final models, different values of β are used to estimate enzyme activities for beef versus dual-purpose and dairy breeds and different values of γ permitted estimations for males versus steers and cows (Figure 3; Table 1).

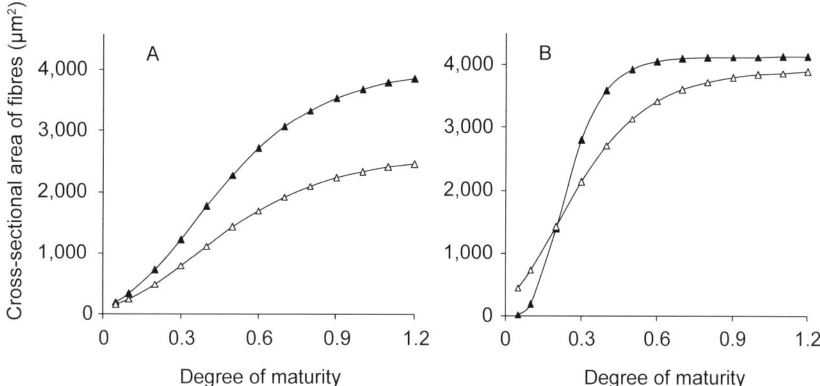

Figure 1. Modelled development of the cross-sectional area of (A) slow-oxidative and (B) fast-glycolytic fibres in longissimus thoracis *in bulls (▲) and cows/steers (△). There was no difference in cross-sectional fibre area between breeds. A degree of maturity greater than 1 occurs when an animal grows to be heavier than the industry standard mature live weight.*

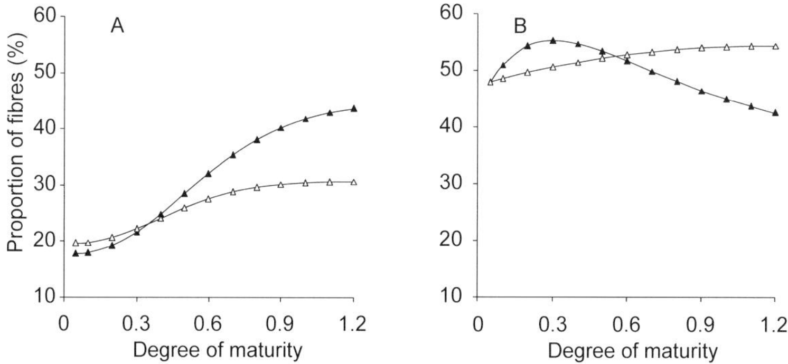

Figure 2. Modelled development of the proportion of (A) slow-oxidative and (B) fast-glycolytic fibres in the longissimus thoracis *muscle in bulls (▲) and in cows and steers (△). There was no difference in the proportion of fibre types between breeds. A degree of maturity greater than 1 is possible when an animal grows to be heavier than the industry standard mature live weight by which the degree of maturity was calculated.*

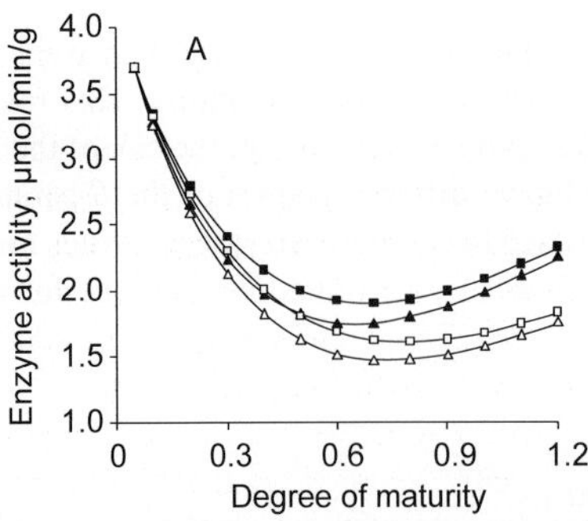

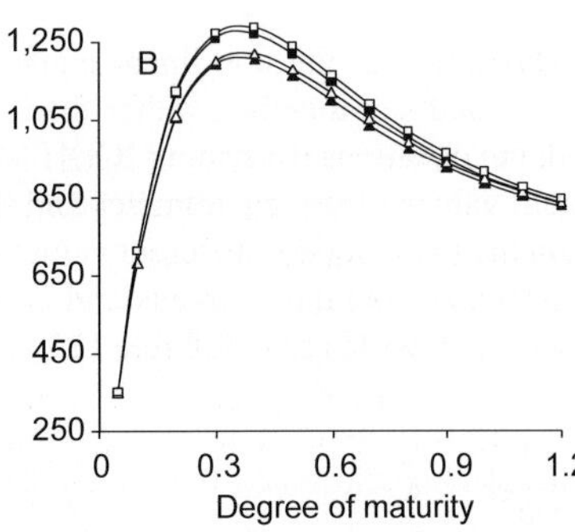

Figure 3. Development of (A) isocitrate dehydrogenase (ICDH) or (B) Lactate dehydrogenase (LDH) activity in longissimus thoracis *muscle in beef breeds e.g. Charolais, Limousin (triangles) and dairy and dual-purpose breeds e.g. Holstein, Aubrac, Salers (squares). The development of enzyme activities was different between bulls (dark shapes) and cows/steers (empty shapes). A degree of maturity greater than 1 is possible when an animal grows to be heavier than the industry standard mature live weight by which the degree of maturity was calculated.*

Table 1. Estimated parameter values (and standard errors of the estimates) for the non-linear equations, as defined by Equations 1-4 in the text, fitted to describe the development of muscle characteristics in growing cattle of different breeds and sexes.

	α	β	γ	δ
Slow-oxidative fibre cross-sectional area				
Bulls	204 (142)	11.0 (4.5)	3.7 (0.8)	-
Steers/Cows	164 (148)	9.4 (5.6)	3.4 (1.0)	-
Fast-glycolytic fibre cross-sectional area				
Bulls	24.9 (66.4)	52.9 (38.5)	10.3 (2.3)	-
Steers/Cows	454 (353)	10.9 (6.5)	5.1 (1.3)	-
Percentage of slow-oxidative fibres				
Bulls	44.7 (3.5)	26.9 (3.1)	2.5 (0.72)	-
Steers/Cows	30.8 (0.9)	11.0 (2.8)	4.2 (1.46)	-
Percentage of fast-glycolytic fibres				
Bulls	48.0	101.2 (24.9)	2.8 (1.22)	37.6 (10.1)
Steers/Cows	48.0	27.1 (15.4)	0.4 (0.25)	10.9 (2.7)
Isocitrate dehydrogenase activity				
Beef breeds				
Bulls	3.7	2.9 (0.22)	1.9 (0.16)	-
Steers/Cows	3.7	2.9 (0.22)	1.4 (0.16)	-
Dual-purpose and dairy breeds				
Bulls	3.7	2.6 (0.16)	1.9 (0.16)	-
Steers/Cows	3.7	2.6 (0.16)	1.4 (0.16)	-
Lactate dehydrogenase activity				
Beef breeds				
Bulls	350	6,333 (574)	4.0 (0.21)	762 (57)
Steers/Cows	350	6,333 (574)	3.9 (0.21)	762 (57)
Dual-purpose and dairy breeds				
Bulls	350	7,069 (582)	4.0 (0.21)	762 (57)
Steers/Cows	350	7,069 (582)	3.9 (0.21)	762 (57)

Checking the fit

From a biological point of view, it is logical that differences between breeds were equalised in the models describing the development of fibre area and fibre type percentages. Using the degree of maturity as the dependant variable equalises breed differences due to comparing the cattle at the same physiological stage thereby, removing age effects. When comparing sexes at the same degree of maturity this will still give differences in muscle characteristics due to a maturity dependant partitioning of nutrients and differences in lean muscle growth that are signalled by sex hormones (Purchas, 1989). For the models of ICDH and LDH activity in growing cattle it is logical that the β parameter which influences the first part of the curve is different between the breeds. The first part of the curve denotes the period in the animals development that occurs before puberty. Prior to puberty the muscle characteristics are predominately affected by different growth rates between different breeds. The γ parameter modifies the curve when the degree of maturity is at stages representing the time after puberty. After puberty the effect of sex is likely to have the predominant effect due to strong signals from the sex hormones influencing the development of the muscle characteristics (Figure 3).

Variation is an undeniable fact of biological systems. Graphical comparison of the predicted values to observations indicated that there was substantial variation in the data which is the major contributor to the MSD values. The difference between the predicted and observed values was minimised by stepwise fitting of parameter values to minimise the MSD (Figure 4). However, large variation in the data exists and to capture this aspect in the model needs further consideration. Additionally, lack of data for newborn cattle (Figure 4B) lead us to fix values of parameter α in Equations 3 and 4 from our published data on foetuses (Gagnière *et al.*, 1997). This need to be validated with new data. Therefore, the non-linear, mathematical descriptions of the development of fibre cross-sectional area, fibre type proportions and the metabolic enzyme activities are congruent with the biological processes that occur in the muscles of cattle but do not capture the variation that occurs within a particular cattle breed or sex.

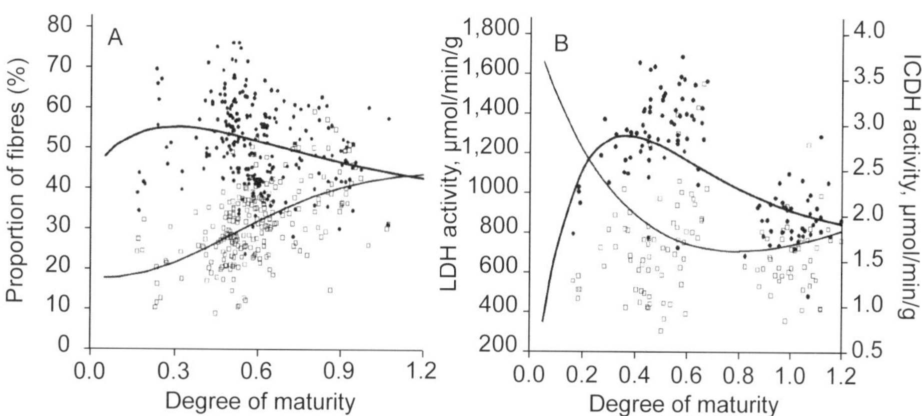

Figure 4. Comparison of the data to the fitted model curve in the longissimus thoracis *of bulls. (A) The proportion of slow-oxidative fibres (squares) and fast-glycolytic fibres (filled circles). (B) The enzyme activities of iscocitrate dehydrogenase (ICDH; squares) and lactate dehydrogenase (LDH; filled circles). A degree of maturity greater than 1 is possible when an animal grows to be heavier than the industry standard mature live weight by which the degree of maturity was calculated.*

Sensitivity analysis was done by altering the fitted values by 10% for each parameter in each model. A sensitivity index (S; Equation 5) was calculated to enable quantitative assessment of the response to a small change in the parameter value (Haefner, 2005).

$$S = \frac{(R_a - R_n)/R_n}{(P_a - P_n)/P_n} \quad (5)$$

In calculating the sensitivity index, R_a and R_n are model responses for altered and nominal parameters, respectively and P_a and P_n are the altered and nominal parameters, respectively. A sensitivity index less than 1 would indicate that the proportional response is less than the proportional change in the parameter value while a value greater than 1 indicates the proportional response is greater than the proportional change in the parameter value. In all cases, the sensitivity index was less than 1 which indicates that the models may be simplified further but in consideration with the MSD it is possible to say that the mathematical formulations and solutions have been applied in a manner that will give logical output values.

Further work

To understand if the non-linear equations fitted to the data are also representative of a larger beef cattle population there needs to be further evaluation of the models to external data. Also to fully understand how the whole muscle develops as the animal matures, there needs to be model fitting for other muscle characteristics such as collagen or fat concentration.

Considering the large variation in the data for some of the muscle characteristics, further modelling procedures may want to consider adding some form of stochasticity to be able to represent the muscle characteristics of the herd rather than just an individual animal.

The animal factors that drive the development of muscle characteristics result from changes in the animal's growth characteristics and body composition. Further work will consider interactions between the muscle characteristics to build a whole muscle model that can then be linked to a growth model. This will enable the daily gain or the existence of discontinuous growth or effects of different nutritional regimes that alter body composition or growth characteristics to be factored into the development of the muscle characteristics associated with meat quality. It will be particularly needed for the estimation of intramuscular fat proportion.

Conclusion

The trend in consumer demands for meat products is oriented towards better control of meat quality for improved sensorial qualities. To understand and predict the responses of on-farm treatments affecting meat quality, models can be developed that describe the dynamics of muscle characteristics that influence meat quality through changes in animal growth. The non-linear equations that have been generated are able to mathematically describe, and provide further understanding of, the changes in the fibre cross-sectional area, fibre type proportions, and the activity of metabolic enzymes that occurs during post-natal growth in different types of cattle.

The ultimate goal will be to produce a model that can predict the development of muscle and provide an indication of meat quality from the growth and composition characteristics of cattle. The model would assess what growth paths of cattle can provide the best quality meat products and also give an indication of which carcasses and cuts are going to be best suited for different markets. This study is a first step toward a dynamic model that would account for the effects of nutrition and growth on the development of muscle characteristics in different classes of cattle.

References

Dransfield, E., Martin, J.F., Bauchart, D., Abouelkaram, S., Lepetit, J., Culioli, J., Jurie, C. and Picard, B., 2003. Meat quality and composition of three muscles from French cull cows and young bulls. Animal Science 76:387-399.

Gauch, H.G., Gene Hwang, J.T. and Fick, G.W., 2003. Model evaluation by comparison of model-based predictions and measured values. Agronomy Journal 95:1442-1446.

Gagnière, H., Picard, B., Jurie, C. and Geay, Y, 1997. Comparative study of metabolic differenciation of foetal muscle in normal and double-muscled cattle. Meat Science 45:145-152.

Haefner, J.W., 2005. Modeling Biological Systems – Principles and Applications. Springer, New York, USA, 475 pp.

Hoch T., Picard B., Jurie C. and Agabriel J., 2002. Modélisation de l'évolution des caractéristiques des fibres musculaires de bovins. 9èmes Journée des Sciences du Muscle et Technologies de la Viande. Clermont-Ferrand, France,15-16 Octobre 2002. Viandes et Produits Carnés, Hors Série, pp.121-122.

Purchas, R.W., 1989. On-farm factors affecting meat quality characteristics. In: Purchas, R.W., Butler-Hogg, B.W. and Davis, A.S. (eds.). Meat Production and Processing. New Zealand Society of Animal Production, Hamilton, New Zealand, pp.159-171.

Renand, G., Picard, B., Touraille, C., Berge, P. and Lepetit, J., 2001. Relationships between muscle characteristics and meat quality traits of young Charolais bulls. Meat Science 59:49-60.

Schreurs, N.M., Garcia, F., Jurie, C., Agabriel, J., Micol, D., Bauchart, D., Listrat, A. and Picard, B., 2008. Meta-analysis of the effect of animal maturity on muscle characteristics in different muscles, breeds, and sexes of cattle. Journal of Animal Science 86:2872-2887.

Thornley, J.H.M. and France, J., 2007. Mathematical Models in Agriculture. CAB International, Wallingford, UK, 906 pp.

Wegner, J., Albrecht, E., Fiedler, I., Teuscher, F., Papstein, H.J. and Ender, K., 2000. Growth- and breed-related changes of muscle fiber characteristics in cattle. Journal of Animal Science 78:1485-1496.

An isotope dilution model for partitioning phenylalanine uptake by the liver of lactating dairy cows

L.A. Crompton[1], C.K. Reynolds[1], J.A.N. Mills[1] and J. France[2]
[1]Animal Science Research Group, School of Agriculture, Policy and Development, University of Reading, Whiteknights, Reading RG6 6AR, United Kingdom; l.a.crompton@reading.ac.uk
[2]Centre for Nutrition Modelling, Department of Animal & Poultry Science, University of Guelph, Ontario N1G 2W1, Canada

Abstract

An isotope dilution model for partitioning phenylalanine uptake by the liver of the lactating dairy cow was constructed and solved in the steady state. If assumptions are made, model solution permits calculation of the rate of phenylalanine uptake from portal vein and hepatic arterial blood supply, phenylalanine release into the hepatic vein, phenylalanine oxidation and synthesis, and degradation of hepatic constitutive and export proteins. The model requires the measurement of plasma flow rate through the liver in combination with phenylalanine concentrations and plateau isotopic enrichments in arterial, portal and hepatic plasma during a constant infusion of [1-^{13}C]phenylalanine tracer. The model can be applied to other amino acids with similar metabolic fates and will provide a means for assessing the impact of hepatic metabolism on amino acid availability to peripheral tissues. This is of particular importance for the dairy cow when considering the requirements for milk protein synthesis and the negative environmental impact of excessive nitrogen excretion.

Keywords: isotope dilution, phenylalanine, liver, dairy cows

Introduction

The ability to predict accurately dietary nitrogen utilisation for production in the ruminant depends on a clear understanding of amino acid flows from the gut to productive tissues. This requires an accurate description of the processes involved, including the regulation of intestinal amino acid supply, absorption of these products from the gut, metabolism by gut and liver tissues and subsequent use by peripheral tissues. Whilst each of these processes is important, a major site of net metabolism of digested and absorbed amino acids (and peptides) in a lactating dairy cow is the liver (Reynolds, 2006). In many cases, up to 80% of the free α-amino nitrogen appearing in the portal vein is removed by ruminant liver metabolism (Reynolds, 2006). Whilst the liver is a major site of oxidation, some of the extracted amino acids can be used for protein synthesis and other synthetic processes. The first step in amino acid oxidation is generally deamination, after which some of the carbon skeletons can be used subsequently for glucose synthesis. The nitrogen arising from deamination can be used for *de novo* synthesis of non-essential amino acids or in urea synthesis. There are a variety of mechanisms that exert a coordinated effort to control the rate of amino acid extraction and deamination in the liver.

Phenylalanine (PHE) and to a lesser extent tyrosine have been widely used as tracer amino acids for the study of protein metabolism in several species. Phenylalanine normally has only two metabolic fates, incorporation into cellular proteins or oxidation via hydrolysis to tyrosine. As a result, PHE catabolism always follows the pathway of tyrosine catabolism and the major site of this breakdown is in the liver. The primary objective of this work was to develop a steady-state model of liver PHE metabolism adequate to analyse *in vivo* isotopic data in the lactating dairy cow.

The model

The scheme adopted is shown in Figure 1A. It contains two intracellular and three extracellular pools. The intracellular pools are free PHE (pool 4) and PHE in export protein (pool 3), whilst the extracellular ones represent portal vein PHE (pool 1), hepatic artery PHE (pool 2) and hepatic vein PHE (pool 5). The flows of PHE between pools and into and out of the system are shown as arrowed lines. The export protein-bound PHE pool has a single inflow: from free PHE, F_{34}, and two efflows: secretion of export protein, F_{03}, and degradation, F_{43}. The intracellular free PHE pool has four inflows: from the degradation of constitutive liver protein, F_{40}, from the extracellular portal vein pool, F_{41}, from the hepatic artery pool, F_{42}, and from degradation of export protein, F_{43}. The pool has four efflows: synthesis of constitutive liver protein, $F_{04}^{(s)}$, oxidation, $F_{04}^{(o)}$, incorporation into export protein, F_{34}, and outflow to the extracellular hepatic vein PHE pool, F_{54}. The extracellular portal vein PHE pool has a single inflow: entry into the pool, F_{10}, and two efflows: uptake by the liver, F_{41}, and release into the extracellular hepatic vein PHE pool, F_{51}. The extracellular hepatic artery PHE pool has a single inflow: entry into the pool, F_{20}, and two efflows: uptake by the liver, F_{42}, and release into the extracellular hepatic vein PHE pool, F_{52}. The extracellular hepatic vein PHE pool has three inflows: bypass from the portal vein PHE pool, F_{51}, bypass from the hepatic artery PHE pool, F_{52}, and release from the intracellular PHE pool, F_{54}, and one efflow out of the system, F_{05}.

The scheme adopted for the movement of label is shown in Figure 1B. [1-^{13}C]PHE was infused into the jugular vein at a constant rate and the enrichment of all five pools monitored. The enrichment of the extracellular pools is measured directly by taking blood samples from the portal vein, hepatic artery and hepatic vein during the isotope infusion. Blood flow rate across the liver is measured by downstream dye dilution using para-aminohippuric acid (PAH). The scheme assumes that the only entry of label into the system is into the PHE portal vein and hepatic artery pools via the effective infusion rates, flows I_1 and I_2 and that the duration of the infusion is such that the enrichment of constitutive protein can be regarded as negligible.

Conservation of mass principles can be applied to each pool in Figure 1A and B to generate differential equations that describe the dynamic behaviour of the system. For total (isotopic plus non-isotopic) PHE these differential equations are (mathematical notation is defined in Table 1):

$$\frac{dQ_1}{dt} = F_{10} - F_{41} - F_{51} \quad (1)$$

$$\frac{dQ_2}{dt} = F_{20} - F_{42} - F_{52} \quad (2)$$

$$\frac{dQ_3}{dt} = F_{34} - F_{03} - F_{43} \quad (3)$$

$$\frac{dQ_4}{dt} = F_{40} + F_{41} + F_{42} + F_{43} - F_{04}^{(o)} - F_{04}^{(s)} - F_{34} - F_{54} \quad (4)$$

$$\frac{dQ_5}{dt} = F_{51} + F_{52} + F_{54} - F_{05} \quad (5)$$

and for [^{13}C] labelled PHE:

$$\frac{dq_1}{dt} = I_1 - e_1\left(F_{41} + F_{51}\right) \quad (6)$$

$$\frac{dq_2}{dt} = I_2 - e_2\left(F_{42} + F_{52}\right) \quad (7)$$

$$\frac{dq_3}{dt} = e_4 F_{34} - e_3\left(F_{03} + F_{43}\right) \quad (8)$$

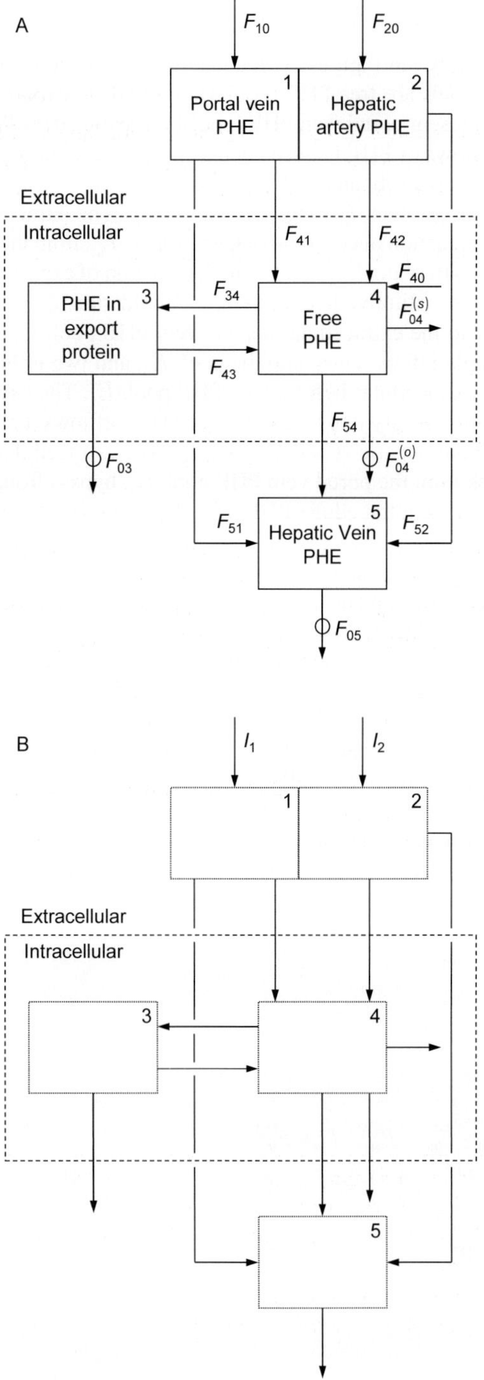

Figure 1. Scheme for the uptake and utilisation of PHE by the liver of lactating dairy cows: (A) total PHE and (B) [^{13}C] labelled PHE. The small circles in Figure 1A indicate flows out of the system which need to be measured experimentally.

$$\frac{dq_4}{dt} = e_1F_{41} + e_2F_{42} + e_3F_{43} - e_4\left(F_{04}^{(o)} + F_{04}^{(s)} + F_{34} + F_{54}\right) \tag{9}$$

$$\frac{dq_5}{dt} = e_1F_{51} + e_2F_{52} + e_4F_{54} - e_5F_{05} \tag{10}$$

When the system is in steady state with respect to both total and labelled PHE, the derivative terms in Equations 1-10 are zero. For the scheme assumed, the enrichment of intracellular export protein-bound pool equalizes with that of the free pool in steady state (i.e. $e_3 = e_4$) otherwise Equations 3 and 8 are inconsistent. After equating intracellular enrichments and eliminating redundant equations, the five differential equations for labelled PHE (Equations 5-10) yield the following 4 identities:

$$I_1 - e_1\left(F_{41} + F_{51}\right) = 0 \tag{11}$$

$$I_2 - e_2\left(F_{42} + F_{52}\right) = 0 \tag{12}$$

$$e_1F_{41} + e_2F_{42} - e_4\left(F_{04}^{(o)} + F_{04}^{(s)} + F_{34} + F_{54} - F_{43}\right) = 0 \tag{13}$$

$$e_1F_{51} + e_2F_{52} + e_4F_{54} - e_5F_{05} = 0 \tag{14}$$

To obtain steady state solutions to the model, it is assumed that PHE secreted in export protein, CO_2 production and PHE removal from the hepatic vein pools (i.e. F_{03}, $F_{04}^{(o)}$ and F_{05}, respectively) can all be measured experimentally. Further, it is mathematically convenient to assume that percentage PHE extraction by the liver is the same from the portal vein and hepatic artery supplies, giving:

$$\frac{F_{51}}{F_{52}}\left(=\frac{F_{41}}{F_{42}}\right)=\frac{F_{10}}{F_{20}} \tag{15}$$

Algebraic manipulation of Equations 1-5 with the derivatives set to zero, together with Equations 11-15, gives the solution to the model:

$$F_{10} = I_1 / e_1 \tag{16}$$

$$F_{20} = I_2 / e_2 \tag{17}$$

$$\overline{F_{34} - F_{43}} = \widetilde{F}_{03} \tag{18}$$

Table 1. Principle symbols.

F_{ij}	Flow of phenylalanine[1] to pool i from j; F_{i0} denotes an external flow into pool i and F_{0j} denotes a flow from pool j out of the system	μmol/min
I_i	Effective constant infusion rate of ^{13}C labelled phenylalanine into pool i	μmol/min
Q_i	Quantity of total phenylalanine in pool i	μmol
q_i	Quantity of ^{13}C labelled phenylalanine tracer in pool i	μmol
e_i	Enrichment of ^{13}C labelled phenylalanine in pool i (= q_i/Q_i)	atoms % excess (μmol label/μmol of total phenylalanine[1])
t	Time	min

[1]Total material (*i.e.* tracee + tracer).

$$F_{51} = \frac{(e_5 - e_4)\widetilde{F}_{05}}{(e_1 - e_4) + (e_2 - e_4)\frac{F_{20}}{F_{10}}} \tag{19}$$

$$F_{52} = \frac{F_{20}}{F_{10}} F_{51} \tag{20}$$

$$F_{54} = \widetilde{F}_{05} - F_{51} - F_{52} \tag{21}$$

$$F_{41} = F_{10} - F_{51} \tag{22}$$

$$F_{42} = F_{20} - F_{52} \tag{23}$$

$$F_{04}^{(s)} = \frac{e_1}{e_4} F_{41} + \frac{e_2}{e_4} F_{42} - \widetilde{F}_{04}^{(o)} - \overline{F_{34} - F_{43}} - F_{54} \tag{24}$$

$$F_{40} = \widetilde{F}_{04}^{(o)} + F_{04}^{(s)} + \overline{F_{34} - F_{43}} + F_{54} - F_{41} - F_{42} \tag{25}$$

where for Equations 16-25 the italics denote steady state values of flows and enrichments, the tilde identifies a measured flow, and the over-lining indicates coupled flows (which cannot be separately estimated).

Application

Application of the model is illustrated using data from experiments conducted at our laboratories with multi-catheterised mid-lactation Holstein-Friesian dairy cows (average live-weight 600 kg) fed total mixed ration (TMR) diets consisting of a 50:50 mixture on a dry matter basis of forage and concentrate with the forage comprised of grass silage and chopped dried Lucerne in a 25:75 ratio on a dry matter basis. Concentrates were formulated to provide crude protein levels of approximately 110 and 200 g per kg concentrate dry matter, such that average TMR crude protein concentration were 128 and 175 g/kg dry matter. Average daily dry matter intake and milk yield for these animals were 22 kg/d and 30 l/d, respectively. The cows were given constant abomasal infusions of water (18 l/d) for 4 d, followed by a buffered mixture of essential amino acids for a further 6 d. The essential amino acids were administered at a daily rate equivalent to the essential amino acids in 800 g milk protein. On the final day of each infusion, the animals received a primed, constant jugular vein infusion of [1-^{13}C]PHE (350 mg/h) for 8 h and 6 hourly blood sample sets were taken simultaneously from catheters in the dorsal aorta and the portal and hepatic veins for the measurement of blood flow rate (by PAH dilution) and nutrient metabolism by the portal drained viscera and liver. Preliminary reports from these experiments have been presented previously (see Reynolds *et al.*, 2000a, b).

The relevant experimental measurements are given in Table 2. They are reported for three animals during the water infusion (2 low protein; 1 high protein diet) and two animals during the amino acid infusion, (1 low protein; 1 high protein). They are based on plasma rather than whole blood values. Phenylalanine measurements are based on free rather than total (i.e. free plus bound) plasma PHE. The effective isotope flows to the liver I_1 and I_2 were obtained from portal/arterial concentration and enrichment of PHE and portal/arterial plasma flow. The flow F_{05} was determined from hepatic measurements of plasma flow and PHE concentration. Two of the measurements shown in Table 2, namely the intracellular enrichment e_4 and the flow F_{03} were not made in the experiment and therefore had to be estimated. Unpublished observations from parallel experiments in our laboratory, using similar animals and isotope infusion rates, have shown that the ratio of hepatic intracellular PHE enrichment to arterial plasma free PHE enrichment is 0.306 with a standard error of 0.037 (Crompton *et al.*, unpublished observations). Therefore the missing intracellular free PHE enrichment, e_4, was

Table 2. Experimental measurements.

Animal/infusion		323/21	323/23[c]	6132/41	6031/42	6132/43[c]
Concentrate protein (%)		11	11	20	11	20
Plateau enrichment (ape)[a]	e_1	2.40	2.21	2.73	2.79	2.10
	e_2	4.05	3.13	4.35	4.31	3.32
	e_3[b]	1.24	0.96	1.33	1.32	1.02
	e_4[b]	1.24	0.96	1.33	1.32	1.02
	e_5	2.33	2.20	2.59	2.70	2.00
Flow (μmol/min)	I_1	53.7	79.3	49.3	53.7	56.9
	I_2	18.3	19.0	8.9	16.5	9.5
	F_{03}[b]	33.8	33.8	33.8	33.8	33.8
	$F_{04}^{(o)}$	316.1	517.3	429.0	350.9	525.2
	F_{05}	2,304.3	3,553.1	1,700.3	1,938.1	2,472.8

[a] Ape: atoms % excess.
[b] Estimated value.
[c] Essential amino acid mixture infused.

estimated at 0.306 times the corresponding arterial enrichment e_2. The flow $F_{04}^{(o)}$ was obtained from labelled CO_2 elevation in plasma flow across the liver and hepatic vein PHE enrichment. The export protein flow F_{03} was assigned the value of 33.8 μmol/min from the study of Raggio *et al.* (2007).

The model calculations obtained from the experimental data are presented in Table 3. For the three animals shown, there was little difference in the supply of PHE to the liver between the high and low protein concentrate diets, but the abomasal infusion of essential amino acids increased the supply of PHE to the liver. Phenylalanine extraction by the liver from the portal vein and hepatic artery was 31.3% on average. The degradation of constitutive liver protein plus undefined sources (F_{40}), and the portal and arterial supply (F_{41} and F_{42}), contribute almost equal amounts of PHE to the intracellular free PHE pool (between 53 and 59% from F_{40}). The hepatic vein efflow of PHE was always less than PHE inflow, showing that the liver is a net utilizer of PHE. The proportion of the intracellular free PHE pool that was oxidised to CO_2 ranged from 15 to 31%. The flow of PHE for constitutive protein synthesis was always much larger than the export protein synthesis flow, the contribution of the former averaged 96.7% of the total protein synthetic flow (Table 3).

Discussion

The model was developed to resolve *in vivo* trans-organ and isotope dilution data obtained from experiments conducted in our laboratories with multi-catheterised lactating dairy cows. Despite some limitations, the model described is a useful tool for partitioning the uptake of PHE by the ruminant liver. The level of representation adopted means that the model could be applied to other amino acids with similar metabolic fates within the liver. If the model is to be rigorously applied, future *in vivo* studies must measure directly the intracellular free amino acid enrichment in conjunction with the measurements of plasma enrichments and flow rates. Kinetic schemes such as the one described herein, could be used as a basis for future process-based simulation models of hepatic metabolism.

Table 3. Model calculations.

Animal/infusion		323/21	323/23[a]	6132/41	6031/42	6132/43[a]
Concentrate protein (%)		11	11	20	11	20
Flow (μmol/min)	F_{10}	2,240.5	3,585.3	1,806.3	1,921.3	2,710.8
	F_{20}	451.3	606.2	205.5	382.3	287.5
	$F_{34} - F_{43}$	33.8	33.8	33.8	33.8	33.8
	F_{51}	1,455.6	2,711.1	1,229.2	1,289.2	1,831.8
	F_{52}	293.2	458.4	139.8	256.5	194.3
	F_{54}	555.6	383.6	331.2	392.4	446.8
	F_{41}	784.9	874.2	577.0	632.1	879.1
	F_{42}	158.1	147.8	65.6	125.8	93.2
	$F_{04}^{(s)}$	1,131.5	1,569.4	605.3	972.4	1,116.0
	F_{40}	1,094.0	1,482.1	756.7	991.6	1,149.5

[a]Essential amino acid mixture infused.

Acknowledgements

The financial support of UK Defra project LS3656 and the Canada Research Chairs Program are gratefully acknowledged.

References

Raggio, G., Lobley, G.E., Berthiaume, R., Pellerin, D., Allard, G., Dubreuil, P and Lapierre, H., 2007. Effect of protein supply on hepatic synthesis of plasma and constitutive proteins in lactating dairy cows. Journal of Dairy Science 90:352-359.

Reynolds, C.K., 2006. Splanchnic metabolism of amino acids in ruminants. In: Sejrsen, K., Hvelplund, T. and Nielsen, M.O. (eds.), Ruminant Physiology. Digestion, metabolism and impact of nutrition on gene expression, immunology and stress. Wageningen Academic Publishers, Wageningen, the Netherlands, pp.225-248,

Reynolds, C.K., Crompton, L.A., Bequette, B.J., France, J., Beever, D.E. and MacRae, J.C., 2000a. Effects of diet protein level and abomasal amino acid infusion on phenylalanine and tyrosine metabolism in lactating dairy cows. Journal of Dairy Science 83: Supplement 1, 298.

Reynolds, C.K., Lupoli, B., Aikman, P.C., Benson, J.A., Humpheries, D.J., Crompton, L.A., France, J., Beever, D.E. and MacRae, J.C., 2000b. Effects of diet protein level and abomasal amino acid infusions on splanchnic metabolism in lactating dairy cows. Journal of Dairy Science 83: Supplement 1, 299.

Modeling the effects of insulin and amino acids on the phosphorylation of mTOR, Akt, and 4EBP1 in mammary cells

J.A.D.R.N. Appuhamy and M.D. Hanigan
Department of Dairy Science, Virginia Polytechnic Institute and State University, 2470 Litton-Reaves Hall, Blacksburg, VA 24061, USA; appuhamy@vt.edu

Abstract

The objective of this work was to develop a mathematical representation of the effects of insulin and essential amino acid (EAA) on phosphorylation of protein kinase B (Akt), mammalian target of rapamycin (mTOR) and eukaryotic initiation factor 4E binding protein 1 (4EBP1), the latter being critical for initiation of protein synthesis. The model included six protein pools (Q) representing phosphorylated (P) and unphosphorylated (U) forms of Akt, mTOR, and 4EBP1. Mass action equations were used to represent kinase ($F_{U,P(X)}$) and phosphatase ($F_{P,U(X)}$) reactions. The $F_{U,P(X)}$ for Akt and mTOR were regulated by extracellular insulin (C_{Ins}) and EAA (C_{EAA}) concentrations, respectively. Exponents were used to adjust the sensitivity of the fluxes to the regulators. Changes in pool size with respect to time were calculated as the difference between $F_{U,P(X)}$ and $F_{P,U(X)}$. The $Q_{U(X)}$ were determined by numerical integration of the differentials starting from specified initial pool sizes. The $Q_{P(X)}$ were calculated by subtracting $Q_{U(X)}$ from the fixed total protein mass ($Q_{T(X)}$). The model was fitted to observed phosphorylation data obtained from a bovine mammary epithelial cell line treated with four C_{Ins} (0, 5, 10, and 100 ng/ml) and four C_{EAA} (0, 0.35, 1.00, and 3.5 mM) arranged in a 4x4 factorial design. Model optimization and sensitivity analyses were carried out in ACSLXtreme. The data were adequate to describe the model parameters as standard deviations of model parameters were <20% of the parameter estimates. Sensitivity exponent estimates were greater than 1 indicating EAA and insulin signal loss associated with transmission down the cascade was partially mitigated. Phosphorylation of Akt was highly sensitive to insulin compared to mTOR and 4EBP1. Phosphorylation of mTOR and 4EBP1 responded similarly to insulin and EAA. The model was able to predict $Q_{P(X)}$ with root mean square prediction errors less than 10% of the observed means. There appeared to be a slight negative slope bias for $Q_{P(Akt)}$, indicating the model tended to overpredict $Q_{P(Akt)}$ as predicted $Q_{P(Akt)}$ increased.

Keywords: amino acid, cellular signals, insulin, mathematical representation

Introduction

Amino acids (AA) and insulin are able to stimulate translation initiation in mammalian cells. These stimulatory signals have been shown to converge on the protein kinase, mammalian target of rapamysin (mTOR) (Proud, 2007). Insulin and AA activate mTOR by stimulating its phosphorylation via two independent pathways (Suryawan *et al.*, 2007). Insulin actions are mediated through phosphorylation of the upstream kinase, Akt (Hinault *et al.*, 2004). The exact pathway through which amino acids stimulate mTOR phosphorylation is still not clear. Eukaryotic initiation factor 4E (eIF4E) binding protein 1 (4EBP1) is one of the best studied direct substrates of mTOR (Avruch *et al.*, 2009). The eIF4E is able to bind the 5'cap of mRNA and thus enhances translation initiation by recruiting methionyl-tRNA to mRNA. The 4EBP1 competitively binds eIF4E thus inhibiting its function. mTOR mediated hyper-phosphorylation of 4EBP1 leads to inactivate 4EBP1 thereby releasing eIF4E and enhancing rates of translation initiation (Richter and Sonenberg, 2005).

Understanding mTOR mediated signals is important for continued development of nutrient requirement systems that allow construction of sound nutritional and feeding management programs for human and livestock species. Since it has been well recognized that amino acids play an important role in controlling protein synthesis in addition to its role as a substrate (Proud, 2007), mathematical representation of these signalling effects is critical to improving amino acid requirement predictions for livestock species. Moreover, quantification of regulatory effects of amino acids and their interactions with other signals such as insulin, could also be important in human health studies given the involvement of mTOR dependent cellular signals in obesity, type 2 diabetes, and cancer (Marshall, 2006). The objective of this work was to develop a mathematical model representing the effects of insulin and amino acids on phosphorylation of Akt, mTOR, and 4EBP1 and to parameterize the model using a data set derived from a bovine mammary epithelial cell line (MAC-T cells).

Materials and methods

Model development

The model was constructed as a dynamic, deterministic, mechanistic representation of the Akt/mTOR/4EBP1 signalling cascade as depicted in Figure 1.

The model considered 3 proteins each in the *P* and *U* forms: Akt ($Q_{P(Akt)}$ and $Q_{U(Akt)}$), mTOR ($Q_{P(mTOR)}$ and $Q_{U(mTOR)}$), and 4EBP1 ($Q_{P(4EBP1)}$ and $Q_{U(4EBP1)}$), respectively for a total of 6 protein pools. Mass action equations were used to represent kinase ($F_{U,P}$) and phosphatase ($F_{P,U}$) reactions for each signalling protein. The time unit for the model was minutes.

For purposes of model development, a reference model state was defined as having inputs of 10 ng/ml insulin concentration (C_{Ins}) and 1 mM total EAA concentration (C_{EA}). In preliminary *in vitro*

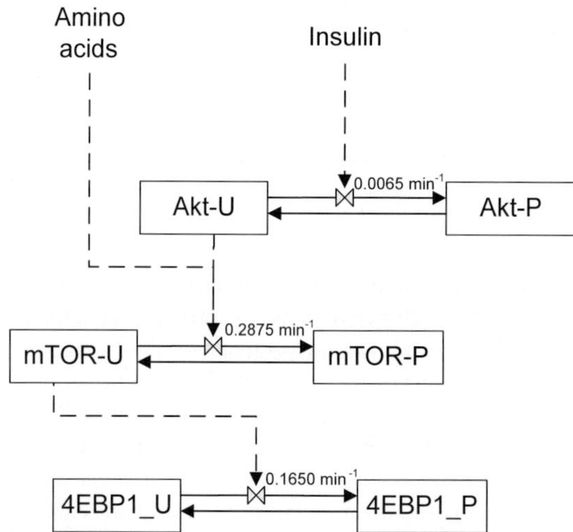

Figure 1. A schematic representation of the model. The arrows represent fluxes. Dashed arrows represent the upstream effects on phosphorylation fluxes. Boxes with solid lines represent protein mass pools. Values of kinase fluxes (min^{-1}) for each protein at reference state (insulin = 10 ng/mland EAA = 1 mM) are given.

experiments using a bovine mammary epithelial cell line, MAC-T, these concentrations fell in the middle of dose response curves of total protein synthesis At this reference state, the total pool size (Q_T) and Q_P of each signalling protein were set to 1 and 0.5 (arbitrary units) respectively assuming 50% of the total protein in each pool was in the phosphorylated state.

Kinase reactions ($F_{U,P}$) for Akt ($F_{U,P(Akt)}$), mTOR ($F_{U,P(mTOR)}$), and 4EBP1 (($F_{U,P(4EBP1)}$) were represented respectively as:

$$F_{U,P(Akt)} = Q_{U(Akt)} \times K_{U,P(Akt)} \times \left(\frac{C_{Ins}}{J_{Ins}}\right) \tag{1}$$

$$F_{U,P(mTOR)} = Q_{U(mTOR)} \times K_{U,P(mTOR)} \times \left(\left(\frac{Q_{P(Akt)}}{J_{QP(Akt)}}\right)^{a} + \left(\frac{C_{EAA}}{J_{EAA}}\right)^{b}\right) \tag{2}$$

$$F_{U,P(4EBP1)} = Q_{U(4EBP1)} \times K_{U,P(4EBP1)} \times \left(\frac{Q_{P(mTOR)}}{J_{QP(mTOR)}}\right)^{c} \tag{3}$$

where a, b, and c represented the sensitivity exponents, and J_{Ins}, J_{EAA},. $J_{QP(Akt)}$, and $J_{QP(mTOR)}$ were constants representing the regulatory effects of C_{Ins}, C_{EAA}, $Q_{P(Akt)}$, and $Q_{P(mTOR)}$ respectively. They were set equal to the reference values of C_{Ins} and C_{EAA}, respectively and the steady state values of $Q_{P(Akt)}$, and $Q_{P(mTOR)}$ respectively when the model was in the reference state. This resulted in the ratios assuming a value of 1 in the reference state which allowed a, b, and c to assume any positive value without affecting model balance in the reference state. $K_{U,P(Akt)}$, $K_{U,P(mTOR)}$, and $K_{U,P(4EBP1)}$ were the mass action rate constants for phosphorylation of Akt, mTOR, and 4EBP1 respectively. Since it has been demonstrated that insulin and AA stimulate kinase activity resulting in the phosphorylation of mTOR via independent pathways in mammalian cells (Hinault *et al.*, 2004 and Suryawan *et al.*, 2007), the effects of insulin and EAA on $F_{U,P(mTOR)}$ were represented as additive factors in Equation 2.

Phosphotase reactions for Akt ($F_{P,U(Akt)}$), mTOR ($F_{P,U(mTOR)}$), and 4EBP1 ($F_{P,U(4EBP1)}$) were represented as:

$$F_{P,U(Akt)} = Q_{P(Akt)} \times K_{P,U(Akt)} \tag{4}$$

$$F_{P,U(mTOR)} = Q_{P(mTOR)} \times K_{P,U(mTOR)} \tag{5}$$

$$F_{P,U(4EBP1)} = Q_{P(4EBP1)} \times K_{P,U(4EBP1)} \tag{6}$$

where $K_{P,U(Akt)}$, $K_{P,U(mTOR)}$, and $K_{P,U(4EBP1)}$ represented the respective reaction rate constants. No regulation of dephosphorylation was represented.

The differential equations defining the rate of change (units/min) for each unphosphorylated signalling protein were calculated from the balance of the phosphorylation and dephosphorylation fluxes:

$$\frac{dQ_{U(Akt)}}{dt} = F_{P,U(Akt)} - F_{U,P(Akt)} \tag{7}$$

$$\frac{dQ_{U(mTOR)}}{dt} = F_{P,U(mTOR)} - F_{U,P(mTOR)} \tag{8}$$

$$\frac{dQ_{U(4EBP1)}}{dt} = F_{P,U(4EBP1)} - F_{U,P(4EBP1)} \tag{9}$$

The Q_U for each signalling protein was calculated by numerical integration of Equations 7, 8, and 9 given specified initial pool sizes for Q_U (iQ_U).

$$Q_{U(Akt)} = Integ\left(dQ_{U(Akt)}, iQ_{U(Akt)}\right) \tag{10}$$

$$Q_{U(mTOR)} = Integ\left(dQ_{U(mTOR)}, iQ_{U(mTOR)}\right) \tag{11}$$

$$Q_{U(4EBP1)} = Integ\left(dQ_{U(4EBP1)}, iQ_{U(4EBP1)}\right) \tag{12}$$

The total mass for each signalling protein (Q_T) which represents the sum of Q_P and Q_U was fixed in size and used to calculate Q_P of each signalling protein.

$$Q_{P(Akt)} = Q_{T(Akt)} - Q_{U(Akt)} \tag{13}$$

$$Q_{P(mTOR)} = Q_{T(mTOR)} - Q_{U(mTOR)} \tag{14}$$

$$Q_{P(4EBP1)} = Q_{T(4EBP1)} - Q_{U(4EBP1)} \tag{15}$$

Because phosphorylated and total quantities of signalling proteins are measured independently in arbitrary units using Western blotting techniques, it is possible that the signal intensity for the phosphorylated protein assay could exceed the signal intensity for the total assay. Of course this simply reflects the relative binding affinity of the two sets of antibodies given that the phosphorylated form of the protein cannot exceed the total form in reality. To accommodate this in the model and provide for a model variable that could be compared to the observed data, a scaled value (rQ_P) was calculated from Q_P using a reference scalar for each Q_P (ref_Q_P) in the same manner that the Western results were scaled:

$$rQ_{P(Akt)} = Q_{P(Akt)} / ref\_Q_{P(Akt)} \tag{16}$$

$$rQ_{P(mTOR)} = Q_{P(mTOR)} / ref\_Q_{P(mTOR)} \tag{17}$$

$$rQ_{P(4EBP1)} = Q_{P(4EBP1)} / ref\_Q_{P(4EBP1)} \tag{18}$$

Model optimization

MAC-T cells were grown to approximately 90% confluence in complete DMEM/F12 media supplemented with 8% serum. Treatments were applied for 1h in serum free DMEM/F12 media containing 3.51 g/lD-glucose. Treatments were four concentrations of EAA (0.0, 0.35, 0.70, and 3.50 mM) and four concentrations of insulin (0, 5, 10, and 100 ng/ml) arranged in a 4×4 factorial design. Protein homogenates obtained from treated cells were subjected to Western immunoblotting analysis. Briefly, proteins were electroporetically separated in 8-14% sodium dodecyl sulfate (SDS)-polyacrylamide gels and transferred onto polyvinylidene fluoride (PVDF) membranes (Bio-Rad, Hercules, CA). The blots were blocked with 5% non-fat dry milk in a tris-base Tween buffer (10 mM tris-base, 150 mM NaCl, and 1% Tween 20) and first probed with antibodies (1:1000 dilutions) specific to the phosphorylated forms (P) of Akt (Ser473), mTOR (Ser2448), and 4EBP1 (Thr37/46). Blots were then stripped and reprobed with antibodies (1:2,000 delusions) that recognized the total form (T) of each. All antibodies were purchased from Cell Signaling Technologies (Beverly, MA). Protein abundance was visualized by chemiluminescence (ECL Plus, Amersham; Piscataway, NJ) and quantified using Un-Scan-It software (Silk Scientific, Inc., Orem, UT). The phosphorylation state of each signaling protein was expressed as a ratio of P:T that was then standardized to the observations from cells treated with media containing highest concentrations of insulin (100 ng/ml) and EAA (3.50 μM). This standardized ratio was equivalent to $rQ_{P(X)}$ in the model.

The model was fitted to the observed P:T ratio data by solving for $K_{U,P(Akt)}$, $K_{U,P(mTOR)}$, $K_{U,P(4EBP1)}$, $K_{P,U(Akt)}$, $K_{P,U(mTOR)}$, $K_{P,U(4EBP1)}$, J_{Ins}, J_{EAA}, $J_{QP(mTOR)}$, $J_{QP(Akt)}$, a, b, c, $Ref\_Q_{P(Akt)}$, $Ref\_Q_{P(mTOR)}$, $Ref\_Q_{P(4EBP1)}$, $iQ_{P(Akt)}$, $iQ_{P(mTOR)}$, and $iQ_{P(4EBP1)}$ while maximizing a log-likelihood function using the Nelder Mead search algorithm in ACSLXtreme (AEgis Technologies, Huntsville, AL). The residual errors of $Q_{P(Akt)}$, $Q_{P(mTOR)}$, and $Q_{P(4EBP1)}$ were regressed on corresponding predicted values using SAS 9.1 (SAS Inc., Cary, NC). Sensitivities of Akt, mTOR, and 4EBP1 to C_{Ins} and C_{EAA} were also determined.

Results and discussion

Parameter estimates are presented in Table 1. As evidenced by standard deviations for parameter estimates that were less than 20% of the estimates in all cases, the data were adequate to describe the model parameters. Correlations among parameters did not appear to significantly contribute to variance inflation in the parameter estimates. The final parameter estimates for J_{Ins} (8.390), J_{EAA}

Table 1. Model parameter estimates and associated standard deviations (SD). The SD are presented as percentages of the parameter estimate (SD%).

Parameter	Estimate	SD	SD%
Kinase reaction rate constants			
$K_{U,P(Akt)}$	0.013	5.2×10^{-6}	0.000
$K_{U,P(mTOR)}$	0.575	1.2×10^{-6}	0.000
$K_{U,P(4EBP1)}$	0.330	9.3×10^{-7}	0.003
Phosphatase reaction rate constants			
$K_{P,U(Akt)}$	1.220	9.5×10^{-5}	0.008
$K_{P,U(mTOR)}$	2.219	4.6×10^{-5}	0.002
$K_{P,U(4EBP1)}$	1.206	2.6×10^{-5}	0.002
Regulatory effect constants			
J_{Ins}	8.390	7.5×10^{-4}	0.001
J_{EAA}	1.070	2.2×10^{-2}	0.002
$J_{QP(mTOR)}$	0.494	8.1×10^{-6}	0.002
$J_{QP(Akt)}$	0.487	7.7×10^{-6}	0.002
Sensitivity exponents			
a	1.450	2.5×10^{-5}	0.002
b	1.490	3.1×10^{-5}	0.002
c	2.370	3.9×10^{-5}	0.002
Reference scalars for phosphorylated protein pools			
$Ref\_Q_{P(Akt)}$	0.910	5.2×10^{-6}	0.001
$Ref\_Q_{P(mTOR)}$	1.037	1.2×10^{-2}	1.157
$Ref\_Q_{P(4EBP1)}$	0.828	9.3×10^{-7}	0.000
Initial phosphorylated protein pools			
$iQ_{P(Akt)}$	0.512	2.7×10^{-6}	0.000
$iQ_{P(mTOR)}$	0.668	6.6×10^{-2}	9.880
$iQ_{P(4EBP1)}$	0.566	9.4×10^{-2}	16.61

(1,070), $J_{QP(Akt)}$ (0.494), and $J_{QP(mTOR)}$ were close to their values at the reference state; 8, 1000, and 0.5 respectively.

The sensitivity exponents, *a* and *b* were 1.450 and 1.490, respectively for the effects of phosphorylated Akt and EAA on mTOR phosphorylation (Equation 2). The sensitivity exponent, c, for the effect of active mTOR on 4EBP1 phosphorylation (Equation 3) was 2.370. A value greater than one is needed to maintain signal intensity as it is transmitted down the cascade. Setting *c* to a value of one results in loss of signal range, i.e. low to high given varying inputs, which was inconsistent with the observed data. This propensity for signal decay contradicts the suggestion that multiple steps in a cascade would propagate and magnify phosphorylation signals as distance from signal source increased (Kholodenko, 2006). However it has also been suggested that that there can be a gradient of kinases and phosphatases within cells where kinases are concentrated near the cell membrane while phosphatases are uniformly distributed. Such a gradient would dampen the range in signal as distance from the kinase location increased (Muñoz-García *et al.*, 2009). Akt has been found to be phosphorylated in close proximity to the plasma membrane (Huang and Kim, 2006). If downstream elements of the pathway are located in close proximity to the Akt phosphorylation site, the range in signal as determined from cell homogenates may be less than the local concentration range, which would explain the need for an exponent greater than 1 in Equations 2 and 3. Unlike phosphorylattion of mTOR and phosphorylation of 4EBP1, phosphorylation of Akt did not necessitate a sensitivity exponent to amplify the effects of insulin (Equation 1).

As mentioned before, the model was fitted to the observed phospho:total ratios that were standardized to values observed for the highest concentrations of insulin (100 ng/ml) and EAA (3500 µM). This resulted in ratios of 1 for those treatments which has an implicit assumption of complete phosphorylation of $Q_{T(X)}$. Additionally, it was assumed that the reference insulin and EAA concentrations of 10 ng/mland 1000 µM, respectively represented 50% phosphorylation. Neither assumption can be tested using Western blotting data given the semi-quantitative nature of the assay. However, the relative responses in signal intensities are representative of concentration changes in the samples. In order to maintain the assumption of 50% phosphorylation in the reference state, an additional parameter was required during fitting to scale the phosphorylation ratios in the same manner the observed data were scaled. This was achieved through the use of the variable $Ref\_Q_{P(X)}$ in the model which has the same role in the model as the scalar applied to standardize the observed ratios. Estimates of $Ref\_Q_{P(X)}$ were very close to 1 supporting the original assumption of complete phosphorylation when both insulin and EAA concentrations are very high.

The sensitivity analysis results are given in Table 2. These results represent the average increase in $Q_{P(Akt)}$, $Q_{P(mTOR)}$, and $Q_{P(4EBP1)}$ in response to unit increases in C_{Ins} (1.0 ng/ml) and C_{EAA} (1.0 µM). The C_{Ins} related sensitivity coefficients of $Q_{P(Akt)}$ (0.0083) that is more than 4 fold greater than that of $Q_{P(mTOR)}$ (0.0019), and $Q_{P(4EBP1)}$ (0.0017) reflect that phosphorylation of Akt is mainly mediated by extracellular insulin compared to phosphorylation of mTOR and phosphorylation of 4EBP1. These results pertained to MAC-T cells.

Similar significant effects of insulin on phosphorylation of Akt have been observed in muscle cells (Alessi *et al.*, 1996). The comparatively lower sensitivity of $Q_{P(mTOR)}$ (0.0019), and $Q_{P(4EBP1)}$ (0.0017) to insulin reflects the degeneration of the signal as it is propagated down the cascade and indicate that mTOR and 4EBP1 have available phosphorylation range to respond to other signals, i.e. insulin does not overwhelm the responsiveness of other effectors. It is widely accepted that additional signals such as energy, regulate mTOR phosphorylation (Sofer *et al.*, 2005). Both $Q_{P(mTOR)}$ and $Q_{P(4EBP1)}$ had similar sensitivities to insulin (0.0019 and 0.0017 respectively) and EAA (0.0008 and 0.0007 respectively) indicating responses are of similar magnitude to unit changes in insulin

Table 2. Average observed phosphorylated Akt ($Q_{P(Akt)}$), mTOR ($Q_{P(mTOR)}$), and 4EBP1 ($Q_{P(4EBP1)}$) masses, root mean square prediction error (RMSPE) and sensitivity coefficients of insulin (S_{Ins}) and EAA (S_{EAA}).

	Observed	RMSPE	S_{Ins}	S_{EAA}
$Q_{P(Akt)}$	0.535	0.043	0.0083	N/A
$Q_{P(mTOR)}$	0.638	0.050	0.0019	0.0008
$Q_{P(4EBP1)}$	0.603	0.053	0.0017	0.0007

and EAA. This observation is not surprising as 4EBP1 has been found to be a direct substrate of mTOR in several mammalian cells (Avruch *et al.*, 2009).

Root mean square prediction errors (RMSPE) associated with $Q_{P(Akt)}$, $Q_{P(mTOR)}$ and $Q_{P(4EBP1)}$ are presented in Table 2. The RMSPE are less than 10% in all cases reflecting good model accuracy and precision in predicting $Q_{P(Akt)}$, $Q_{P(mTOR)}$ and $Q_{P(4EBP1)}$. Residual errors are plotted in Figure 2. Mean and slope bias analyses indicated non significant mean (intercept) bias and slope bias for $Q_{P(mTOR)}$ ($P<0.47$ and $P<0.18$ respectively) and $Q_{P(4EBP1)}$ ($P<0.32$ and $P<0.25$ respectively). However model predictions of $Q_{P(Akt)}$ appeared to have slope bias ($P<0.07$) indicating the model tended to over predict $Q_{P(Akt)}$ as C_{Ins} increased.

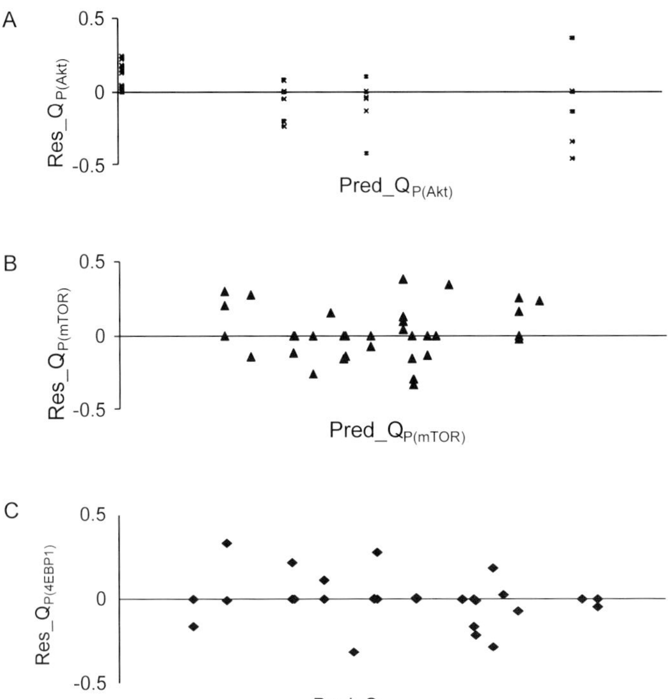

Figure 2. Residuals plots representing relationships between residual error (Res_$Q_{P(X)}$) and predicted values of phosphorylated masses (Pred_$Q_{P(X)}$) of Akt (A), mTOR (B), and 4EBP1 (C).

The effects of insulin and EAA on mTOR phosphorylation were represented additively in the model which supports the hypothesis that AA and insulin stimulate protein synthesis by independent signaling through mTOR. As the model was well defined by the data and appeared to reasonably predict the effects of insulin and EAA on phosphorylation of AKT, mTOR, and 4EBP1, this modeling work further strengthens the idea of the differential effects of insulin and EAA on protein synthesis. Since the model was fitted to data from a bovine mammary cell line, the estimates are more applicable to dairy and beef cattle. The work also supports the idea of independent effects of insulin and EAA on protein synthesis.

Model described in this work could be used to improve mathematical representations of amino acid requirements for milk protein synthesis in dairy cows. Since the mTOR pathway is highly conserved, this representation may also be directly applicable to other species.

Acknowledgements

This study was partially supported by Land O'Lakes/Purina Feed, LLC. Travel was supported by the Maryland Department of Agriculture. General departmental support by the Virginia State Dairymen is acknowledged.

References

Alessi, D.R., Andjelkovic, M., Caudwell B, Cron, P., Morrice, N., Cohen, P. and Hemmings, B.A., 1996. Mechanism of activation of protein kinase B by insulin and IGF-1. EMBO Journal 15:6541-6551.

Avruch, J., Long, X, Ortiz-Vega, S., Rapley, J., Papageorgiou, A. and Dai, N., 2009. Amino acid regulation of TOR complex 1. American Journal of Physiology Endocrynology and Metabolism 296:E592-602.

Hinault, C., Mothe-Satney, I., Gautier, N., Lawrence, J.C. Jr. and Van Obberghen, E., 2004. Amino acids and leucine allow insulin activation of the PKB/mTOR pathway in normal adipocytes treated with wortmannin and in adipocytes from db/db mice. FASEB Journal 18:1894-1896.

Huang, B.X. and Kim, H.Y., 2006. Interdomain conformational changes in Akt activation revealed by chemical cross-linking and tandem mass spectrometry. Molecular Cell Proteomics 5:1045-1053.

Kholodenko, B.N., 2006. Cell-signalling dynamics in time and space. Nature Reviews Molecular Cell Biology 7:165-176.

Marshall, S., 2006. Role of insulin, adipocyte hormones, and nutrient-sensing pathways in regulating fuel metabolism and energy homeostasis: a nutritional perspective of diabetes, obesity, and cancer. Science STKE 346:re7.

Muñoz-García, J., Neufeld, Z. and Kholodenko, B.N., 2009. Positional information generated by spatially distributed signaling cascades. PLoS Computational Biology 5:e1000330.

Proud, C.G., 2007. Amino acids and mTOR signalling in anabolic function. Biochemical Society Transactions 35:1187-1190.

Richter, J.D. and Sonenberg, N., 2005. Regulation of cap-dependent translation by eIF4E inhibitory proteins. Nature 433:477-480.

Sofer, A., Lei, K., Johannessen, C.M. and Ellisen, L.W., 2005. Regulation of mTOR and cell growth in response to energy stress by REDD1. Molecular Cell Biology 25:5834-5845.

Suryawan, A., Orellana, R.A., Nguyen, H.V., Jeyapalan, A.S., Fleming, J.R. and Davis, T.A., 2007. Activation by insulin and amino acids of signaling components leading to translation initiation in skeletal muscle of neonatal pigs is developmentally regulated. American Journal of Physiology Endocrynology and Metabolism 293:E1597-1605.

From metabolisable energy to energy of absorbed nutrients: quantitative comparison of models

C. Loncke[1,5], P. Nozière[1], S. Amblard[1,2], S. Léger[2], J. Vernet[1], H. Lapierre[3], D. Sauvant[4] and I. Ortigues-Marty[1]
[1]INRA, UR1213 Herbivores, Site de Theix, 63122 Saint Genès Champanelle, France; isabelle.ortigues@clermont.inra.fr
[2]Laboratoire de Mathématiques, Université de Clermont Ferrand II, 63177 Aubière, France
[3]Agriculture and Agri-Food Canada, STN Lennoxville, Sherbrooke, QC, Canada
[4]INRA-AgroParisTech, UMR 791, Physiologie de la Nutrition et Alimentation, 75231 Paris, France
[5]Present address: Union Invivo ets INZO SAS, rue de l'église, B.P. 50019, 02407 Chierry cedex, France

Abstract

In ruminant nutrition, metabolisable energy (ME) is an aggregated unit, whereas feed evaluation systems for ruminants have to evolve toward nutrient-based systems. Empirical models of net portal appearance (NPA) of nutrients have been established from dietary characteristics. This work firstly extended the meta-analysis to energy expenditure of portal drained viscera (PDV-EE) in relation with changes in dry matter intake (DMI) and dietary composition and, second, assessed the quantitative consistency between dietary ME defined according to INRA and the amount of absorbed energy (AE), calculated as the sum of the predicted NPA and PDV-EE. Publications presenting data on sheep and cattle were selected from the FLORA database and diets were described according to INRA tables of feedstuffs. For animals consuming on average 20.2 ± 7.2 g DM/d/kg BW and containing 25.6 ± 31.7% concentrate, the best adjustment of the PDV-EE prediction model was obtained with DMI. Adding parameters of diets composition did not improve the model. Concerning the assessment of the coherence between ME and AE, since urinary energy contributes to AE but not heat of fermentation, AE was compared by GLM to Y = ME + urinary energy – fermentation heat. Factors potentially responsible for this difference were investigated by testing the influence of DMI and dietary characteristics. For animals eating 18.1 ± 9.7 g DM/d/kg BW of diets containing 34 ± 34% concentrates, the amount of AE was linearly related to Y independently of the unit expression. On a DM basis, no difference was obtained between the two variables. When expressed in MJ/d/kg BW the difference ranged from -0.03 to 0.13 and was negatively correlated to DMI and to the concentrate proportion in the diet, suggesting that differences could be partly due to digestive interactions. This work demonstrates that a quantitative coherence exists between models of ME and AE, and that energy systems can evolve towards systems having a more physiological basis.

Keywords: meta-analysis, energy, net portal appearance, oxygen consumption, ruminant

Introduction

The feeding systems developed for ruminants were designed to adjust feed allowances to the production potential of the animals. Major progress has been made in feed characterisation but energy is still evaluated as an aggregate unit, such as metabolisable energy (ME). The new challenges of ruminant nutrition require that feed evaluation systems evolve towards nutrient-based systems (Reynolds, 2000) while remaining compatible with the existing systems, widely used in practice. Improvements to diet formulation depend on our capacity to quantitatively decompose ME into individual nutrients. We previously showed (Loncke *et al.*, 2008, 2009) that the net portal appearance (NPA) of energy sources such as acetate, propionate, butyrate, β-hydroxybutyrate, glucose, lactate and total amino acids could be predicted from simple dietary characteristics derived from the INRA feed evaluation system (2007) and valid for both sheep and cattle.

The major objective of the present work is to assess the quantitative consistency between dietary ME calculated according to INRA (ME_{INRA}) and the concept of absorbed energy (AE) defined as the sum of the predicted nutrient-NPA and the energy expenditure by the PDV (PDV-EE). Also, a prediction equation of the PDV-EE will be established.

Material and methods

The FLORA data base, choice of units and meta-analysis

The FLORA (FLuxes across Organs and tissues in Ruminant Animals) database, which is an exhaustive base from approximately 250 international publications reporting results of net splanchnic nutrient fluxes in ruminants, was used. The structure of Flora has been described in detail by Vernet and Ortigues-Marty (2006). Briefly, animals were encoded according to species (sheep, cattle) and physiological status (non-productive adult, growing, gestating or lactating animals). Chemical composition and nutritional value of feeds and diets, plus catheter location, blood sampling, metabolite analysis methods, and results for blood or plasma flows, metabolite concentrations and fluxes, were entered in the FLORA database as reported in the publications but after systematic verification steps. To fully and homogeneously characterise feeds and diets from different publications, a consistent description of the chemical composition and nutritional values of all feeds and diets included in the FLORA database was subsequently made using INRA Feed Tables (2007), assuming additivity, as described by Loncke *et al.* (2009). All intake and nutrient flux data were expressed per unit body weight. This unit allowed normalizing distributions of the explanatory (X) variables and total overlapping of cattle and sheep data, contrary to the metabolic body weight (Vernet *et al.*, 2006). Digestive processes are better expressed by the kg body weight (Sauvant *et al.*, 2006) and present results confirmed the relevance of this unit also for PDV-EE (Figure 1).

In this work, statistical methods dealing with the analysis of literature data, known as meta-analysis, were used as described by Sauvant *et al.* (2008). Briefly, publications and studies within publications were selected based on the specific objectives of the study. Statistical models were applied which allow to dissociate inter- and intra-study variations, considering study as a fixed

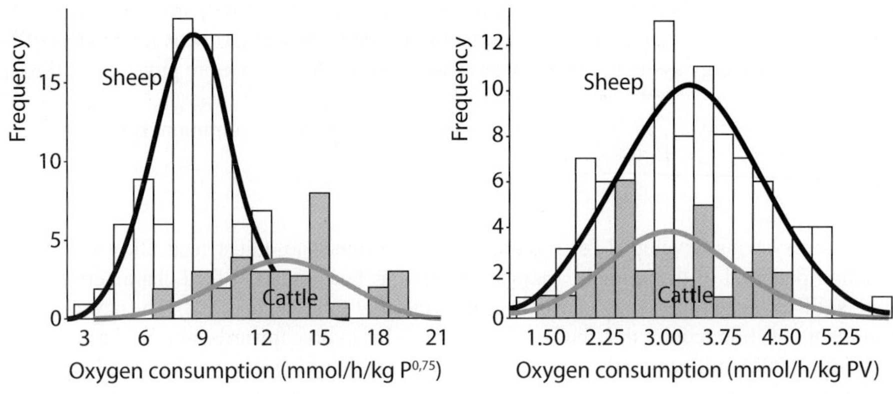

Figure 1. Oxygen consumption in sheep and cattle expressed by kilo body weight or by metabolic body weight.

factor. Once model parameters were determined, extensive analyses of potential interfering factors, i.e. discrete (qualitative) or continuous (quantitative) variables were performed on slopes, LSMeans and residuals, to detect potential additional covariables to be included in the model. A graphical examination of the data was performed throughout.

Concept of absorbed energy

The concept of AE was defined as the traditional concept of ME using digestible energy as a starting point, but on more physiological bases (Figure 2). It is equal to digestible energy minus energy lost either as methane or as fermentation heat from the gut. A fraction of AE is then metabolised in the PDV by oxidation (leading to PDV-EE) or to be deposited as net energy in PDV tissues (assumed here to be negligible on a daily basis). Another fraction of AE appears in the portal blood as nutrients. Energy-NPA was calculated as the energy supplied by nutrient NPA (i.e. acetate, propionate, butyrate, glucose, lactate, β-hydroxybutyrate, total amino acids). NPA of peptides, nucleic acids and aceto-acetate were not considered because of lack of data, and their contribution was considered as minor. A last fraction of AE is in the form of fat which appears in the portal vein or in lymph. Energy absorbed as fatty acids was estimated from the dietary lipids assuming an apparent digestibility of 80% (Glasser *et al.*, 2008). The amount of AE was defined as 'energy-NPA + absorbed fatty acids + PDV-EE'.

Prediction of energy expenditure by the PDV

PDV-EE was shown to vary with dietary bulk and composition (Han *et al.*, 2002). Consequently, variables reflecting the ruminal digesta bulk (expressed by DMI), the amount of nutrients (digestible organic matter intake and ME intake) or the nutrient pattern [dietary digestible NDF (dNDF) intake, crude protein intake] were first tested as the predictors of PDV-EE. Based on the results obtained, other predictors were also considered: the dietary dNDF and crude protein contents. As part of the heuristic process of meta-analysis, 3 sub-data bases were finally selected from FLORA. Publications reporting PDV O_2 consumption were selected if rations varied only for DMI (DMI database), the dietary dNDF content (dNDF database) or crude protein content (crude protein database). PDV energy expenditure (EE, in MJ/d) was calculated as 0.460 O_2 consumption (mol/d).

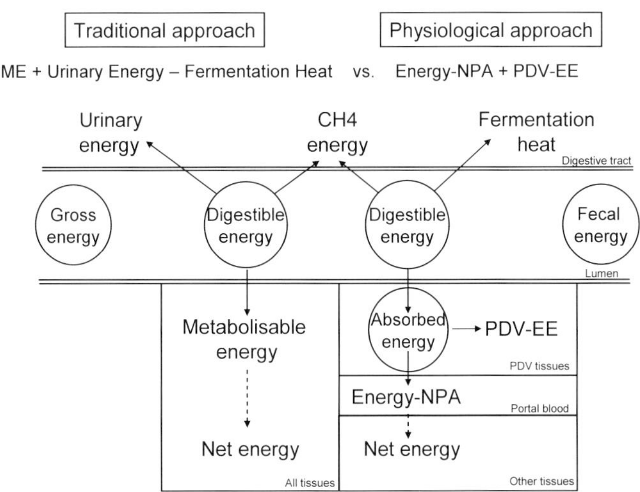

Figure 2. Comparison between the traditional and a more physiological definition of digestible energy.

When applying meta-analytical methods (Sauvant *et al.*, 2008), calculation of reliable within-experiment responses requires a minimum variation on each explanatory variable within that experiment, therefore, the minimum acceptable variation was determined from all the selected publications for each target variable, as described by Loncke *et al.* (2009) as 2 g DMI/d/kg BW, 15 g dNDF/kg DM and 8 g crude protein /kg DM. Variance covariance models ($Y = \alpha + \beta X +$ species $+ \alpha i$ (species) + species $\times X +$ e, with α = overall intercept; αi = effect of the experimental group i on the intercept α, nested within species; β = slope; e = error), were then adjusted using the GLM procedure to separate within and inter-experiment effects. When a model was significant, relevant additional variables, including intake level, dietary composition, site and extent of digestion, were tested using analysis of interfering factors on residuals, LSMeans and slopes.

Prediction of the absorbed energy

All publications presenting results on nutrient-NPA were selected from FLORA. If diets included feed additives which modified rumen fermentation, they were not included in the study. All diets had previously been described according to INRA feed characteristics, including their ME_{INRA} content which assumed additivity. The NPA of each energy nutrient was systematically predicted according to Loncke *et al.* (2008, 2009 summarised in Table 1). Thus, the acetate, propionate and butyrate NPA were predicted from the combination of the intake of ruminally fermented organic matter (RfOM) and the nature of RfOM (defined by Ruminally dNDF/RfOM), the β-hydroxybutyrate NPA from the RfOM intake, the lactate NPA from the starch digested in the rumen, and the glucose NPA from the starch digested in the small intestine. Finally, the alpha amino nitrogen NPA was predicted from the protein digested in the small intestine, and the total amino acids NPA was calculated as 2.35×NPA α-amino-nitrogen (Martinau *et al.*, 2009). Energy absorbed as fatty acids was estimated as mentioned above. For each publication and each treatment, AE (MJ/d/kg BW) was then calculated as described above and the contribution of each nutrient to AE was calculated as: (energy absorbed as nutrient)/AE×100.

Comparison of the metabolisable energy and the absorbed energy models

The difference between ME_{INRA} and AE was assessed. Since urinary energy contributes to absorbed energy but not heat of fermentation, we first compared, by GLM procedure including study effect (Sauvant *et al.*, 2008), AE to $Y = ME_{INRA} +$ urinary energy – fermentation heat with urinary energy and fermentation heat calculated from Sauvant and Giger-Reverdin (2007). Subsequently, factors responsible for the calculated difference were investigated by testing the influence of level intake and dietary characteristics on Δ = AE-Y.

Results

Prediction of the portal drained viscera oxygen consumption

Thirty-eight publications (93 experimental treatments on sheep and 32 on cattle) presented results on PDV-EE. The rations covered a large range of intake levels (0 to 41.5 g DM/d/kg BW) and diet composition (13 to 74% NDF and 4.5 to 22% CP). The derived response equations are reported in Table 2.

The PDV-EE was significantly related to the digestible organic matter, dry matter, NDF, crude protein and ME intake. Nevertheless, only the DMI explained the inter-study variability. It was thus considered as the best predictor of the PDV-EE.

Table 1. Equations (Loncke et al., *2008, 2009) used to predict total and molar proportion of volatile fatty acids (VFA), β-hydroxybutyrate (BHBA), glucose, lactate (mmol/h/kg BW) and α-amino nitrogen (α-amino N, g N/h/kg BW) from variations (g/d/ kg BW) in ruminal fermented organic matter (RfOM), starch digested in the small intestine (SdSI), rumen digested starch (RdS), protein digestible in the small intestine (PDI) and to ruminal digestible NDF (RdNDF).*

Nutrients	Nt	Predictors	Intercept	SE	Slope	SE	RMSE	R^2
Total VFA	36	RfOM	0.06^{NS}	0.20	0.247^{***}	0.022	0.243	0.95
% Acetate	20	RdNDF/RfOM	58.7^{***}	2.5	25.1^{***}	5.1	2.23	0.85
% Propionate	20	RdNDF/RfOM	34.0^{***}	2.5	-18.9^{**}	5.0	2.18	0.76
% Butyrate	22	RdNDF/RfOM	7.9^{***}	1.1	-7.3^{**}	2.1	1.10	0.76
BHBA	38	RfOM	0.114^{***}	0.033	0.0105^{**}	0.0035	0.036	0.91
Glucose	41	SdSI	-0.103^{***}	0.0094	0.0913^{***}	0.0088	0.035	0.89
Lactate	53	RdS	0.156^{***}	0.015	0.0136^{***}	0.0031	0.39	0.87
α-amino N	76	PDI	$-0.030^{\dagger}$	0.016	0.570^{***}	0.046	0.026	0.88

$^*P<0.05$; $^{**}P<0.01$; $^{***}P<0.001$; $^{\dagger}P<0.1$; NS: $P>0.1$.
SE: standard error, RMSE: root mean squared error, Nt: number of treatments.

Table 2. Response equations of portal drained viscera oxygen consumption (kJ/d/kg BW) to variations in dry matter intake (DMI, g/kg BW) and to the diet proportion of NDF and crude protein (CP, g/100g DM).

Variable	Nt	Intercept	Slope	R^2	RMSE	Species	Interfering factors		
							Residues	Slope	LSMeans
DMI	40	$19.8^{***} \pm 2.49$	$0.89^{***} \pm 0.14$	0.78	4.41	NS	none	dNDF *	CP *
NDF	28	$26.2^{**} \pm 6.62$	$0.37^{\dagger} \pm 0.22$	0.80	4.19	NS	ND	ND	ND
CP	40	$43.4^{***} \pm 4.75$	$-0.36^{NS} \pm 0.37$	0.76	4.30	NS	ND	ND	ND

$^*P<0.05$; $^{**}P<0.01$; $^{***}P<0.001$; $^{\dagger}P<0.1$; NS: $P>0.1$; ND: not determined; Nt: number of treatments.

In the DMI database, PDV-EE varied from 16.6 to 51.9 kJ/d/kg BW, for DMI varying from 0 to 32 g DM/kg BW. PDV-EE was significantly and linearly related to DMI (Figure 1), accounting for 78% of the variance, without a species effect. The RMSE was relatively small (4.41 kJ/d/kg BW). This model (Table 2) suggests that when the DMI increases by 1 g DM/d/kg BW, the PDV-EE increases of 0.89 kJ/d/kg BW (Figure 3). The intercept was significantly different from 0. The interfering factors study showed the individual slopes were positively ($P = 0.019$) related to the dNDF content and that the LSMeans of the model were positively ($P = 0.035$) related to the crude protein content. Nevertheless, the inclusion of the dNDF and/or the crude protein contents in the model was not significant.

Considering the interfering factors that were identified, different prediction models were investigated, using the other sub-data bases. In the NDF database, the PDV-EE varied from 18.8 to 55.5 kJ/d/kg BW for diets ranging in dNDF contents from 200 to 460 g/kg DM. In the crude protein database, the dietary crude protein content ranged from 44.7 to 177 g/kg DM and PDV-EE varying from 18.7

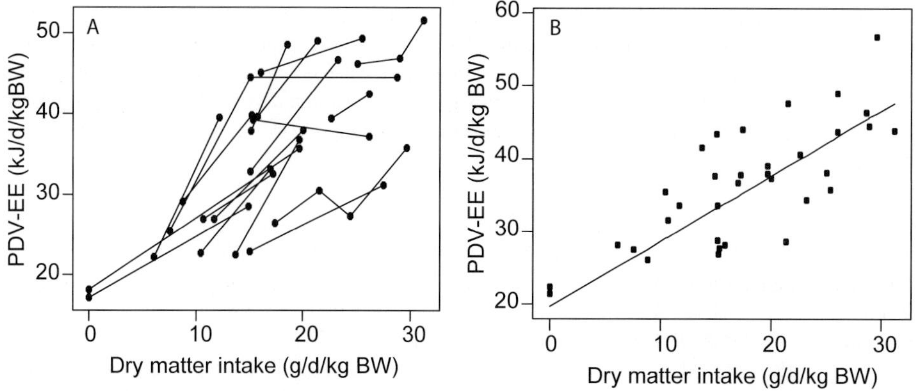

Figure 3. Within-experiment relationship (A) and adjusted response model (B) between the dry matter intake and the energy expenditure of the portal drained viscera (PDV-EE).

to 55.5 kJ/d/kg BW. No significant within- or inter-experiment relationship could be established between the PDV-EE and the digestible NDF or crude protein content.

The model based on DMI was thus considered as the best model and further evaluated. No interfering factor was observed on the residues. Comparative analysis of the PDV-EE measured by the authors (*Y*) vs. predicted by our model (*X*) using a GLM analysis (Nt = 125) showed an intercept not different from 0 (-0.38 ± 0.38, $P = 0.3$) and a slope not different from 1 (1.064 ± 0.11, $P = 0.6$). The R^2_{adj} was high (0.82) and the RMSE rather low (4.19 kJ/d/kg BW).

The absorbed energy model

One hundred and twenty-three publications (209 experimental treatments on cattle and 205 on sheep) were selected from FLORA. Animals ate 18.1 ± 9.7 g DM/d/kg BW (ranging from 0 to 47.5) of diets containing 34 ± 34 g concentrate/100 g DM (ranging from 0 to 100). The dietary contents averaged 235 ± 90.5 g ruminally digestible NDF /kg DM, 510 ± 52.8 g RfOM/kg DM. The meta-design is graphically described in Figure 4.

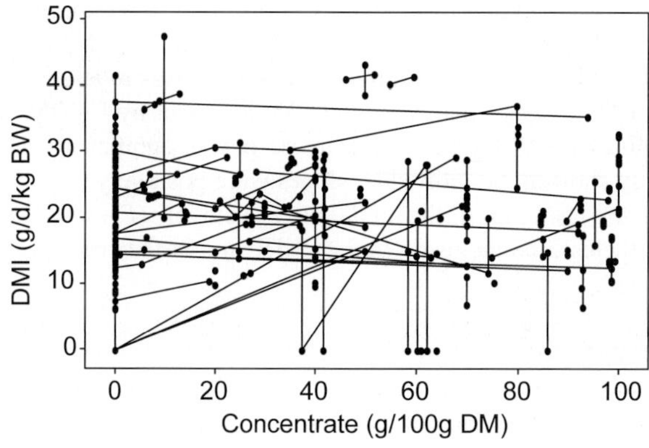

Figure 4. Meta design of the data set used to compare the ME_{INRA} and absorbed energy models.

As predicted, the nutrients which contributed the most to the average AE were the volatile fatty acids (43.6 ± 5.58% of the AE ranging from 30.5 to 65.5%). Other contributions averaged 8.05 ± 3.46% (from 3.86 to 17.1%) for the β-hydroxybutyrate, 4.31 ± 1.51% (from 1.76 to 16.2%) for the lactate, 0.38 ± 0.91% (from 0 to 6.37%) for the glucose, 6.95 ± 2.49% (from 0.96 to 14.5%) for the total amino acids, 11.1 ± 4.0% (from 4.0 to 27.9%) for the lipids and 25.6 ± 3.92% (from 6.72 to 33.6%) for PDV-EE.

Fermentation heat and urinary energy contributed respectively to 7.10 ± 1.84% (from 1.51 to 10.8%) and 7.85 ± 2. 39% (from.4.34 to 13.8%) of the ME_{INRA}, and thus the Y value was lower than ME_{INRA} by an average of 0.75 ± 1.25 percentage units (= urinary energy – fermentation heat).

Comparison of the ME_{INRA} and absorbed energy models.

The results indicated that the ME_{INRA} and the energy of absorbed nutrients were closely and linearly linked, independently of the unit expression. More precisely, the amount of AE (MJ/d/kg BW) was linearly related to Y (MJ/d/kg BW, Figure 5A) as:

$$Y = -0.04^{***} \pm 0.0008 + 1.38^{***} \pm 0.004\ AE$$

$N_{exp} = 113$; $N_t = 399$; $R^2_{adj} = 0.996$; RMSE = 0.007

The slope obtained indicated that the amount of AE was lower than Y. The Δ ranged from -0.03 to 0.13 MJ/d/kg BW, corresponding to about 9% of the average ME_{INRA}.

It was negatively correlated to DMI (g/d/kg BW, Figure 6A) and to the proportion of concentrate (conc, g/100 g DM, Figure 6B), in a within study relationship as:

$$\Delta = 0.031^{***} \pm 0.005 + 0.0023^{***} \pm 0.00023\ DMI + 0.00036^{***} \pm 0.00004\ conc$$

$N_{exp} = 113$; $N_t = 399$; $R^2_{adj} = 0.96$; RMSE = 0.007

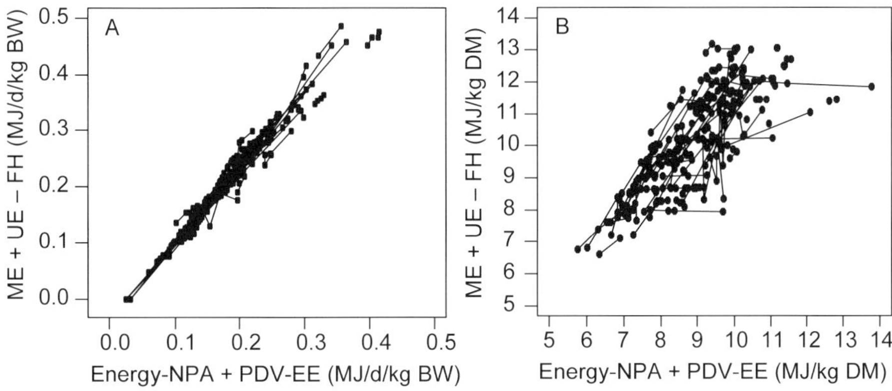

Figure 5. Within-experiment relationships between the Y = ME + Urinary energy – Fermentation heat and Absorbed Energy models expressed by kilo BW (A) or by g DMI (B).

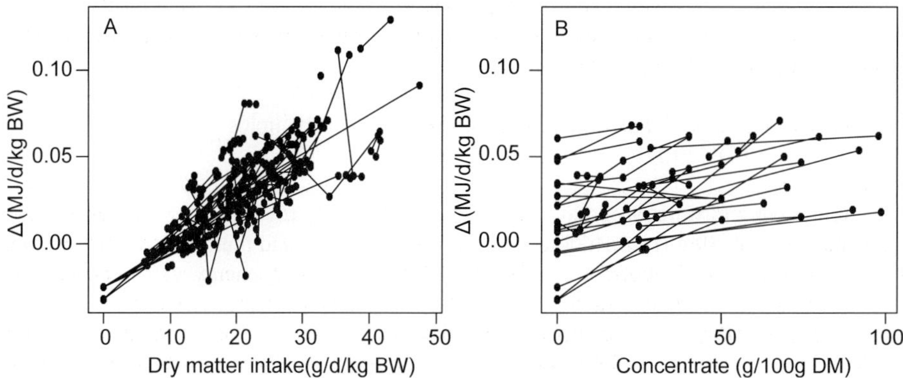

Figure 6. Within-experiment relationships between Δ = ME - AE and DM intake (A) or proportion of concentrate (B).

Additionally, the amount of AE, expressed by DMI (MJ/d/kg MDI) was linearly related to Y (MJ/d/kg DMI, Figure 5B):

$$Y = 1.72^{**} \pm 0.59 + 0.95^{***} \pm 0.06 \text{ AE}$$

$N_{exp} = 120$; $N_t = 380$; $R^2_{adj} = 0.903$; RMSE = 0.461

The slope obtained was not different from 1, suggesting a strong relationship between the two variables.

Discussion

The data published on NPA of energy nutrients in ruminants proved relevant to establishing prediction models in both sheep and cattle at different physiological stages. Most of the data available for this work was obtained in animals fed at intakes lower than 35 g DMI/d/kg BW and with concentrate proportions of below 70 g/100 g DM. This reflects most real-world feeding situations except for high-producing dairy females receiving up to 40 g DMI/d/kg BW, or for feeding management strategies based on high-concentrate dry feeds (Loncke *et al*., 2009). Despite the wide range of intakes and diets covered by the meta design, it should also be stressed that models developed in the current study do not account for specific feed additive effects (buffers, essential oils, ionophores, etc.), which may affect the profile of digestion end-products and absorbed nutrients.

Prediction of energy expenditure of portal drained viscera

The best model was obtained with DMI. It is the first published model which predicts PDV-EE in both sheep and cattle consuming a large range of diets and intake levels. Interestingly, the intercept was significantly different from 0, suggesting that PDV metabolism has an energy cost in absence of DMI. The intercept obtained was equal to 19.8 ± 2.49 kJ/d/kg BW) consistently with the PDV-EE obtained (22.6 ± 6.39 kJ/d/kg BW) in 6 trials on fasted animals. The RMSE of the model represented about 10% of the average PDV-EE. Digestible NDF or crude protein contents of the diet increased PDV-EE independently of the intake level, but their effect was too low to significantly improve the model. This model is consistent with Han *et al*. (2002) who showed that digesta mass and nutrients were the main factors of variations of PDV-EE and with Goetsch's (1997), who predicts variability in PDV-EE from digestible energy intake (explaining 89% of the variance) and fecal NDF excretion in sheep fed exclusively forage diets.

The absorbed energy model and comparison with the ME_{INRA} model

With the AE model, energy actually recovered in the portal vein averaged 86.0 ± 9.5% of ME_{INRA}. This is higher than the range of recovery (68-74% of ME) reported in sheep by Lindsay (1993). Reasons for a recovery lower than 100%, as discussed by Lindsay (1993), include technical limits in the estimation of blood flow and A-V differences, lack of data for nutrients such as nucleic acids, peptides or nutrients deposited in adipose tissues of PDV. In the present study, since all nutrient NPA were predicted, uncertainties of predictions cannot be excluded. Nevertheless, when the variables X and Y were expressed per kg DMI, the slope was not significantly different from 1, reflecting the close quantitative coherence between both 2 concepts. Another reason could be assumed from the present work from the comparison between Y and AE expressed per kg BW. First, the linear relationship suggested a systematic discrepancy. Second, the fact that Δ increased with DM intake and with the percentage of concentrate of the diet, suggested that differences between the Y and the AE models could be partly attributed to digestive interactions, which are not accounted for in ME_{INRA}. Uncertainties on ME_{INRA}, calculated assuming additivity, might therefore contribute to the discrepancy between AE and ME. Additionally, the error associated to calculated urinary energy and fermentation heat cannot be excluded but it was not possible to evaluate it.

Conclusion

This work demonstrates that a global quantitative coherence exists between the classical energetic models of ME and the more physiological based model of AE. Energy feeding systems will be able to evolve towards systems closer to physiological bases, which represents a major step forward in the development of new rationing tools for ruminants.

Moreover, this work stressed the importance of taking into account the digestive interactions when defining the energy value of diets.

Acknowledgements

The authors wish to thank INZO and LIMAGRAIN as well as the Association Nationale de la Recherche Technique for funding this project.

References

Han, X. T., Nozière, P., Rémond, D., Chabrot, J., and Doreau, M., 2002. Effects of nutrient supply and dietary bulk on O2 uptake and nutrient net fluxes across rumen, mesenteric- and portaldrained viscera in ewes. Journal of Animal Science 80:1362-1374.

Glasser, F., Ferlay, A., Doreau, M., Schmidely, P., Sauvant, D. and Chilliard, Y., 2008. Long-chain fatty acid metabolism in dairy cows: A meta-analysis of milk fatty acid yield in relation to duodenal flows and de novo synthesis. Journal of Dairy Science 91:2771-2785.

Goetsch, A.L., Patil, A.R., Wang, Z.S., Park, K.K., Galloway, D.L., Rossi, J.E. and Kouakou, B., 1997. Net flux of nutrients across splanchnic tissues in wethers consuming grass hay with or without and alfalfa. Animal Feed Science and Technology 66:271-282.

INRA, 2007. Alimentation des bovins, ovins et caprins – Besoins des animaux – Valeurs des aliments – Tables INRA 2007. In: INRA (ed.) Versailles, 307 pp.

Lindsay, D.B., 1993. Making the sums add up – the importance of quantification in nutrition. Australian journal of Agricultural Research 44:479-493.

Loncke, C., Ortigues-Marty, I., Vernet, J., Lapierre, H., Sauvant, D. and Nozière, P., 2008. Capacité du système PDI à prédire les quantités d'azote alpha-aminé absorbées en veine porte chez les ruminants. Rencontre Recherches Ruminants 15:285.

Loncke C, Ortigues-Marty, I., Vernet, J., Lapierre, H., Sauvant, D. and Nozière, P., 2009. Empirical prediction of net portal appearance of volatile fatty acids, glucose and their secondary metabolites (β-hydroxybutyrate, lactate) from dietary characteristics in Ruminants: a meta-analysis approach. Journal of Animal Science 87:253-268.

Martineau, R., Ortigues-Marty, I., Vernet, J., and Lapierre, H., 2009. Technical note: Correction of net portal absorption of nitrogen compounds for differences in methods: First step of a meta-analysis. Journal of Animal Science 87:3300-3303.

Reynolds, C.K., 2000. Feed evaluation for animal production. In: Theodorou, M. K. and France, J. (eds.) Feeding systems and feed evaluation models. CABI Publishing, London, UK, pp 87-108.

Sauvant, D., and Giger-Reverdin, S., 2007. Empirical modelling by meta-analysis of digestive interactions and CH4 production in ruminants. In: Ortigues-Marty, I., Miraux, N. and Brand-Williams, W. (eds.). Energy and Protein Metabolism and Nutrition, EAAP Publication No 124, Wageningen Academic Publishers, Wageningen, the Netherlands, pp.561-562.

Sauvant D, Schmidely, P, Daudin, J.J. and St-Pierre, N.R., 2008. Meta-analyses of experimental data in animal nutrition. Animal 2:1203-1214.

Sauvant, D., Assoumaya, C., Giger-Reverdin, S. and Archimède, H., 2006. Etude comparative du mode d'expression du niveau d'alimentation chez les ruminants. Rencontre Recherches Ruminants 13:103.

Vernet, J., Nozière, P., Sauvant, D., Léger, S., and Ortigues-Marty, I., 2005. Regulation of hepatic blood flow by feeding conditions in sheep: a meta-analysis. In: Priolo, A., Biondi, L., Ben Salem, H. and Morand-Fehr, P. (eds.). Proceeding of the 11th Seminar of the Sub-Nutrition FAI-CIHEAM on sheep and goat nutrition. September 8-10 2005. Catalia, Italy, pp 435-440.

Vernet, J., and Ortigues-Marty I., 2006. Conception and development of a bibliographic database of blood nutrient fluxes across organs and tissues in ruminants: Data gathering and management prior to meta-analysis, Reproduction Nutrition Development 46:527-546.

Development of a heat balance model for cattle under hot conditions

V.A. Thompson[1], L.G. Barioni[2], J.W. Oltjen[1], T. Rumsey[1], J.G. Fadel[1] and R.D. Sainz[1]
[1]University of California, Department of Animal Sciences, One Shields Avenue, Davis, CA 95616-8521, USA, vathompson@ucdavis.edu
[2]Embrapa Cerrados, BR 020 Km 18, Caixa Postal 08223, CEP 73310-970 Planaltina-DF, Brasil

Abstract

We developed a model that simulates heat production and heat flow between the animal and its environment. Heat production is predicted using the Davis Growth Model (DGM), which in turn is based on the maintenance functions in NRC (2000). Heat flows between the animal and the environment are based largely on models by Turnpenny *et al.* (2000a) and McGovern and Bruce (2000). Heat flows are calculated between the body, the skin+coat and the environment. The heat flow is based on four types of flow: solar radiation, long wave radiation, convection and evaporative heat loss. Solar radiation flows from the environment to the animal, while long wave radiation and convection can flow both ways depending on the environmental conditions. Evaporative heat loss includes respiratory and skin latent heat loss. The animal increases heat losses through increased vasodilation, respiration and sweating. Vasodilation increases blood flow to the skin surface which increases heat losses through convection. Evaporative heat losses through respiration and sweating directly decrease the internal body temperature and the skin temperature, respectively. These animal responses are modeled through mechanistic equations and have parameters which can be fitted to specific breeds of cattle. The inputs needed to run this model are divided into two parts, the environmental inputs (radiation, temperature, wind and humidity) and animal inputs (body weight, diet composition and breed). The model predicts that as body heat content increases, metabolic rate and body temperature both increase, and subsequently feed intake decreases. This can be useful in heat stress situations to calculate the quantitative decrease in feed intake and the concomitant decrease in performance. This model still needs to be fitted and tested against more extensive actual datasets in order to get breed- and system-specific parameter sets. Once completed, this model should be a useful tool for researchers and livestock producers.

Introduction

Thermal stress (both hot and cold) in cattle may result in large losses in production. Physiological responses to heat stress include vasodilation, sweating, increased respiration and decreased feed intake. One of the first reactions is vasodilation which allows an increased blood flow to the surface of the skin in order to bring the warm blood in close proximity to the air which would increase heat losses through convection. The animal also can increase its sweating rate and respiration rate to increase latent heat loss. Lastly, if the animal cannot compensate for the excess body heat, it can decrease its intake in order to decrease heat production. While the first three methods can increase the animal's maintenance requirement, decreasing intake further limits the amount of nutrients directed towards growth. Due to this, the animal's production decreases, resulting in an economic loss.

Because there are economic losses associated with heat stress, temperature-humidity indices (THI) were created that allow for the prediction of when an animal will be stressed. The THI initially proposed by Thom (1959), takes into account temperature and humidity to determine the level of heat stress within the animal. However, these indices have limited utility to predict the dynamic change in body temperature and to evaluate different mitigation measures. Therefore, there is a need for a model that will simulate the actual heat flow between the animal and its environment, the heat production by the animal and any potential loss in animal performance.

The objectives of this research were to create a dynamic, mechanistic model to:
- Estimate major flows of heat into and out of the animal.
- Use heat balance to calculate changes in body temperature in response to air temperature, relative and absolute humidity, radiation/shade and wind.
- Model changes in intake and performance; analyze potential management strategies.

Materials and methods

The heat balance model has two state variables: body heat content and skin + coat heat content (Joules). Two layers are necessary because temperatures (i. e., heat contents) vary between compartments and it is this temperature gradient that determines the heat flows between each layer and the environment. The model with all the flows is shown in Figure 1; arrow sizes are not scaled to the magnitude of flow, as these vary widely. The heat flows are through Convection (lungs and skin), Long Wave Radiation, Solar Radiation and Evaporation (lungs and skin). Convection and long wave radiation can have either positive or negative net flows between the animal and the environment whereas solar radiation and evaporation each only flow in one direction, into and out of the animal, respectively. The differential equations representing these heat flows are as follows:

$$dHC_{body}/dt = HE - Con_l - E_r - G_s \quad (1)$$

$$dHC_{skin}/dt = G_s + Solar - Con_s - E_s - LWR \quad (2)$$

where dHC/dt is the change in heat content with respect to time, HE is heat production, Con_l and Con_s are convection in the lungs and skin, G_s is the heat flow from the body to the skin, Solar is solar radiation, E is the evaporative heat loss (through respiration, E_r, and sweating, E_s) and LWR is long wave radiation at a given time.

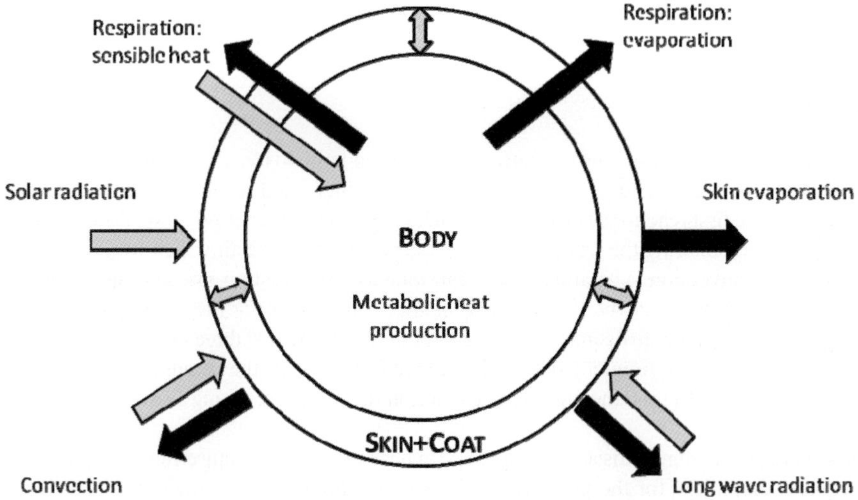

Figure 1. Overview of the model. The two black circles represent the state variables: Body and Skin. The arrows leaving the circles represent heat lost from the animal to the environment and the arrows entering the circles represent heat gained by the animal. The two sided arrows represent the heat flow between the body and the skin.

Flows into and out of the body

Heat production is modeled using the NRC equations (2000) where HE is heat energy or heat production, H_eE is endogenous heat production, H_dE is heat of digestion, absorption and assimilation, and H_rE is heat associated with production (growth, lactation, reproduction; all in joules):

$$HE = H_eE + H_dE + H_rE \quad (3)$$

$$H_eE = 0.077 \times EBW^{0.75} \times 2^{(Tb-39)/10} \quad (4)$$

$$H_dE = DMI_{maint} \times (MEC - NE_m) \quad (5)$$

$$H_rE = DMI_{gain} \times (MEC - NE_g) \quad (6)$$

where EBW is empty body weight (kg), T_b is body temperature, DMI_{maint} and DMI_{gain} are dry matter intake for maintenance and gain (kg DM/d), MEC is metabolizable energy concentration (J/kg DM) and NE_m and NE_g are dietary net energy contents for maintenance and gain, respectively (both in J/kg DM). Temperature is always in degrees Celsius. The animal's maintenance energy requirement is the sum of H_eE and H_dE, and all costs of production (including additional heat of digestion associated with higher intakes) are included in H_rE. The cubic relationships between MEC, NE_m and NE_g used in NRC (2000) implicitly increase H_dE and H_rE with lower quality feeds. Whilst heat is certainly lost as excreta leave the body at body temperature, this would not in itself alter body temperature. Ingestion of solid or liquid at temperatures different from the body would change overall temperature; however this is not presently included in the model.

While DMI is currently an input, DMI_{maint} and DMI_{gain} can be calculated as follows (kg DM):

$$DMI_{maint} = H_eE / NE_m \quad (7)$$

$$DMI_{gain} = DMI - DMI_{maint} \quad (8)$$

Convection in the lungs (Con_l, W/m^2) is modeled as:

$$Con_l = V_t \times RR \times \rho_a \times \rho_{cp} \times (T_{vb} - T_{va}) \quad (9)$$

where V_t is tidal volume (m^2), RR is respiration rate (breaths/min), ρ_a is the density of moist air (kg/m^3), ρ_{cp} is the specific heat of air (1220 J/(m^3 × K)) and T_{vb} and T_{va} are the virtual temperatures of the body and air respectively. The calculation uses virtual temperatures because they adjust for humidity using saturation vapor pressure (esT, kPa), atmospheric air pressure (P, kPa) and temperature (T) (Turnpenny *et al.*, 2000b):

$$T_v = (T + 273.15) \times (1 + 0.38 \times esT / P) \quad (10)$$

Tidal volume and respiration rate are calculated based on equations by McGovern and Bruce (2000) which have a working temperature range from 15 to 40 °C.

$$V_t = -0.10 \times T_a + 6.22 \quad (11)$$

$$RR = 6.25 \times T_a - 80.47 \quad (12)$$

Evaporation from the lungs is modeled as (McGovern and Bruce, 2000):

$$E_r = V_t \times RR \times \rho_a \times \lambda \times (\chi_b - \chi_a) \quad (13)$$

where λ is the latent heat of vaporization (J/g) and χ_b and χ_a are the absolute humidities at body and air temperatures (kg/kg), respectively.

The flow of heat from the body to the skin, G_s (W/m^2), is modeled using a standard heat flow equation (McArthur, 1987):

$$G_s = (T_b - T_s) \times \rho_{cp} / r_s \quad (14)$$

where T_b and T_s are the temperatures of the body and skin (°C) and r_s is the resistance to heat transfer (s/m). The resistance, r_s, has a linear relationship to the difference between the reference body temperature (39 °C) and the actual body temperature (Finch, 1985).

$$r_s = (T_b - T_{bref}) \times (-30.833) + 50 \quad (15)$$

Body and skin temperatures (T) are calculated using the following equation:

$$T = HC / (W \times Ca) \quad (16)$$

where W is weight (of the skin or body, g), HC is the heat content (J) and Ca is the specific heat capacity ($3.4\ J/g^1/K^1$; Monteith, 1973).

Flows into and out of the skin+coat layer

Solar radiation absorbed by the animal ($Solar_{abs}$, W/m^2) is from three sources: direct radiation, diffuse radiation and radiation reflected from the ground surface (Monteith and Unsworth, 2008).

$$Solar_{abs} = f_c\,(1 - \rho_c) \times R_{dir} + 0.5 \times [(1 - \rho c) \times (R_{diff}) + \rho_g \times (R_{dir} + R_{diff})] \quad (17)$$

The form factor of the animal, f_c, takes into account its angle in relation to the sun and its body dimensions and is given by the following equation (Monteith and Unsworth, 2008):

$$f_c = A_h / A \quad (18)$$

where A_h and A are the areas of the shadow cast by the animal and the animal's total surface area (m^2). Direct and diffuse radiation are given by R_{dir} and R_{diff} respectively (W/m^2), while ρ_c and ρ_g are the reflection coefficients (J radiation reflected / J radiation intercepted) of the animal and the ground, respectively. Gomes da Silva (2008) found the reflection coefficient of short grass to be 0.25 and Monteith (1973) found white and black cattle to have coat reflection coefficients of 0.51 and 0.12, respectively.

The latent heat lost from the surface of the animal is modeled using a minimizing function. Heat lost by the animal through evaporation is limited by either its sweating rate or by the rate that the water can evaporate, which depends on environmental conditions. Therefore, to model the latent heat lost through sweating, E_s (W/m^2), a minimizing function is used (Turnpenny *et al.*, 2000b):

$$E_s = \min(SR, ER_{max})\ \lambda \quad (19)$$

where SR is the sweating rate and ER_{max} is the maximum evaporation rate due to environmental conditions. The equation for ER_{max} accounts for temperature, wind and humidity:

$$ER_{max} = \rho_{cp} / \gamma \times (esT_s - eT_a) / r_v \tag{20}$$

where γ is the psychrometer constant (kPa/K), esT_s is the saturation vapor pressure of the skin (kPa), eT_a is the vapor pressure of the air (kPa), and r_v is the resistance to water vapor transfer (s/m). The equation for r_v accounts for coat length l (m) and body diameter d (m).

$$r_v = (l - l_w) / (D \times (1 + 1.54 \times ((l - l_w) / d) \times T_v dif^{0.7})) \tag{21}$$

where l_w (m) is the wind penetration of the coat (given by Turnpenny *et al.*, 2000a), D is the diffusion coefficient of water vapor in air (m^2/s) at 20 °C and $T_v dif$ is the difference in virtual temperatures between the skin and air.

The sweating rate (SR, $g/m^2/h$) of the animal is modeled using an equation presented by McArthur (1987):

$$SR = c\,(T_{st} - T_s) + E_{cmin} \tag{22}$$

where c is the rate of sweat deposition per unit area per degree rise in skin temperature ($g/m^2/h/°C$), T_{st} is the threshold skin temperature above which sweating increases linearly (°C) and E_{cmin} is the minimum sweating rate ($g/m^2/h$).

Long wave radiation (LWR, W/m^2) is described by the following equation:

$$LWR = \sigma \times \varepsilon \times (T_s^4 - T_r^4) \tag{23}$$

where ε is the emissivity in the infra-red of the surface of the animal (0.99, Blaxter 1962), σ is the Stefan-Boltzmann constant (5.67×10^{-8} $W/m^2/K^4$) and T_s and T_r are the skin temperature and radiant temperature of the surroundings, respectively. T_r is calculated as the average of the sky (T_{sky}; Duffie and Beckman, 1980) and ground (T_g; Turco *et al.*, 2008) radiant temperatures (°C).

Convection at the skin surface, Con_s (W/m^2), accounts for the effects of temperature and wind speed:

$$Con_s = (T_b - T_a) \times \rho_{cp} / r_H \tag{24}$$

The resistance to convection, r_H (s/m), is dependent on the animal size and decreases as wind speed increases:

$$r_H = \rho_{cp} \times d / (k_a \times Nu) \tag{25}$$

The Nusselt number, Nu, is calculated based on Reynold's (Re) and Grashof's (Gr) numbers, which are dimensionless numbers associated with forced and natural convection, respectively (Incopera and DeWitt, 1990; McGovern and Bruce, 2000; Monteith 1973). For natural convection, $Gr > 16 \times Re^2$, $Nu = 0.48 \times Gr^{0.25}$. With forced convection, $Gr < 0.1 \times Re^{0.6}$, $Nu = 0.24 \times Re^{0.6}$. Otherwise, if wind speed is small, then Nu will be the maximum value between the two previous equations.

$$Nu = \max(0.48 \times Gr^{0.25}, 0.24 \times Re^{0.6}) \tag{26}$$

The variables Con_s, Con_l, LWR, $Solar_{abs}$, E_s, E_r and G_s can be converted from W/m^2 into J/h/animal by multiplying by 3,600 × A, where A (m^2) is the surface area given by the following formula (Brody 1945):

$$A = 0.14 \times BW^{0.57} \tag{27}$$

This surface area equation works for *Bos taurus*. Johnston *et al.* (1958) found that *Bos indicus* have approximately 12% more surface area than *Bos taurus*, so the equation would then be multiplied by 1.12.

Model simulation

The model is run using Matlab 7.8.0 (R2009a) using a variable-step integration algorithm (Matlab solver: ode23). The climatic input data: temperature, relative humidity, solar radiation and wind speed are obtained from regional weather stations. The run time is set for 96 hours (4 days), the air temperature varies between 29 and 36 °C, and the wind speed varies between 2 and 4 m/s. The humidity drops from 1.0 to 0.67 at the hottest part of the day and the solar radiation reaches 850 W/m^2 at noon.

Results and discussion

The simulated evaporation potential (Equation 20) behaves as expected where at both high temperatures and high humidity (which is commonly seen in the humid tropics) the potential sharply decreases (Figure 2). At low relative humidity, the potential increases as air temperature increases. This is because as air temperature increases, the absolute humidity potential also increases. On the other hand, at a high relative humidity, the evaporation potential decreases as air temperature increases. At a high relative humidity, a higher air temperature will have a much higher absolute humidity potential than a lower air temperature.

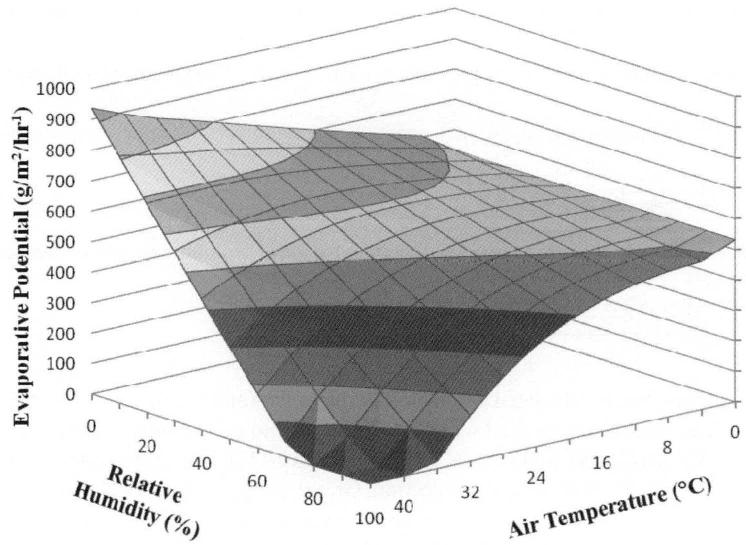

Figure 2. Simulated evaporation potential of sweat on the skin of cattle versus air temperature and relative humidity.

The long wave radiation and convection losses from the animal (Figure 3) are dependent on air and body temperatures and independent of humidity. As air temperature decreases, the temperature difference between the animal and its environment increases, which increases both long wave radiation and convection losses from the animal. Above 36 °C, at zero wind speed, both long wave radiation and convection become negative indicating a net sensible heat gain by the animal. Zhang *et al.* (2007) found that humidity does have an effect on convection, but they measured objects at temperatures ranging from 77 to 177 °C. The effect of humidity is due to the fact that a wet surface, with evaporation, will have a lower temperature than a dry surface. This lower temperature will affect the temperature gradient which impacts the convection and long wave radiation losses. In the figure, the lines are representative of climatic conditions at zero wind speed. Gebremedhin (1987) found that with no wind, long wave radiation and convection will contribute equally to heat loss, but under windy conditions convection losses can be up to six times the value of long wave radiation losses.

The simulation outputs of the model are shown in Figure 4. As expected, the skin temperature lies between the body and air temperature. This is consistent with findings by Gebremedhin *et al.* (2008) who showed that in this air temperature range, skin temperature lies between body and air temperature. Future work on the model may reveal the need to increase the number of layers between the body core and the outside air, for example, by addition of a separate coat compartment, or by explicitly modeling the influence of subcutaneous fat on heat transfer between the core and skin. The fluctuation in daily skin temperature is slightly offset from the air temperature. This is due to the impact of solar radiation. Although peak air temperature is reached around 15:00, peak solar radiation is reached at midday. Due to this, the animal's solar load increases its skin temperature slightly ahead of the air temperature and subsequently, as soon as air temperature begins to decrease, the solar load is no longer significant so the skin temperature can decrease along with it.

Conclusion

Though the many parts of the model can be independently tested and compared, the full model has yet to be tested. This is due in part to the lack of a complete data set against which to test the model. While this data set may not be available, further testing should result in components being tested simultaneously. This model integrates concepts and data regarding heat flows between animals

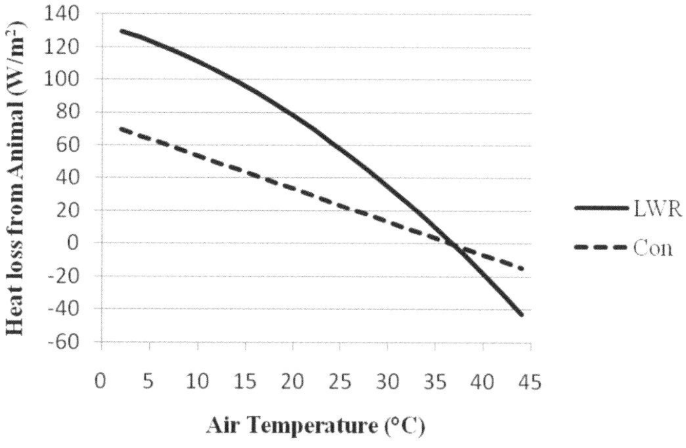

Figure 3. Long Wave Radiation (LWR) and Convection (Con) losses from cattle at zero wind speed.

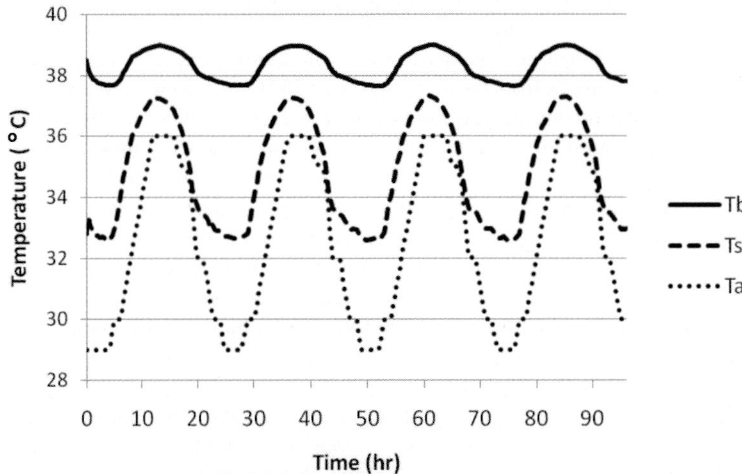

Figure 4. The model simulation outputs. The model simulated four days (96 hours). The outputs are body, T_b, skin, T_s, and air temperature, T_a. The skin temperature lies between the body and air temperature as air temperature never exceeds 36 °C.

and their environment. Equation forms and parameters are largely based on generally accepted physical, anatomical and physiological principles. While this model appears to behave acceptably, it still needs to be further tested against more complete, extensive datasets to evaluate its ability to predict body temperature in response to different environmental conditions.

References

Blaxter, K.L., 1962. The Energy Metabolism of Ruminants. Hutchinson & Co., Ltd., London, UK.

Brody, S., 1945. Bioenergetics and Growth with Special Reference to the Energetic Efficiency Complex in Domestic Animals. Reinhold Publishers, New York, NY, USA.

Duffie, J.A. and Beckman, W.A., 1980. Solar Engineering of Thermal Processes. Wiley, New York, NY, USA.

Finch, V.A., 1985. Comparison of Non-Evaporative Heat Transfer in Different Cattle Breeds. Australian Journal of Agricultural Research 36:497-508.

Gebremedhin, K.G., 1987. Effect of animal orientation with respect to wind direction on convective heat loss. Agricultural and Forest Meteorology 40:199-206.

Gebremedhin, K.G., Hillman, P.E., Lee, C.N., Collier, R.J., Willard, S.T., Arthingon, J. and Brown-Brandl, T.M., 2008. Livestock Environment VIII. Proceedings of the 31 Aug-4 Sep 2008 Conference. Iguassu Falls, Brazil. ASABE Publication Number 701P0408.

Gomes da Silva, R. 2008. Biofisica Ambiental: Os animais e seu ambiente. Funep, Jaboticabal, SP, Brazil, 393 pp.

Johnston, J.E., Hamblin, F.B. and Schrader, G.T., 1958. Factors Concerned in the Comparative Heat Tolerance of Jersey, Holstein and Red Sindhi-Holstein (F1) Cattle. Journal of Animal Science 17:473-479.

Incopera, F.P. and DeWitt, D.P., 1990. Fundamentals of Heat and Mass Transfer. John Wiley and Sons, New York, NY, USA.

McArthur, A.J., 1987. Thermal interaction between animal and microclimate: a comprehensive model. Journal of Theoretical Biology 126:203-238.

McGovern, R.E. and Bruce, J.M., 2000. A model of the thermal balance of cattle in hot conditions. Journal of Agricultural Engineering Research 77:81-92.

Monteith, J.L., 1973. Principles of Environmental Physics. Contemporary Biology. Edward Arnold, London, UK.

Monteith, J.L. and Unsworth, M.H., 2008. Principles of Environmental Physics, 3rd Edition. Edward Arnold, London, UK.

NRC (National Research Council), 2000. Nutrient Requirements of Beef Cattle. 7th rev. ed. National Acadamies Press, Washington, DC, USA.

Thom, E.C., 1959. The discomfort index. Weatherwise 12:57-59.

Turco, S.H.N., da Silva, T.G.F., de Oliveira, G.M., Leitão, M.M.V.B.R., de Moura, M.S.B., Pinheiro, C. and da Silva Padilha, C.V., 2008. Estimating Black Globe Temperature based on Meteorological Data. Proceedings of Livestock Environment VIII. 31 Aug – 4 Sep, 2008. ASABE 701P0408. Available at: http://asae.frymulti.com/azdez.asp?JID=1&AID=25593 &CID=lenv2009&T=2. Accessed 10 Nov 2009.

Turnpenny, J.R., McArthur, A.J., Clark, J.A. and Wathes, C.M., 2000a. Thermal balance of livestock 1. A parsimonious model. Agricultural and Forest Meteorology 101:15-27.

Turnpenny, J.R., McArthur, A.J., Clark, J.A. and Wathes, C.M., 2000b. Thermal balance of livestock 1. Applications of a parsimonious model. Agricultural and Forest Meteorology 101:29-52.

Zhang, J., Gupta, A. and Baker, J. 2007. Effect of Relative Humidity on the Prediction of Natural Convection Heat Transfer Coefficients. Heat Transfer Engineering 28:335- 342.

An interactive, mechanistic nutrition model to determine energy efficiency of lactating dairy cows

L.O. Tedeschi[1], *D.G. Fox*[2] *and D.K. Roseler*[3]
[1]*Texas A&M University, Department of Animal Science, 133 Kleberg, 2471 TAMU, College Station, TX 77843-2471, USA; luis.tedeschi@tamu.edu*
[2]*Cornell University, Department of Animal Science 124 Morrison Hall Ithaca, NY 14853, USA*
[3]*Land O Lakes Purina Feed, 505 Palmer St. Wooster OH 44691, USA*

Abstract

Selection for milk production alone may result in some cows that have lower energy efficiency (increased energy efficiency index – EEI – Mcal of metabolisable energy (ME) required per kg of milk) on the diets available due to limitations in dry matter intake (DMI), increased body size relative to the amount of milk produced, or inadequate ability to mobilize and replete energy reserves. Ideally, an energy efficient dairy cow would produce the same amount of milk with less DMI by mobilizing reserve tissues during negative energy balance and replenishing reserve tissues during positive energy balance in a more efficient and faster manner. The objective of this paper is to describe a nutrition model that identifies differences in ME required (MER) for the observed milk production under farm conditions where individual DMI is not known. This model is comprised of four submodels that account for energy requirements for maintenance, growth, pregnancy, and lactation. A fifth submodel accounts for the dynamic mobilization and replenishment of body reserves throughout the reproductive cycle. The model was evaluated with a database containing records of 241 Holstein cows individually fed for a complete lactation at four locations in the US (New York, Utah, Florida, and Arizona). A Monte Carlo simulation was used to assess the distribution and variation. Influential variables on the simulation output were accomplished by using standardized regression coefficients. Model predicted MER accounted for up to 83% of the variation in actual DMI consumed. We conclude this model can be used to identify differences in feed requirements for the observed milk production when individual cow DMI is not known.

Keywords: modelling, simulation, lactation, dairy

Introduction

Extraordinary increases in milk production have been achieved by the dairy industry due to improvements in nutrition and feeding, reproduction and genetics, and management and welfare of the cows. Much of this increase is directly related to widespread use of sires whose daughters have increased genetic potential for milk yield. However, the use of selection for milk production alone may result in some cows that have lower energy efficiency (increased Mcal of metabolisable energy (ME) required per kg of milk) on the diets available due to limitations in dry matter intake (DMI), increased body size relative to the amount of milk produced or inadequate ability to mobilize and replete energy reserves. Berry *et al.* (2003) indicated that despite the moderate genetic correlation between body condition score (BCS), milk yield (MY), body weight (BW), and fertility traits in dairy cows, selection strategies to increase MY is possible without any negative consequences for fertility and BCS. Veerkamp (1998) suggested that significant improvements in economic efficiency can be achieved when BW is decreased (maintenance requirements are likely decreased), resulting in an increase in feed efficiency (FE, kg of milk per kg of DMI). The intrinsic inter-relationships among BW, MY, BCS, milk composition, and DMI are nonlinear and continuous and any attempt to improve production efficiency would have to account for their interactions to prevent unintended consequences.

Knaus (2009) argued that the ever-increasing MY of dairy cows during the last half century has reduced cow longevity. Furthermore, an increase in milk production at the expense of more ME required per kg of milk is not sustainable for a long period of time mainly because of competition for resources (including human edible feeds and arable land), environmental pollution, and costs associated with maintaining energy inefficient cows. A successful management example of application of technology in support of production efficiency with reduced carbon footprint of dairy production is the recombinant bovine somatotropin (rbST) (Capper *et al.*, 2008). These authors suggested a decrease of 1.9 Tg CO_2/y at the same level of MY when rbST was used. A historical comparison indicated that modern dairy systems require only 21% of animals, 23% of feedstuffs, and only 10% of the land to produce the same amount of milk with less manure (24%) and methane (43%) production (Carro *et al.*, 1980). Therefore, identification of efficient animals may decrease even more these statistics, mitigating further the environmental impact of ruminant production.

Ideally, an energy efficient dairy cow would produce the same amount of milk with less dry matter (DM) by mobilizing reserve tissues during negative energy balance and replenishing reserve tissues during positive energy balance in a more efficient and faster manner. Linn *et al.* (2006) suggested that efficiency of lactating dairy cows should be performed on a fat-corrected milk basis because it accounts for milk composition and compares cows at the same level of production.

The identification of efficient lactating cows requires accurate individual cow measurements of BW and BCS changes, DMI, and ration content of ME and metabolisable protein (MP) at the same physiological stage. However, not all of these measurements are possible under current practical conditions or they require extra costs that would be prohibitive. The use of accumulated scientific knowledge of nutrition, including fluxes of energy and nutrients, and growth, pregnancy, and lactation requirements may assist in determining efficient animals via computer modelling.

The objective of this paper is to describe a dynamic nutrition model that can be applied on dairy farms to predict ME required (MER) for the observed milk production for individual lactating cows where individual cow DMI is not known. This model is comprised of four submodels that account for energy requirements for maintenance, growth, pregnancy, and lactation, and a fifth submodel that accounts for the dynamics of mobilization and replenishment of body reserves throughout the reproductive cycle.

Material and methods

Model development

An interactive, mechanistic, dynamic model was developed to account for dietary MER by a dairy cow throughout the reproductive cycle given the milk production and composition, changes in BW and BCS, expected calf birth weight and dietary ME. The model computes MER on a daily basis and summarizes the calculations into indexes based on physiological status of the cow. Most of the calculations of MER were based on the equations used in the Cornell Net Carbohydrate and Protein System (CNCPS) version 5 (Fox *et al.*, 2004) and the National Research Council publication Nutrient Requirements of Dairy Cattle (NRC, 2001). Figure 1 depicts a flow chart that describes the model. The following sections describe specific calculations of the model.

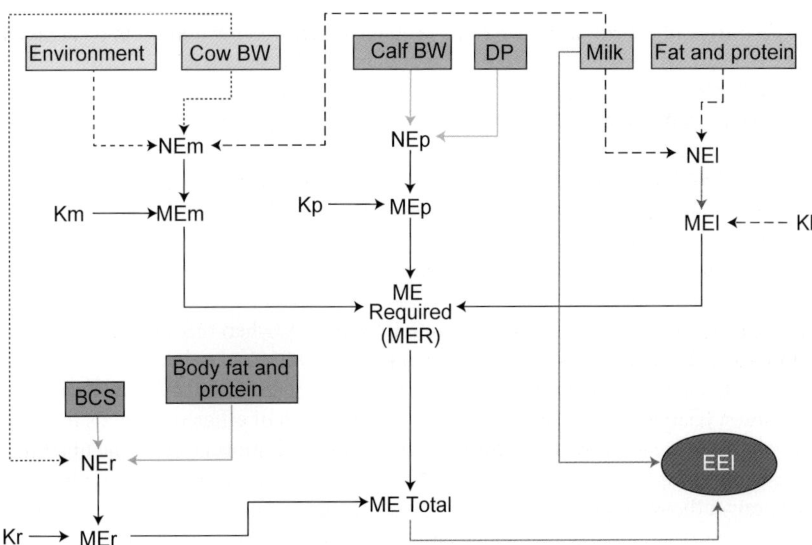

Figure 1. Schematic representation of the relationships among inputs needed to predict the energy efficiency index for lactating dairy cows (BCS: body condition score, EEI: energy efficiency index, BW: body weight, ME: metabolisable energy, DP: days pregnant).

Dietary factors. The partial efficiency of ME to net energy (NE) for maintenance (km) was computed based on ME as described by Fox *et al.* (2004). The partial efficiencies of ME to NE for lactation (kl) and pregnancy (kp) were assumed to be 64.4 and 14%, respectively.

Maintenance. The maintenance requirement of ME was estimated based on metabolic BW and km. This calculation does not account for composition of the body (e.g. fat content). No adjustment was performed for environmental effects on maintenance requirement. As discussed by Tedeschi *et al.* (2004), the recurrent daily adjustment for environmental effects may overpredict energy requirement by the animal when using a dynamic model. The environment submodel used by CNCPS-based models (Fox *et al.*, 2004) was developed for static conditions. Adverse environmental effects must be accounted for (Delfino and Mathison, 1991; Fox *et al.*, 1988; Fox and Tylutki, 1998), but further work is needed for continuous simulations.

Lactation. The MY was either predicted using Wood's equation (Equation 1) or observed values were used. When observed values were used, values were interpolated using Cubic Spline curves to estimate the missing values. Milk fat and(or) protein values were entered to compute milk energy (Equation 2 or 3) content using the NRC (2001) equation. Fat corrected milk for 4% milk fat ($FCMY_{4\%}$) was computed using Equation 4 and ME for lactation (MEl) was computed using Equation 5.

$$MY = a \times (\frac{DIM}{7})^{b} \times e^{-c \times \frac{DIM}{7}} \quad (1)$$

$$MkE = 0.97 \times MkFAT + 0.361 \quad (2)$$

$$MkE = 0.0929 \times MkFAT + 0.0563 \times MkPROT + 0.192 \quad (3)$$

$$FCMY_{MkFAT} = MY \times MkE / MkE_{MkFAT} \quad (4)$$

$$MEl = MY \times MkE / kl \quad (5)$$

Where MY is milk yield, kg/d; a, b, and c are parameters; DIM is days in milk, e is exponential, MkE is milk energy, Mcal of NE/kg; MkFAT is milk fat, %; MkPROT is milk protein, %, FCMY is fat-corrected MY for a given milk fat energy (MkE_{MkFAT}), kg/d, MEl is metabolisable energy required for lactation, Mcal/d.

Pregnancy. The ME required for pregnancy was computed assuming the equation proposed by Bell *et al.* (1995) when days pregnant was greater than 190 d as shown in Equation 6.

$$MEp = \left((2 \times 0.00159 \times DP) - 0.0352\right) \times (\frac{CBW}{45})/kp \quad (6)$$

Where MEp is metabolisable energy required for pregnancy, Mcal/d; DP is days pregnant, d; CBW is calf birth weight, kg; and kp is partial efficiency of ME to NE for pregnancy.

Body reserves. Changes in BW reflect the use of energy reserves either to supplement ration deficiencies during early lactation or to store energy consumed above the requirements (Moe *et al.*, 1972; NRC, 2001). The body energy was assessed using BCS throughout the reproductive cycle. The adjustments discussed by Tedeschi *et al.* (2006) were applied in this model. Briefly, empty BW (EBW) of the cows was adjusted for changes in BCS assuming a fixed change of 6.85% of EBW per change in BCS (ΔBCS) using Equations 7 and 8 and total energy was computed as shown in Equations 9 to 11.

$$WAF_t = 1 - 0.0685 \times (5 - BCS_{[1-9],t}) \quad (7)$$

$$aEBW_t = \frac{EBW_{t=1}}{WAF_{t=1}} \times WAF_t \quad (8)$$

$$TF_t = 0.037683 \times BCS_{[1-9],t} \times aEBW_t \quad (9)$$
$$TP_t = \left(0.200886 - 0.0066762 \times BCS_{[1-9],t}\right) \times aEBW_t \quad (10)$$
$$TE_t = 9.367 \times TF_t + 5.554 \times TP_t \quad (11)$$

Where WAF is weight adjustment factor; aEBW is adjusted empty BW, kg; TF is total body fat, kg; TP is total body protein, kg; and TE is total energy, Mcal.

Daily changes in body energy were computed as the difference of TE between two consecutive days as shown in Equation 12.

$$\Delta TE_t = TE_t - TE_{t-1}; t \geq 2 \quad (12)$$

The ME for reserves (MEr) was computed based on physiological stages of the cow and the ΔTE value, assuming different partial efficiencies for mobilization and repletion. When ΔTE was negative (a negative energy balance from one day to another – energy reserves were used to support milk production), Equation 13 was used to compute MEr if cow was lactating or Equation 14 was used if cow was not lactating. When ΔTE was positive (a positive energy balance from one day to another – energy reserves were replenished), Equation 15 was used if the cow was lactating.

$$MEr = \Delta TE_t \times 0.84/0.644; \text{ if lactating cow and ΔTE is negative} \quad (13)$$
$$MEr = \Delta TE_t/0.6; \text{ if dry cow and ΔTE is negative} \quad (14)$$
$$MEr = \Delta TE_t/0.7; \text{ if lactating cow and ΔTE is positive} \quad (15)$$

As discussed by Tedeschi *et al.* (2006), the partial efficiency of use of ME and NE changes, depending on the physiological stage of the cow. When ΔTE <0, a partial efficiency for NE for reserves (NEr) to NE for lactation (NEl) of 84% was assumed based on the Commonwealth Scientific and Industrial Research Organisation (CSIRO, 1990, 2007) and the Agricultural Research Council (ARC, 1980). Then, NEl was converted to MEl using kl of 64.4%. When $\Delta TE > 0$, NEr was converted to MEr assuming a partial efficiency of 72.6% based on data of Moe *et al.* (1970) and calculations as shown by Tedeschi *et al.* (2006).

Metabolisable energy required. The daily MER was computed as the sum of ME for maintenance, lactation, and pregnancy. Because computation of MER accounts for the composition of milk, further adjustment to FCMY is not necessary.

Energy efficiency index. Figure 1 depicts the scheme of calculation of energy efficiency index (EEI) for lactating dairy cows. Several EEI were computed for different physiological stages. The comparison among cows within the same physiological stage (e.g. days in milk) might be more adequate due to changes in BCS that is intrinsic to each cow. The EEI was computed as MER plus any energy change in body reserves (MET) divided by milk production within a given period.

Model evaluation

The database used in the evaluation of the dairy nutrition efficiency model contained records of 241 lactating Holstein cows from four locations in the US (New York, Utah, Florida, and Arizona). The database was described by Roseler *et al.* (1997). Simulations were performed for a reproductive cycle (i.e. calving interval) for each cow. Calving interval varied from 339 to 648 d. Calf birth weight was assumed to be 45 kg. Gestation interval was assumed to be 280 d. Because observations were typically obtained on a weekly basis, linear interpolation was used to maintain a continuous shape in estimating daily BCS, MY, DMI, BW, milk fat (MkFAT), and milk protein (MkPROT) based on weekly measured data. The relationship between predicted EEI and observed FE or predicted MER and observed DMI with and without adjustments to MER for ΔBCS during the complete lactation were evaluated, using the complete data set in which the data were adjusted for the random effects of locations and treatments.

Monte Carlo simulation

Monte Carlo simulation was used to assess the distribution and variation of key output variables. Monte Carlo simulations, distribution fitting and sensitivity analysis were performed with @Risk v. 5.5 (Palisade Corp., Ithaca, NY). The Monte Carlo technique has been used frequently to understand relationships among important variables in biology (Tedeschi *et al.*, 2009; Waller *et al.*, 2003). It is based on probability density functions (PDF) from which samples are drawn to perform specific calculations and solutions are presented as distributions (Law, 2007). Spearman correlations are used during the simulations to account for non-independency among variables. The Monte Carlo simulation was obtained from 1,000 iterations and Latin hypercube sampling. Distribution fitting was used to obtain the PDF of input variables. The determination of most influential variables on the simulation output was accomplished by using standardized regression coefficients (SRC) as discussed by Kutner *et al.* (2005) and tornado charts. The SRC reflects the change in the standard deviation of the dependent (output) variable associated with one unit change in the standard deviation of the independent (input) variable at a ceteris paribus condition, i.e. when all other input variables are fixed and unchanged (Helton and Davis, 2002).

Statistical analyses

Statistical analyses were done with SAS v. 9.2. The PROC MIXED was used to evaluate the impact of locations and treatments as random effects on selected variables. Treatments were levels of a biweekly sustained-release form of rbST (0, 250, 500, and 750 mg) initiated at 60 d post-partum. The PROC REG was used to develop linear equations and the PROC NLIN was used to converge parameters (*a*, *b*, and *c*) of Equation 1 for milk yield data of the 241 cows.

Results and discussion

Figure 2 depicts the prediction of MER during the reproductive cycle for a representative cow in the database (630 kg of BW, average MY of 27.6 kg/d, and lactation length and calving intervals of 305 and 365 days, respectively). The MY was predicted throughout the lactation using Equation 1 assuming the average value for the parameters *a* (23.4), *b* (0.308), and *c* (0.0291) obtained from the 241 data points. Similarly, the average of the 241 data points for BW, milk fat (3.56%) and protein (3.21%), and BCS (3.07) were assumed constant throughout the reproductive cycle. The diet ME was assumed to be 2.6 Mcal ME/kg of dry matter.

Figure 2B shows the partition of MER among maintenance, lactation, and pregnancy as simulated by the model for this representative cow. The model assumed 0.082 Mcal of NE/kg$^{0.75}$ for maintenance (basal metabolism plus 15% for physical activity) and an average partial efficiency of 65%, resulting in an average of 126 Kcal of ME/kg$^{0.75}$. This value is less than the 143 Kcal of ME/kg$^{0.75}$ reported by Agnew *et al.* (2003) even though the partial efficiency was the same. Variations in ME or NE for maintenance among breeds and breed-types have been well documented for cows (Solis *et al.*, 1988), but more work is needed to accurately identify traits that can be used to account for individual variations. Lactation requirement was greater than 50% of the MER throughout the lactation. As expected this is affected by the amount and composition of milk. Pregnancy accounted for up to 30% at the end of the gestation. Therefore, identification of efficient dairy cows is likely to be related to milk production and maintenance. VandeHaar and St-Pierre (2006) suggested that a cow producing 45 kg/d needs 4 times more NEl to support lactation than maintenance. This number would be 7 times more for an elite high-producing cow of 90 kg/d. Therefore, we have been able to double the amount of milk production with only 75% increase in energy requirement for lactation.

The distribution (not shown) and descriptive statistics of MET, FE, and EEI with and without adjustment for ΔBCS for early lactation (1 to 120 DIM) obtained for the 241 cows indicated the 90% confidence interval ($CI_{90\%}$) was 35.2 to 61.2 Mcal/d of MET. The average was 49.1 Mcal/d. Even though the study variance was not different than zero (P=0.116), cows from Florida had a significantly lower MET (ΔMET of -7.3 Mcal/d, P=0.009) than the average, suggesting that Utah, Arizona, and New York locations had on average 51.6 Mcal/d whereas cows from Florida had 41.8 Mcal/d. This is likely because cows from Florida produced less milk (ΔMY of -5.3 kg/d, P=0.01) than the average (31.5 kg/d). The milk production of treatment 0 rbST (control) tended to differ (P=0.07) from the 750 mg of rsBT, but because treatments were considered random effects we accounted for treatment effects by removing them from the variables of interest.

The $FCMY_{4\%}$ FE $CI_{90\%}$ ranged from 1 to 1.65 with an average of 1.37. Cows in the Florida data base were less efficient (ΔFE of -0.165, P=0.0178) than the average, and cows in the New York data base converted better than the average (ΔFE of 0.153, P=0.029). This is in agreement with Britt *et al.* (2003), who reported that $FCMY_{3.5\%}$ FE was less in warm than in cold climates. Our average was similar to that reported by Britt *et al.* (2003) (1.40), but our variation was 1.5 times greater. Linn *et al.* (2006) reported a variation in FE from 1.3 to 1.8 for mid-lactation cows (158

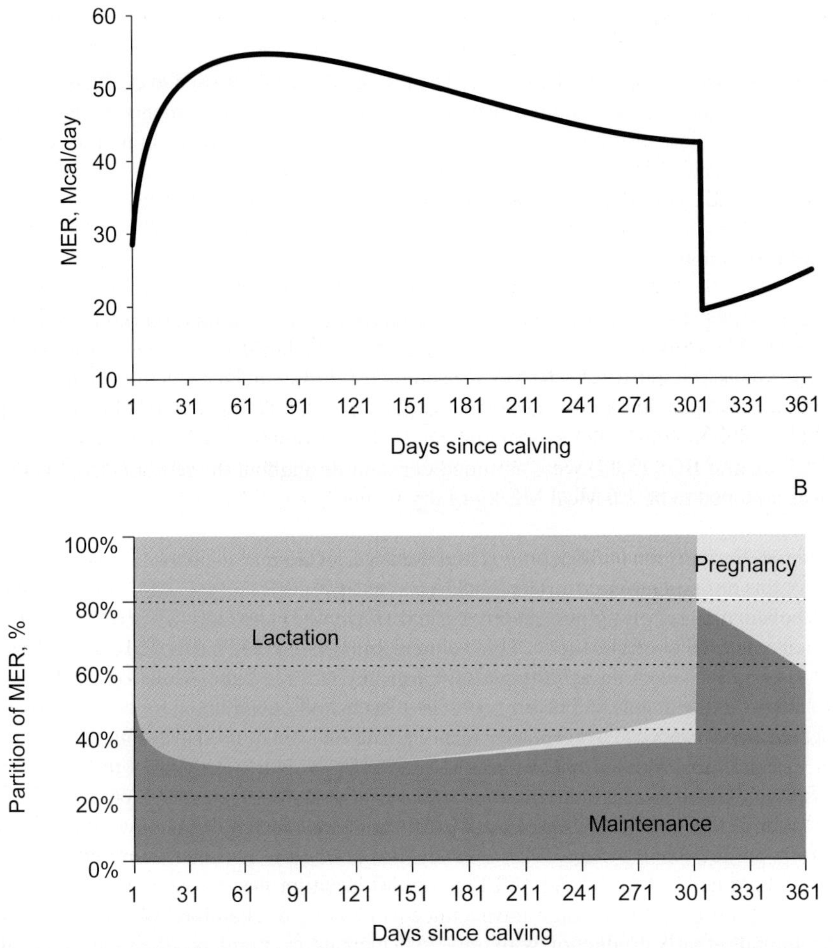

Figure 2. Simulated daily (A) metabolisable energy required (MER, Mcal/d) and (B) partition of MER into percentage used for maintenance, pregnancy and lactation, assuming a body weight of 630 kg, average milk yield of 23 kg/d, and lactation and calving intervals of 305 and 365 days, respectively. Pregnancy requirement was computed with the equation of the NRC (2000).

to 231 DIM). Linn *et al.* (2006) further suggested that during early lactation, a greater FE (>1.8) might indicate excessive mobilization of body reserves, leading to ketosis. A lesser FE (<1.2) might suggest health problems such as acidosis or, *in facto*, a low efficient animal.

When no adjustment was made for changes in BCS, EEI $CI_{90\%}$ varied from 1.38 to 1.78 Mcal/kg with an average of 1.56 Mcal/kg. The calculation of EEI based on FCMY alone is not accurate because of not accounting for changes in BCS. When EEI was adjusted for ΔBCS, the average EEI increased to 1.58 Mcal/kg, and the $CI_{90\%}$ was 1.38 to 1.89, suggesting variation in BCS occurred. The average and SD values for EEI using the complete database for mid-, late-, and complete lactation are 1.64±0.19, 1.96±0.35, and 1.65±0.14; respectively, and the average and SD values for $EEI_{\Delta BCS}$ for mid-, late-, and complete lactation are 1.68±0.22, 2.08±0.49, and 1.70±0.16; respectively. The reason that $EEI_{\Delta BCS}$ increased for mid- and late-lactation is likely because MY decreased and cows were depositing body reserves, which on average is less efficient – partial efficiency of ME to NEl is 0.767 (Equation 13) versus 0.726 (Equation 15) when cows are in negative or positive energy balances, respectively. This likely explains the larger difference between EEI and $EEI_{\Delta BCS}$ for mid- and late lactation.

Therefore, adjustment for ΔBCS is needed to adequately compute energy efficiency of dairy cows. A comparison of different systems (Australia, France, the Netherlands, the United Kingdom, and the United States) to predict energy and nutrients requirements for lactating dairy cows indicated that all systems had a poor performance in accounting for BW changes of lactating cows (Yan *et al.*, 2003) in which the prediction error increased with increasing BW changes. Furthermore, BW changes were deemed to be inaccurate for accounting for effects of changes in energy reserves on energy balance in lactating dairy cows (Yan *et al.*, 2003), likely because changes in BW are related to temporary changes in gastro-intestinal tract contents (Andrew *et al.*, 1994; Gibb *et al.*, 1992; Komaragiri and Erdman, 1997). Models have been developed to use changes in BCS rather than BW (Tedeschi *et al.*, 2006), but individual variation is yet to be accounted for. Some information about the relationship of adipose tissue and leptin single-nucleotide polymorphism has been documented for beef cattle (Kononoff *et al.*, 2005) and dairy cattle (Buchanan *et al.*, 2003; Liefers *et al.*, 2002).

Figure 3 shows the relationship between predicted EEI and observed FE or predicted MER and observed DMI with and without adjustments to MER for ΔBCS during the complete lactation, using the complete data set in which the data were adjusted for the random effects of locations and treatments. As expected, there was a negative correlation between EEI and FE in which efficient cows had high FE and low EEI. Figures 3A and 3B show that EEI explained 77.5 and 76.4% of the variation in FE when EEI was uncorrected or corrected for changes in BCS, respectively. Figures 3C and 3D show that when MER was uncorrected or corrected for changes in BCS, MER explained 82.9 and 82.2% of the variation in observed DMI; respectively. The slightly lower r^2 for Figure 3B and 3D are likely a result of FE not accounting for changes in BCS whereas EEI and MER do. For example, in Figure 3B, cows mobilizing body energy could have a high FE but not a correspondingly lower EEI, because the EEI includes the mobilized or ME reserves gained in the ME requirement whereas FE does not. A possible calculation approach to account for ΔBCS on FE calculation is to (1) convert the ΔBCS energy reserves to ME, (2) convert the change in energy reserves ME to diet DM equivalent (k = energy reserves ME/diet ME concentration), (3) add the diet DM equivalent to DMI if cows lost BCS or to subtract it from the DMI if cows gained BCS, and (4) compute FE using the adjusted DMI.

Given the high correlation between MER and MET with actual individual cow DMI (r=0.91), the model presented in this paper can be used to account for differences in feed requirements among individual lactating dairy cows. This information can be used to monitor how well the ration being fed is meeting their individual requirements. The ME being fed to each cow is estimated by applying each cow's proportional share of the pen MET (sum of the individual MET) to the DM actually fed to a pen and the ME content of the ration.

There are additional variables that need to be accounted for in evaluating and ranking cows for energy efficiency. To be accurate for the entire reproductive cycle, the ME required over the reproductive cycle should be adjusted for differences between beginning and ending BCS. Because of the different partial efficiencies of use of energy for body reserves mobilization and deposition, cows that begin and end a reproductive cycle (calving interval) with the same BCS (ΔBCS = 0) may not be efficient, even though they have an accumulated energy balance equal to zero. As shown above, the efficiency during early lactation might be greater than that during late lactation. Reproduction efficiency needs to be accounted for when determining production efficiency of dairy

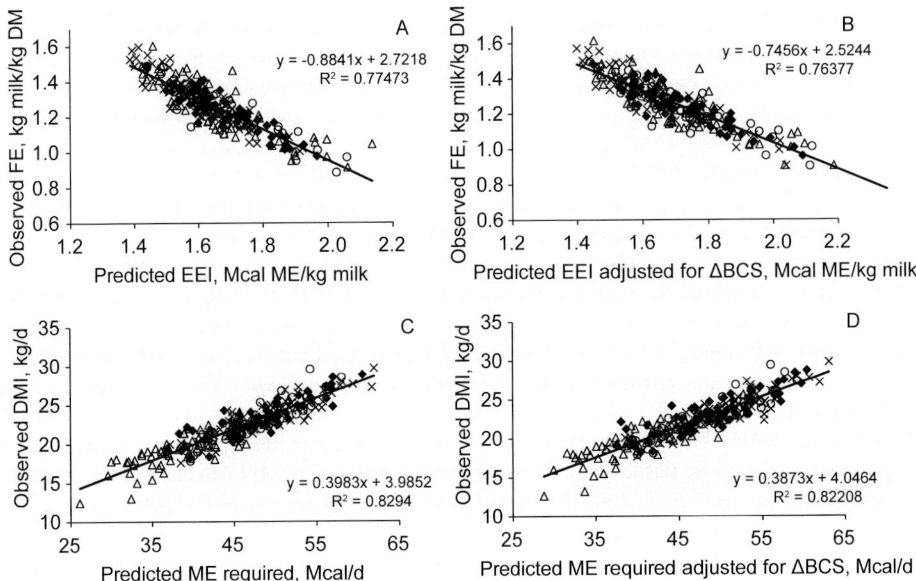

Figure 3. Regression between observed feed efficiency (FE) on energy efficiency index (EEI) either (A) unadjusted or (B) adjusted for ΔBCS and observed dry matter intake (DMI) on metabolisable energy (ME) required either (C) unadjusted or (D) adjusted for ΔBCS for 4 locations (NY, ○; AZ, ◆; UT, ×; FL, △) throughout the lactation. Observed FE and DMI were adjusted for random effects of treatment and location.

cows. Knaus (2009) has indicated that the increase in MY achieved during the last 60 years has dramatically impaired fertility and longevity of dairy cows. This suggests that increased genetic potential for milk yield has not been fully matched with increased MEI in early lactation, resulting in an increased time for cows to reach NADIR.

References

Agnew, R.E., Yan, T., Murphy, J.J., Ferris, C.P. and Gordon, F.J., 2003. Development of maintenance energy requirement and energetic efficiency for lactation from production data of dairy cows. Livestock Production Science 82:151-162.

Andrew, S.M., Waldo, D.R. and Erdman, R.A., 1994. Direct analysis of body composition of dairy cows at three physiological stages. Journal of Dairy Science 77:3022-3033.

ARC (Agricultural Research Council), 1980. The Nutrient Requirements of Ruminant Livestock. The Gresham Press, London, UK.

Bell, A.W., Slepetis, R. and Enrhardt, R.A., 1995. Growth and accretion of energy and protein in the gravid uterus during late pregnancy in Holstein cows. Journal of Dairy. Science 78:1954-1961.

Berry, D.P. Buckley, F., Dillon, P. Evans, R.D., Rath, M. and Veerkamp, R.F., 2003. Genetic relationships among body condition score, body weight, milk yield, and fertility in dairy cows. Journal of Dairy Science 86:2193-2204.

Britt, J.S., Thomas, R.C., Speer, N.C. and Hall, M.B., 2003. Efficiency of converting nutrient dry matter to milk in Holstein herds. Journal of Dairy Science 86:3796-3801.

Buchanan, F.C., Van Kessel, A.G. Waldner, C. Christensen, D.A. Laarveld, B. and Schmutz, S.M., 2003. Hot Topic: An association between a leptin single nucleotide polymorphism and milk and protein yield. Journal of Dairy Science 86:3164-3166.

Capper, J.L., Castañeda-Gutiérrez, E., Cady, R.A. and Bauman, D.E., 2008. The environmental impact of recombinant bovine somatotropin (rbST) use in dairy production. Proceedings of the National Academy of Sciences. 105:9668-9673.

Carro, O., Hillaire-Marcel, C. and Gagnon, M., 1980. Detection of adulterated maple products by stable carbon isotope ratio. Journal of AOAC International 63:840-844.

CSIRO (Australian Commonwealth Scientific and Industrial Research Organization), 1990. Feeding Standards for Australian Livestock. Ruminants. Commonwealth Scientific and Industrial Research Organization, Melbourne, Australia.

CSIRO (Australian Commonwealth Scientific and Research Organization), 2007. Nutrient Requirements of Domesticated Ruminants. Commonwealth Scientific and Industrial Research Organization, Collingwood, Victoria, Australia.

Delfino, J.G. and Mathison, G.W., 1991. Effects of cold environment and intake level on the energetic efficiency of feedlot steers. Journal of Animal Science 69:4577-4587.

Fox, D.G., Sniffen, C.J. and O'Connor, J.D., 1988. Adjusting nutrient requirements of beef cattle for animal and environmental variations. Journal of Animal Science 66:1475-1495.

Fox, D.G. and Tylutki, T.P., 1998. Accounting for the effects of environment on the nutrient requirements of dairy cattle. Journal of Dairy Science 81:3085-3095.

Fox, D.G., Tedeschi, L.O., Tylutki, T.P. Russell, J.B., Van Amburgh, M.B., Chase, L.E., Pell, A.N. and Overton, T.R., 2004. The Cornell Net Carbohydrate and Protein System model for evaluating herd nutrition and nutrient excretion. Animal Feed Science Technology 112:29-78.

Gibb, M.J., Irvings, W.E., Dhanoa, M.S. and Sutton, D.J., 1992. Changes in body composition of autumn-calving Holstein Friesian cows over the first 29 weeks of lactation. Animal Production 55:339-360.

Helton, J.C. and Davis, F.J., 2002. Illustration of sampling-based methods for uncertainty and sensitivity analysis. Risk Analysis 22:591-622.

Knaus, W., 2009. Dairy cows trapped between performance demands and adaptability. Journal of Science and Food Agriculture 89:1107-1114.

Komaragiri, M.V.S. and Erdman, R.A., 1997. Factors affecting body tissue mobilization in early lactating dairy cows: 1.Effect of dietary protein on mobilization of body fat and protein. Journal of Dairy Science 80:929-937.

Kononoff, P.J., Deobald, H.M., Stewart, E.L., Laycock, A.D. and Marquess, F.L.S., 2005. The effect of a leptin single nucleotide polymorphism on quality grade, yield grade, and carcass weight of beef cattle. Journal of Animal Science 83:927-932.

Kutner, M.H., Nachtsheim, C.J., Neter, J. and Li, W., 2005. Applied Linear Statistical Models (5th ed.). McGraw-Hill Irwin, New York, NY, USA.

Law, A.M., 2007. Simulation modeling and analysis (4th ed.). McGraw-Hill, Boston, MT, USA.

Liefers, S.C., Te Pas, M.F.W., Veerkamp, R.F. and Van der Lende, T., 2002. Associations between leptin gene polymorphisms and production, live weight, energy balance, feed intake, and fertility in Holstein heifers. Journal of Dairy Science 85:1633-1638.

Linn, J., Raeth-Knight, M., Fredin, S. and Bach, A., 2006. Feed efficiency in lactating dairy cows. In: Proceedings of Cornell Nutrition Conference for Feed Manufacturers, Syracuse, State College of Agriculture & Life Sciences, Cornell University, New York, NY, USA, pp.141-150.

Moe, P.W., Flatt, W.P. and Tyrrell, H.F., 1972. Net energy value of feeds for lactation. Journal of Dairy Science. 55:945-958.

Moe, P.W., Tyrrell, H.F. and Flatt, W.P., 1970. Partial efficiency of energy use for maintenance, lactation, body gain and gestation in the dairy cow. In: Proceedings of Energy Metabolism of Farm Animals, 5, EAAP, Vitznau, Switzerland, pp.65-68.

NRC (National Research Council), 2000. Nutrient Requirements of Beef Cattle (updated 7th ed.). National Academy Press, Washington, DC, USA.

NRC (National Research Council), 2001. Nutrient Requirements of Dairy Cattle (7th ed.). National Academy Press, Washington, DC.

Roseler, D.K., Fox, D.G.,. Pell, A.N and Chase, L.E., 1997. Evaluation of alternative equations for prediction of intake for Holstein dairy cows. Journal of Dairy Science 80:864-877.

Solis, J.C., F.M. Byers, F.M., Schelling, G.T., Long, C.R. and Greene, L.W., 1988. Maintenance requirements and energetic efficiency of cows of different breed types. Journal of Animal Science 66:764-773.

Tedeschi, L.O., Kononoff, P.J., Karges, K. and Gibson, M.L., 2009. Effects of chemical composition variation on the dynamics of ruminal fermentation and biological value of corn milling (co)products. Journal of Dairy Science 92:401-413.

Tedeschi, L.O., Fox, D.G. and Guiroy, P.J., 2004. A decision support system to improve individual cattle management. 1. A mechanistic, dynamic model for animal growth. Agricultural Systems 79:171-204.

Tedeschi, L.O., Seo, S., Fox, D.G. and Ruiz, R., 2006. Accounting for energy and protein reserve changes in predicting diet-allowable milk production in cattle. Journal of Dairy Science 89:4795-4807.

VandeHaar, M.J. and St-Pierre, N., 2006. Major advances in nutrition: Relevance to the sustainability of the dairy industry. Journal of Dairy Science 89:1280-1291.

Veerkamp, R.F., 1998. Selection for economic efficiency of dairy cattle using information on live weight and feed intake: A review. Journal of Dairy Science 81:1109-1119.

Waller, L.A., Smith, D., Childs, J.E. and Real, L.A., 2003. Monte Carlo assessments of goodness-of-fit for ecological simulation models. Ecological Modeling 164:49-63.

Yan, T., Agnew, R.E., Murphy, J.J., Ferris, C.P. and Gordon, F.J., 2003. Evaluation of different energy feeding systems with production data from lactating dairy cows offered grass silage-based diets. Journal of Dairy Science 86:1415-1428.

The development and evaluation of the Small Ruminant Nutrition System

A. Cannas[1], L.O. Tedeschi[2], A.S. Atzori[1] and D.G. Fox[3]
[1]Dipartimento di Scienze Zootecniche, Università di Sassari, Via E. De Nicola, 9, 07100 Sassari, Italy; cannas@uniss.it
[2]Texas A&M University, Department of Animal Science, 133 Kleberg, 2471 TAMU, College Station, TX 77843-2471, USA
[3]Cornell University, Department of Animal Science 124 Morrison Hall Ithaca, NY 14853, USA

Abstract

A mechanistic model that predicts nutrient requirements and biological values of feeds for sheep and goats (Small Ruminant Nutrition System, SRNS) was developed based on the Cornell Net Carbohydrate and Protein System for sheep. The SRNS uses animal and environmental factors to predict metabolisable energy (ME) and protein requirements. This model has been subjected to an extensive evaluation. In particular, evaluation of the SRNS for sheep using published papers indicated good accuracy and precision in the prediction of organic matter and CP digestibility, while NDF digestibility was underpredicted. In addition, the SRNS accurately predicted daily ME intake (mean bias (MB) = 0.04 Mcal/d; root mean square error of prediction (RMSEP) = 0.24 Mcal/d; r^2 = 0.99) of lactating goats and goat wethers. The SRNS also accurately predicted the ADG of lambs (n = 42; MB = 1 g/d; RMSEP = 37 g/d; r^2= 0.84) and kids (n = 31; MB = -6.4 g/d; RMSEP = 32.5 g/d; r^2= 0.85) and the gains and losses of shrunk body weight of adult sheep (MB = -5.8 g/d; RMSEP = 30 g/d; r^2= 0.73) and the energy balance (n = 21; RMSEP = 0.20 Mcal/d; r^2 = 0.87) of lactating goats and wethers. In conclusion, based on our accumulated evaluation of the SRNS with literature data, the SRNS accurately predicts nutrient supply and requirements of sheep and goats. Recent unpublished evaluations, however, suggested that the SRNS may underpredict ADG when compensatory growth occurs.

Keywords: modelling, requirements, sheep, goats

Introduction

Mathematical models have been proven to be powerful tools for improving animal performance while reducing nutrient excretion (Tedeschi *et al.*, 2005). Several mathematical models based on feeding standards or feed evaluation systems for sheep and goats have been developed by different countries (Cannas, 2002; Cannas *et al.*, 2008). Recently, new requirements for sheep and goats were published by NRC (2007). This system uses the equations proposed by Cannas *et al.* (2004) for sheep requirements and the research published by the E(Kika) de la Garza Institute for Goat Research at Langston University (OK, USA) for goats requirements. The Small Ruminant Nutrition System (SRNS) model was developed, based on the Cornell Net Carbohydrate and Protein System (CNCPS) for sheep (CNCPS-S) framework (Cannas *et al.*, 2004), to account for energy and protein requirements of sheep and goats under diverse practical conditions. The objectives of this paper are to briefly document the SRNS model and to show accumulated and recent evaluations. More specific description of the SRNS model was provided by Tedeschi *et al.* (2010).

Energy and protein requirements

Maintenance

Maintenance requirements comprise the largest portion of energy and protein needed by farm animals and are influenced by many different variables (Cannas, 2002; Cannas *et al.*, 2008). Most sheep and goats feeding systems estimate them by using only few variables. The SRNS attempts to overcome this limitation by including additional variables previously not considered for these species.

The basal metabolic rate (BMR) is defined as the amount of energy needed to maintain life processes of an animal (e.g. vital cellular activity, respiration, and blood circulation) under the fasting and resting state in a thermo neutral environment (NRC, 1981). The intake of energy below or above the BMR would result in a decrease or increase in body weight (BW), respectively. Cannas *et al.* (2004) modified the maintenance submodel of the CNCPS (Fox *et al.*, 2004) to account for the requirement of metabolisable energy (ME) for maintenance (ME_M) and net energy (NE) for maintenance (NE_M) for sheep. The modifications were based on recommendations of the ARC (1980) and CSIRO (1990, 2007) feeding standards (Box 1, Equation 1). As discussed by Cannas *et al.* (2004), a1 is the thermo neutral maintenance requirement, which is assumed to be 0.06214 Mcal of $NE_M/kg^{0.75}$ of shrunk BW (SBW) for all types of sheep. Based on the work of Sahlu *et al.* (2004), a1 was assumed to be 0.0777 and 0.0652 Mcal of $NE_M/kg^{0.75}$ of SBW for dairy goats and other goat breeds, respectively (Cannas *et al.*, 2007b). For goats, Fernandes *et al.* (2007) obtained 0.067 Mcal of $NE_M/kg^{0.75}$ of SBW for Boer crossbred kids; they found values of NE_M for goats in the literature varying from 0.0625 to 0.0851 Mcal of $NE_M/kg^{0.75}$ of SBW (Fernandes *et al.*, 2007). In the SRNS, the k_M was assumed to be constant and equal to 0.644. The ACT in Equation 1 (Box 1) accounts for extra physical activity of grazing animals. Alternatively, CSIRO (1990) computed allowances of energy expenditure for grazing animals based on dry matter digestibility and availability of herbage of the pasture. This calculation was revised in the CSIRO (2007) publication. The SRNS calculates the energy cost to eliminate excess of N as urea and the recycled N as done by the CNCPS.

The metabolisable protein (MP) for maintenance (MP_M) requirements are the sum of dermal (wool) protein, urinary endogenous protein, and faecal endogenous crude protein (CP) losses (CSIRO, 1990; 2007). Equation 11 (Box 1) is used to compute MP_M for sheep. For goats, the first term in Equation 11 was modified (Equation 12; Box 1) to account for hair and scurf (Cannas *et al.*, 2007c). The hair requirement was based on the AFRC (1993) recommendation for scurf MP for goats (and cattle). The faecal endogenous CP requirement is variable; it increases as the dry matter intake (DMI) increases. This is consistent with the CNCPS model (Fox *et al.*, 2004); however, the INRA (1989) and AFRC (1993) suggested that faecal endogenous was based on BW. The efficiency of conversion of MP to net protein (NP) used for urinary and faecal endogenous CP is 0.67 (Cannas *et al.*, 2004).

Pregnancy and lactation

The ME for pregnancy (ME_P) requirements are estimated using the ARC (1980) equation as adapted by CSIRO (1990, 2007). The same equation was also adopted by the AFRC (1993). The ME_P for sheep and goats from 63 d after conception to delivery is shown in Equation 13 (Box 2). Similarly, the MP for pregnancy (MP_P) requirements are computed in Equation 14 (Box 2) using the recommendations of CSIRO (1990, 2007), which were derived from the ARC (1980), and assuming an efficiency of conversion of MP to NP of 0.70.

Box 1. Energy and protein requirements for maintenance.

$$ME_M = \frac{SBW^{0.75} \times a1 \times a2 \times e^{(-0.03 \times AGE)} + 0.09 \times MEI \times k_M + ACT + NE_{MCS} + UREA}{k_M} \quad (1)$$

$$ACT = (0.00062 \times \text{Flat Distance} + 0.00669 \times \text{Sloped Distance}) \times FBW \quad (2)$$

$$ACT = (0.000836 \times \text{Flat Distance} + 0.00669 \times \text{Sloped Distance}) \times FBW \quad (3)$$

$$NE_{MCS} = \frac{SA \times (LCT - Tc) \times k_M}{IN} \quad (4)$$

$$SA = 0.09 \times SBW^{0.66} \quad (5)$$

$$LCT = 39 + E \times EI - \frac{HP \times IN}{SA} \quad (6)$$

$$HP = ME_M + (MEI - ME_M) \times (1 - k_G) \quad (7)$$

$$IN = TI + \left[1 - 0.3 \times \left(1 - e^{\frac{-1.5 \times \text{Rainfall}}{\text{Wool Depth}}}\right)\right] \times \text{EI} \quad (8)$$

$$EI = WF \times \frac{4.184}{0.481 + 0.326 \times \sqrt{Wind}} + 4.184 \times r \times Log\left(\frac{1}{WF}\right) \times \left(z - 0.017 \times \sqrt{Wind}\right) \quad (9)$$

$$WF = \frac{r}{r + \text{Wool Depth}} \quad (10)$$

Where shrunk BW (SBW) is defined as 96% of full BW (FBW), kg; ME_M is ME required for maintenance, Mcal/d; MEI is ME intake Mcal/d; a2, an adjustment for the effects of previous temperature, is $(1 + 0.0091 \times C)$, where $C = (20 - Tp)$ and Tp is the daily temperature of the previous month; k_M is the efficiency of utilization of ME for NE for maintenance; AGE is the age of the animals, months; ACT is in Mcal of NE_M/d; flat and sloped distances are in km/d; UREA is NE required to eliminate excess of nitrogen, kcal/g, as estimated for each diet by the SRNS supply submodel; NE_{MCS} is NE required for maintenance due to cold stress, Mcal/d; SA is body surface area, m^2; LCT is lower critical temperature, °C; Tc is current temperature, °C; IN is total insulation, °C×m^2×d/Mcal; SBW is shrunk body weight, kg; E is evaporative losses, Mcal/(m^2×d); EI is external insulation, °C×m^2×d/Mcal; HP is heat production, Mcal/d; MEI is ME intake, Mcal/d; k_G is partial efficiency of ME to NE for growth; TI is tissue insulation, °C×m^2×d/Mcal; WF is wool (sheep) or hair (goats) factor; Wind is km/h; Rainfall is mm/d; Wool or hair depth is mm; r and z are factors for sheep and goats.

$$MP_M = \frac{Clean Wool}{0.6 \times 365} + \frac{0.147 \times FBW + 3.375}{0.67} + \frac{15.2 \times DMI}{0.67} \quad (11)$$

$$MP_M = 0.1125 \times FBW^{0.75} + \frac{0.147 \times FBW + 3.375}{0.67} + \frac{15.2 \times DMI}{0.67} \quad (12)$$

Where MP_M is the MP required for maintenance, g/d; the first term is scurf and wool CP requirement, g/d; the second term is the urinary endogenous CP, g/d; the third term is the faecal endogenous CP, g/d; Clean Wool is the clean wool produced per head, g/yr; FBW is full BW, kg; and DMI is dry matter intake, kg/d.

The ME for lactation (MP_L) requirement is computed based on the gross energy of milk and a constant efficiency of conversion of ME to NE for lactation (k_L = 0.644), as suggested by the CNCPS for cattle (Fox *et al.*, 2004). The gross energy of milk is computed as shown by Cannas

Box 2. Energy and protein requirements for pregnancy and lactation.

Pregnancy

$$ME_P = \frac{9.2438 \times LBW \times e^{\left(-11.465 \times e^{(-0.00643 \times t)} - 0.00643 \times t\right)}}{k_P} \tag{13}$$

$$MP_P = \frac{1428.06 \times LBW \times e^{\left(-11.22 \times e^{(-0.00601 \times t)} - 0.00601 \times t\right)}}{0.7} \tag{14}$$

Where ME_P is the ME required for pregnancy, Mcal/d; t is days of pregnancy; and LBW is the expected total lamb or lambs birth weight, kg. The efficiency of utilization of ME for pregnancy (k_P) is constant at 0.13; MP_P is MP required for pregnancy, g/d. The efficiency of utilization of MP for pregnancy is constant at 0.7.

Lactation

$$ME_L = \frac{(251.73 + 89.64 \times PQ + 37.85 \times (PP / 0.95)) \times Yn}{1000 \times k_L} \tag{15}$$

$$ME_L = \frac{(289.72 + 71.93 \times PQ + 48.28 \times (PP / 0.92)) \times Yn}{1000 \times k_L} \tag{16}$$

$$MP_L = \frac{10 \times PP \times Yn}{k_{PL}} \tag{17}$$

Where ME_L is ME required for lactation, Mcal/d; Yn is actual milk yield at a particular day of lactation, kg/d; PQ is milk fat, %; and PP is true milk protein, %; MP_L is MP required for lactation, g/d; and k_{PL} is the efficiency of MP to NP for lactation (0.58 for sheep and 0.64 for goats).

---

et al. (2004) for sheep (Equation 15, Box 2) and by Pulina *et al.* (1992) for goats (Equation 16, Box 2). The MP for lactation (MP_L) requirement for sheep and goats are computed from true milk protein content as shown in Equation 17 (Box 2). The coefficient for conversion of MP to NP (k_{PL}) is suggested specifically by the INRA (1989) to be 0.58 for sheep and 0.64 for goats. Both values were adopted by the SRNS model.

Growth and body reserves

The original requirements for growth of the CNCPS-S were described by Cannas *et al.* (2004). They were based on the requirements proposed by the CSIRO (1990; 2007) and modified by Freer *et al.* (1997). This model was evaluated and then revised by Cannas *et al.* (2006a). Box 3 reports the equations of the modified growth model for lambs and kids of the SRNS and Box 4 lists the whole energy and protein reserve submodel of the SRNS.

The original energy and protein reserve model was presented by Cannas *et al.* (2004), while its current SRNS version is discussed by Tedeschi *et al.* (2010).

Body condition score (BCS), BW, and body composition are used to calculate changes in energy and protein reserves after first lambing, applying the same approach developed by the NRC (2000). The proportion of fat in the empty body (AF) is computed for sheep as shown in Equation 25 of Box 4 (Cannas *et al.*, 2004) and for goats as shown in Equation 26 of Box 4, based on Ngwa *et al.*

Box 3. Energy and protein requirements for growth.

$$ADG = \frac{NE_G \times 1000}{0.92 \times EVG} \quad (18)$$

$$EVG = 0.239 \times \left\{6.7 + 2 \times (L-1) + \frac{16.5 - 2 \times (L-1)}{1 + e^{-6 \times (P-0.4)}}\right\} \quad (19)$$

$$NE_G = ME_G \times k_G \quad (20)$$

$$k_G = \frac{18.36}{27 + 41 \times RE_p} \quad (21)$$

$$RE_P = \frac{RE_{Prot}}{RE_{Prot} + RE_{Fat}} \quad (22)$$

$$RE_{Prot} = \left[212 - 8 \times (L-1) - \frac{120 - 8 \times (L-1)}{1 + e^{-6 \times (P-0.4)}}\right] \times 5.7 \quad (23)$$

$$RE_{Fat} = \left[43 + 56 \times (L-1) + \frac{490 - 56 \times (L-1)}{1 + e^{-6 \times (P-0.4)}}\right] \times 9.4 \quad (24)$$

Where ADG is average daily gain, g/d; NE_G is NE available for growth, Mcal/d; EVG is NE content of gain, Mcal/kg; ME_G is ME available for growth calculated as MEI (estimated by the supply submodel) and ME_m (Equation 1), Mcal/d; k_G is efficiency of ME to NE_G; RE_P is the proportion of retained energy as protein; RE_{Prot} is retained energy as protein, Mcal/kg; RE_{Fat} is retained energy as fat, Mcal/kg; L is the intake above ME_M requirement factor (ME intake/ME_M - 1); and P is degree of maturity = current BW/ standard BW, where the latter is the FBW at maturity when BCS is 3 (scale 0-5).

Box 4. Variations in energy and protein body reserves.

$AF_{sheep} = 0.0269 + 0.0868 \times BCS$ (25)

$AF_{goats} = 0.0289 + 0.0708 \times BCS$ (26)

$AP = -0.0039 \times BCS^2 + 0.0279 \times BCS + 0.1449$ (27)

$EBW = 0.851 \times 0.96 \times FBW$ (28)

$FBW = (0.594 + 0.163 \times BCS) \times FBW_{BCS\,2.5}$ (29)

$TF = AF \times EBW$ (30)

$TP = AP \times EBW$ (31)

$TE = 9.4 \times TF + 5.7 \times TP$ (32)

$FBW_c = (EB \times k_R)/((TE \text{ of } FBW \text{ at next } BCS - TE \text{ of } FBW \text{ at current } BCS)/ (FBW \text{ at next } BCS - FBW \text{ at current } BCS)) \times 1000$ (33)

Where BCS is body condition score, 0 to 5 scale; AF is proportion of empty body fat in sheep (AF_{sheep}) or goats (AF_{goats}); AP is proportion of empty body protein; EBW is empty body weight (0.851 × SBW), kg; SBW is shrunk body weight (0.96 × FBW), kg; FBW is the current full body weight, kg; FBW_c is FBW change, g/d; $FBW_{BCS2.5}$ is FBW at BCS 2.5, kg; TF is total body fat, kg; TP is total body protein, kg; and TE is total body energy, in Mcal of NE; EB = MEI - ME_m (Equation 1) - ME_L (Equation 15) - ME_p (Equation 13), Mcal/d, where MEI is ME intake as estimated by the SRNS, Mcal/d; k_R is the reserves NE to ME ratio and is 0.6 for all categories.

(2007) data. Due to difficulties in developing a model to account for breed differences, Equations 25 and 26 simply describe the average proportion of AF observed in the two species.

Dietary supply of energy and nutrients

Equations used to predict dietary supply of energy and nutrients in the SRNS are those used for the CNCPS-S, as described by Cannas *et al.* (2004). The main differences between the SRNS and the CNCPS for cattle (Fox *et al.*, 2004) are passage rates of forages and concentrates that are based on Cannas and Van Soest (2000) and the passage rate of liquids developed by Cannas *et al.* (2004). Another modification in the SRNS compared to the CNCPS (Fox *et al.*, 2004) is the calculation of faecal CP, fat and ash (Cannas *et al.*, 2004).

Model evaluations

Table 1 summarizes the evaluations performed with the CNCPS-S and SRNS. The statistics used for the evaluations are described in detail by Tedeschi (2006).

Prediction of nutrient supply

The predictions on feed digestibility were evaluated by using 14 published *in vivo* total tract digestibility experiments, in which 56 different diets were tested. For the diets for which the CNCPS-S predicted no rumen N shortage, predicted DM, OM, and CP digestibility did not differ from observed values, while NDF digestibility was underpredicted ($P<0.01$; Table 1). When the model predicted negative rumen N balance, OM and CP digestibility was predicted without significant bias, while DM and NDF digestibility were significantly underpredicted ($P<0.01$; Table 1). The accuracy of OM digestibility prediction is of particular importance, because it is used by the SRNS to predict dietary ME (Cannas *et al.*, 2004). More recently, Cannas *et al.* (2007b) evaluated the SRNS by using 5 published studies in which balance experiments including indirect calorimetric measurements on lactating does (15 treatment means) and wethers (6 treatment means) were performed. The prediction of daily MEI was very accurate (root of the mean squared error of prediction (RMSEP) = 0.24 Mcal/d; r^2 =0.99) when DMI was known (Table 1).

Prediction of animal performances

Cannas *et al.* (2006b) carried out the first of a series of evaluations on growth performances of lambs and kids. The predictions of the CNCPS-S on the average daily gain (ADG) of lambs were evaluated by using 42 treatment means from 9 published studies. The CNCPS-S markedly underpredicted ADG (mean bias, i.e. observed-predicted values (MB) = 74 g/d; RMSEP = 85, r^2 = 0.79), due to an overprediction of ME requirements for maintenance (ME_M). One factor that was causing this overprediction was a multiplier used by Cannas *et al.* (2004) that increases the BMR of males by 15%. This multiplier is no longer used in the current SRNS. In addition, the factor that accounts for the visceral organ component of the ME_M on requirement ($0.09 \times MEI \times k_M$ term in Equation 1) was removed. Despite these improvements, the underprediction was still large (MB = 34 g/d; RMSEP = 53 g/d, r^2 = 0.80). Thus, based on the equation form proposed by Tedeschi *et al.* (2004), Cannas *et al.* (2006a) derived a new partial efficiency of use of ME to NE for growth, assuming mean deposition efficiency of 27% for protein and 68% for fat for growing lambs. The evaluation of this new equation (Cannas *et al.*, 2006a) explained 84% of the variation, with a MB of 1 g/d and RMSEP of 37 g/d, suggesting an improvement in growth predictions when the efficiency of use of ME to NE for growth was based on the new equation instead of the empirical equations proposed by NRC (2000, 2001), ARC (1980), and CSIRO (1990, 2007) systems. The NRC (2007)

Table 1. Summary of evaluations performed with the CNCPS-S and the SRNS.

Species	Stage	Variable, units[5]	N	Obs.	Pred.	MB[5]	RMSEP[5]	r^2	Ref.[6]
Nutrient supply									
Sheep[1]	All	OM dig., N+, %	19	60.2	61.3	-1.1	3.6	0.83	A
Sheep[2]	All	OM dig., N–, %	12	56.6	53.3	3.3	6.5	---	A
Sheep[1]	All	CP dig., N+., %	23	69.6	67.7	1.9	7.2	0.34	A
Sheep[2]	All	CP dig., N–, %	7	42.7	45.4	-2.7	12.8	---	A
Sheep[1]	All	NDF dig., N+, %	7	54.6	50.3	4.3	6.9	0.51	A
Sheep[2]	All	NDF dig., N–, %	7	54.3	45.2	9.1	12.0	---	A
Goats	Lactating + wethers	MEI, Mcal/d	21	4.04	4.00	0.04	0.24	0.99	B
Animal performances									
Sheep[1]	Lactating	SBW changes, g/d	151	21.8	27.6	-5.8	30	0.73	B
Sheep[2]	Lactating	SBW changes, g/d	14	-56.9	-110.3	53.4	84.1	0.84	B
Sheep	Lactating	NE_L, Mcal/d	19	--	--	0.174	--	0.82	C
Sheep[3]	Growing	ADG, g/d	42	198	197	1	37	0.84	D
Sheep[3]	Growing	ADG, g/d	156	189	179	10	--	0.70	C
Sheep[3]	Growing	ADG, g/d	8	285	282	2.4	21.4	0.76	E
Sheep[3]	Growing	Fat, g/kg EBG	8	275	295	-19.8	59.2	0.30	E
Sheep[3]	Growing	Protein, g/kg EBG	8	146	152	-5.8	19.3	0.10	E
Sheep[4]	Growing	MP for gain, g	48	--	--	0.9	--	0.88	C
Sheep[4]	Growing	ME_M, Mcal/d	23	--	--	0.005	0.02	0.96	F
Sheep	Growing	DMI, g/d	8	1,327	1,291	35	45.5	0.95	E
Goats	Lactating	NE_L, Mcal/d	21	1.664	1.746	-0.082	0.102	0.99	B
Goats[4]	Lactating + wethers	NE balance, Mcal/d	21	0.361	0.286	0.075	0.201	0.87	B
Goats[3]	Growing	ADG, g/d	31	136.1	142.5	-6.4	32.5	0.85	G

[1] Ruminal N balance positive (N+).
[2] Model adjusted for a negative ruminal N balance (N–).
[3] Using k_G as described by Cannas *et al.* (2006a) and no adjustment for ME intake.
[4] No adjustment for ME intake.
[5] OM: organic matter, SBW: shrunk body weight, NE_L: NE for lactation, ADG: average daily gain, EBG: empty body gain, MP: metabolisable protein, DMI: dry matter intake, MEI: ME intake, ME_M: ME for maintenance, NE: net energy, (MB) mean bias: observed – predicted values, RMSEP: root of the mean square of prediction error.
[6] References: A = Cannas *et al.* (2004), B = Cannas *et al.* (2007b), C = NRC (2007), D = Cannas *et al.* (2006a,b), E = Cannas *et al.* (2009) , F = Galvani *et al.* (2008), and G = Cannas *et al.* (2007b).

conducted an evaluation of the CNCPS-S as published by Cannas *et al.* (2004). For growing and finishing sheep, the NRC (2007) developed a database containing 156 treatment means (N=1,876 sheep) from 31 references. When the feed biological value was not provided by the references, the NRC (2007) assumed the values reported by the NRC (2000, 2001). The CNCPS-S accounted for 70% of the variation of the observed ADG with MB of 37 g/d. When the term $0.09 \times MEI \times k_M$ (Equation 1) was excluded, the CNCPS-S had a lower MB (10 g/d) than the original equation, but similar r^2 (0.70) (Table 1). Based on these results, the current SRNS (Tedeschi *et al.*, 2010) excludes the term $0.09 \times MEI \times k_M$ (Equation 1) for growing animals.

The evaluation of the SRNS for growing kids was performed by Cannas *et al.* (2007a) using eight publications with goat breeds for milk, meat, and indigenous (n = 31). The evaluation showed that the mean and the systematic bias were very small and the RMSEP was smaller (MB of -6.4 g/d; RMSEP of 32.5 g/d; r^2 of 0.85) than that reported when the SRNS predictions of the ADG of lambs were evaluated (Cannas *et al.*, 2006a) (Table 1). When the predictions were based on the SRNS estimates of MEI and on the ME requirements for maintenance and gain of the NRC (2007) goat model (based on Sahlu *et al.*, 2004), the RMSEP was increased, mainly due to a fairly large and significant systematic bias, which made the regression line statistically different from the equivalence line (MB of 7.2 g/d; RMSEP of 34.9 g/d; and r^2 of 0.86) (Cannas *et al.*, 2007a). In particular, the NRC (2007) model overpredicted the ADG at high observed ADG and underpredicted the ADG at low observed values. This is likely because the SRNS uses a fixed value for the energy content of gain, regardless of the BW or the relative size of the kids. To evaluate the MP calculations of the CNCPS-S for growing or finishing sheep, the NRC (2007) collected information on 48 treatment means, representing 335 sheep, from 14 references. A regression between MP available and used yielded an r^2 of 0.88 and MB of 0.9 g/d (Table 1). The NRC (2007) recommended the CNCPS-S for calculating MP requirements of growing sheep. Cannas *et al.* (2009) presented the results of an evaluation of the SRNS based on a experiment specifically designed to evaluate this model in South-African conditions (Table 1). Male and female lambs of two breeds (Dorper and South African Merino) and two slaughter ages were used. Predictions of feed intake were accurate and precise with a low systematic bias (MB = 35 g/d; RMSEP = 50 g/d; r^2 = 0.95). Predictions on the ADG were also accurate and precise (MB = 2.4 g/d; RMSEP = 21.4 g/d; r^2 = 0.76), while the EBG was overpredicted (MB = 31 g/d; RMSEP = 38 g/d; r^2 = 0.69). The model slightly overpredicted the fat and protein content of the EBG (by 7.2% and 3.9%, respectively), showing high accuracy but very low precision, probably because the range of variation of fat and protein content of gain was narrow.

The NRC (2007) also tested the predictions of ADG of the NRC (1985) system and concluded that the CNCPS-S and the NRC (1985) were both accurate and precise in predicting energy requirements for growing sheep. However, because the CNCPS-S included the calculation for protein requirements and supplies, calculations for BW changes, and several classes of sheep, the NRC (2007) adopted it to develop its nutrient requirement tables.

Regarding mature animals, the CNCPS-S accurately predicted variations in SBW when diets had a positive rumen N balance (MB = -5.8 g/d; RMSEP = 30.0 g/d), while when rumen N balance was negative, SBW variations were markedly underpredicted (MB = 53.4 g/d; RMSEP = 84.1 g/d) (Table 1). In the study previously reported in which balance experiments including indirect calorimetric measurements on lactating does and wethers were performed, the SRNS underpredicted the energy balance of goats (MB = 0.31 Mcal/d; RMSEP of 0.40 Mcal/d; and r^2 of 0.79). Similarly to what observed for growing animals, Cannas *et al.* (2007b) reported that when the term $0.09 \times MEI \times k_M$ of Equation 1 was removed for the calculation of ME_M, the predictions of energy balance of goats by the SRNS were remarkably improved (MB = 0.075 Mcal/d; RMSEP of 0.20 Mcal/d; and r^2 of 0.87). The NRC (2007) evaluation of the CNCPS-S predictions for lactating sheep contained 6 publications with 19 observations. The CNCPS-S predicted NE_L accurately when changes in ADG were accounted for. The r^2 was 0.82 and MB was 0.174 Mcal/d. This supports the need to adjust for changes in BCS or BW to accurately predict ME or MP available for lactation (Tedeschi *et al.*, 2006).

Discussion and conclusions

Overall, the evaluation carried out on the SRNS showed that it has consistently been able to predict nutrient supply and animal performances with moderate to high accuracy and precision.

Some limitations of the SRNS have been identified during the development phase. One limitation regards the uncertainties of using a factor that accounts for the visceral organ component of the ME_M requirements. Another limitation is concerned to the use of the age correction factor in the prediction of ME_M requirements for growing animals. Freetly *et al.* (2002) reported that the effect of age on ME_M should be better accounted for not by using the chronological age but the physiological age, i.e. the degree of maturity of the growing animals. Recent unpublished work (A. Cannas and I.A.M.A. Teixeira, personal communication) on growing kids confirms this finding. A third issue is related to the lack of correlation between maintenance costs and body fatness, despite evidence suggesting they are correlated (Tedeschi *et al.*, 2006). A fourth problem is associated to the prediction of ADG in animals under compensatory growth. Unpublished evaluations based on an experiment carried out in Sweden (A. Cannas and G. Bernes, personal communication) and using a dataset from California, US (A. Cannas and J. W. Oltjen, personal communication), clearly indicated that when compensatory growth occurred the SRNS markedly underpredicted the ADG. Since the SRNS is inherently unable to account for compensatory growth, an effort is underway to integrate the SRNS with the dynamic California growth model (A. Cannas and J. W. Oltjen, personal communication).

References

AFRC (Agricultural and Food Research Council), 1993. Energy and Protein Requirements of Ruminants. Agricultural and Food Research Council. CAB International, Wallingford, UK, 159 pp.

ARC (Agricultural Research Council), 1980. The Nutrient Requirements of Ruminant Livestock. Agricultural Research Council. The Gresham Press, London, UK, 351 pp.

Cannas, A., 2002. Energy and protein requirements. In: G. Pulina (ed.). Dairy Sheep Feeding and Nutrition. Avenue Media, Bologna, Italy, pp.55-81.

Cannas, A., Atzori, A.S., Boe, F., and Teixeira, I.A.M.A., 2008. Energy and protein requirements of goats. In: A. Cannas and G. Pulina (eds.). Dairy Goats, Feeding and Nutrition. CAB International, Cambridge, MA , USA, pp.118-146.

Cannas, A., Linsky, A., Erasmus, L.J., Tedeschi, L.O., van Niekerk, W.A., and Coertze, R., 2009. Evaluation of performance predictions of the Small Ruminant Nutrition System model using growth and body composition data of South African Mutton Merino and Dorper. Journal of Animal Science 87 E-Suppl. 2:522.

Cannas, A., Tedeschi, L.O., Atzori, A.S., and Fox, D.G., 2006a. Prediction of energy requirements for growing sheep with the Cornell Net Carbohydrate and Protein System. In: J. Dijkstra (ed.). Nutrient Digestion and Utilization in Farm Animals. CABI Publishing, Cambridge, MA, USA, pp.99-113.

Cannas, A., Tedeschi, L.O., and Fox, D.G., 2006b. Small Ruminant Nutrition System: a computer model to develop feeding programs for sheep and goats. Journal of Animal Science, 84 (Suppl. 1), pp.376.

Cannas, A., Tedeschi, L.O., Atzori, A.S., and Fox, D.G., 2007a. Prediction of the growth rate of kids with the Small Ruminant Nutrition System. Anais da 44ª Reunião Anual da Sociedade Brasileira de Zootecnia, pp.1-3. (CD-ROM).

Cannas, A., Tedeschi, L.O., and Fox, D.G., 2007b. Prediction of metabolizable energy intake and energy balance of goats with the Small Ruminant Nutrition System. In: Ortigues-Marty I. (ed.). Energy and protein metabolism and nutrition. EAAP publication No 124, Wageningen Academic Publishers, Wageningen, the Netherlands, pp.569-570.

Cannas, A., Tedeschi, L.O., Atzori, A.S., and Fox, D.G., 2007c. The Small Ruminant Nutrition System: development and evaluation of a goat submodel. Italian Journal of Animal Science 6 (Suppl. 1):609-611.

Cannas, A., Tedeschi, L.O., Fox, D.G., Pell, A.N., and Van Soest, P.J., 2004. A mechanistic model for predicting the nutrient requirements and feed biological values for sheep. Journal of Animal Science 82:149-169.

Cannas, A., and Van Soest P.J., 2000. Allometric models to predict rumen passage rate in domestic ruminants. In: J.P. McNamara, J. France and D.E. Beever (eds.). Modeling Nutrient Utilization in Farm Animals. CABI Publishing, Wallingford, Oxon, UK, pp.49-62.

CSIRO (Australian Commonwealth Scientific and Industrial Research Organization), 1990. Feeding Standards for Australian Livestock. Ruminants. Commonwealth Scientific and Industrial Research Organization, Melbourne, Australia, 266 pp.

CSIRO (Australian Commonwealth Scientific and Industrial Research Organization), 2007. Nutrient Requirements of Domesticated Ruminants. Commonwealth Scientific and Industrial Research Organization, Collingwood, Australia, 270 pp..

Fernandes, M.H.M.R., Resende, K.T., Tedeschi L.O., Fernandes, J.S., Silva, Jr., H.M., Carstens, G.E., Berchielli, T.T., Teixeira, I.A.M.A. and Akinaga, L., 2007. Energy and protein requirements for maintenance and growth of Boer crossbred kids. Journal of Animal Science 85:1014-1023.

Fox, D.G., Tedeschi, L.O., Tylutki, T.P., Russell, J.B., Van Amburgh, M.E., Chase, L.E., Pell, A.N., and Overton, T.R., 2004. The Cornell Net Carbohydrate and Protein System model for evaluating herd nutrition and nutrient excretion. Animal Feed Science and Technology 112:29-78.

Freer, M., Moore A.D., and Donnelly, J.R., 1997. GRAZPLAN: decision support systems for Australian grazing enterprises-II. The animal biology model for feed intake, production and reproduction and the GrazFeed DSS. Agricultural Systems 54:77-126.

Freetly, H.C., Nienaber, J.A. and Brown-Brandl, T., 2002. Relationships among heat production, body weight, and age in Finnsheep and Rambouillet ewes. Journal of Animal Science 80:825-832.

Galvani, D.B., Pires, C.C., Kozloski, G.V. and Wommer, T.P., 2008. Energy requirements of Texel crossbred lambs. Journal of Animal Science 86:3480-3490.

INRA (Institut National de la Recherche Agronomique), 1989. Ruminant nutrition. Recommended allowances and feed tables. John Libbey Eurotext, Montrouge, France, p. 389.

Ngwa, A.T., Dawson, L.J., Puchala, R., Detweiler, G., Merkel, R.C., Tovar-Luna, I., Shalu, T., Ferrell, C.L. and Goetsch, A.L., 2007. Urea space and body condition score to predict body composition of meat goats. Small Ruminant Research 73:27-36.

NRC (National Research Council), 1981. Nutritional Energetics of Domestic Animals and Glossary of Energy Terms. National Academy Press, Washington, DC, USA.

NRC (National Research Council), 1985. Nutrient Requirements of Sheep (6th ed.). National Academy Press, Washington, DC, USA, 99 pp.

NRC (National Research Council), 2000. Nutrient Requirements of Beef Cattle (updated 7th ed.). National Academy Press, Washington, DC, USA, 248 pp.

NRC (National Research Council), 2001. Nutrient Requirements of Dairy Cattle (7th ed.). National Academy Press, Washington, DC, USA, 381 pp.

NRC (National Research Council), 2007. Nutrient Requirements of Small Ruminants: Sheep, Goats, Cervids, and New World Camelids (6th ed.). National Academy Press, Washington, DC, USA 384 pp.

Pulina, G., Cannas, A., Serra, A., and Vallebella, R., 1992. Determinazione e stima del valore energetico di latte di capre di razza Sarda. Atti della Società Italiana di Scienze Veterinarie 45:1779-1781.

Sahlu, T., Goetsch, A.L., Luo, J., Nsahlai, I.V., Moore, J.E., Galyean, M.L., Owens, F.N., Ferrell, C.L. and Johnson, Z.B., 2004. Nutrient requirements of goats: developed equations, other considerations and future research to improve them. Small Ruminant Research 53:191-219.

Tedeschi, L.O., 2006. Assessment of the adequacy of mathematical models. Agricultural Systems 89:225-247.

Tedeschi, L.O., Cannas, A. and Fox, D.G., 2010. A nutrition mathematical model to account for dietary supply and requirements of energy and nutrients for domesticated small ruminants: the development and evaluation of the Small Ruminant Nutrition System. Small Ruminant Research, in press: doi.org/10.1016/j.smallrumres.2009.12.041.

Tedeschi, L.O., Fox, D.G., and Guiroy, P.J., 2004. A decision support system to improve individual cattle management. 1. A mechanistic, dynamic model for animal growth. Agricultural Systems 79:171-204.

Tedeschi, L.O., Fox, D.G., Sainz, R.D., Barioni, L.G., Medeiros, S.R. and Boin, C., 2005. Using mathematical models in ruminant nutrition. Scientia Agricola 62:76-91.

Tedeschi, L.O., Seo, S., Fox, D.G. and Ruiz, R., 2006. Accounting for energy and protein reserve changes in predicting diet-allowable milk production in cattle. Journal of Dairy Science 89:4795-4807.

A model of phosphorus metabolism in growing sheep

R.S. Dias[1], T. Silva[2], R.M.P. Pardo[3], J.C. Silva Filho[4], D.M.S.S. Vitti[2], E. Kebreab[5], S. López[6] and J. France[1]
[1]Centre for Nutrition Modelling, Department of Animal and Poultry Science, University of Guelph, Guelph, Ontario N1G 2W1, Canada; jfrance@uoguelph.ca
[2]Animal Nutrition Laboratory, Centro de Energia Nuclear na Agricultura, Caixa Postal 96, CEP 13400-970, Piracicaba, SP, Brazil
[3]Facultade de Ciências Agropecuarias, University of Sucre, Carrera 28 5-267, Sincelejo, Sucre, Colombia
[4]Federal University of Lavras, Animal Research Laboratory, CEP 37200-000, Lavras, MG, Brazil
[5]Department of Animal Science, University of California, Davis, CA 95616, USA
[6]Instituto de Ganadería de Montaña (IGM), Universidad de León – Consejo Superior de Investigaciones Científicas (CSIC), Departamento de Producción Animal, Universidad de León, 24007 León, Spain

Abstract

A compartmental model was developed to provide information on P metabolism in ruminants. The model is an extension of that proposed by Vitti *et al.* (2000) and amended by Dias *et al.* (2006). Compartments representing saliva and rumen were added to the original model. The data used to develop the extended model were generated in an experiment using 24 male sheep, initial live weight 34.5 kg, aged 8 mo, fed 0.14; 0.31; 0.49 and 0.65% P in DM. Animals received a single dose of 7.4 MBq of ^{32}P injected into the jugular vein and subsequently samples of plasma and faeces were collected daily for one week. Mixed saliva samples and rumen fluid samples were taken only at 4 and 6 d after injection to prevent animal stress. Specific activities and inorganic P measurements from bone, soft tissues, plasma, rumen, saliva and faeces were used to calculate flows between compartments. Increased dietary P levels affected positively total P and endogenous P excretion in faeces and P flow from rumen to lower gastrointestinal tract (GIT). The latter flow was also related to both total and endogenous P excreted in faeces. Phosphorus flow from GIT to plasma was correlated with endogenous P excretion and P recycled to GIT. Phosphorus flow from saliva to rumen was related to P flow from rumen to GIT. Phosphorus mobilized from bone was addressed to soft tissues as shown in the correlations between P flow from bone to plasma and net P flow in soft tissue, and P flow from plasma to soft tissue and net P flow in bone. The flows from the model resulted in appropriate biological description of P metabolism in ruminants.

Keywords: bone, plasma, rumen, saliva, soft tissues

Introduction

Mathematical models have been used as a tool to study P metabolism in ruminants for many years (e.g. Schneider *et al.*, 1987; Vitti *et al.*, 2000; Dias *et al.*, 2006) providing much valuable information. Vitti *et al.* (2000) proposed a model with the following 4 compartments: total gastrointestinal tract (GIT), plasma, soft tissues and bone. Dias *et al.* (2006) amended this model by incorporating new measurements that allowed a more accurate approach to calculating P flows between compartments and into and out of the system.

In ruminants, endogenous P is particularly important due to its role in the maintenance of a healthy rumen environment. Provision of endogenous P to the rumen is achieved mainly through salivary secretion which is regulated by plasma P levels. Endogenous P that is not used for an animal's metabolic function is excreted in faeces and can represent a large proportion of total P excreted. Hence, a study of whole body P metabolism in ruminants needs to consider the endogenous P fraction. In addition, knowledge of endogenous P is a crucial step in determining the P requirement for maintenance of these animals. However, endogenous P determination is not achieved using a simple direct technique. Instead, a more refined approach as such isotopic dilution needs to be used if this is the experimental objective. This technique provides data that can be used for mathematical model development and kinetics studies.

Therefore, the present study aimed to use an isotopic dilution technique to generate data to allow the addition of saliva and the rumen as two new compartments to the Vitti model in order to provide more precise representation of whole body P metabolism in ruminants.

Material and methods

Animal and diets

Twenty-four male sheep, of Santa Inês breed, initial live weight 34.5 kg, aged 8 months, were fed the experimental diets for 10 d. Treatments consisted of the basal diet supplemented with different amounts of dicalcium phosphate and limestone to provide increasing P (0, 2, 4 and 6 g/animal/d) and Ca levels, respectively. The basal diet comprised roughage (coast cross hay), a concentrate mixture (cassava bran, soya bean bran, and urea) and a mineral mixture (Table 1). Treatments corresponded to diets providing 0.14, 0.31, 0.49 and 0.65% P in DM respectively. Vitamins A, D and E were supplemented in the diets (10^6; 250000 and 6250 UI/kg respectively).

After a 10 d adaptation period the animals received a single dose of 7.4 MBq of ^{32}P injected into the jugular vein, and subsequently samples of plasma were collected daily for one week. Mixed saliva and rumen fluid samples were taken only at 4 and 6 d after injection to minimize stress caused to the animals. Before feeding the animals, saliva was collected using a pair of pincers and small pieces of plastic sponge, put directly in the animal's mouth without impairing chewing movements. Rumen content was collected using an oesophageal tube and a syringe extracted 50 ml of the rumen sample each time of collection. Rumen fluid was obtained after filtering the rumen sample using a gauze, diluting it with distilled water and then centrifuging the filtrate (Sorval/ model RC2B) at 3,000 rpm for 15 min. Feed was given twice daily after samples were collected and feed refusals were weighed to determine DM intake (DMI).

Table 1. Nutrient analysis of feedstuffs (g/kg DM).

	DM	CP	Ash	NDF	ADF	Ca	P
Cassava bran	880	15	11	65	51	2	0.6
Soya bean bran	886	431	64	140	100	2.9	6.8
Limestone	1000	0	1000	-	-	340	0.2
Dicalcium phosphate	1000	0	1000	-	-	340	160
Coast cross hay	890	60	76	801	434	3.8	1.3
Mineral mixture[1]	1000	0	1000	0	0	0	0

[1] Mineral mixture: Mg, 30 g/kg; S, 30 g/kg; Cu, 3,000 mg/kg; Mn, 5,000 mg/kg; Zn, 12,000 mg/kg; Se, 80 mg/kg and I, 180 mg/kg.
DM: dry matter, CP: crude protein, NDF: neutral detergent fibre, ADF: acid detergent fibre, Ca: calcium, P: phosphorus.

The experiment was conducted using a protocol approved by the Centre of Nuclear Energy in Agriculture (CENA) Animal Care Committee.

Analysis of samples

Samples of feed and feed refusals were analysed for DM, CP, P, and ADF following recommendations of the Association of Official Analytical Chemists (1995). Crude protein was determined by the Kjeldahl method, and NDF according to Mertens (2002), without using amylase or sodium sulphite.

Samples of saliva and rumen fluid were filtered and centrifuged (Sorval, model RC2B) and then an aliquot of 0.5 ml of supernatant was used for protein precipitation. Inorganic P levels in plasma, saliva and rumen were determined by colorimetric analysis (Fiske and Subbarow, 1925). Radioactivity of ^{32}P was measured in a Packard Liquid Scintillation Spectrometer (model 2450B) using Cerenkov radiation.

Model development

The model of whole body phosphorus metabolism is an extension of that proposed by Vitti *et al.* (2000) for use in ruminants. The extended model is illustrated in Figure 1. The six compartments of P are: (1) rumen, (2) lower GIT, (3) saliva, (4) plasma, (5) bone and (6) soft tissues. Phosphorus entry to the system is via intake (F_{10}), and exit via faeces (F_{02}) and urine (F_{04}) while ^{32}P is administered as a single dose into systemic blood at time zero, and specific activity (SA, µCi) of all pools monitored (s_1, SA in rumen; s_2, SA in faeces; s_3, SA in saliva; s_4, SA in plasma; s_5, SA in bone and s_6, SA in soft tissues). The scheme assumes there is no re-entry of label from external sources.

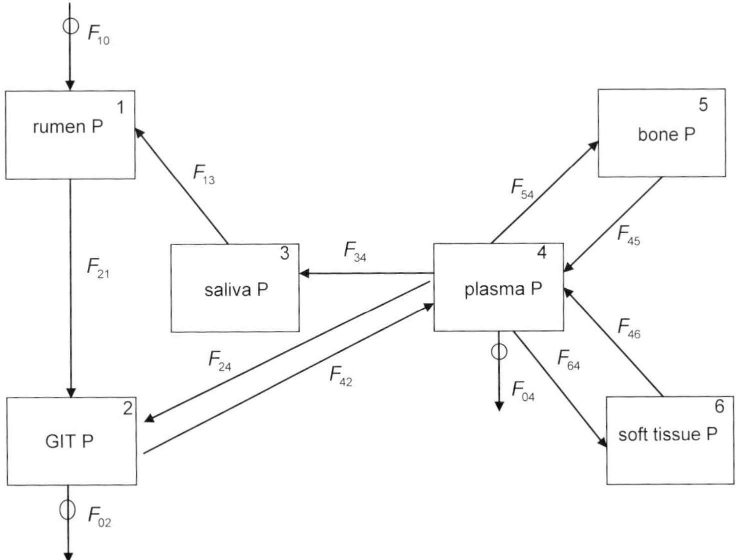

Figure 1. The model of whole body P metabolism in ruminants. The circles identify the measured flows: F_{10} - P intake; F_{02} - P excreted in faeces and F_{04} - P in urine. The arrows represent the flows between compartments: from rumen to GIT - F_{21}; from GIT to plasma - F_{42}; from plasma to GIT - F_{24}; plasma to soft tissues - F_{64}; from soft tissues to plasma - F_{46}; from plasma to bone - F_{54}; from bone to plasma - F_{45}; from plasma to saliva - F_{34} and from saliva to rumen - F_{13}. Flows are in g/d.

Conservation of mass principles can be applied to each compartment to generate differential equations which yield algebraic formulae for the following flows (described in Figure 1):

$$F_{13} = s_1 F_{10} / (s_3 - s_1)$$

$$F_{21} = F_{10} + F_{13}$$

$$F_{24} = (s_1 - s_2) F_{21} / (s_2 - s_4)$$

$$F_{42} = F_{21} + F_{24} - F_{02}$$

$$F_{34} = F_{13}$$

Assuming that after several hours the relative rates of change of specific activity in plasma (*-k*, per d), bone and soft tissue have the same magnitude, i.e.

$$\frac{1}{s_5}\frac{ds_5}{dt} = \frac{1}{s_6}\frac{ds_6}{dt} = k$$

then flows between bone and plasma likewise soft tissues and plasma can be calculated as follows:

$$F_{54} = k\, s_5\, Q_5 / (s_4 - s_5)$$

$$F_{64} = k\, s_6\, Q_6 / (s_4 - s_6)$$

$$|F_{45} + F_{46}| = F_{04} + F_{24} + F_{34} + F_{54} + F_{64} - F_{42}$$

$$F_{46} = (s^* - s_5) \times |F_{45} + F_{46}| / (s_6 - s_5)$$

$$F_{45} = |F_{45} + F_{46}| - F_{46}$$

where $s^* = (s_5 Q_5 + s_6 Q_6)/(Q_5 + Q_6)$ and Q_5 and Q_6 are the quantities of P (g) in compartments 5 and 6, respectively (France *et al.*, 2010).

Statistical analysis

The results from each treatment were aggregated in order to perform a correlation analysis using PROC CORR of SAS (SAS, 1999). Linear regression using PROC REG of SAS was used to provide intercept and slope for the relationships selected based on Pearson product-moment correlations.

Results and discussion

The P flow values from each treatment were assembled in a single dataset with the aim of finding relationships between flows to further assess their biological meaning. The linear equations are represented by numbers in Table 2 to facilitate the discussion.

Total P excreted in faeces was well correlated with P intake (1) supporting the well established feeding strategy which proposes to mitigate P in the environment by reducing dietary P levels. Endogenous P excreted in faeces was also well correlated with P intake (2) and hence with total P in faeces (4), reflecting the increasing levels of P used for animal metabolism. In agreement, Dias

Table 2. Flow correlations, where numbers in brackets represent the standard error.

Flows†	no.	Linear equations*	r^2
F_{10} vs. F_{02}	1	F_{02} = 0.87 (0.047) F_{10} – 0.12 (0.241)	0.94
F_{10} vs. F_{02end}	2	F_{02end} = 0.26 (0.035) F_{10} + 0.24 (0.177)	0.72
F_{10} vs. F_{21}	3	F_{21} = 1.40 (0.090) F_{10} + 0.59 (0.455)	0.92
F_{02end} vs. F_{02}	4	F_{02} = 2.35 (0.342) F_{02end} + 0.37 (0.555)	0.68
F_{02} vs. F_{21}	5	F_{21} = 1.52 (0.129) F_{02} + 1.17 (0.558)	0.86
F_{21} vs. F_{02end}	6	F_{02end} = 3.64 (0.624) F_{21} + 1.65 (1.011)	0.61
F_{42} vs. F_{02end}	7	F_{02end} = 3.55 (0.594) F_{42} + 1.14 (0.963)	0.62
F_{21} vs. F_{13} (F_{34})	8	F_{13} (F_{34}) = 2.17 (0.260) F_{21} + 1.63 (0.723)	0.76
F_{42} vs. F_{24}	9	F_{24} = 0.62 (0.073) F_{42} – 0.81 (0.518)	0.77
Net P in soft tissue vs. Net P in bone	10	Net P in bone = -0.98 (0.041) Net P in soft tissue + + 0.55 (0.208)	0.55
F_{45} vs. Net P in soft tissue	11	Net P in soft tissue = 0.22 (0.036) F_{45} – 5.36 (1.450)	0.64
F_{64} vs. Net P in bone	12	Net P in bone = -0.58 (0.054) F_{64} + 2.23 (0.511)	0.85
F_{64} vs. Net P in soft tissue	13	Net P in soft tissue = 0.59 (0.038) F_{64} – 1.76 (0.376)	0.91
F_{46} vs. Net P in soft tissue	14	Net P in soft tissue = 1.05 (0.190) F_{46} – 2.19 (1.105)	0.58

*For all relationships $P<0.0001$.
†Endogenous faecal P calculated as $F_{02end} = F_{02} \times$SA in faeces/SA in plasma.

et al. (2009) also reported an increased flow of endogenous P into the rumen when increasing levels of P were provided in the diet of growing sheep.

Phosphorus intake also affected P flow from the rumen to GIT (3) which was also correlated with total P excreted in faeces (5). The increase in ruminal P flow may be related to the increase in ruminal P concentration as a response to increased P intake, as reported by Witt and Owens (1983) and Dias *et al.* (2009) studying steers and sheep fed increasing dietary P levels respectively. Phosphorus absorption occurs mainly in the upper small intestine portion of the tract (Poppi and Ternouth, 1979) thus any undigested P entering the GIT is excreted in faeces as shown in the relationship between P from the rumen entering the GIT and P excreted in faeces (5).

Endogenous P secreted in saliva is necessary for microbial growth and to maintain a healthy rumen environment. Consequently, P in rumen is mainly a result of salivary P secretion (Tomas, 1973). Thus salivary flow is strongly related to P flow from rumen to GIT as demonstrated by Equation 8. Endogenous P coming from the rumen is either absorbed in the GIT or excreted in faeces (6) if this P fraction is not used for animal metabolic functions. The model does not discriminate which portion of absorbed P is of endogenous or dietary origin, however, the model demonstrated that higher P absorption (F_{42}) leads to higher endogenous P recycling in the GIT (9). Further, the model indicates a relationship between P flow from GIT to plasma, which encompasses endogenous and dietary P, with endogenous P excreted in faeces (7).

The removal of P from plasma through salivary secretion is an important mechanism of P regulation in whole body metabolism in ruminants. The lack of relationship between F_{34} and F_{42} is probably related to the fact that P levels in plasma are also influenced by P mobilization from bone. According to Knowlton and Herbein (2002), P balance is regulated by P absorption from the small intestine, mobilization from bone and salivary secretion. Although soft tissue does not represent an important contribution to the plasma P pool, P mobilization from bone plays a role in the changes in plasma P that in turn seem to affect P channelling to soft tissues. In fact, the model clearly shows that net P flow in soft tissues is dependent on the P exchanges between these and the plasma pool (13, 14), as well as P mobilized from bone and net P flow in bone (11, 12). In other words, net P flow in soft tissue is strongly related to changes in net P flow in bone (10). Feaster *et al.* (1953) working with rats fed vitamin D free and low P diets suggested that in the rats fed the low P diet soft tissue P was maintained by P withdrawal from the bones. These findings are in agreement with the relationship between P flow from bone and net P flow in soft tissues generated by the model.

In conclusion, the main relationships found between the flows in each pool are in agreement with existent knowledge on P metabolism in ruminants, demonstrating that the proposed model is an appropriate representation of whole body P metabolism in ruminants. This model will be useful in developing feeding strategies that aim to mitigate P excretion in the environment without compromising the soft tissue and bone demands for this mineral. Moreover, the model will help elucidate additional knowledge on the mechanism underlying P metabolism in ruminants.

References

AOAC (Association of Official Analytical Chemists), 1995. Official methods of analysis of AOAC, 16th ed. Association of Official Analytical Chemists International. Arlington, VA, USA, 1200 pp.

Dias, R.S., Kebreab, E., Vitti, D.M.S.S., Roque, A.P., Bueno, I.C.S. and France, J., 2006. A revised model for studying phosphorus and calcium kinetics in growing sheep. Journal of Animal Science 84:2787-2794.

Dias, R.S., Lopez, S., Silva, T., Pardo, R.M.P., Silva Filho, J.C., Vitti, D.M.S.S., Kebreab, E. and France, J., 2009. Rumen phosphorus metabolism in sheep. Journal of Agricultural Science 147:391-398.

Feaster, J.P., Shirley, R.L., McCall, J.T., and Davis, G.K., 1953. 32P distribution and excretion in rats fed vitamin D-free and low phosphorus diets. Journal of Nutrition 51:381-392.

Fiske, C.H. and Subbarow, Y., 1925. The colorimetric determination of phosphorus. Journal of Biological Chemistry 66:375-400.

France, J., Dias, R.S., Kebreab, E., Vitti, D.M.S.S., Crompton, L.A. and Lopez, S., 2010. Kinetic models for the study of phosphorus metabolism in ruminants and monogastrics. In: Vitti, D.M.S.S. and Kebreab, E. (eds.) Phosphorus and calcium utilization and requirements in farm animals, CAB International, Wallingford, UK, p. 18-44.

Knowlton, K.F. and Herbein, J.H., 2002. Phosphorus partition during early lactation in dairy cows fed diets varying in phosphorus content. Journal of Dairy Science 85:1227-1236.

Mertens, D.R., 2002. Gravimetric determination of amylase-treated neutral detergent fiber in feeds using refluxing in beakers or crucibles: collaborative study. Journal of AOAC International 82:1217-1240.

Poppi, D.P. and Ternouth, J.H., 1979. Secretion and absorption of phosphorus in the gastrointestinal tract of sheep fed on four diets. Australian Journal of Agricultural Research 30:503-512.

SAS Institute Inc., 1999. SAS Macro Language: Reference version 8. SAS Institute Inc. Cary, NC, USA, 1256 pp.

Schneider, K.M., Boston, R.C. and Leaver, D.D., 1987. Quantification of phosphorus excretion in sheep by compartmental analysis. American Journal of Physiology 252:R720-R731.

Tomas, F.M., 1973. Parotid salivary secretion in sheep: its measurement and influence on phosphorus in rumen fluid. Quarterly Journal of Experimental Physiology 58:131-138.

Vitti, D.M.S.S., Kebreab, E., Lopes, J.B., Abdalla, A.L., De Carvalho, F.F.R., De Resende, K.T., Crompton, L.A. and France, J., 2000. A kinetic model of phosphorus metabolism in growing goats. Journal of Animal Science 78:2706-2712.

Witt, K.E. and Owens, F.N., 1983. Phosphorus: ruminal availability and effects on digestion. Journal of Animal Science 56:930-937.

A generic stoichiometric model to analyse the metabolic flexibility of the mammary gland in lactating dairy cows

S. Lemosquet[1,4], *O. Abdou Arbi*[2], *A. Siegel*[2,3], *J. Guinard-Flament*[4,1], *J. Van Milgen*[5,6] *and J. Bourdon*[7,2]
[1]*INRA UMR1080 Dairy Production, 35590 Saint-Gilles, France; sophie.lemosquet@rennes.inra.fr*
[2]*INRIA Rennes Bretagne Atlantique, Symbiose project, 35042, Rennes, France*
[3]*CNRS Université de Rennes 1, UMR 6074 IRISA, 35042 Rennes, France*
[4]*Agrocampus Ouest, UMR1080 Dairy Production, 35000 Rennes, France*
[5]*INRA UMR1079 SENAH, 35590 Saint-Gilles, France*
[6]*Agrocampus Ouest, UMR1079 SENAH, 35000 Rennes, France*
[7]*LINA UMR 6241, Université de Nantes, 44322 Nantes, France*

Abstract

The stoichiometry model tool of Van Milgen (2002) was adapted to study the metabolism of the ruminant mammary gland. It was used to study the uptake of nutrients in several metabolic reactions using two studies of mammary gland net balance (net nutrient uptake and milk yield and composition) in dairy cows receiving casein infusions in the duodenum. The rules applied to find the partitioning were based on the hypotheses that there is no accumulation of intermediary metabolites in the mammary gland (rule 1) and that there is no deficiency of ATP or cofactors (rule 2). The system was solved manually using a spreadsheet, while respecting the 2 rules. A new automatic tool using Flux Balance Analysis theory was developed. The stoichiometry of the model appeared to be appropriate since the predicted CO_2 production was close to the measured CO_2 production. The nutrient partitioning obtained by manual solving corresponded to one solution among possibly infinite number of solutions. The manual solving of the equations typically corresponded to the maximisation of ATP production while respecting nutrient output for milk synthesis. This maximization of ATP production is a third, explicit rule required to obtain the same partitioning with the automatic tool based on Flux Balance Analysis theory.

Keywords: nutritional model, biochemistry, flux balance analysis, ATP control

Introduction

The main biochemical pathways in the different metabolisms were recently integrated in a generic stoichiometric model (Van Milgen, 2002; Van Milgen *et al.*, 2003) called the 'metabolic calculator'. The structure of this calculator is based on a restricted number of intermediary metabolites called carbon-chain 'pivots', which correspond to important cross points between metabolic pathways (i.e. glucose, pyruvate, acetyl-CoA, oxaloacetate, α-ketoglutarate and serine). Each biochemical reaction is then described by the variations of these intermediary metabolites and also integrates the variation in ATP and other cofactors. The user then has to manually combine these reactions while respecting the rule that there will be no accumulation or deficit of pivots and cofactors. This approach allows the computation of the global cost of product synthesis by a biological entity (e.g. cell, tissue, whole organism). For example, the calculator allows comparison of the efficiency of ATP production from direct glucose oxidation or from glucose oxidation via (temporary) storage as glycogen (Van Milgen, 2002). This tool was also used to check the consistency of the overall nutrient balance in the mammary gland of the sow (Reneaudeau *et al.*, 2003; Van Milgen *et al.*, 2003). A major characteristic of the calculator is its large flexibility, since the choice of reactions to be equilibrated is left entirely to the user.

From a mathematical viewpoint, the metabolic calculator can be seen as a computational environment for studying the solution set of a system of linear equations. Each reaction is associated to a variable, estimating how often the reaction is used in the biological process. Equilibrating the concentration of given nutrients (the pivots) gives rise to linear equations. Thus, performing a computation with the metabolic calculator involves solving a system of linear equations. Unfortunately, the linear system is generally over-dimensioned, since the number of nutrients to be equilibrated (i.e. the number of equations) is much smaller than the number of reactions (i.e. variables). Therefore, from a mathematical viewpoint, the system has an infinite number of solutions, even though the user of the metabolic calculator usually finds what is often claimed to be a unique solution after computation. It appears that the user usually follows an *a priori* 'deterministic' choice when performing the computation.

Therefore, one can question which unconscious hypotheses are assumed by the user to navigate in the infinite space of solutions, and how these hypotheses can be formalised as equations and constraints over this space of solutions to design a completely automatic tool. This question will be examined through an example of data of mammary metabolism in dairy cows (Guinard *et al.*, 1994; Guinard and Rulquin, 1994; Lemosquet *et al.*, 2009; Raggio *et al.*, 2006).

Materials and methods

The metabolic calculator: reaction as a function of carbon chain pivots and of cofactors

In the generic stoichiometric model of Van Milgen (2002), reactions were mainly described by a restricted number of intermediary metabolites, called carbon-chain 'pivots', at cross points between metabolic pathways presented in Figure 1. These pivots were chosen to describe the main possible conversions between metabolites. As mentioned above, reactions also integrate the variations of ATP, other cofactors and metabolites (i.e. NADH, NADPH, $FADH_2$ NADH, O_2, CO_2 and NH_3).

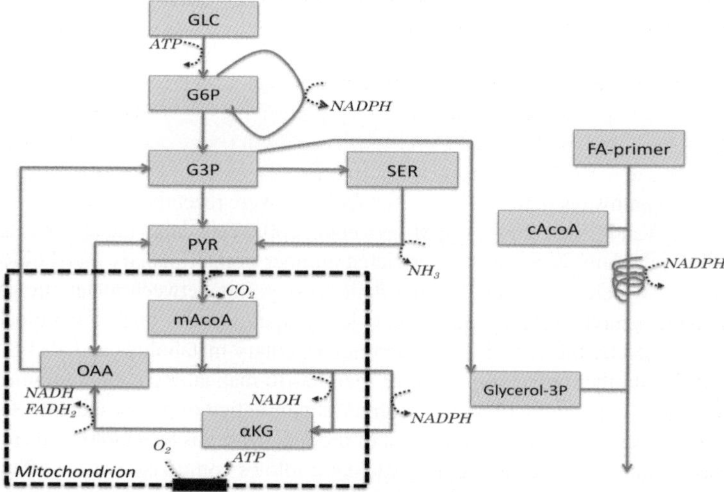

Figure 1. Possible conversions between intermediary pivots (11) in the example of the ruminant mammary gland (GLC: glucose; G6P: glucose 6-phosphate; G3P: triose phosphate; SER: serine; PYR: pyruvate; ACoA: mitochondrian (m) and cytosolique (c) acetyl CoA; αKG: α-ketoglutarate; OAA: oxaloacetate; FA primer: Fatty acid primer: the first two or four carbon chain of a fatty acid; Glycerol 3-P: glycerol 3-Phosphate).

The calculator is provided as a spreadsheet which computes the nutrient balance resulting from the combination of metabolic reactions. As shown in Figure 2, columns contain both cofactors (columns E to K) and products to be equilibrated (pivots – columns M to T). Rows correspond to available reactions and include the stoichiometry of the reactions. Each reaction will be used with a given rate (column C), which will be used as a multiplicative factor for the stoichiometry reactions. The final balance (last row) is obtained by summation of the balances of the selected reactions.

For example, the user may have measured experimentally the uptake of glucose by the mammary gland and lactose output and may have observed that the glucose uptake was greater than the glucose use for lactose synthesis. The user then has to select the appropriate equation (column B) for lactose synthesis and glucose oxidation and manually compute the reaction rates (column C), so that the balance of each pivot equals zero. This computation has two main interests: first, it allows the calculation of the theoretical balance of non-measured nutrients, such as ATP. Second, the value of the reaction rates permit the estimation of the way each nutrient is consumed by the different pathways. This is used to find, analyse and compare scenarios of nutrients partitioning within the biological entity from balance studies, and as mammary net balance studies.

Mathematical modelling and flux balance analysis

Studying metabolic networks is crucial for detecting products that play key roles in biochemical processes. There exists a large body of literature describing tools that address this problem from different viewpoints, included in the large domain of Flux Balance Analysis (Schuster *et al.*, 2000). One of the most important hypothesis of Flux Balance Analysis theory relies in the mass conservation hypothesis (i.e. all the pivots are balanced). It allows one to revisit metabolic networks features as problems of linear algebra in which the stoichiometric matrix plays a central role, tackled with optimization algorithms.

Let us revisit the stoichiometric model tool in this framework. Determining reactions rates in column C of Figure 2 is formalised by setting a variable for each coefficient rate. The vector that gathers all these rate variables is denoted by F. We also introduce a vector B which describes the balance between outputs and inputs of each nutrient. Experimentally measured data allow the computation of the coefficients of B which correspond to unbalanced measured nutrients (such as cofactors). Assuming that there is no accumulation of pivots in the cell (commonly referred to as the balance hypothesis) means that the coefficients of B corresponding to pivots equal zero. Independently,

A	B	C	D	E	F	G	H	J	K	L	M	N	O	P	Q	R	S	T
1	eq.	Flux value.	reaction	ATP	cNADH	mNADH	FADH2	CO2	O2	NH3	OAA	αKG	PYR	mACoA	cACoA	GLC	G6P	G3P
100	4	F_4= 1,0	lactose synthesis	-1	0	0	0	0	0	0	0	0	0	0	0	-1	-1	0
101	2	F_2= 3,0	glucose entry	0	0	0	0	0	0	0	0	0	0	0	0	3	0	0
102	1	F_1= 3,0	GLC → G6P	-2	0	0	0	0	0	0	0	0	0	0	0	-2	2	0
103	5	F_5= 1,0	G6P → G3P	-1	0	0	0	0	0	0	0	0	0	0	0	0	-1	2
104	6	F_6= 2,0	G3P → PYR	4	2	0	0	0	0	0	0	0	2	0	0	0	0	-2
105	8	F_8= 0,0	G3P → G6P	0	0	0	0	0	0	0	0	0	0	0	0	0	0	0
106	9	F_9= 2,0	PYR → mACoA	0	0	2	0	2	0	0	0	0	-2	2	0	0	0	0
107	10	F_{10}= 2,0	OAA + mACoA → αKG	0	0	2	0	2	0	0	-2	2	0	-2	0	0	0	0
108	11	F_{11}= 2,0	αKG → OAA	2	0	4	2	2	0	0	2	-2	0	0	0	0	0	0
109	13	F_{13}= 0,0	PYR → OAA	0	0	0	0	0	0	0	0	0	0	0	0	0	0	0
110	15	F_{15}= 0,0	OAA → G3P	0	0	0	0	0	0	0	0	0	0	0	0	0	0	0
111																		
112	17	F_{17}= 2,0	cNADH ↔ mNADH	0	-2	2	0	0	0	0	0	0	0	0	0	0	0	0
113	16	F_{16}= 10,0	mNADH oxidation	30	0	-10	0	0	-5	0	0	0	0	0	0	0	0	0
114	18	F_{18}= 2,0	FADH2 oxidation	4	0	0	-2	0	-1	0	0	0	0	0	0	0	0	0
			balance	36	0	0	0	6	-6	0	0	0	0	0	0	0	0	0
				B_{ATP}	B_{cNADH}	B_{mNADH}	B_{FADH2}	B_{CO2}	B_{O2}	B_{NH3}	B_{OAA}	$B_{\alpha KG}$	B_{PYR}	B_{mACoA}	B_{cACoA}	B_{GLC}	B_{G6P}	B_{G3P}

Figure 2. The metabolic calculator spreadsheet: three moles of glucose (GLC) were used and 1 mol of lactose was produced (eq: equations; Coef: coefficient; OAA: oxaloacetate; αKG: α-ketoglutarate; PYR: pyruvate; mAcoA and cACoA: mitochondrian (m) and cytosolique (c) acetyl CoA; G6P: glucose 6-phosphate; G3P: triose phosphate).

the stoichiometry of reactions is described by a matrix S, called the stoichiometric matrix ($S_{i,j}$ is the stoichiometric coefficient of the *i*-th metabolite in the *j*-th reaction). Computing the reaction rates (i.e. the coefficients of the vector *F*) needed to explain how the balance vector *B* may be equilibrated by using the stoichiometry *S* is then performed by solving the matrix equation $S.F=B$, that is studying the solutions of a system of linear equations. Figure 3A presents the mathematical modelling of the metabolic calculator for the simple example depicted in Figure 2.

Unfortunately, this system commonly has an infinite number of solutions. Flux Balance Analysis is a set of mathematical methods that rely on additional constraints over the space of solutions to explore it and enhance solutions with a biological relevance. Reactions rates (called flux coefficients in this theory) are constrained by both thermodynamic constraints (that impose some reactions to be irreversible) and by Michaelis-Menten dynamics (yielding bounds for the productions of metabolites in the cell). All together, such bounds and constraints imply that the space of solutions has a very particular shape: it is a simplex (i.e. a convex polyedra) as depicted in Figure 3B.

The very specific shape of the space of solution allows a better understanding of the biological strategy used by the entity to adapt to uptakes. By taking benefit of the linear structure and optimisation algorithms, it becomes possible to compute the solution satisfying a global objective function. For instance, as stated by Edwards and Palsson (2000) and Edwards *et al.* (2001), it was shown that bacteria select reactions that maximise biomass growth. Similarly, other species have been shown to evolve to use the smallest quantity of an exogenous nutrient such as glucose (Fell and Small, 1986).

From these examples, it turns out that exploring the infinite space of solution of linear equations yielded by stoichiometry can be handled in a formalised framework, by solving a linear optimisation problem based on three types of information: (1) A stoichiometric model that defines the equations (this is exactly what appears in the stoichiometric model tool of Van Milgen); (2) Constraints deduced from thermodynamics or obtained from the literature, allowing to restrict the combinations between reactions; (3) An objective function, which is the main phenotype to globally optimise the system to select a unique solution within the space of solutions.

Notice that the objective function defines an optimisation direction. The optimal solution is thus placed at a vertex node of the simplex. The simplex method is an efficient method to solve such a problem (Chvatal, 1983). It consists in a walk on the vertices of the simplex passing from one vertex to its best neighbour according to the objective function. The algorithm ends when it is not possible to improve the objective function.

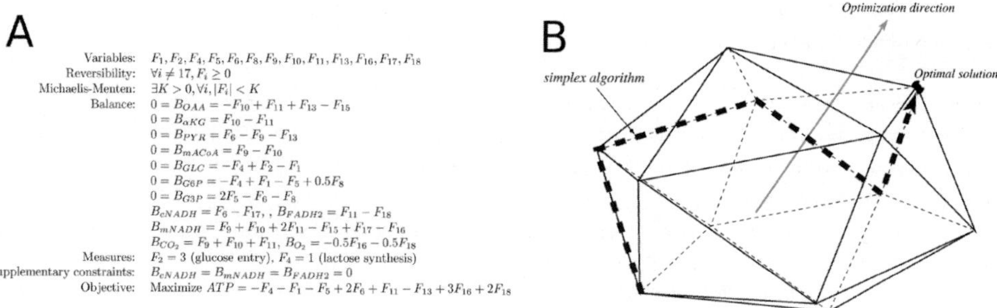

Figure 3. (A) a mathematical viewpoint for the metabolic calculator (same example as Figure 2) and (B) the space of solutions of the stoichiometric model has a polyhedra shape.

With this Flux Balance Analysis theory at hand, a natural question is now to understand whether the computations performed with the metabolic calculator can be fully included in this framework, by exhibiting some constraints and objective functions which can model the implicit search of a solution. We studied this question on a specific example related to metabolism of the mammary gland.

Adaption of the stoichiometry to ruminant mammary gland

To adapt the original model to mammary gland metabolism of ruminants, five additional intermediary metabolites (i.e. pivots) were introduced (Figure 1). Mitochondrial (m) and cytosolic (c) acetyl-CoA (ACoA) were used to distinguish between utilisation of cACoA for fatty acid (FA) synthesis and utilisation of mACoA in the tricarboxylic acid cycle (TCA). In addition, glucose-6-phosphate (G6P) and triose-phosphate (G3P) were introduced to account more specifically for the partitioning of glucose (in this case, the use of G6P) between the pentose phosphate pathway, lactose synthesis and glycerol utilisation. For lipid metabolism, an intermediary metabolite called fatty acid primer (FA primer) was introduced. It corresponds to the primers of two or four carbon-units necessary to initiate fatty acid synthesis before elongation. In ruminants, this primer can be acetate or β-hydroxybutyrate (BHBA). Additional reactions were introduced to describe milk synthesis. For example, NADPH synthesis through the isocitrate deshydrogenase pathway was included. The final spreadsheet included a total of 11 pivots and 92 reactions. However, some reactions were considered as non-existing in the ruminant mammary gland, in particular the NADPH synthesis through malic enzyme. Overall, the reactions taken into account were close to those retained in other models of mammary metabolism (Waghorn and Baldwin, 1984; Hanigan and Baldwin, 1994).

Dataset and input data

Two experiments studying mammary metabolism in response to casein infusions in the duodenum in supplement to diet were analysed. In the first experiment (Exp. A: Guinard *et al.*, 1994; Guinard and Rulquin, 1994), treatments corresponding to 0 (Ctrl), 177 g/d, 762 g/d of casein (CN) infused were compared to control (Ctrl: 0 g/d of CN infused) and CN (743 g/d) in the second experiment (Exp. B: Lemosquet *et al.*, 2009; Raggio *et al.*, 2006). The input data corresponded to half-udder uptakes of major nutrients (GLC, all individual amino acids (AA), acetate, BHBA, glycerol, lactate) and of half-udder milk output (lactose, AA in protein synthesised in the mammary gland and all FA synthesised within the mammary gland, glycerol-3-P required for milk triglycerides), and mammary blood CO_2 production. Precise calculations of milk component output were described in Lemosquet *et al.* (2009). We used similar hypotheses to those adopted by Hanigan and Baldwin (1994) for AA composition of milk protein (Raggio *et al.*, 2006) and to those adopted by Waghorn and Baldwin (1984) for milk fatty acid synthesised in the mammary gland. It was assumed that all FA from C4 to C12, 85% of C14 and C15, and 60% of C16 and C17 were synthesized in the mammary gland as in Waghorn and Baldwin (1984).

In the present computation, protein and long chain FA directly taken up by the mammary gland and not synthesised were not taken into account. In addition, if the net balance (uptake – output in milk protein) of a non-essential AA (NEAA) was significantly different from zero and negative then this NEAA was considered to be synthesized within the mammary gland. Similarly, an AA was considered to be catabolised (including NEAA synthesis) if its net balance (uptake-output in milk protein) was significantly different from zero and positive.

Results

Performing computations with the metabolic calculator

Several hypotheses were used to select the reactions and finally compute all reactions rates describing the nutrient partitioning between metabolic pathways. Two types of rules could be distinguished: 'strong' rules absolutely necessary to find a solution and 'weak' rules. The first two strong rules were based on the hypotheses that there is no accumulation or deficiency of intermediary metabolites in the mammary gland (rule 1) and that there is no deficiency of ATP or cofactors (rule 2). Most of the weak rules were deduced from the literature from ruminant mammary metabolism. These rules are open for debate and may be modified in future computations. Among these weak rules, glucose was assumed to be the sole precursor of lactose, uptake of glycerol by the mammary gland was not used for glycerol-3P synthesis but only oxidised. For NADPH production, it was assumed that two-third of NADPH was produced in the pentose phosphate pathway (Hanigan and Baldwin, 1994) and the rest by the isocitrate deshydrogenase pathway.

One important point, in terms of the mathematical problem, is that these weak rules allowed selection of a set of reactions in the spreadsheet from which to compute coefficient rates such that intermediary metabolites and cofactors (in columns) and reactions (rows) respect the two strong rules.

Biological validation of dataset and results of 'metabolic calculator' computations

This model allowed us to test easily the validity of the datasets for several aspects.

First, it analysed carbon and nitrogen balances through CO_2 and NH_3, respectively (Table 1). With respect to the stoichiometry of the reactions, CO_2 calculated by the model corresponded to carbon uptake minus carbon output in milk and NH_3 production to nitrogen balance. In Exp. B, carbon (Lemosquet *et al.*, 2009) and nitrogen balances (Raggio *et al.*, 2006) were quite well equilibrated and close to the model value. In both experiments, CO_2 production was measured by a blood gas apparatus method, and the model gave very similar results, thereby indirectly validating the measurement. The differences in nitrogen balance were linked to differences in AA uptake retained. In Exp. A, N balances were not equilibrated. For example, in the CN (743 g/j) treatment of Exp. A, NH_3 was positive. This positive nitrogen balance was also highlighted by Guinard and Rulquin (1994) and they hypothesised that peptide production may occur, which was not accounted for in the present model.

Second, it allowed the determination of whether enough nutrients were taken up to explain milk synthesis based on other criteria than carbon or nitrogen balances. In the Ctrl treatment of Exp. A, in addition to the deficit in NH_3 group (-29.7 mmol/h) to synthesise NEAA, it was not possible to respect the rule of no deficit in pivots since intermediary metabolites such as G3P were negative. These observations may indicate an underestimation of the mammary blood flow in Ctrl treatment of Exp. A.

Third, in this example by applying the same rules of computation, the partitioning of nutrient utilisation can be analysed and compared between experiments. As an illustration, ATP balance was very similar in high CN treatments in both experiments (in Table 1: 1,941 vs. 1,835 in Exp. A and B, respectively). In Exp. A, ATP balance and ATP gain from AA catabolism greatly increased in response to increasing CN infusions. However, in Exp. B, ATP balance decreased but ATP gain by BHBA oxidation greatly increased in response to CN infusions.

Table 1. Example of the results obtained after manual computation of net mammary balances of CO_2, NH_3 and ATP in two experiments (Exp. A and Exp. B) in dairy cows.

	Exp. A[1]			Exp. B[2]	
	Ctrl[3]	CN[4]: 177 g/d	CN[4]: 743 g/d	Ctrl[3]	CN[4]: 762 g/d
(mmol/h/half udder)					
CO_2 measured	628	760	864	930	1,030
CO_2 by the model	526	720	841	1,047	988
N balance calculated					
NH_3 by the model	-29.7	10.1	61.4	-3.7	-2.2
Balance of ATP[5]	232	1,438	1,941	2,579	1,835
ATP from BHBA[6]	827	940	911	505	1,887
ATP from AA[7]	60	298	990	280	391

[1] Guinard *et al.* (1994) and Guinard and Rulquin (1994).
[2] Lemosquet *et al.* (2009) and Raggio *et al.* (2006).
[3] Control treatment.
[4] Casein treatments (duodenal infusions).
[5] Obtained by the model.
[6] ATP produced during BHBA oxidation through isocitrate dehydrogenase pathways.
[7] ATP produced during amino acids (AA) catabolism.

Automatically recovering identical results with Flux Balance Analysis together with an ATP optimisation objective

The computations with the metabolic calculator presented above might suggest a unique solution of partition, which is not the case. Indeed, the stoichiometric model adapted to the ruminant mammary gland generates a linear system $S.F = B$ composed by 23 equations and 40 variables (only a subset of all the 92 possible reactions described by Van Milgen model are used). This leads to a rectangular stoichiometric matrix S of size 23×40, a flux vector F with 40 coefficients rates to be estimated and a balance vector B with 23 nutrients to be balanced (11 of these corresponds to pivots). For all irreversible reactions i, the constraint F_i *strictly positive* is added. The resulting linear problem is clearly under-parameterised and has an infinite number of solutions. Several additional constraints are simply added to formalise the weak rules used to perform the computation with the stoichiometric tool (see above). For instance, one of these constraints concerns lactose synthesis. In our study, we impose that the unique origin of lactose is glucose (for GLC and G6P parts of lactose). Thus equation 8 (G3P → G6P) is not active leading to $F_8 = 0$. Unfortunately, even with these new constraints, the system of equations is still under-parameterised: it admits an infinite number of solutions.

As shown in the Flux Balance Analysis theory, focusing on a particular point of the space of solutions requires the definition of a general objective function that formalises the general phenotype that is optimised during milk synthesis. To find a unique solution, it would turn the mathematical problem into a linear optimisation problem. Notice that linear means that both constraints defining the space of solutions and the objective function are linear combinations of the variables F_i, such as balances of products that we are currently considering.

As mentioned before, several objective functions have been proposed in the literature depending on biological applications. However, they are often related to the optimisation of biomass production. In our case, the consumption of every nutrient is known and appears in the balance. Furthermore, the quantity of nutrients is exactly the same for all solutions. Therefore, the biomass growth depends only on the consumption of ATP. Thus, maximising the biomass growth is equivalent here to a minimisation of ATP consumption.

Following this idea, we used the simplex algorithm (see the Materials and Methods section) to estimate reaction rates that maximise the production of ATP while satisfying the stoichiometric constraints given by the model and the literature constraints discussed above. Performing this computation automatically resulted in the same reaction rates as those obtained with the manual computation of the 'metabolic calculator'. This suggests that computations with the metabolic calculator spreadsheet tool actually follow an optimisation strategy just as those used in Flux Balance Analysis. In the context of ruminant mammary gland, the optimisation is performed such that ATP production globally reaches its maximum.

Discussion

The formalisation as equations and constraints by the completely automatic tool based on Flux Balance Analysis revealed that an objective function was required to find a unique solution of nutrient partition in each treatment. In the present case, manually using the metabolic calculator appears to correspond to the hypothesis of a maximisation of ATP production (i.e. the objective function). This means that the usual choice of pathways for milk synthesis corresponded to a minimisation of ATP consumption. This part was not a conscious choice. However, this also means that when nutrient uptake is higher than the amount required for milk synthesis, the remainder of the nutrients is catabolised, which is necessary to satisfy the rule (constraint) of non-accumulation of pivots and cofactors. For example, it is evident that mammary glucose uptake should always exceed lactose synthesis. The question remains whether it is relevant to take the maximisation of the ATP as the objective function?

At a cell scale this objective function seemed to be relevant. In fact, several objective functions, mainly based on maximisation of biomass growth, have been proposed in the literature, especially for bacterial metabolism (Fell and Small, 1986; Edwards and Palsson, 2000). In our context, as discussed in the result section, maximising the biomass growth is equivalent here to a minimisation of ATP consumption.

At the scale of the mammary gland metabolism, the question concerning the appropriate objective function remains open. To our knowledge, the ruminant mammary metabolism model of Hanigan and Baldwin (1994) is based one other constraints than the maximisation of ATP. It is a model of simulation of the literature knowledge on mammary metabolism. It was validated by using *in vitro* measurements of partitioning through tracer studies. However, Hanigan *et al.* (2009) seemed to find similar quantities of ATP used in milk component (3.3 mol/kg of milk) and of ATP generated by mammary gland in their model compared with the present computation. With the Ctrl treatment of Exp. B, the quantity of ATP used for milk synthesis was also 3.3 mol/kg of milk and, the quantity of ATP generated for a whole udder (7.8 mol/kg of milk) exceed by 2-fold the quantity of ATP used for milk as reported by Hanigan *et al.* (2009). The similarity of the results between both models suggests that the biological rule of maximisation of ATP may be acceptable to explore mammary metabolism. This concept related to ATP will be also interesting to explore since mammary nutrient uptake is regulated by ATP availability in the model proposed by Cant and McBride (1995).

At a level of the whole organism, the rule of maximisation of ATP seems to be less obvious. In Molly cow model, the efficiency of ATP in several metabolic pathways was assumed to decrease to avoid a too high increase of the cow body weight in high fat diet treatments (Johnson *et al.*, 2009). Another example which occurred at whole body metabolism is the variation of efficiencies of ATP production from GLC (Van Milgen, 2002). It can vary from 38 ATP/mol of GLC (direct oxidation) to 31 ATP/mol of GLC, if GLC is converted in glutamate and if glutamate is incorporated. However, the maximisation rule may still hold if additional constraints will be considered. For example, glutamate is extensively catabolised in the gut, and the oxidation of glucose to ATP via glutamate can be seen as a spatial organisation of metabolism (Van Milgen, 2002). Similarly, glucose oxidation via lipid can be interpreted as a temporal organisation of metabolism and the constraint (glucose to ATP via lipid) may be included as a constraint. The rule of maximisation of ATP, under this constraint, may still hold.

In conclusion, the mathematical formalisation using the flux balance analysis theory of the stoichiometric tool of Van Milgen (2002, 2003) required an objective function to find a unique solution of nutrient partitioning. The validity of an objective function based on the maximization of ATP as a rule of nutrient partition needs further confirmation.

References

Cant, J.P. and McBride, B.W, 1995. Mathematical analysis of the relationship between blood flow and uptake of nutrients in the mammary glands of a lactating cow. Journal of Dairy Research 62:405-422.

Chvatal, V., 1983. Linear Programming. W.H. Freeman and Co. New York, NY, U.S., 478 pp.

Edwards, J.S., Ibarra, R.U. and Palsson, B.O., 2001. *In silico* predictions of *Escherichia coli* metabolic capabilities are consistent with experimental data. Nature Biotechnology 19:125-130.

Edwards, J.S. and Palsson, B.O., 2000. The *Escherichia coli* MG1655 *in silico* metabolic genotype: its definition, characteristics, and capabilities. Proceedings of the National Academy Sciences 97:5528-5533.

Fell, D.A. and Small, J.R., 1986. Fat synthesis in adipose tissue. An examination of stoichiometric constraints. Biochemical Journal 238:781-786.

Johnson, H.A., Calvert, C.C. and Fadel, J.G., 2009. Impact of adjusting energy values on predictions of energy metabolism using molly. In: Sauvant, D., Agabriel, J., Faverdin, P., Lescoat, P. and Van Milgen, J. (eds.) 7e International workshop. Modelling Nutrient Digestion and Utilization in Farm Animals, September 10-12, 2009. Paris, France, p 42.

Guinard, J., Rulquin, H. and Vérité. R., 1994. Effect of graded levels of duodenal infusions of casein on mammary uptake in lactating cows. 1. Major nutrients. Journal of Dairy Science 77:2221-2231.

Guinard, J. and Rulquin, H., 1994. Effect of graded levels of duodenal infusions of casein on mammary uptake in lactating cows. 2. Individual amino acids. Journal of Dairy Science 77:3304-3315.

Hanigan, M.D. and Baldwin, R.L., 1994. A mechanistic model of mammary gland metabolism in the lactating cow. Agricultural Systems 45:369-419.

Hanigan, M.D., France, J., Mabjeesh, S.J., Mcnabb, W.C. and Bequette. B.J., 2009. High rates of mammary tissue protein turnover in lactating goats are energetically costly. Journal of Nutrition 139:1118-1127.

Lemosquet, S., Raggio, G., Lobley, G.E., Rulquin, H., Guinard-Flament, J. and Lapierre, H, 2009. Whole-body glucose metabolism and mammary energetic nutrient metabolism in lactating dairy cows receiving digestive infusions of casein and propionic acid. Journal of Dairy Science 92:6068-6082.

Raggio, G., Lemosquet, S., Lobley, G.E., Rulquin, H. and Lapierre, H, 2006. Effect of casein and propionate supply on mammary protein metabolism in lactating dairy cows. Journal of Dairy Science 89:4340-4351.

Renaudeau, D. Noblet, J. and Dourmad, J.Y., 2003. Effect of ambient temperature on mammary gland metabolism in lactating sows. Journal of Animal Science 81:217-231.

Schuster, S., Fell, D.A. and Dandekar, T., 2000. A general definition of metabolic pathways useful for systematic organization and analysis of complex metabolic networks. Nature Biotechnology 18:326-332.

Van Milgen, J., 2002. Modeling biochemical aspects of energy metabolism in mammals. Journal of Nutrition 132:3195-3202.

Van Milgen, J., Gondret, F. and Renaudeau, D., 2003. The use of nutritional models as a tool in basis research. In: Souffrant, W. B. and Metges, C. C. (eds.) Progress in research on energy and protein metabolism. September 13-18, 2003. Rostock, Germany, pp.259-263.

Waghorn, G.C. and Baldwin, R.L., 1984. Model of metabolite flux within mammary gland of the lactating cow, Journal of Dairy Science 67:531-544.

Model development of nutrient utilization to represent poultry growth dynamics: application to the turkey

V. Rivera-Torres[1,2], P. Ferket[3] and D. Sauvant[4]
[1]Techna, B.P. 10, 44220 Couëron, France; virginie_rivera@techna.fr
[2]AgroParisTech, 16 rue C. Bernard, 75231 Paris cedex 5, France
[3]North Carolina State University, NC State Dept. of Poultry Science, Box 7608, Raleigh, NC 27695-7608, USA
[4]INRA, UMR 791 Modélisation Systémique Appliquée aux Ruminants, 16 rue C. Bernard, 75231 Paris cedex 5, France

Abstract

To ensure a sustainable development of animal farming systems, relating growth performance to production cost has become a major issue. The use of models represents a useful means to assess the nutritional response of growing animals and to relate it to the cost of production. A mechanistic poultry growth model was developed to describe the response of poultry to nutrition in a controlled environment. The model is based on the representation of the homeorhetic and homeostatic regulations associated with energy and amino acid utilization. Protein, ash and water in carcass, viscera and feathers, and lipid in carcass and viscera constitute body compartments, whereas amino acids, fatty acids, glucose and acetyl-coenzyme A constitute circulating compartments. Protein and lipid turnover are at the origin of protein and lipid retention and are driven by homeorhetic regulations. Ash and water retention are defined through allometric relations with protein retention. The circulating compartments are controlled by homeostatic regulations to maintain a constant concentration of amino acids, glucose and fatty acids in plasma. The model allows simulating the growth dynamics and the composition of body weight gain and was evaluated using data from growing turkeys. The flexibility of the model may enable an application to commercial facilities to relate the nutritional response of growing poultry to production costs.

Keywords: turkey, growth, model

Introduction

The understanding of nutrient utilization in farm animals has become a challenge for better management of farming systems from both a technical and economical point of view. Mechanistic models simulate the metabolism of ingested nutrients associated with energy and amino acid utilization for growth in several species (Pettigrew *et al.*, 1992, Gerrits *et al.*, 1997, Halas *et al.*, 2004), but very few such models have been developed for poultry. Most growth simulation models in poultry are based on empirical responses to nutrient intake. To our knowledge, no existing poultry model is currently capable of simulating the metabolic processes associated with the regulation of body growth dynamics. Such a mechanistic simulation model would be a step towards the determination of the regulation of metabolic processes in poultry nutrition.

The goal of our work is to develop a mechanistic model that simulates growth dynamics in poultry subjected to different feeding and environmental conditions, so that it can be used as a practical decision-making tool for adapting nutritional strategies to the outcome of potential field or industry applications, or socio-economic scenarios. The model describes the regulation of energy and amino acid metabolism during growth. Turkeys reared in a controlled environment are used to evaluate the model.

Homeorhetic and homeostatic regulations drive the growth trajectory to maturity

Growth is defined as the trajectory to a desired mature body weight

For the first step towards the development of the model, we simulate the growth of meat poultry fed standard diets in a controlled environment (e.g. experimental conditions). The genetic response of the animal to this unrestricted environment results in a constrained feed intake that corresponds to the desired feed intake, as described by Kyriazakis and Emmans (1999). Thus, the desired feed intake is an input to the model, and it conditions growth as a trajectory towards mature body weight (BW).

The turkey is taken as the basis for the development of the poultry model because turkeys are characterized by a high sexual dimorphism and a long production in comparison with other poultry species, which makes it possible to run the simulations on the basis of a greater BW range for further evaluations. Because the experimental unit in experiments often corresponds to a group of turkeys, we use the average growth performance to run the simulations representing an individual. The data used for the simulations are the average performance of the experimental unit (BW, feed intake and feed efficiency of the group of turkeys), the digestible nutrient contents of total amino acids, fat, and starch, and the metabolizable energy in the dietary feed.

Regulatory processes are based on the interaction between homeorhetic and homeostatic regulations

We assumed homeorhetic (HR) and homeostatic (HS) regulations to be the driving forces for growth. The HR regulations correspond to long-term processes, whereas HS regulations refer to short-term controlled events. The HR regulations (mostly genetic) interact at different levels of organization and determine the composition of BW. While protein and lipid deposition are considered as the determinant components of BW gain, homeorhesis regulates protein and lipid turnover, thereby establishing a decisional system (Sauvant, 1992) that drives the accretion of muscle (mostly through protein deposition) and adipose tissue (mostly through lipid deposition). At a lower level of organization of the animal, HS regulations ensure that the HR regulations are accomplished. The HS regulations are driven by an operative system that is associated with the decisional system. These two systems describe the quantitative and qualitative effects of nutrient intake on daily BW gain through the regulation of the concentration of circulating compartments that act as intermediary metabolites in the transportation and transformation flows of nutrients.

The simulation of growth dynamics is based on protein and lipid turnovers

The factorial method describes body growth dynamics

The factorial method (Hurwitz *et al.*, 1983) is used for the description of the HR regulations. The whole body is partitioned in 3 components defined as auxiliary variables: (1) carcass, the main component of energy storage; (2) viscera, the component for the transformation and transportation flows of nutrients; and (3) feathers, a component that helps maintaining thermoregulation, and mainly associated with an irreversible exportation flow of amino acids. Each component is composed of protein, ash, water, and lipid (except for feathers). Protein and lipid compartments are driven by their respective turnover rates. Because water and ash are usually not-limiting for growth, water and ash intake are not specifically represented and changes in body ash and water are described by allometric relations of carcass, viscera or feathers. The result is that the whole body is partitioned into 11 body compartments that mostly depend on the protein and lipid compartments in carcass, viscera and feathers. Also, 4 compartments are represented in the model to represent circulating nutrients (i.e. amino acids, glucose, fatty acids and acetyl-coenzyme A) in plasma (see later). The partitioning of the body into the 11 body compartments and the 4 circulating nutrients enables us to independently simulate the differences in growth dynamics of carcass, viscera and feathers, and the transformation of nutrients from the feed to body tissues.

Protein and lipid turnovers are components of HR regulations

The deposition rate of the protein and lipid compartments is the difference between the anabolic and catabolic flows of protein and lipid in carcass and viscera. The anabolic and catabolic flows follow mass action laws (Lovatto and Sauvant, 2003) with the mass of the protein or lipid compartment (g) and the fractional rates of anabolism (FrA) and catabolism (FrC) in carcass and viscera (%/d). In feathers, no FrC of protein is represented because the anabolic flow of protein in feathers is considered as an exportation flow. The anabolic and catabolic flows of lipid are also differentiated between carcass and viscera. Because of lack of data on lipid turnover in poultry, the FrA and FrC of lipid were assumed to be similar in carcass and viscera.

The FrA and FrC of protein and lipid follow decreasing exponential functions as defined by Danfaer (1991) and Lovatto and Sauvant (2003) in growing pigs. They constitute the HR engine of the model. At early ages, FrA and FrC are high and the positive difference between these determines the fractional rate of deposition (FrD; Figure 1a). As the turkey grows, FrA and FrC decrease towards a constant value corresponding to a state of maturity, and the difference becomes zero, resulting in zero growth. The FrA and FrC for protein are different for carcass and viscera, and are greater in the latter (Lovatto and Sauvant, 2003). The partitioning of the protein into the feathers is defined by a FrA, which is lower than those in the viscera and the carcass. Values taken from the literature (Figure 1b) confirmed that the optimised FrA and FrC are in the range of the theoretical values determined in broilers and turkeys.

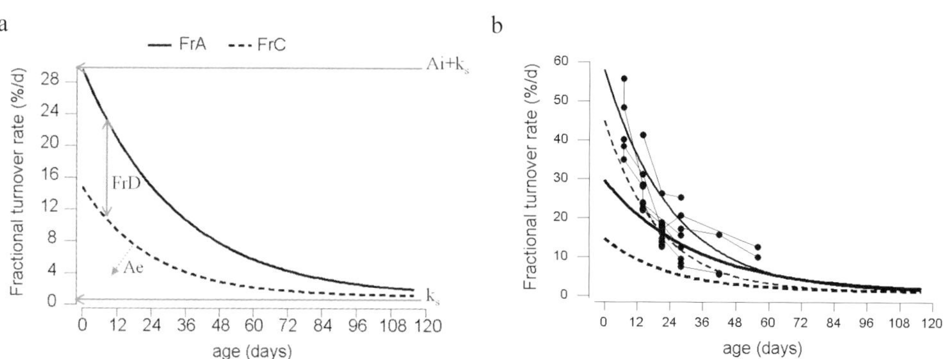

Figure 1. Change in the fractional rates of protein anabolism (FrA, solid line) and catabolism (FrC, dotted line) as a function of age (t). (a) FrA = k_s + Ai exp(-Ae × t) and FrC = k_s + Ai exp(-Ae × t), where k_s is the fractional turnover rate at maturity (asymptote), Ai is the difference between the fractional turnover rate at maturity and the initial state (t=0), and Ae is the fractional rate of change that defines the inflexion of the curve (Lovatto and Sauvant, 2003). The fractional rate of deposition (FrD) is determined by difference between FrA and FrC. (b) Points in gray represent data from the literature in turkeys (Kang et al., *1985b) and broilers (Kang* et al., *1985a; Tesseraud* et al., *1996, 2000; Pym* et al., *2004; Urdaneta-Rincon and Leeson, 2004). Intra-experiment points are joined. The black lines represent the fractional rates of protein anabolism and catabolism of carcass (bold lines) and viscera (thin lines).*

The adjustment of the fractional turnover rates to the observed protein and lipid deposition enables the determination of the anabolic and the catabolic flows of protein and lipid in carcass and viscera. Figure 2 illustrates the change in anabolic and catabolic flows of protein in carcass and viscera over time. Despite a greater fractional turnover rate of protein in viscera compared with carcass, the turnover flows of protein are greater in carcass than in viscera. This is due to the greater size of the protein compartment in carcass than in viscera. Moreover, these anabolic and catabolic flows indicate that the maximum deposition rates in carcass and viscera are not reached at the same age, which justifies the differentiation between these two compartments. Maximum protein deposition in viscera is reached at 42 days of age, whereas it is reached at 68 days of age in the carcass. The protein and lipid deposition in the whole body are presented in Figure 3.

Nutrient utilization depends on HR and HS regulations

Feed intake is an input to circulating nutrients

We considered feed intake to be the result of HR regulations. Feed intake is associated with the incoming flows of N and N-free digestible nutrients. Digestible crude protein, crude fat, starch, fibre, and sugar are inputs for the model and these nutrients enter the circulating compartments. The N-free nutrients are associated with digestible crude fat (tryglycerides) and carbohydrates (digestible sugar, starch and crude fibre).

Body tissue accretion results from transaction flows between circulating nutrients

Once transported to the circulating compartments, the digestible nutrients are subject to HS regulations that attempt to maintain constant nutrient concentrations in the plasma. All transformation flows are subject to stoichiometry relations, as described by Gerrits *et al.* (1997). The HS regulations are based on glucose, fatty acid, and amino acid metabolism. Also, they drive the acetyl-coenzyme

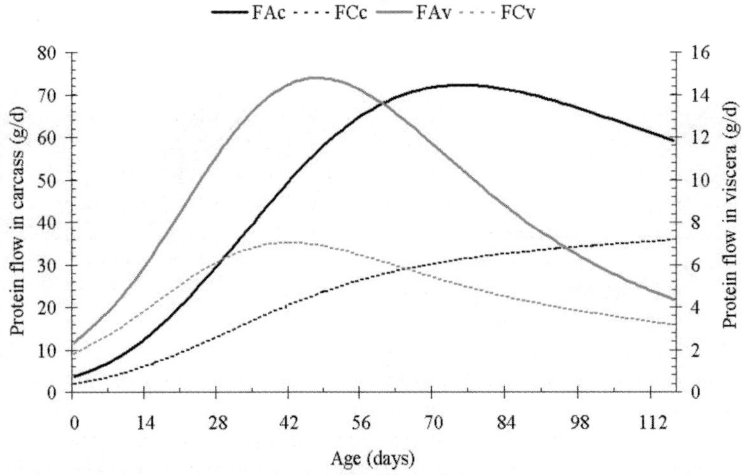

Figure 2. Kinetics of the anabolic and catabolic flows of protein in carcass (FAc and FCc, respectively) and viscera (FAv and FCv, respectively). The anabolic (bold lines) and catabolic (plain lines) flows of carcass (in black) and viscera (in gray) result from a mass action law of the protein pool (P, in g) in carcass or viscera component (i): FAi = FrAi × Pi, where FrAi follows a decreasing exponential law (i.e. Figure 1).

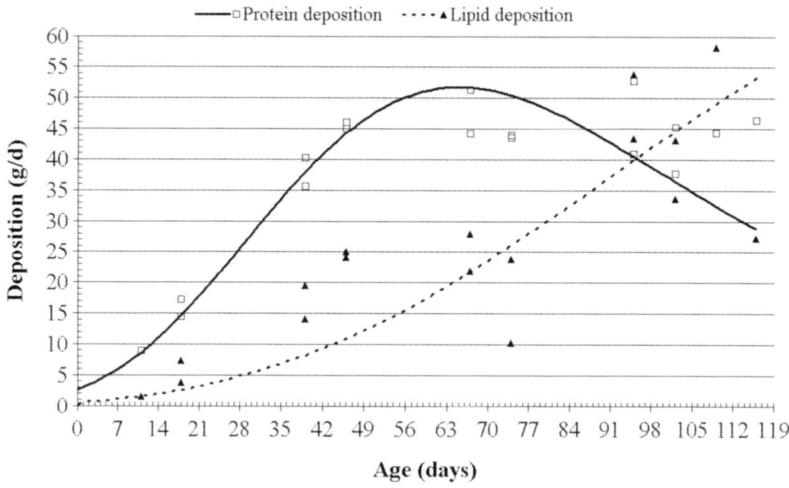

Figure 3. Evolution of daily protein and lipid deposition with age, comparison between observations and simulations. Each observation is represented with squares (protein deposition) and triangles (lipid deposition). The simulation outputs are represented by a straight (protein deposition) and a dotted (lipid deposition) line. The data were taken from Rivera-Torres et al. *(2010) in male medium-size growing turkeys from 0 to 17 weeks of age.*

A (Ay) pool, which is a circulating intermediary metabolite representing the only common intermediate for the oxidation of glucose, fatty acids and amino acids. The circulating glucose pool is directly linked to the lipid compartments. The lipid compartment is expressed as triglycerides equivalents, which correspond to the sum of 3 moles of fatty acids (linoleic acid equivalents) and 0.5 mole of glucose. The excess glucose is either used for fatty acid synthesis or oxidized *via* Ay. The fatty acid pool is an input for the lipid compartments for lipid deposition, and for the Ay pool for fatty acid oxidation. The Ay is also an input to the fatty acid pool for a transformation into tryglycerides.

Finally, the amino acid pool affects the protein turnover of carcass and viscera, and the exportation of amino acids through HS regulations. The irreversible export flows correspond to inevitable oxidation losses, to oxidation losses that result from the amino acids supplied in excess, to endogenous losses, and to protein deposition in the feathers. The endogenous losses correspond to a loss of protein that results from the difference between anabolic and catabolic flows of protein in the gut. The amino acids supplied in excess are converted to glucose (ketogenic amino acids) or Ay, and ammonia in droppings. Depending on the concentration of the circulating nutrients, the inputs to the Ay are differently distributed between amino acids, glucose, and fatty acids.

Homeostatic regulations are at the origin of constant nutrients concentrations

The Ay pool acts as a buffer in the HS regulations of glucose, amino acid, and fatty acid concentrations in plasma. An excess in the concentration of a circulating nutrient results in an increase in its outflow and a decrease in the inflow. In such a situation, the major increase in outflow is observed with the oxidation flow into Ay. This HS regulation enables the transformation of the nutrients into molecules with 2 carbon atoms, which are either reintegrated into the fatty acid compartment or oxidized to provide the energy required for endogenous transactions and for maintenance energy expenditure.

The HS regulations are represented by exponential laws as described by Lovatto and Sauvant (2003; Figure 4). They were preferred to the saturation kinetics (Danfaer, 1991; Gerrits *et al.*, 1997; Halas *et al.*, 2004) because they refer to a return force originating from the HR regulations. In other words, the HS regulations interfere with the HR regulations at different events. They aim at adapting the metabolic flows so that they remain in the desired 'HR trajectory'. The exponential laws are proportional to the relative difference in the concentration of the circulating nutrient.

Homeostatic regulations are primarily controlled by the circulating glucose and amino acid compartments. The relative difference between the actual and the balanced concentration of amino acids is the core of HS regulations of the amino acid oxidation flow, and of protein turnover. An increase in the relative difference between actual and balanced amino acid concentration is associated with increased amino acid oxidation and protein anabolic flows, along with a decrease in the protein catabolic flows of carcass and viscera. The relative difference in glucose concentration acts on the HS regulation of glucose and fatty acid oxidation, fatty acid synthesis from Ay, and lipid synthesis from glucose and fatty acids. An increase in the relative difference increases fatty acid synthesis and lipid synthesis, whereas it reduces glucose and fatty acid oxidation.

Energy utilization conditions the nature of BW gain composition

Feed intake is determined from the desired ME intake

No dietary nutrient is considered limiting in this version of the model, and the animal regulates feed intake on the basis of the ME content of the diet. The calculation of the desired ME intake is based on the equation of Kielanoswki (1965). The desired ME intake also corresponds to the sum of heat production (HP) and the retained energy as protein and lipid, where HP is the sum of the energy expenditure for maintenance and energy retention. The energy expenditure for maintenance

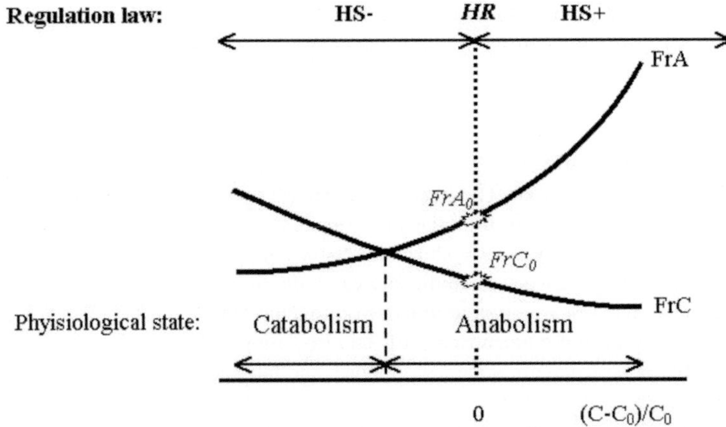

Figure 4. Schematic representation of the interaction between homeorhetic (HR) and homeostatic (HS) regulations (based on Lovatto and Sauvant, 2003) through changes in the fractional anabolic and catabolic rates (FrA and FrC, respectively). When the relative difference between nutrient concentration (C) and balance concentration (C_0) in plasma equals zero at a given age, HR regulations drive the dynamic evolution of FrA and FrC (FrA_0 and FrC_0 in these conditions). When the relative difference is greater than zero (HS+), HS regulation laws increase FrA and decrease FrC, resulting in a greater deposition ($FrA-FrC < FrA_0-FrC_0$). On the contrary, a negative relative difference (HS-) results in a decreased deposition relative to the HR state ($FrA-FrC < FrA_0-FrC_0$).

is assumed to be 641 kJ/kgBW$^{0.75}$/d, and the energy cost of protein deposition is assumed to be 23.7/0.65 = 39.5 kJ/g, while the energy cost of lipid deposition equals 39.6/1.00 = 39.6 kJ/g (Rivera-Torres *et al.*, 2010). The desired feed intake is estimated as the ratio between the desired ME intake and the ME content of the diet. The daily glucose, fatty acid and amino acid intakes are determined from the feed intake and the nutrient content in lipid and amino acids in the diet. The glucose intake is determined by difference between the ME intake and the energy content in the digestible amino acid and lipid intake.

Daily heat production affects nutrient partitioning

The flows are partitioned between endogenous and exogenous reactions at the origin of energy consumption and energy expenses, respectively. The exogenous reactions correspond to the transformation of amino acids, glucose and fatty acids into Ay, and to the lipid degradation into fatty acids and glucose, whereas the endogenous transaction flows correspond to protein and lipid synthesis, and protein degradation. All these transactions are assimilated with ATP yields as described by Gerrits *et al.* (1997) and Halas *et al.* (2004), assuming an energy content of 79.5 kJ per mole of ATP. The difference between the daily HP and the exogenous reactions corresponds to the energy provided by the oxidation of Ay (Halas *et al.*, 2004), assuming an energy yield of 12 moles of ATP per mole of Ay oxidized.

Conclusions

This study aimed at describing the regulatory processes upon which the construction of a poultry growth model was based. The regulation processes are based on HR and HS principles, which enable flexibility of the model in simulating the growth profile of turkeys, depending on different HR driving forces (protein and lipid deposition potential) and on different feeding diets. The HS regulations enabled simulating the metabolic adaptation of the turkey to different diets. The simulation of the regulation principles provides the model with a flexibility that may result in an application to the poultry industries to simulate the effect of nutrient intake on growth and the composition of body weight gain.

References

Danfaer, A., 1991. Mathematical modeling of metabolic regulation and growth. Livestock Production Science 27:1-18.

Gerrits, W.J.J., Dijkstra, J. and France, J., 1997. Description of a model integrating protein and energy metabolism in preruminant calves. Journal of Nutrition 127:1229-1242.

Halas, V., Dijkstra, J., Babinszky, L., Verstegen, M.W.A. and Gerrits, W.J.J., 2004. Modelling of nutrient partitioning in growing pigs to predict their anatomical body composition. 1. Model description. British Journal of Nutrition 92:707-723.

Hurwitz, S., Plavnik, I., Bengal, I., Talpaz, H. and Bartov, I., 1983. The amino acid requirements of growing turkeys. 2. Experimental validation of model calculated requirements for sulfur amino acids and lysine. Poultry Science 62:2387-2393.

Kang, C.W., Sunde, M.L. and Swick, R.W. 1985a., Growth and protein turnover in the skeletal muscles of broiler chicks. Poultry Science 64:370-379.

Kang, C.W., Sunde, M.L. and Swick, R.W., 1985b. Characteristics of growth and protein turnover in skeletal muscle of turkey poults. Poultry Science 64:380-387.

Kielanowski, J., 1965. Estimates of the energy cost of protein deposition in growing animals. In: Baxter, K.L. (ed.) Energy metabolism. Proceedings of the 3rd symposium. Academic Press, London, UK, pp.13-20

Kyriazakis, I. and Emmans, G.C., 1999. Voluntary food intake and diet selection. In: Kyriazakis, I. (ed.) A quantitative biology of the pig. CABI Publishing, Wallingford, UK, pp.229-248.

Lovatto, P.A. and Sauvant, D., 2003. Modeling homeorhetic and homeostatic controls of pig growth. Journal of Animal Science 81:683-696.

Pettigrew, J.E., Gill, M., France, J. and Close, W.H., 1992. A mathematical integration of energy and amino acid metabolism of lactating sows. Journal of Animal Science 70:3742-3761.

Pym, R.A.E., Leclercq, B., Tomas, F.M. and Tesseraud, S., 2004. Protein utilisation and turnover in lines of chickens selected for different aspects of body composition. British Poultry Science 45:775-786.

Rivera-Torres, V., Noblet, J., Dubois, S., and van Milgen, J., 2010. Energy partitioning in male growing turkeys. Poultry Science 89:530-538.

Sauvant, D., 1992. Systemic modeling in nutrition. Reproduction Nutrition Development 32:217-230.

Tesseraud, S., Peresson, R., and Chagneau, A.M., 1996. Age-related changes of protein turnover in specific tissues of the chick. Poultry Science 75:627-631.

Tesseraud, S., Chagneau, A.M. and Grizard, J., 2000. Muscle protein turnover during early development in chickens divergently selected for growth rate. Poultry Science 79:1465-1471.

Urdaneta-Rincon, M. and Leeson, S., 2004. Muscle (Pectoralis major) protein turnover in young broiler chickens fed graded levels of lysine and crude protein. Poultry Science 83:1897-1903.

Modelling energy utilization in poultry

N.K. Sakomura, J.B.K. Fernandes, K.T. Resende, C.B.V. Rabello, F. Longo and R. Neme
Universidade Estadual Paulista, UNESP, Via de Acesso Prof. Paulo Donato Castellane s/n Jaboticabal, 14 884-900, São Paulo, Brazil; sakomura@fcav.unesp.br

Abstract

Models of daily energy requirement can help to establish better and more profitable feeding programs for poultry. Studies have been conducted at UNESP-Jaboticabal-Brazil with the aim of studying energy utilization in broiler breeders, laying hens, and broilers, and to establish metabolisable energy requirement models. The factorial approach was used to partition the energy requirements into maintenance, growth, and production components. The resulting models consider body weight, weight gain, egg production, and environmental temperature for the determination of the energy requirements of poultry. These models were evaluated in performance trials and provided good estimates. Therefore, they can be used to establish nutritional programs. The aim of this chapter is to describe the development of these models and to outline the results of our studies at UNESP.

Keywords: factorial approach, energy metabolism, energy requirement, poultry

Introduction

Over the past 20 years, growth modelling has become an important activity in animal science. Scientists have developed concepts and intellectual frameworks to allow description of complex biological systems. Simulation models have also played an important role in commercial poultry production, where they are used to design economically optimum feeding production regimens (Fisher and Gous, 2009).

An accurate prediction of nutrient requirements is important to establish profitable nutritional programs in poultry production. Among models used to predict nutrient requirement, the factorial or partitioning model has proved to be the most used approach. Such models can be applied in growth models to establish better and more profitable feeding programs for poultry.

The nutritional recommendations for swine by the NRC (1998) are based on mathematical models for growth and reproduction. These models use a simple structured method to develop factorial estimations of nutrient requirements. They estimate the amount of a nutrient used for each major function of the body (e.g. maintenance, protein accretion, and milk production). In poultry, however, complete factorial energy models have been developed only for laying hens (Emmans, 1974; NRC, 1994; Peguri and Coon, 1988; Sakomura and Rostagno, 1993).

Several studies have been conducted at the Universidade Estadual Paulista-UNESP, Jaboticabal, São Paulo, Brazil, with the aim of studying the energy utilization in broiler breeder, laying hens, and, broilers, and also to determine models of energy requirements in poultry. The aim of this chapter is to describe the development of these models and to outline the results of these studies.

Energy partitioning for maintenance, growth, and production

The conventional method to represent energy utilization has been to partition metabolisable energy intake. According to Birkett and Lange (2001), the simplest approach to partition the ME intake (ME_i) is in terms of its use by animal for production, as retained energy (RE), and an amount associated with maintenance (ME_m): $ME_i = ME_m + (1/k_g)RE$. However, different energy efficiencies are associated with production, for instance for growth (k_g), as compared with maintenance purposes. However, the simple model does not take into account the differences in energy retention efficiency as fat or protein. To model energy utilization more accurately for varying fat:protein ratios in RE, Kielanowski (1966) subdivided the RE into RE in fat (REF) and in protein (REP), as $ME_i = ME_m + (1/k_f)REF + (1/k_p)REP$, where k_f and k_p are energy efficiencies for fat and protein deposition, respectively.

The ME requirements for laying and breeder hens can be partitioned differently into maintenance, growth, and production, as expressed by the model: $ME_i = aW^b(T) + c\Delta W + dEM$, where ME_i is daily ME intake, W^b is a power function of W (i.e. it is equal to metabolic body weight when $b \cong 0.75$), ΔW is body weight change per day, EM is egg mass output, T is environmental temperature (°C), a, c, and d are the maintenance, growth and production requirement coefficients (Emmans, 1974; NRC, 1994; Peguri and Coon, 1988; Sakomura and Rostagno, 1993). The accuracy of these models depends on the estimation of those coefficients. Unfortunately, they vary greatly in the literature. Besides the difference in genetics and environmental conditions, the limitations in the methodologies employed also explain the wide-range of estimated coefficients in the literature (Chwalibog, 1991, 1992). The models to determine ME requirement developed at UNESP were based on the partitioning of ME intake into maintenance, growth, and production according to the approach just described.

Maintenance energy requirement

The classic definition of maintenance is the state 'in which there is neither gain nor loss of energy by the body' (Blaxter, 1972). Therefore, the ME requirement for maintenance (ME_m) has been defined as the amount of energy required to balance anabolism and catabolism, giving an energy retention around zero. According to Chwalibog (1991), this definition is acceptable for adults but not for producing animals, because these animals sometimes face negative energy balance. In this case, Chwalibog (1985) defines the ME maintenance requirement as being the amount of ME to maintain a dynamic equilibrium of protein and fat turnover, to maintain body temperature and a normal level of locomotion. Studies were conducted at UNESP to determine the effect of temperature on maintenance requirements (ME_m and NE_m), and also the efficiencies of energy utilization (k_g) in broiler breeders, laying hens and broiler chickens. The trials were carried out in environmental controlled rooms at temperatures below, at, and above the thermal neutral zone. Birds were fed below, close to, and above maintenance intake levels. Energy balance components were obtained by comparative slaughter method. Retained energy (RE) was measured by slaughtering a representative group of birds at the beginning and at the end of each trial. Heat production (HP) was calculated by difference between ME intake (ME_i) and RE. ME_m was estimated by the linear regression of RE on ME_i; the intercept in the x-axis (i.e. when RE = 0) provides the estimated ME_m. The slopes represent the efficiencies of ME utilization (k_g). The regression of the log of HP on ME_i provides an estimate of net energy for maintenance (NE_m), (i.e. the HP when $ME_i = 0$). The efficiency of energy utilization for maintenance k_m was calculated by NE_m/ME_m.

As shown in Table 1, the ME_m requirements of poultry vary widely. Assuming a constant requirement across genotypes and rearing T appears to be incorrect. Some of this variation could be attributed to genetic differences between broilers and laying hens, which exhibit different growth potential and body composition. The efficiency of ME utilization for maintenance (k_m) varied between 0.67 and 0.80, while its efficiency for gain (k_g) ranged from 0.57 to 0.69. Temperature, genotype and age of poultry did not affect these efficiencies. Because the same diet (based on corn and soybean meal)

Table 1. Maintenance metabolizable energy (MEm) and net energy requirements (NEm), efficiency of energy utilization above maintenance (k_g), and for maintenance (k_m) according to ambient temperatures, and genotype[1].

Genotype	Temperature (°C)	Requirements ($kcal/kg^{0.75}/day$)		Efficiency	
		ME_m	NE_m	k_g	k_m
Laying-type pullets (cage)	12	142	-	-	-
Laying hens (cage)	12	138	100	0.66	0.72
Broiler breeder pullet (ground)	15	158	119	0.69	0.75
Broiler (ground)	13	158	119	0.63	0.76
Broiler breeder hen (ground)	13	131	-	-	-
Broiler breeder hen (cage)	13	111	78	0.61	0.70
Laying-type pullets (cage)	24	94		0.59	0.67
Laying hens (cage)	22	112	80	0.62	0.71
Broiler breeder pullet (ground)	22	144	109	0.69	0.76
Broiler breeder hen (ground)	21	113	-	-	-
Broiler breeder hen (cage)	21	91	65	0.60	0.71
Broiler (ground)	23	112	90	0.59	0.80
Laying-type pullets (cage)	30	109	-	-	-
Laying hens (cage)	31	93	69	0.69	0.74
Broiler breeder pullet (ground)	30	128	92	0.62	0.72
Broiler breeder hen (ground)	30	111	-	-	-
Broiler breeder hen (cage)	30	88	59	0.57	0.67
Broiler (ground)	32	127	96	0.66	0.76

[1]Longo *et al.*, 2006; Neme *et al.*, 2005; Rabello *et al.*, 2004; Sakomura *et al.*, 2003; Sakomura *et al.*, 2005.

was utilized in the assays with broiler breeders, laying hens and broiler chicks, and diet composition is the main factor that affects the efficiency of ME utilization, no variation in k_g was observed.

Another important result is the difference between ME_m for broiler breeder raised in cages vs. those on the ground. The requirement of breeders raised on the ground was 20% higher than that in cages. This is likely due to more energy being spent for physical activity. These results are important because most maintenance energy requirement for broiler breeders have been determined in metabolic chambers or cages, which underestimate the requirements for breeders raised on the ground. Johnson and Farrell (1983) and Spratt *et al.* (1990) reported ME_m for broiler breeder hens in metabolic chambers at 21 °C to be 87 and 88 $kcal/kg^{0.75}/day$, respectively. A similar result (91 $kcal/kg^{0.75}/day$) was reported by Rabello *et al.* (2004) for breeders in cages at 21 °C, while for those raised on the ground ME_m was 113 $kcal/kg^{0.75}/day$.

Nesheim *et al.* (1990) reported that the energy requirement for activity is about 50% of basal metabolism and is influenced by raising conditions. Birds in cages have lower activity and HP is about 30% of basal metabolism. The expenditure of energy for activity of laying hens is about 20 to 25% of HP (MacLeod *et al.*, 1982). According to Wenk (1997), the physical activity in growing farm animals kept under practical conditions accounts for almost 20% of the maintenance requirements.

Under restricted housing conditions, often in respiration chambers, lower ME_m values are found. The ME_m must be corrected to account for different levels of activity.

Measurement of total HP includes the energy required for maintenance as well as energy spent in response to changes in the environment. The major environmental factor that influences HP is temperature. The critical lower temperature is the point below which the animal must increase HP to maintain body temperature; the critical higher temperature is the point where the animal needs to expand more energy to dissipate heat.

Using the data shown in Table 1, regression of ME_m on ambient temperature (T) was performed. The differences due to genotype and the bird's age are probably due to variations in body weight and body composition. A linear decrease of ME_m with increase of T was observed for laying hens ($ME_m = BW^{0.75}$ (166 – 2.37 T), r^2=0.92) and broiler breeder pullets ($ME_m = BW^{0.75}$ (174 – 1.88 T), r^2=0.99). On the other hand, a quadratic effect was observed for broiler chicks ($ME_m = BW^{0.75}$ (308 – 15.63 T + 0.31 T^2), r^2=0.93) and broiler breeder hens ($ME_m = BW^{0.75}$ (193 – 6.32 T + 0.12 T^2), r^2=0.72). There was a decrease in ME_m when the temperature increased up to 26 °C; above that temperature ME_m increased. According to Leeson and Summers (1997), a small variation is observed in HP of birds from 19 to 27 °C, but below the lower critical limit temperature, birds need to produce heat to maintain their body temperature, whereas above 27 °C, energy is spent to dissipate heat. However, these temperature limits are not the same for all birds because body weight, feed intake, feathering, and activities can affect response to T changes.

Because feathering is an important factor modulating maintenance energy requirements, a study was carried out at UNESP to evaluate the effect of temperature (12, 18, 24, 30, and 36 °C) and the degree of feathering (0, 50, and 100% of body surface) on ME_m requirements of laying type pullets. Minimal critical temperature (MCT) was established based on the degree of feathering (F, expressed as a proportion ranging between 0 and 1). The resulting equation (MCT = 24.54 – 5.65 F) indicates that MCT is greater for completely de-feathered birds, as they loose heat more easily and therefore respond faster to a reduction in environmental temperature as compared to completely feathered birds. According to the equation obtained, MCT changes as a function of the degree of feathering; MCT increases as feathering is reduced. That is, MCT = 19, 22, and 24 °C for 100, 50, and 0% feathered birds, respectively. The relationship between feathering degree and MCT indicates changes in the temperature at which the birds are in the comfort zone, and consequently spend less energy for basal metabolism maintenance. According to the estimated model, $ME_m = 92.4\ BW^{0.75} + 0.88(T-MCT)$. Thus, for each 1 °C increase above the MCT of each feathering degree, there is an increase in maintenance requirement of 0.88 kcal/$kg^{0.75}$/day. On the other hand, as temperature decreases below the MCT of each feathering degree, there is an increase in maintenance requirements of 6.73 kcal/$kg^{0.75}$/day for each 1 °C, according to the model $ME_m = 92.4\ BW^{0.75} + 6.73\ (MCT - T)$.

Energy requirement for growth

Although maintenance and growth are parallel and continuous in the process of nutrient metabolism, the two processes have traditionally been considered separately (Black, 2000). Most metabolism simulation models consider a hierarchy in requirements. The maintenance requirements are first to be, followed by protein growth and, finally, by fat growth. Tissue deposition only occurs if there are sufficient nutrients available for growth above the maintenance requirements. When the readily available nutrients are not sufficient to supply the maintenance requirements, body tissue must be catabolised to supply the balance (Pomar *et al.*, 1991).

In the UNESP studies, trials were conducted with broilers, laying type pullets, and breeder pullets in order to determine energy body composition. Body weights were recorded weekly, and groups of birds were slaughtered to determine whole body energy. Net energy requirement per gram of body weight gain (NE_g) was determined by regressing whole body energy on body weight, the slope representing NE per gram of weight gain).

Table 2 shows that body energy composition NE/gram of gain differs according to genotype and age, and it is related to fat and protein body composition. There was an increase in body energy content as bird aged, due to higher body fat deposition. The ME requirement for growing was determined by considering NE_g and k_g.

In laying hens and broiler breeder hens, it is difficult to measure the efficiency of energy utilization in body and egg deposition due to difficulties in partition energy intake into body and egg energy deposition. In our lab, Rabello *et al.* (2006) determined the energy efficiency for body deposition in broiler breeder hens by inhibiting egg production using a low calcium diet (0.466% Ca) and from light restriction. The efficiency of energy utilization for body deposition was calculated by the formula: $k_g=RE_b/(ME_i-ME_m)$, where RE_b is energy retained in the body, ME_i is ME intake, and ME_m is ME for maintenance. The efficiencies of ME for energy deposition shown in Table 2 varied from 47% for broiler breeder hens to 69% for growing breeders. These differences resulted in variation in the estimated ME requirements, ranging from 2.50 to 9.49 kcal ME/gram of body, according to genetics and ages.

The ME utilization above maintenance depends on the partitioning of energy between protein and lipid synthesis and their respective efficiencies, which can be estimated by the factorial approach, where the ME intake is partitioned into protein and lipid deposition, according to the model suggested

Table 2. Net energy (NE_g) and metabolizable energy (ME_g) requirements for growth, and efficiencies of energy utilization (k_g)[1].

Laying-type pullets	Age (weeks)	NE_g (kcal/g BW)	k_g	ME_g (kcal/g BW)
White-Egg-Laying strain	1 to 7	2.03	0.59	3.44
	8 to 12	3.06		5.19
	13 to 18	5.60		9.49
Brown-Egg-Laying strain	1 to 7	2.03	0.63	3.22
	8 to 12	3.11		4.94
	13 to 18	3.98		6.32
Broiler breeder pullet	3 to 8	1.950	0.69	2.83
	9 to 14	1.725		2.50
	15 to 20	2.239		3.24
Male broiler	1 to 3	2.190	0.59	3.72
	4 to 6	2.479		4.21
	7 to 8	2.657		4.51
Female broiler	1 to 3	2.341	0.59	3.97
	4 to 6	2.316		3.93
	7 to 8	4.148		7.04
Laying hen	20 to 36	4.340	0.65	6.68
Broiler breeder hen	26 to 33	3.580	0.47	7.62

[1]Longo *et al.*, 2006; Neme *et al.*, 2005; Rabello *et al.*, 2004; Sakomura *et al.*, 2003; Sakomura *et al.*, 2005a,b.

by Kielanowski (1965). This model is important because body composition changes with genetic, age, body weight, and diet. With this approach, it is possible to consider body composition when determining energy requirements. Sakomura *et al.* (2003) and Sakomura *et al.* (2005b) used this approach to determine the efficiencies of energy deposition as protein (k_p) and fat (k_f) in broiler breeder pullets and broilers. The k_p obtained in broilers (0.45) and broiler breeder pullets (0.46) were similar. Based on the k_p and the net energy content of body protein (5.66 kcal/g), the ME requirements per gram of protein deposited were estimated at 12.59 kcal/g for broiler and 12.57 kcal/g for breeder pullets. However, the k_f was numerically much greater for broiler breeder (1.04) than that for broiler chick (0.69). This can be explained by the difference on the feeding programs; broilers were fed *ad libitum* whereas broiler breeders received controlled feeding. Based on the net energy of body fat (9.37 kcal/g) and k_f, the ME requirements per gram of fat deposited were 13.52 kcal/g for broilers and 9.04 kcal/g for breeder pullets.

Energy requirements for egg production

ME requirements of egg production (kcal/g of egg) were determined by taking into account the energy content per gram of egg and the efficiency of energy deposition in the egg (k_o). Studies were performed at UNESP with laying hens (27-61wks) and broiler breeder hens (31-37wks) to determine ME requirement for egg production. Egg samples were collected every 15 days for energy and dry matter determination. The gross energy of eggs did not change with age. Thus, the means obtained for layer (1.49±0.11kcal/g) and for broiler breeder (1.54±0.10 kcal/g) represent the net energy requirement per gram of egg (NE_e). The efficiency of ME utilization for deposition in the egg (k_o) was calculated by: $k_o = RE_e/[ME_i-(ME_m + ME_g)]$, where RE_e= energy retained in the eggs. The k_o for layers was 0.62, compared to 0.64 for breeders. The ME coefficients for egg production were determined by $ME_e=NE_e/k_o$. The ME_e for layers was similar as those of breeders, resulting in the same ME requirement for egg production (2.40 kcal ME/g of egg (Sakomura *et al.*, 2005a; Rabello *et al.*, 2006).

Models to estimate ME requirements

Models to estimate the ME requirements of broiler breeders, layers, and broilers were developed based on the factorial approach and on the coefficients previously determined (Table 3).

Models evaluation

Aiming to evaluate the models developed at UNESP, performance trials were carried out with broiler breeder pullets (Sakomura *et al.*, 2003), broiler breeder hens (Rabello *et al.*, 2006), and laying hens (Sakomura *et al.*, 2005c). The birds were fed according to the ME requirements calculated using as inputs to the models the data of body weight, daily weight gain, egg mass produced, and environmental temperature as measured during the experiments. The models were compared to the energy supply recommended for the genetic strains used in the trials. According to the results obtained in these trials, the models were accurate in their estimation of ME requirements. In summary, the UNESP models can be used for determining feeding programs. The coefficients estimated for maintenance, growth and production can be used in computer models designed to optimize feeding programs in poultry.

Application of models in feeding programs

Models aim at determining the nutritional requirements, taking into considerations the differences in weather, strains, weight, and body composition. Table 4 shows a simulation of ME requirements

Table 3. Metabolisable energy requirement models for broiler breeders, laying hens, and broilers.

Birds age (weeks)	Models
Broiler breeder pullet[1]	
3 to 8	$ME = BW^{0.75} (174.15 - 1.88\ T) + 2.83\ BWG$
9 to14	$ME = BW^{0.75} (174.15 - 1.88\ T) + 2.50\ BWG$
15 to 20	$ME = BW^{0.75} (174.15 - 1.88\ T) + 3.24\ BWG$
Broiler breeder hen[2]	
	$ME = BW^{0.75} (192.8 - 6.32\ T + 0.12\ T^2) + 7.62\ BWG + 2.40\ EM$
Laying-type pullet[3]	
White-egg-laying strains	
1 to 6	$ME = BW^{0.75}\ 92.40 + 6.73\ (LCT–T) + 3.44\ BWG\ (T<LCT)$
	$ME = BW^{0.75}\ 92.40 + 0.88\ (T–LCT) + 3.44\ BWG\ (T\geq LCT)$
7 to 12	$ME = BW^{0.75}\ 92.40 + 6.73\ (LCT–T) + 5.19\ BWG\ (T<LCT)$
	$ME = BW^{0.75}\ 92.40 + 0.88\ (T–LCT) + 5.19\ BWG\ (T\geq LCT)$
13 to 18	$ME = BW^{0.75}\ 92.40 + 6.73\ (LCT–T) + 9.49\ BWG\ (T<LCT)$
	$ME = BW^{0.75}\ 92.40 + 0.88\ (T–LCT) + 9.49\ BWG\ (T\geq LCT)$
Brown-egg-laying strains	
1 to 6	$ME = BW^{0.75}\ 92.40 + 6.73\ (LCT–T) + 3.22\ BWG\ (T<LCT)$
	$ME = BW^{0.75}\ 92.40 + 0.88\ (T–LCT) + 3.22\ BWG\ (T\geq LCT)$
7 to 12	$ME = BW^{0.75}\ 92.40 + 6.73\ (LCT–T) + 4.94\ BWG\ (T<LCT)$
	$ME = BW^{0.75}\ 92.40 + 0.88\ (T–LCT) + 4.94\ BWG\ (T\geq LCT)$
13 to 18	$ME = BW^{0.75}\ 92.40 + 6.73\ (LCT–T) + 6.32\ BWG\ (T<LCT)$
	$ME = BW^{0.75}\ 92.40 + 0.88\ (T–LCT) + 6.32\ BWG\ (T\geq LCT)$
Laying hen[4]	
	$ME = BW^{0.75} (165.74 - 2.37\ T) + 6.68\ BWG + 2.40\ EM$
Broiler[4]	
1 to 8	$ME = BW^{0.75} (307.87 - 15.63\ T + 0.3105\ T^2) + 13.52\ FG + 12.59\ PG$

ME = metabolizable energy requirement (kcal/bird/day); $BW^{0.75}$ = metabolic body weight (kg); T = ambient temperature (C); BWG = daily body weight gain (g); EM = daily egg mass (g); FG = daily fat weight gain (g); PG = daily protein weight gain (g); LCT = 24.54 – 5.65F, F is feathering score (0 to 1).
[1] Sakomura *et al.*, 2003.
[2] Rabello *et al.*, 2006.
[3] Neme *et al.*, 2005.
[4] Sakomura *et al.*, 2005a,b.

calculated for broiler breeders, using the UNESP model according to the environmental temperature, body weight, weight gain, and production data recommended for Cobb. The definition of a feeding program fit to the strain objectives, as well as to local climatic conditions is essential for the success of the flock, particularly in case of broiler breeders that are submitted to a controlled feeding program and energy supply is essential for a good performance.

Acknowledgements

The authors is grateful to Fundação de Amparo a Pesquisa do Estado de São Paulo for financial support to UNESP studies, and Dr. Normand St-Pierre from The Ohio State University, Columbus, OH, USA, and visiting professor in UNESP-Jaboticabal for suggestions and the manuscript English proofreading.

Table 4. Simulation of broiler breeder hens ME requirements by applying the Cobb performance in the model[1].

Age (weeks)	Body weight mean (g)	Weight gain (g/b/d)	Egg mass (g/b/d)	ME intake[2] (kcal/b/d)	Feed intake[3] (g/b/d)
26	3,235	12.86	7.9	383	134
28	3,373	6.43	39.4	418	147
30	3,450	2.86	48.4	417	146
32	3,490	2.86	50.4	424	149
34	3,530	2.86	50.7	427	150
36	3,570	2.86	50.1	428	150
38	3,610	2.86	49.0	428	150
40	3,650	2.86	48.5	429	151
42	3,683	2.14	47.7	424	149
44	3,713	2.14	46.9	424	149
46	3,743	2.14	46.0	423	148
48	3,773	2.14	45.0	423	148
50	3,803	2.14	44.0	422	148
52	3,825	1.43	42.9	415	146
54	3,845	1.43	41.8	414	145
56	3,865	1.43	40.7	412	145
58	3,885	1.43	39.4	410	144
60	3,905	1.43	38.1	408	143
62	3,925	1.43	36.8	406	143
64	3,945	1.43	35.5	405	142

[1] Calculated by the model: $ME = W^{0.75} (192.76 - 6.32.T + 0.12.T^2) + 7.62.WG + 2.40.EM$.
[2] Ambient temperature at 24 °C.
[3] Feed intake determined considering the dietary energy of 2,850 kcal ME/kg.

References

Birkett, S. and Lange K., 2001. Limitations of conventional models and a conceptual framework for a nutrient flow representation of energy utilization by animals. British Journal of Nutrition 86:647-659.

Black, J.L., 2000. Modelling growth and lactation in pigs. In: Theodorou, M.K. and France, J. (eds.) Feeding systems and feed evaluation models. CABI Publishing, Oxon, UK, pp.363-391.

Blaxter, K.L., 1972. Fasting metabolism and energy required by animals for maintenance. In: Festskrift til Knud Breirem, Mariendals Boktrykkeri A.S., Gjovik, Norway, pp.19-36.

Chwalibog, A., 1985. Studies on energy metabolism in laying hens. Institute of Animal Science, Denmark, 190 pp.

Chwalibog, A., 1991. Energetics of animal production. Acta Agriculturæ Scandinavica 41:147-160.

Chwalibog, A., 1992. Factorial estimation of energy requirements for egg production. Poultry Science 71:509-515.

Emmans, G.C., 1974. The effect of temperature on performance of laying hens. In: Morris, T.R. and Freeman, B.M. (eds.), Energy requirements of poultry. British Poultry Science, Edinburgh, UK, pp.79-90.

Fisher, C. and Gous, R.M., 2009. Protein and amino acid responses. In: Hocking, P. (ed.), Biology of breeding poultry. British Poultry Science Symposium Series, CABI Publishing, Wallingford, UK, pp.331-360.

Johnson, R.J. and Farrell, D.J., 1983. Energy metabolism of groups of broiler breeders in open-circuit respiration chambers. British Poultry Science 24:439-453.

Kielanowski, J., 1965. Estimates of the energy cost of protein deposition in growing animals. In: Blaxter K.L. (ed.), Proceedings of the 3rd symposium on energy metabolism. Academic Press, London, UK, pp.13-20.

Kielanowski, J., 1966. Conversion of energy and the chemical composition of gain bacon pigs. Animal Production 8:121-128.

Leeson, S. and Summers, J.D., 1997. Commercial poultry nutrition, 2nd ed., Guelph, Canada, 350 pp.

Longo, F.A., Sakomura, N.K., Rabello, C.B., Figueiredo, A.N. and Fernandes, J.B.K., 2006. Exigências energéticas para mantença e para o crescimento de frangos de corte. Revista Brasileira de Zootecnia 35:119-125.

MacLeod, M.G., Jewitt, T.R., White J., Verbrugge, M. and Mitchell, M.A., 1982. The contribution of locomotion activity to energy expenditure in the domestic fowl. In: Ekern, A. and Sundstol S. (eds.), Energy metabolism of farm animals. The Agricultural University of Norway, pp.297-300.

Neme, R., Sakomura, N.K., Fialho, F.B., Freitas, E.R. and Fukayama, E.H., 2005. Modelling energy utilization for laying type pullets. Brazilian Journal of Poultry Science 7:39-46.

Nesheim, M.C., Austic, R.E. and Card, L.E., 1990. Poultry production. Lea and Febiger, Philadelphia, USA, 399 pp.

NRC (National Research Council), 1994. Nutrient requirements of poultry. 9th ed. National academy of sciences, Washington, DC, USA.

NRC (National Research Council), 1998. Nutrient requirements of swine. 10th edition. National academy of sciences, Washington, DC, USA.

Peguri, A. and Coon, C.N., 1988. Development and evaluation of prediction equations for metabolizable energy and true metabolizable energy intake for Dekalb XL-Link White Leghorn hen. In: Proceeding of Minnesota nutrition conference and Degussa technology symposium. Bloming, USA, pp.199-211.

Pomar, C.D., Harris, L. and Minvielle, F., 1991. Computer simulation model of swine production systems. 1. Modeling the growth of young pigs. Journal of Animal Science 69:1468-1488.

Rabello, C.B., Sakomura, N.K., Longo, F.A. and Resende, K.T., 2004. Efeito da temperatura ambiente e do sistema de criação sobre as exigências de energia metabolizável para mantença de aves reprodutoras pesadas. Revista Brasileira de Zootecnia 33:382-390.

Rabello, C.B., Sakomura, N.K., Longo, F.A., Couto, H.P., Pacheco, C.R. and Fernandes, J.B.K., 2006. Modelling energy utilization in broiler breeder hens. British Poultry Science 47:622-631.

Sakomura, N.K. and Rostagno, H.S., 1993. Determinação das equações de predição da exigência nutricional de energia para matrizes pesadas e galinhas poedeiras. Revista Brasileira de Zootecnia 22:723-731.

Sakomura, N.K., Silva, R., Couto, H.P., Coon, C. and Pacheco, C.R., 2003. Modelling metabolizable energy utilization in broiler breeder pullets. Poultry Science 82:419-427.

Sakomura, N.K., Basaglia, R., Sa-Fortes, C.M.L. and Fernandes, J.B.K., 2005a. Modelos para estimar as exigências de energia metabolizável para poedeiras, Revista Brasileira de Zootecnia 34:575-583.

Sakomura, N.K., Longo, F.A., Oviedo-Rondon, E.O., Boa-Viagem, C. and Ferraudo, A., 2005b. Modeling energy utilization and growth parameter description for broiler chickens. Poultry Science 84:1363-1369.

Sakomura, N.K., Basaglia, R., Fernandes, J.B.K., Resende, K.T. and Sa-Fortes, C.M.L., 2005c. Avaliação dos modelos para determinar exigências energéticas de poedeiras, Acta Scientiarum 27:349-354.

Spratt, R.S., Bayley, H.S., Mcbride, B.W. and Leeson, S., 1990. Energy metabolism of broiler breeder hens. 1. The partition of dietary energy intake. Poultry Science 65:1339-1347.

Wenk, C., 1997. Summary of the discussion. Energy metabolism of farm animals. In: McCracken, K.J., Unsworth, E.F. and Wylie, A.R.G. (eds.), Proceedings of the 14th symposium on energy metabolism, Newcastle, Co. Down, Northern Ireland. University Press, Cambridge, UK, pp.265-267.

Model to estimate lysine requirements of broilers

J.C. Siqueira[1], N.K. Sakomura[1], R.M. Gous[2], I.A.M.A. Teixeira,[1] J.B.K. Fernandes[1] and E.B. Malheiros[1]
[1]Universidade Estadual Paulista, UNESP, Via de Acesso Prof. Paulo Donato Castellane s/n, Jaboticabal 14 884-900, São Paulo, Brazil; sakomura@fcav.unesp.br
[2]University of KwaZulu-Natal, Private bag X01, Scottsville 3209, South Africa

Abstract

The purpose of this study is to describe the development of a model to predict the digestible lysine requirements of broilers using the factorial approach, and to evaluate the model using as reference the model presented in Brazilian Tables for Poultry and Swine. The model partitions the requirement for maintenance and growth for feather-free body protein and feather protein, in which the inputs are body and feather protein weight and the daily rates of protein deposition in the feather-free body and feathers. The parameters that express the lysine requirement for maintenance were obtained in metabolism trials with roosters, and those for the efficiency of lysine utilization in experiments with broilers from 1 to 42 d. Based on these results the model proposed was: $Lys (mg/d) = [(151 \times BP_m^{-0.27} \times BP_t) + (0.01 \times P_t \times 18)] + [(75 \times BPD/0.77) + (18 \times FPD/0.77)]$, where Lys = digestible lysine requirement (mg/d), BP_m=body protein weight at maturity (kg), BP_t= body protein weight at time t (kg), FP_t= feather protein weight at time t (kg), BPD=body protein deposition (g/d), FPD = feather protein deposition (g/d). The model yields sensible predictions of the digestible lysine requirements of broilers of different strains and ages growing at their potential, and suggests a lower lysine requirement after 27 d than does the Brazilian model. The proposed model is the first step in the development of a simulation model that predicts the food intake of a broiler and hence the dietary amino acid content that would optimise performance.

Keywords: amino acids, maintenance, modeling, performance, poultry

Introduction

Amino acid requirements of broilers have most commonly been estimated from the results of dose-response experiments. Estimates obtained in this way vary widely not only because of the method of interpretation of such results but also because of differences in genotype, diet composition and environmental conditions, since these latter factors have a direct influence on performance and body composition of birds. Therefore, when amino acid requirement is estimated it is necessary to take into account performance and body composition.

In this sense the factorial approach is more useful for the estimation of amino acid requirements of chickens than is direct experimentation. This method estimates the requirement in two stages: (1) the net requirements are calculated, based on estimates of amino acid deposition during growth and/or egg production, and on endogenous losses (Owens and Pettigrew, 1989; Sakomura and Rostagno, 2007); (2) the nutrient requirements are calculated by dividing the net requirements by the efficiency of utilization of each amino acid, these efficiencies being determined from the results of metabolism trials or from studies measuring the response of the broiler or laying hen to each amino acid. The factorial approach accounts for differences in the state of the bird (body weight and composition), potential growth rate and rate of laying in the case of laying hens. Thus, the factorial approach has the potential to predict the nutrient requirements of birds of different genotypes and body weights kept under different environmental conditions.

In poultry research, a model can be considered as a mathematical description of a given biological phenomenon (growth and egg production, for example) and this is obtained by combining equations in which the qualitative and / or quantitative variables are taken to represent the factors that affect such phenomena (Oviedo-Rondon *et al.*, 2002). According to Zoons *et al.* (1991), modeling involves three distinct steps. The first concerns the identification of inputs (independent variables) and outputs (dependent variables), taking into account the objectives of the model. The second step involves the establishment of the structure of mathematical relationships that will comprise the model, and the final step involves model evaluation, this step playing an important role in checking the applicability of the model. According to Sakomura and Rostagno (2007) model evaluation can be carried out by performance trials, computer simulation or by comparing values predicted by the model with observed values reported in the literature.

The use of models to predict the amino acid requirements for poultry is still quite limited in Brazil. In the latest edition of the Brazilian Tables for Poultry and Swine, Rostagno *et al.* (2005) suggested lysine requirements for broilers based on data from 30 studies that used a dose-response approach, their recommendation of the amount of lysine required (mg) per g of body weight gain being based on mean body weight (g). Thus, the lysine requirement for growth (mg/d) can be obtained by multiplying the value predicted by the equation (mg/g) with the expected body weight gain (g/d). Those authors established the lysine requirement for maintenance according to the recommendations of Fisher (1998) and Edwards *et al.* (1999), being 100 $mg/kg^{0.75}.d$. The digestible lysine requirements (mg/d) were estimated as the sum of the requirements for maintenance and growth. Although this model has contributed greatly to the establishment of the amino acid requirements of broilers in Brazil, its value is limited because it is based on total body weight gain. The amount of lipid in body weight gain changes as the bird grows and can be affected by many biological and environmental factors, and amino acids do not contribute directly to the growth of body lipid.

The aim of this study is to describe the development of a model to predict the digestible lysine requirements of broilers using the factorial approach, and to evaluate the model using as reference the model of Brazilian Table of Poultry and Swine (Rostagno *et al.*, 2005).

Model structure and estimation of parameters

The model is structurally similar to that developed by Martin *et al.* (1994) to estimate the amino acid requirements of laying pullets, which was in turn based on the principles of the growth model proposed by Emmans (1981). According to Emmans and Fisher (1986) the amino acid composition of the protein in the feather-free body and in feathers differs markedly, as do their relative proportions during growth due to differences in their growth rates, and therefore it is necessary when calculating the amino acid requirements of the bird to consider separately the requirements for these two components for maintenance and growth. The structure on which the model is based is presented in Figure 1.

Lysine requirement for maintenance

Three nitrogen balance (NB) trials were carried out to determine lysine requirements for maintenance. Roosters of three different genotypes and body composition were used, namely Leghorn (1.98±0.18 kg, n=30), ISA Label (4.33±0.15 kg, n=60) and Cobb 500 (5.11±0.44 kg, n=60). The diets were based on corn, starch and crystalline amino acids formulated to contain all the essential amino acids at 50% of their recommended level (Rostagno *et al.*, 2005), and lysine was first limiting in the test feed. A dilution technique was used to produce the dietary treatments, which resulted in digestible lysine intakes ranging from 10.3 to 93.2 $mg/kg^{0.75}/d$. The roosters were fed for an adaptation

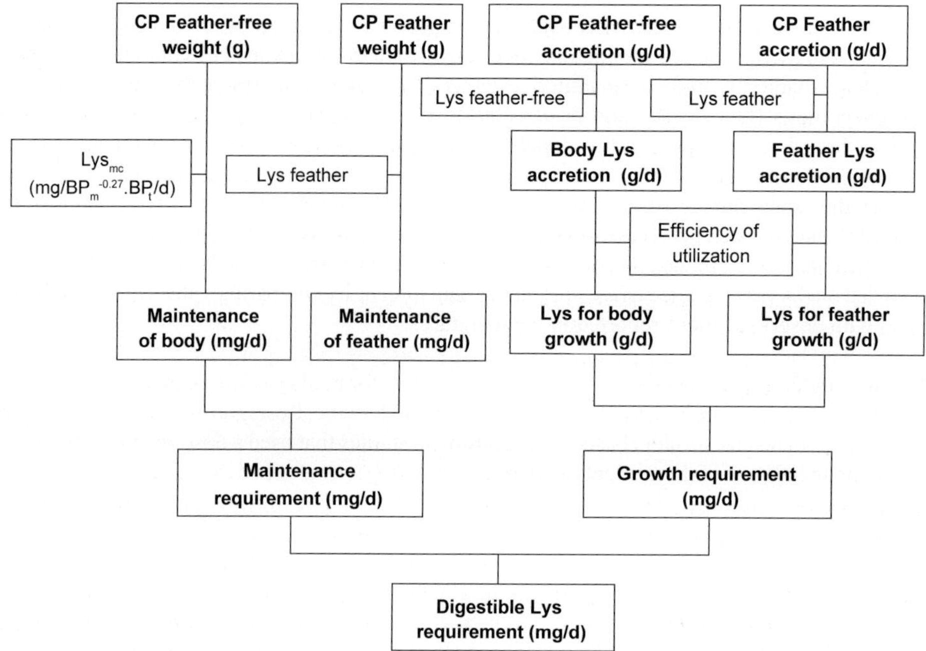

Figure 1. Schematic diagram of the model used to calculate the lysine requirement in broilers.

period of 120 h, followed by total excreta collection for the following 96 h. The nitrogen excreted was subtracted from that consumed and the resultant NB was regressed against lysine intake (mg/kg$^{0.75}$/d). Digestible lysine required for maintenance was estimated as the intake of lysine (LysI) required for NB = 0.

The linear regressions obtained in these trials were:

NB= -107.42 (±20.44) + 2.39 (±0.39)LysI (r^2 =0.60; P<0.0001) for Leghorn roosters (1)
NB= -120.95 (±18.69) + 2.72 (±0.30)LysI (r^2=0.74; P<0.0001) for ISA Label roosters (2)
NB= -140.55 (±24.86) + 2.99 (±0.41)LysI (r^2=0.66; P<0.0001) for Cobb 500 roosters (3)

The data were combined and subjected to a parallelism test which indicated that the linear regression coefficients for the three strains did not differ significantly, and the resulting general equation over all three strains was:

NB = -122.16 (±12.09) + 2.71 (±0.21) LysI (r^2=0.67; P<0.0001) (4)

Based on Equation 4, the lysine requirement for maintenance (i.e. the lysine intake required to obtain a zero NB) was calculated to be 45 mg/kg$^{0.75}$/d, or 32 mg/kg/d, independent of genotype. Several studies relate the amino acids requirements for maintenance with the body weight (mg/kg/d) or metabolic weight (mg/kg$^{0.75}$/d). However, Emmans and Fisher (1986) showed that amino acid requirements for maintenance are related more closely to the body protein content, because body lipid has no maintenance requirement (Emmans and Oldham, 1988; Burnham and Gous, 1992; Gous, 2007). Therefore a more appropriate way to express the requirements is to relate it with the body protein contents (Emmans and Fisher, 1986; Nonis and Gous, 2008). Lysine requirement for the maintenance of body protein was calculated using the composition of the feather-free body protein of roosters, according to Emmans and Fisher (1986): MP = mp.BPm$^{0.73}$.u; MP = maintenance protein requirement (g/d), mp = 0.008 kg ideal protein/unit/d, BP_m = mature protein weight (kg), and u = degree of protein maturity (BP_t/BP_m), equal to one in this study because adult birds were used. To exemplify the application of this equation were used the mean data: Lys_{mb} = 123/0.754$^{0.73}$ = 151 mg/$BP_m^{0.73}$.u/d (Table 1).

Protein required for the maintenance of feathers was considered to be directly proportional to the feather lost, according to Emmans (1989) to be 0.01 g of protein/g of feathers per d, and hence the lysine required for maintenance of feathers was Lys_{mf} (mg/d) = 0.01×FP_t×Lys_f, where FP_t = feather protein weight (g), Lys_f = lysine content of protein in feathers (18 mg/g) (Emmans and Fisher, 1986; Emmans, 1989). Total lysine requirement for maintenance (Lys_m) was calculated as the sum of Lys_{mb} and Ly_{mf}.

Lysine requirement for growth

The amount of lysine required for growth was calculated by considering the rates of protein deposition in the feather-free body and in feathers and the efficiency of lysine utilization.

To determine the efficiency of utilization dietary lysine (k_{Lys}), four dose-response trials were conducted using 4,800 (1,200 in each phase) male Cobb 500 chicks from one to eight, eight to 22, 22 to 35 and 35 to 42 d of age, fed a range of feeds varying in lysine level obtained using the dilution technique. The lysine contents in the feeds were: 9.75, 10.82, 11.89, 12.96 and 14.03 g/kg in the pre-starter phase; 8.40, 9.32, 10.24, 11.16 and 12.08 g/kg in the starter phase; 7.86, 8.72, 9.58, 10.44 and 11.30 g/kg in the growth phase and 7.45, 8.26, 9.07, 9.89 and 10.70 g/kg in the finisher phase.

By sampling birds at the start and end of each phase, whole body protein and lysine contents, calculated by summing the lysine contents of the feather-free body and feathers, were measured and used to calculate protein and lysine deposition. The k_{Lys} (%) was calculated as lysine deposited ×100/(lysine intake – lysine for maintenance).

The averages of efficiencies determined for each phase were: 75.6 for pre-starter, 79.4 for starter, 73.5 for growth and 79.1 for finish, and they were statistically similar ($P<0.05$) by SNK test. Therefore the average efficiency (76.9%) represents the efficiency of lysine utilization (k_{Lys}) measured in this study, independent of phase of growth.

Table 1. Lysine requirement for maintenance of feather-free body protein (Lys_{mb}) based on metabolic body weight ($BW^{0.75}$), body weight (BW), body protein weight ($BP_m^{0.73}$·u), and bird per d.

Genotype	BW_m (kg)	CPC (g/kg)	BP_m (kg)	Lys_{mb} (mg/unit/d)			
				Bird	BW	$BW^{0.75}$	$BP_m^{0.73}$·u[1]
Leghorn	1.99	160	0.318	75.2	37.8	44.9	174
ISA Label	4.33	174	0.753	133	30.8	44.4	164
Cobb 500	5.11	233	1.191	160	31.3	47.1	141
Mean	3.81	198	0.754	123	32.3	45.1	151

BW: body weight at maturity, CPC: crude protein contents in the body, BP_m: body protein weight at maturity.
[1]u=BP_t/BP_m; If $BP_t<BP_m$ (growing birds), then the unit is: $BP_m^{-0.27}$·BP_t.

The calculation of the lysine requirements for growth of the feather-free body and feathers was based on the predicted lysine deposition in these components and above k_{Lys}.

Model to estimate lysine requirement for broilers

The digestible lysine requirements for growth were obtained as the sum of the quantities required for the growth of feather-free body and feathers. The total digestible lysine requirements were obtained as the sum of lysine requirements for maintenance and growth. The equation to estimate the lysine requirement of a broiler, based on the parameters Lys_{mb}, Lys_{mf} and k_{Lys} determined in this study was:

$$Lys\ (mg/d) = [(151 \times BP_m^{-0.27} \times BP_t) + (0.01 \times FP_t \times 18)] + [(75 \times BPD/0.77) + (18 \times FPD/0.77)] \quad (5)$$

where, Lys = digestible lysine requirement (mg/d), BP_m=body protein weight at maturity (kg), BP_t= body protein weight at time t (kg), FP_t= feather protein weight at time t (g), BPD= body protein deposition (g/d), FPD = protein deposition in feathers (g/d). The lysine contents in feather-free body and feathers were considered to be 75 and 18 mg/g, respectively (Emmans and Fisher, 1986; Emmans, 1989). A k_{Lys} of 77% was considered to be the same for the growth of both feather-free body and feathers as there is no specific information in the literature concerning the efficiency of utilization of lysine for feather protein growth.

Application and evaluation of model

The model was evaluated by comparing the predicted estimates with those in the Brazilian Tables for Poultry and Swine model (Rostagno *et al.*, 2005), calculated as:

$$Lys\ (mg/d) = [(0.1 \times BW^{0.75}) + (g\ Dig.Lys/kg\ BWG) \times BWG] \times 1000 \quad (6)$$

where BW = body weight (kg), (g Dig.Lys / kg BWG) = 14.28 + 2.0439×BW, BWG = body weight gain (kg/d).

Daily values to 42 d of age for the independent variables BW, BWG, BP_t, BPD, FP_t and FPD, used in the models to calculate the lysine requirements, were obtained using Gompertz equations for Cobb 500 and Ross 308 broilers, values for the parameters of which are given in Table 2.

The digestible lysine requirements for broilers, at weekly intervals based on the equation developed in this paper and that by Rostagno *et al.* (2005), were then calculated using the daily values of the variables calculated above. These requirements, together with the values of the five independent variables at weekly intervals during the growing period of both strains of broiler, are given in Table 3.

Discussion

It is widely accepted that the optimum amino acid contents in feeds for broilers will vary due to factors inherent in the bird such as genetic potential, state and sex, to environmental and nutritional conditions, as well as economic conditions. For these reasons digestible amino acid requirements estimated by experimentation have limited applicability outside of the conditions under which the studies were conducted (Oviedo-Rondón and Waldroup, 2002; Sakomura and Rostagno, 2007). In this sense, there is great interest in developing mathematical models to predict the nutritional requirements of poultry of different genetic potential, reared under different conditions.

Table 2. Values for the Gompertz parameters mature protein weight (P_m), rate of maturing (B) and age at point of inflection (t^*) for body weight, feather-free body protein weight and feather protein weight of Cobb 500 and Ross 308.

Genotype	Variables	Parameters[1]		
		P_m (kg)	B (per d)	t* (d)
Cobb 500	Body weight	7.592	0.0410	40.6
	Body protein feather-free	1.042	0.0470	37.2
	Feather protein weight	0.193	0.0630	31.1
Ross 308	Body weight	7.832	0.0380	42.0
	Body protein feather-free	1.309	0.0370	44.0
	Feather protein weight	0.328	0.0380	45.6

[1] Gompertz equation: $P_t = P_m \cdot \exp(-\exp(-B.(Age_t - t^*)))$ (Marcato, 2007).

Table 3. Lysine requirements calculated using the model developed in this study and that by Rostagno et al. *(2005) for two broilers strains.*

Genotype	Age (d)	Independent variables[1]						Requirements	
		BW (kg)	BWG (kg/d)	BP (kg)	BPD (g/d)	FP (g)	FPD (g/d)	Lys[2] (mg/d)	Lys[3] (mg/d)
Cobb 500	1	0.048	0.010	0.0043	1.12	0.24	0.10	112	153
	7	0.144	0.023	0.0167	3.25	1.99	0.57	333	365
	14	0.388	0.047	0.0532	7.44	10.17	1.88	780	762
	21	0.814	0.075	0.1226	12.33	29.04	3.46	1,307	1,274
	28	1.421	0.098	0.2233	16.17	57.04	4.38	1,723	1,808
	35	2.159	0.111	0.3439	17.92	88.06	4.35	1,916	2,259
	42	2.954	0.114	0.4693	17.59	116.45	3.70	1,894	2,548
Ross 308	1	0.068	0.012	0.0097	1.75	1.41	0.29	180	190
	7	0.179	0.026	0.0257	3.73	4.29	0.71	385	403
	14	0.432	0.048	0.0630	7.06	11.81	1.49	736	774
	21	0.850	0.072	0.1258	10.89	25.66	2.48	1,145	1,237
	28	1.427	0.092	0.2147	14.35	46.53	3.45	1,522	1,719
	35	2.124	0.105	0.3244	16.73	73.42	4.18	1,792	2,137
	42	2.881	0.109	0.4460	17.76	104.14	4.54	1,925	2,429

[1] BW: body weight, BWG: daily body weight gain (g), BP: body protein weight, BPD: body protein deposition, FPD: protein deposition in feathers, Lys: digestible lysine requirement.
[2] Model-generated.
[3] Model of Rostagno *et al.* (2005).

Because the amino acid composition of feather-free body protein differs from that of feather protein, more accuracy can be achieved in estimating amino acid requirements if both components are considered separately (Emmans and Fisher, 1986). Accordingly, in the proposed model the requirements for maintenance and growth are partitioned into specific proportions according to the contents of feather-free body protein and feather protein.

The digestible lysine requirements estimated with both equations were generally higher for the Cobb strain after the first week than those for the Ross strain, reflecting the large differences between the two strains in their mature body and feather protein weights and the rate of feather growth. The greatest difference between the two models was in the estimated digestible lysine requirements of both strains after 27 d, (Figure 2) where the estimates using the proposed model were considerably lower than those using the equation of Rostagno *et al.* (2005). These differences may be justified based on the principles used for the construction of the models. The recommendations obtained using the equation of Rostagno *et al.* (2005), were estimated to meet a response of weight gain, in which the main components are protein and lipid. On other hand, the proposed model was developed from data of body protein deposition. Whereas the protein:fat ratio, the main components of the gain reduces systematically with increasing age (Emmans and Fisher, 1986; Gous and Burnham, 1992), and considering that the lysine requirements are limited to maintenance and growth of body protein (Emmans and Fisher, 1986; Emmans and Oldham, 1988; Gous and Burnham, 1992; Gous, 2007; Nonis and Gous, 2008), the findings were no surprising, showing the sensitivity of the proposed model to detect these differences.

The lower lysine requirements after 28 d of age suggested by the model proposed here has important implications. Under the concept of 'ideal protein' the concentrations of essential amino acids in feeds for broilers are based on fixed ratios between lysine and the other amino acids (Baker and Han, 1994, Baker *et al.*, 2002; Rostagno *et al.*, 2005), so where this practice is used to calculate the requirements for all amino acids other than lysine, any reduction in the lysine requirement would result in reductions of all other essential amino acids. Before such adjustments are made it would be prudent to determine whether the published ideal protein ratios are correct or relevant for a given strain. Practical and experimental observations have demonstrated that approximately 75% of feed intake of broilers occurs between 21 and 42 d and during this period the production of waste increases dramatically. In this sense, the reduction of lysine in the feed during this period, if the model estimates are correct, would bring both economic and environmental benefits to poultry production.

The theory applied in this paper to predict the lysine requirements for broilers is based on the potential growth of feather-free body protein and feather protein. These requirements change according to the potential growth rate of the broiler, defined by the Gompertz growth curve, which is under both genetic and environmental control. The approach used here can predict the requirements only where the bird is growing at its potential, although this potential can be altered

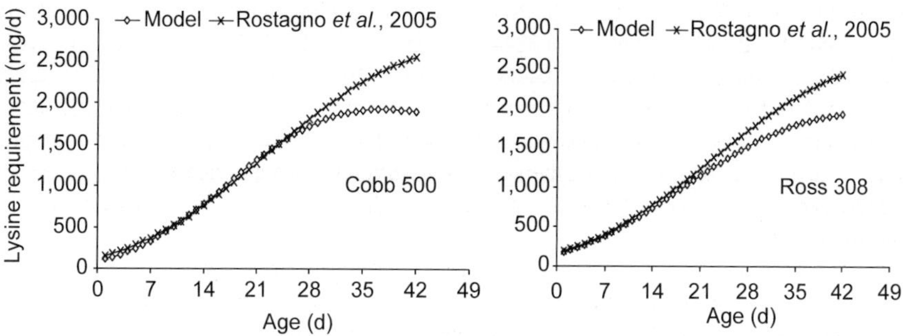

Figure 2. A comparison of the digestible lysine requirements of two broiler strains estimated using the model developed in this paper and that of Rostagno et al. *(2005).*

to simulate environmental insults such as disease, vaccination stress, high temperature stress, etc. However, the extent to which the potential is reduced by these factors is difficult to predict and requires a more sophisticated approach than that used here. Also, this approach does not predict the concentration of lysine required in the feed to ensure that the bird consumes the desired amount each day. This requires the prediction of the amount of food the bird will consume each day, which is dependent on many interacting factors including genotype, feed composition and environmental conditions. Such a simulation model is available (EFG Software, 2009), which makes use of the theory elaborated here to determine the daily amino acid requirements of growing broilers, but goes further by predicting the actual feed intake of the bird given the specific circumstances under which it is housed and fed, and from this the dietary amino acid contents during each phase of growth are predicted. Because food intake is predicted by such models the optimization of feeds and feeding programs is possible.

In conclusion, the requirement model elaborated here is the key step in developing a model that accurately predicts the amino acid requirements of broilers during growth, that predicts food intake and hence the amino acid contents of feeds used during the growing period, and ultimately, that optimizes the feeding program for growing broilers.

Acknowledgements

The authors thank Fundação de Amparo a Pesquisa do Estado de São Paulo/FAPESP for financial support to this project.

References

Baker, D.H. and Han, Y., 1994. Ideal amino acid profile for chickens during the first three weeks posthatching. Poultry Science 73:1441-1447.

Baker, D.H., Batal, A.B., Parr, T.M., Augspurger, N.R. and Parsons, C.M., 2002. Ideal ration (relative to lysine) of tryptophan, threonine, isoleucine and valine for chicks during the second and third weeks posthatch. Poultry Science 81:485-494.

Burnham, D. and Gous, R.M., 1992. Isoleucine requirements of the chicken: requirement for maintenance. British Poultry Science 33:59-69.

Edwards, H.M. Fernandez, S.R. and Baker, D.H., 1999. Maintenance lysine requirement and efficiency of using lysine for accretion of whole-body lysine and protein in young chicks. Poultry Science 78:1412-1417.

EFG Software, 2009. Available at: www.efgsoftware.net. Accessed October 2009.

Emmans, G.C., 1981. A model of the growth and feed intake of ad libitum fed animals, particularly poultry. In: Hillyer, G.M.; Whittemore, C.T. and Gunn, R.G (eds.), Computers in Animal Production, Occasional Publication. n.5, British Society of Animal Production, UK, pp.106-110.

Emmans, G.C., 1989. The growth of turkey. In: Nixey, C. and Grey, T.C. (eds.), Recent Advances in Turkey Science. Butterwordths, UK, pp.135-166.

Emmans, G.C. and Fisher, C., 1986. Problems in nutritional theory. In: Fisher, C. and Boorman, K.N. (eds.), Nutrient requirements of poultry and nutritional research. Butterwordths, UK, pp.9-39.

Emmans, G.C. and Oldham, J.D., 1988. Modelling of growth and nutrition in different species. In: Korver, S. and Van Arendonk, J.A.M. (eds.), Modelling of livestock production systems. Kluwer Academic Publishers, the Netherlands, pp.13-21.

Fisher, C., 1998. Amino acid requirements of broiler breeders. Poultry Science 77:124-133.

Gompertz, B., 1825. On the nature of the function expressive of the law of human mortality and on a new method of determining the value of life contingencies. Philosophical Transactions of the Royal Society of London 115:513-585.

Gous, R.M., 2007. Methodologies for modeling energy and amino acid responses in poultry. Brazilian Journal of Animal Science 36:263-274.

Marcato, S.M., 2007. Características do crescimento corporal, dos órgãos e tecidos de duas linhagens comerciais de frangos de corte. 207 f. Tese (Doutorado) – Faculdade de Ciências Agrárias e Veterinárias, Universidade Estadual Paulista, Jaboticabal, Brazil.

Martin, P.A., Bradford, G.D. and Gous, R.M., 1994. A formal method of determining the dietary amino acid requirements of laying type pullets during their growing periods. British Poultry Science 35:709-724.

Nonis, M.K. and Gous, R.M., 2008. Threonine and lysine requirements for maintenance in chickens. South African Journal of Animal Science 38:75-82.

Oviedo-Rondón, E.O., Murakami, A.E. and Sakaguti, E.S., 2002. Modelagem computacional para produção e pesquisa em avicultura. Revista Brasileira de Ciência Avícola 4:199-207.

Oviedo-Rondón, E.O. and Waldroup, P.W., 2002. Models to estimate amino acid requirements for broiler chickens: A Review. International Journal of Poultry Science 5:106-113.

Owens, F.N. and Pettigrew, J.E., 1989. Subdividing amino acid requirements into portions for maintenance and growth. In: Friedman, M. (ed.), Absorption and utilization of amino acids. Boca Raton: CRC Press, USA, pp.15-30.

Rostagno, H.S., Albino, L.F.T., Donzele, J.L., Gomes, P.C., Oliveira, R.F.M., Lopes, D.C., Ferreira, A.S. and Barreto, S.L.T., 2005. Tabelas brasileiras para aves e suínos: composição de alimentos e exigências nutricionais. 2. ed. Viçosa, MG: Departamento de Zootecnia, Universidade Federal de Viçosa, Brazil, 186 pp.

Sakomura, N.K. and Rostagno, H.S., 2007. Métodos de pesquisa em nutrição de monogástricos. Jaboticabal, Brazil, Funep. pp.283.

Zoons, J., Buyse, J. and Decuypere, E., 1991. Mathematical models in broiler raising. World's Poultry Science Journal 47:243-255.

Prediction of nellore empty body composition using indirect measurements

M.I. Marcondes[1,3], *M.L. Chizzotti*[2], *S.C. Valadares Filho*[1] *and L.O. Tedeschi*[3]
[1]*Universidade Federal de Viçosa, Department of Animal Science, Viçosa MG 36571-000, Brazil; marcosinaciomarcondes@gmail.com*
[2]*Universidade Federal do Vale do São Francisco, Department of Animal Science, Petrolina, PE 56304-917, Brazil*
[3]*Department of Animal Science, Texas A&M University, College Station, TX 77843-2471, USA*

Abstract

The study proposed to develop equations to determine crude protein (CP), ether extract (EE) and water contents of the empty body weight (EBW) (CP_{EBW}, EE_{EBW} and W_{EBW}, respectively) of pure and crossbred Nellore cattle using measurements of the 9-11th rib section. Nine experiments conducted with a total of 272 animals were used. Data were coded by sex (120 bulls, 37 heifers, and 115 steers) and breed (196 Nellore purebreds and 76 Nellore crossbreds, including 20 F_1 Nellore × Simmental, 56 F_1 Nellore × Red Angus). In all experiments, the EBW composition was individually determined and several indirect measurements were taken, including EBW, carcass weight (CW), visceral weight (VW), visceral fat weigh (VFW), VW and VFW as a percentage of EBW (V and VF, respectively), carcass yield (CY), and EE, CP and water content of the 9-11th rib section (EE_{Rib}, CP_{Rib}, and W_{Rib} respectively). A stepwise procedure to estimate EE_{EBW}, CP_{EBW}, and W_{EBW} using multiple regressions with EE_{Rib}, CP_{Rib}, V, EBW, CY, VF, and CW was conducted. After defining the most significant variables, a random coefficients model analysis was conducted to evaluate the fixed effects of gender and breed, assuming the random effects of studies. The model analysis indicated that there was sex effect on EE_{Rib} (P=0.004) and VF (P=0.002). Therefore one equation for each gender was developed as follows: $EE_{EBW}=2.378\pm0.722 + 0.392\pm0.0511\times EE_{Rib} + 1.311\pm0.325\times VF$, for bulls; $EE_{EBW}=2.378\pm0.722 + 0.217\pm0.061\times EE_{Rib} + 2.397\pm0.345\times VF$, for steers; and $EE_{EBW}=2.378\pm0.722 +0.287\pm0.0703\times EE_{Rib}+ 2.001\pm0.396\times VF$, for heifers. The analysis of CP_{EBW} and W_{EBW} indicated no breed or sex effects; thus, a single equation was developed for them ($CP_{EBW}=14.849\pm0.890 + 0.289\pm0.046\times CP_{Rib} -0.400\pm0.071\times VF$; $W_{EBW} = 53.666\pm2.310 + 0.241\pm0.033\times W_{Rib} - 1.606\pm0.136\times VF$). Our results suggested that it is possible to estimate the empty body composition using measurements of 9-11th rib section and the VF in EBW.

Keywords: crude protein, ether extract, rib section, visceral fat

Introduction

Experiments to determine nutrient requirements using direct determination of empty body composition are expensive and laborious. Indirect methods, such as the composition of the 9-11th rib section, have been widely used to predict carcass composition, but accurate prediction of the empty body composition is lacking.

The object of this study was to develop equations using the composition of the 9-11th rib section to determine crude protein (CP) ether extract (EE), and water contents of the empty body weight (EBW) (CP_{EBW}, EE_{EBW}, and W_{EBW} respectively) of pure and crossbred Nellore cattle.

Material and methods

Model development

The data from 9 experiments (Marcondes *et al.*, 2009a; Chizzotti *et al.*, 2007; Paulino *et al.*, 2005; Paulino *et al.*, 2008; Marcondes *et al.*, 2009b; Paixão, 2008; Machado, 2009; Moraes *et al.*, 2009; Porto, 2009) conducted in Brazil, total of 272 animals, were used in this study. The data were coded by sex (120 bulls, 37 heifers and 115 steers) and breed (196 Nellore purebreds and 76 Nellore crossbreds, including 20 F_1 Nellore × Simmental, 56 F_1 Nellore × Red Angus).

In all experiments, the EBW composition was directly and individually determined and several indirect measurements were taken, including EBW, carcass weight (CW), visceral weight (VW), visceral fat weight (VFW), VW and VFW as a percentage of EBW (V and VF, respectively), carcass yield (CY), and EE, CP and water content of the 9-11th rib section (EE_{Rib}, CP_{Rib}, and W_{Rib}, respectively). The mesentery fat was physically and manually separated from guts and combined with the kidney, pelvic, and heart fat to compose total visceral fat.

A stepwise procedure to estimate EE_{EBW} CP_{EBW}, or W_{EBW} using multiple regression analysis with EE_{Rib}, CP_{Rib}, W_{Rib}, V, EBW, CY, VF, and CW as independent variables was conducted.

A procedure similar to that used by Seo *et al.* (2006) was used to develop the predictive equations. Briefly, after defining the most significant variables in each model, a random coefficients model analysis was conducted to evaluate the fixed effect of gender and breed assuming the random effect of studies. The variance-(co)variance matrix structures analyzed were the variance components (VC) and the unstructured (UN) forms. The Akaike Information Criteria (AIC) was used to select the model with the best fit. The analysis were conducted using PROC MIXED (SAS Inst. Inc., Cary, NC) and fixed effects were considered significant at *P*-values lower than 0.05, and random effects were significant at *P*-values lower than 0.20. When non significance was observed for a variable, it was removed from the model.

We used the equations developed by Hankins and Howe (1946) to predict carcass composition from 9-11th rib section composition. Subsequently, we used the predicted carcass composition to estimate empty body composition using the equations proposed by Garrett and Hinman (1969). We compared the predicted EE_{EBW} ($HH\text{-}GH_{EE}$), CP_{EBW} ($HH\text{-}GH_{CP}$), and W_{EBW} ($HH\text{-}GH_{W}$) with the observed values.

Valadares Filho *et al.* (2006) also developed equations to predict empty body composition from 9-11th rib section. However, they used *Bos indicus* animals to develop their models; thus, these equations were also used to predict EE_{EBW}, CP_{EBW} and W_{EBW} and compared with the observed values.

Model evaluation

Model evaluation was assessed using the concordance correlation coefficient (CCC), mean square error (MSE), mean bias, determination coefficient of the equation of the observed on predicted values, and simultaneous *F*-test for intercept=0 and slope=1 as discussed by Tedeschi (2006). The Model Evaluation System (MES; http://nutritionmodels.tamu.edu) was used to perform the calculations.

A bootstrap also was used to evaluate the robustness of the equations developed. This procedure consists in to build a sampling distribution by re-sampling the database. A cross-validation was conducted to evaluate the accuracy of the new models. The database was split in two subsets and the first one was use to fit the model and the second was used to test the previous equation. This procedure was performed 1000 times to reduce variability and the validation results (mean square error, MSE) are averaged over all the rounds. These procedures were conducted according to Davison and Hinkley (1997) using the statistical software R (R Development Core Team, 2009).

Results and discussion

The summary of database (n = 259) is shown in Table 1. One study was excluded of the database (Moraes, 2006; 13 Nellore bulls under grazing conditions) due to the lack of information on visceral fat. There was no breed effect in any of the statistical models analyzed. The database accounts for a broad range of EBW (145 to 506 kg), EE_{EBW} (4.15 to 30%), and CP_{EBW} (14.29 to 23.4%). The variation of the diets fed to these animals was high in which the concentrate level ranged from 0.20 to 2% of BW and that 27% of the animals were raised on pasture while the remaining 73% were finished on feedlot conditions. This variation is important to account for diverse feeding conditions.

Ether extract

There was no breed effect on the intercept (P=0.357), EE_{Rib} coefficient (P=0.854) and VF coefficient (P=0.338) for EE_{EBW}. In addition, there was no sex effect on the intercept (P=0.061), but there was sex effect on EE_{Rib} (P=0.004) and VF coefficients (P=0.002). Thus, one equation for each gender was developed using a common intercept as follows.

$$\text{Bulls: } EE_{EBW} = 2.378 \pm 0.722 + 0.392 \pm 0.0511 \times EE_{Rib} + 1.311 \pm 0.325 \times VF \quad (1)$$
$$\text{Steers: } EE_{EBW} = 2.378 \pm 0.722 + 0.217 \pm 0.061 \times EE_{Rib} + 2.397 \pm 0.345 \times VF \quad (2)$$
$$\text{Heifers: } EE_{EBW} = 2.378 \pm 0.722 + 0.287 \pm 0.0703 \times EE_{Rib} + 2.001 \pm 0.396 \times VF \quad (3)$$

where EE_{EBW} is the ether extract in empty body weight, EE_{Rib} is the percentage of ether extract in the 9-11th rib section and VF is the percentage of visceral fat (% of EBW).

The bootstrap analysis indicated biases of 0.381% for the intercept and 0.049, 0.034, and 0.052% for EE_{Rib} coefficient and 0.294, 0.179, and 0.291% for VF coefficient, for bulls, steers and heifers, respectively.

Hankins and Howe (1946) developed equations to estimate carcass composition based on the composition of the 9-11th rib section. Several modelling papers have used Hankins and Howe (1946) equation to predict carcass composition from the rib section composition and then the Garrett and Hinman (1969) equation to compute empty body composition from carcass composition (Tedeschi *et al.*, 2004). However, Hankins and Howe (1946) used soft tissue to develop the equation; therefore, because the fat and protein content of body bones were not included, the prediction of carcass composition may be overpredicted.

The comparison between the observed EE_{EBW} and the $HH\text{-}GH_{EE}$ yielded a CCC of 0.909 and a root of MSEP of 2.62% of EE_{EBW}. Even though the simultaneous test for the intercept equals to zero and slope equals to unity was significant (P<0.0001; Mayer *et al.*, 1994), which means they were different from zero and unity, respectively, simultaneously, the equations had a good precision (r^2 = 0.84) and over-predicted EE_{EBW} by 0.57%, suggesting the equations from Hankins and Howe (1946) and Garrett and Hinman (1969) were adequate to estimate EE_{EBW} for these animals under these feeding conditions.

Table 1. Mean, standard deviation (sd), maximum (Max), minimum (Min) and coefficient of variation (CV) of the variables used to develop the models for bulls, steers, heifers, and the data altogether[1].

Sex	EBW	CW	CD	VW	V	VFW	VF	EE_{Rib}	CP_{Rib}	EE_{EBW}	CP_{EBW}
Bulls											
Mean	312.36	193.99	61.98	44.81	14.32	10.15	3.11	15.44	17.56	12.40	18.50
sd	68.02	44.27	2.10	11.04	1.11	5.17	0.96	6.93	1.33	4.63	1.31
Max	449.80	285.70	66.69	73.63	17.64	27.70	6.46	43.57	21.84	24.09	22.62
Min	145.86	87.00	50.49	22.25	12.28	3.23	1.48	6.07	14.54	5.30	15.45
CV%	21.78	22.82	3.39	24.63	7.73	50.92	30.77	44.88	7.56	37.31	7.08
Steers											
Mean	346.38	218.44	63.03	52.78	15.02	16.44	4.45	21.84	17.13	17.47	18.14
sd	79.44	50.87	1.80	15.66	1.33	8.40	1.52	8.37	1.90	5.57	1.46
Max	506.08	322.45	66.43	87.13	17.69	33.54	7.17	50.85	23.38	27.36	23.38
Min	200.53	117.32	56.97	26.84	12.17	3.18	1.40	4.85	11.38	4.15	15.05
CV%	22.93	23.29	2.86	29.68	8.88	51.10	34.15	38.33	11.07	31.87	8.07
Heifers											
Mean	286.85	179.43	62.57	46.07	15.88	14.82	4.85	26.31	16.25	20.27	17.32
sd	63.77	40.17	2.61	13.57	1.64	7.93	1.73	10.30	2.13	6.29	1.57
Max	397.48	251.80	71.86	71.03	19.76	31.80	8.75	49.54	22.12	29.95	20.72
Min	175.47	99.70	56.82	25.82	12.65	3.92	2.10	9.89	12.25	6.57	14.29
CV%	22.23	22.39	4.17	29.45	10.34	53.52	35.63	39.15	13.10	31.04	9.07
Total											
Mean	323.82	202.77	62.53	48.53	14.85	13.61	3.95	19.83	17.18	15.77	18.17
sd	75.72	48.92	2.11	14.10	1.39	7.73	1.53	9.02	1.77	6.07	1.47
Max	506.08	322.45	71.86	87.13	19.76	33.54	8.75	50.85	23.38	29.95	23.38
Min	145.86	87.00	50.49	22.25	12.17	3.18	1.40	4.85	11.38	4.15	14.29
CV%	23.38	24.13	3.37	29.05	9.39	56.77	38.66	45.46	10.30	38.51	8.06

[1] Variables used: EBW = empty body weight (kg), CW = carcass weight (kg), CD = carcass dressing (% of EBW), VW = visceral weight (kg), V = percentage of visceral weight, VFW = visceral fat weight (kg), VF = percentage of visceral fat, EE_{Rib} = ether extract in 9-10-11th rib cut (%), CP_{Rib} = crude protein in 9-10-11th rib cut (%), EE_{EBW} = ether extract in EBW (%), CP_{EBW} = crude protein in EBW (%).

It is well documented the visceral fat is not readily mobilized when the animals is under feed restriction, because it is mainly used in splanchnic metabolism. On the other hand, the carcass fat is mostly related to body reserves and its rate of mobilization is higher than the visceral fat (Butterfield, 1966). Therefore, as the different fats discussed had different mobilizations rates, visceral fat is important to predict empty body fat because for underfed animal the EE_{EBW} will have underpredicted using HH-GH_{EE} approach or Equation 4. In the actual model, the adding of visceral fat represented an improvement of 8.27% in the precision of the model.

Crude protein

Similarly, the random coefficients model analysis analysis for CP_{EBW} indicated no breed effect on the intercept (P=0.358), CP_{Rib} (P=0.322), and VF (P=0.265) coefficients. There was no sex effect on the intercept (P=0.968), CP_{Rib} (P=0.875), and VF (P=0.291) coefficients. Thus, a single equation was developed (Equation 4).

$$CP_{EBW} = 14.849{\pm}0.890 + 0.289{\pm}0.046 \times CP_{Rib} - 0.400{\pm}0.071 \times VF \quad (4)$$

Equation 4 also showed the importance of visceral fat in predicting body protein composition. The crude protein in EBW is highly and negatively correlated with EE in the EBW (r=-0.714; $P<0.0001$). The crude protein content was less variable in the EBW (CV = 10.3%). The contribution of visceral fat to the model was greater (10.10%) than that for $\%EE_{EBW}$.

The bootstrap analysis showed a bias of -0.014 on the intercept and no biases on the CP_{Rib} and VF coefficients. Even though Equation 4 had a moderate precision ($r^2 = 0.61$), it had a reasonably good accuracy as shown by the MSE of 0.86% and RMSE of 0.93%, using cross-validation analysis.

Differences on the N content of meat protein might explain some of that variation. Benedict and Ellis (1987) found that some meat proteins might have more than 16%. These authors suggested that in meat product containing 50% of collagen would produce a factor of 5.583 in contrast to the mean value of 6.25 commonly used. According to these authors, the factor would be a function of the percentage of collagen, described by a linear equation ($6.25 - 0.085 \times A$, where A is the percent of collagen in the total protein). Paulino *et al.* (2005) used a factor lower than 5.88 while others (Marcondes *et al.*, 2009a,b; Machado, 2009; and Paixão, 2008) used the factor 6.25. This finding indicates that the variation in the factor to convert N to protein might increase the source of variation in the protein content in the EBW.

The analyses of observed and predicted values have shown an r^2 of 0.576. The estimated CCC was only 0.667 and the RMSEP was 1.00% of CP in EBW. The partition on RMESP indicated that 90.28% of the RMSEP is due random variation and only 5.18% due biases, Therefore, using Hankins and Howe (1946) along with Garrett and Hyman (1969) did not estimate well the CP in EBW for Nellore purebreds and crossbreds. Even though the breed effect was no significant in predicting CP_{EBW}, only differences between *Bos indicus* purebreds and crossbreed were evaluated, and further analysis should be done in order test differences between *Bos indicus* and *Bos taurus* cattle.

Water

The stepwise procedure determined that the content of water in the 9-11th rib section (W_{Rib}) and VF were the most important variables in predicting what W_{EBW}. Furthermore, the random coefficients model analysis showed no breed effect on the intercept (P=0.369), and on the W_{Rib} (P=0.405) and VF (P=0.179) coefficients. Similarly, there was no sex effect on the same parameters (*P-values* of 0.377, 0.429, and 0.211, respectively). Therefore one single equation was developed (Equation 5).

$$W_{EBW} = 53.666{\pm}2.310 + 0.241{\pm}0.033 \times W_{Rib} - 1.606{\pm}0.136 \times VF \quad (5)$$

As expected, VF was significant in predicting W_{EBW} because of the high correlation between water and ether extract. The bootstrap analysis showed biases of 0.045, -0.001, and -0.004 on the intercept, and W_{Rib} and VF coefficients. The cross-validation indicated a RMSE of 1.90% of W_{EBW}, indicating a good fitness of the model, with high precision and accuracy.

Using Hankins and Howe (1946) and Garrett and Hinman (1969) to estimate W_{EBW} yielded good predictions, with an CCC of 0.853 and RMSE of 2.75% of W_{EBW}. Valadares Filho *et al.* (2006) also developed an equation to predict W_{EBW} for *Bos indicus* cattle. Their equation had an CCC of 0.816 and RMSE of 2.60% of W_{EBW} when compared to our observed values.

Analysis of the fat-free matter

Protein to water ratio: Reid *et al.* (1955) discussed the importance of body water when determining the corporal composition of cattle. The development and testing phases of techniques that predict water or fat in a live animal can allow for predicting other components in the body.

Reid *et al.* (1955) proposed that the ratio between protein and ash (RPA) would be constant in a fat-free matter (dry matter basis), and would be equal to 4.1 (80.26/19.74) with no effect of breeds or sexes. Therefore, by computing W_{EBW}, the remaining components could be estimated. Reid *et al.* (1955) also suggested that age might influence the RPA. They presented equations to predict protein and ash content in the fat-free matter (dry matter basis). The same principle was used with our database. The stepwise procedure selected the logarithm of W_{EBW} and VF in predicting EE_{EBW} ($P<0.0001$). No breed effect was detected on the intercept ($P=0.924$), and log(EBW) ($P=0.902$) and VF ($P=0.400$) coefficients. Similarly, no sex effect was observed on the intercept ($P=0.746$), and log(EBW) ($P=0.791$) and VF ($P=0.074$) coefficients. Therefore, one single equation was developed as shown below.

$$EE_{EBW} = 236.21 \pm 8.853 - 126.25 \pm 4.792 \times \log(W_{EBW}) + 1.114 \pm 0.114 \times VF \quad (6)$$

The equation showed a pattern very similar to that suggested by Reid *et al.* (1955), however, the visceral fat might introduce a fine adjustment to possible feed restriction conditions. The r^2 of the model was 0.957 and the cross-validation indicated a MSE of 1.585% of EE on EBW. The bootstrap showed small bias on intercept (0.118) and log (EEBW) coefficient (-0.066), therefore the model fitness was accurate and precise.

The RPA was computed as 3.37 (77.12/22.88) with no breed and sex effects (*P*-values of 0.524 and 0.4603, respectively). The interval of confidence ($P=0.05$) indicated that this ratio was different from that reported by Reid *et al.* (1955). The database used by Reid *et al.* (1955) was mostly composed of *Bos taurus* cattle whereas ours is mostly comprised of *Bos indicus*. Therefore, it is possible the difference in the RPA is due to different *Bos* species.

We observed a correlation between protein and ash content in the fat-free matter and EBW (0.287, $P<0.001$). It is well known that bone development ceases before muscle development (Berg and Butterfield, 1968), consequently, increasing the proportion of protein when the animal is heavier in relation to ash content. This fact may explain the differences between RPA found in this work and that determined by Reid *et al.* (1955). The average EBW reported in this work was 324 kg.

Protein in the fat-free matter: The random coefficients model analysis of the percentage of crude protein in the fat-free matter of the EBW ($FFMCP_{EBW}$) showed no sex or breed effects on the intercept (*P-values* of 0.194 and 0.435, respectively) and on the EBW coefficient (*P-values* of 0.061 and 0.062, respectively). The ash content in the fat-free matter can be computed by difference ($100 - FFMCP_{EBW}$).

$$FFMCP_{EBW} = 74.089(\pm 0.668) + 0.00976(\pm 0.00193) \times EBW \quad (7)$$

The r^2 of the models was only 0.0089 (Figure 1) because the slope of the equation is nearly zero; however the proximity of the points to the mean indicated the model had a good precision, with an RMSE of 2.32% of CP in EBW.

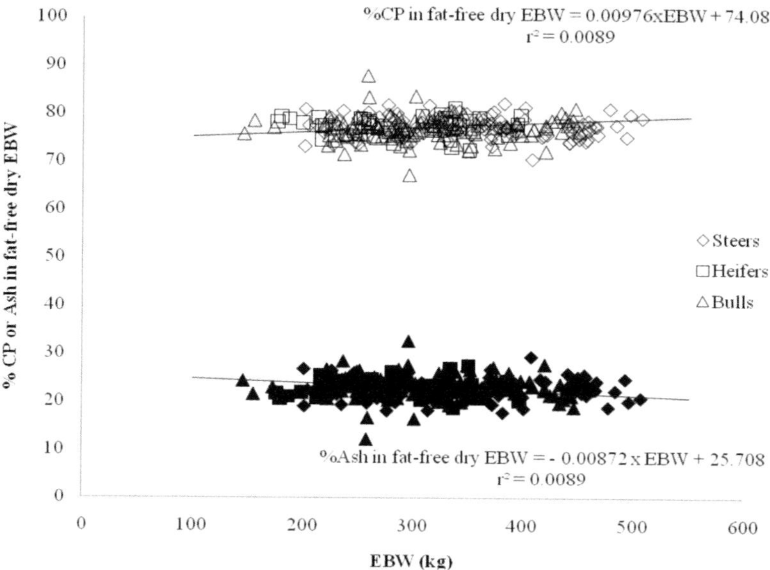

Figure 1. Relationship between protein and ash content in fat-free dry empty body weight (EBW), and the EBW. Solid dots = ash, empty dots = protein.

Using Equation 7 it is possible to predict that an animal would have RPA of 80.32:19.68, suggested by Reid *et al.* (1955), with 632 kg of EBW. This weight is commonly found in animals *Bos taurus* finished on feedlot in United States. Therefore, these facts prove the acceptance of both models.

In the review of literature equations performed by Valadares Filho *et al.* (2006) to predict body composition of *Bos indicus* cattle, no equation included a component which could indicate the nutritional status of the animals or the composition of non-carcass parts.

Alleoni *et al.* (2001) and Lanna (1988) suggested that protein could have a constant ratio with the water content in empty body weight. The ratio protein to water found in this work (0.299) was similar to the values reported by Alleoni *et al.* (2001) and Lanna (1988). However, the coefficient of determination between protein and water in our study was low ($r^2 = 0.103$), suggesting that predictions of water or protein compositions based on that ratio may be imprecise.

Conclusions

Our results suggested it is possible to estimate the empty body composition using indirect measurements of the chemical composition of the 9-11th rib section and the percentage of visceral fat in the EBW.

The ratio protein to ash is constant in fat free matter EBW, and body composition could be easily predicted only by knowing the water percentage in EBW.

References

Alleoni, G.F., Boin, C., and Leme, P.R., 2001. Determinação da composição do corpo vazio, do ganho de peso e das exigências de energia e proteína para mantença e ganho de peso em bovinos da raça Nellore, canchim e brangus. Nova Odessa: Instituto de Zootecnia de Nova Odessa. Relatório Científico Fapesp: Processo 97/02590-5.

Benedict, R. C. and Ellis, R. L., 1987. Determination of nitrogen and protein content of meat and maet products. Journal Association of Official Analytical Chemists 70:69-74.

Berg, R.T., and Butterfield, R.M., 1968. Growth patterns of bovine muscle, fat and bone. Journal Animal Science 27:611-679.

Butterfield, R.M. 1966. The effect of nutritional stress and recovery on the body composition of cattle. Research in Veterinary Science 7:168-179.

Chizzotti, M.L., Valadares Filho, S.C., Tedeschi, L.O., Chizzotti, F.H.M. and Carstens, G.E., 2007. Energy and protein requirements for growth and maintenance of f1 Nellore × red angus bulls, steers, and heifers. Journal of Animal Science 85:1971-1981.

Davidson, A.C. and Hinkley, D.V., 1997. Bootstrap methods and their applications. Cambridge University Press, Cambridge, UK.

Garrett, R.P. and Hinman, H., 1969. Re-evaluation of the relationship between carcass density and body composition of beef steers. Journal of Animal Science 28:1-5.

Hankins, O.G. and Howe, P.E., 1946. Estimation of the composition of beef carcasses and cuts. 926, USDA, Washington, DC, USA.

Lanna, D.P.D., 1988. Estimativa da composição química do corpo vazio de tourinhos nelore através da gravidade específica da carcaça e da composição química do corte das costelas. Dissertation, Escola superior de Agrocultura Luiz de Queiroz, Piracicaba, SP, Brazil.

Machado, P.A.S., 2009. Desempenho produtivo e exigências nutricionais de bovinos de corte em pastagem de *brachiaria decumbens*, suplementados no período de transição águas-seca. Thesis, Universidade Federal de Viçosa, Viçosa-MG, Brazil.

Marcondes, M.I., Valadares Filho, S.C., Paulino, P.V.R., Valadares, R.F.D., Paulino, M.F., Nascimento, F.B. and Fonseca, M.B., 2009a. Nutrient requirements of protein, energy and macrominerais of Nellore cattle of three genders. Revista Brasileira de Zootecnia 38:1587-1596.

Marcondes, M.I., Valadares Filho, S.C., Oliveira, I.M., Oliveira, D.M.S., Santos, T. R., Nascimento F. B. and Mariz, L.D.S., 2009b. Net requirements for pure and crossbred zebu cattle. In: Reunião anual da sociedade brasileira de zootecnia, June 14-17, 2009. Maringá, MG, Brazil. CD-ROM.

Mayer, D.G., Stuart, M.A. and Swain, A.J., 1994. Regression of real-world data on model output: An appropriate overall test of validity. Agricultural Systems 45:93-104.

Moraes, E.H.B.K., Paulino, M.F., Moraes, K.A.K., De Figueiredo, D.M., Valadares Filho, S.C., Paulino, P.V.R., Couto, V.R.M., 2009. Energy requirements of beef cattle at pasture. Revista Brasileira de Zootecnia 38:933-940.

Paixão, M.L., 2008. Desempenho produtivo e exigências nutricionais de bovinos de corte em pastagens de brachiaria decumbens, com suplementação protéica. Thesis, Universidade Federal de Viçosa, Viçosa-MG, Brazil.

Paulino, P.V.R., Costa, M.A.L., Valadares Filho, S.C., Paulino, M.F., Valadares, R.F.D., Magalhães, K.A., Detmann, E., Porto, M.O. and Moraes, K.A.K., 2005. Validation of the equations proposed by hankins and howe for estimating the carcass composition of zebu cattle and development of equation to predict the body composition. Revista Brasileira de Zootecnia 34:327-339.

Paulino, P.V.R., Valadares Filho, S.C., Detmann, E., Valadares, R.F.D., Fonseca, M.A., Véras, R.M.L. and Oliveira, D.M., 2008. Productive performance of Nellore cattle of different gender fed diets containing two levels of concentrate allowance. Revista Brasileira de Zootecnia 37:1079-1087.

Porto, M.O., 2009. Suplementos múltiplos para bovinos de corte nas fases de cria, recria e terminação em pastagens de *brachiaria decumbens*. Thesis, Universidade Federal de Viçosa, Viçosa-MG, Brazil.

R Development Core Team, 2009. R: A language and environment for statistical computing. R Foundation for Statistical Computing, Vienna, Austria. ISBN 3-900051-07-0, Available at: http://www.R-project.org. Accessed October 2009.

Reid, J.T., Wellington, G.H. and Dunn, H.O., 1955. Some relationships among the major chemical components of the bovine body and their application to nutritional investigations. Journal of Dairy Science 38:1344-1359.

Seo, S., Tedeschi, L.O., Schwab, C.G. and Fox, D.G., 2006. Development and evaluation of empirical equations to predict feed passage rate in cattle. Animal Feed Science and Technology 128:67-83.

Tedeschi, L.O., Fox, D.G. and Guiroy, P.J., 2004. A decision support system to improve individual cattle management. 1. A mechanistic, dynamic model for animal growth. Agricultural Systems 79:171-204.

Valadares Filho, S.C., Paulino, P.V.R. and Magalhães, K.A., 2006. Exigências nutricionais de zebuínos e tabelas de composição de alimentos – BR CORTE. 1 ed. Suprema Grafica Ltda, Viçosa, MG, Brazil, 142 pp.

[illegible]

Part 5 – Extrapolating from the animal to the herd

Precision feeding can significantly reduce feeding cost and nutrient excretion in growing animals

C. Pomar[1], L. Hauschild[1,2], G.H. Zhang[1,3], J. Pomar[4] and P.A. Lovatto[2]
[1]Agriculture et Agroalimentaire Canada, Sherbrooke, QC, J1M 1Z3, Canada; candido.pomar@agr.gc.ca
[2]Universidade Federal de Santa Maria, Department of Animal Science, Campus Camobi, Santa Maria, RS 97105-900, Brazil
[3]Northwest A&F University, Yangling, Shaanxi Province 712100, P. R. China
[4]Universidat de Lleida, Department of Agricultural enigeering, 25198 Lleida, Spain

Abstract

Precision feeding is an agricultural concept that relies on the existence of between-animal variation and involves the use of feeding techniques that allow the right amount of feed with the right composition to be provided at the right time to each pig of the herd. Precision feeding is proposed as an essential approach to improve nutrient utilization and thus reduce feeding cost and nutrient excretion. The potential impact of using precision feeding techniques on feeding cost and nitrogen (N) and phosphorus (P) excretion was evaluated in this study. The growing pig module of InraPorc® was slightly modified and used to estimate lysine requirements and simulate growth of individual pigs and the overall population. Detailed information of body composition of individual pigs, feed intake and growth performance was used to calibrate the model. Simulated pigs were fed either according to a typical three-phase feeding program or individually with a daily tailored feed as could be provided using precision feeding techniques. Feeding cost was determined according to common commercial pig feeds sold in Quebec, Canada. As expected, simulated N and P retention was not affected ($P>0.05$) by the feeding method. However, feeding pigs with daily tailored diets reduced ($P<0.001$) N and P intake by 25% and 29%, respectively and nutrient excretions were reduced both by more than 38%. Feed cost was 10.5% lower for pigs fed daily tailored diets. In fact, population protein and P requirements were established to optimize average daily gain of the population. Estimated nutrient requirements did not include safety margins as normally used in commercial feeds and therefore, the simulated N and P reductions are probably underestimated. Phosphorus excretion should be interpreted with caution because actual models simulating P retention seldom take into account the effect of P intake in P retention and bone mineralisation. Precision feeding can be an efficient approach to significantly reduce feeding cost and the excretion of N and P in animal production systems.

Keywords: mathematical model, pigs, precision farming

Introduction

In industrial and semi-industrial swine production areas, feeding cost represents more than 60% of the production cost. In growing-finishing pig production systems, feeding programs are proposed to maximize population responses at minimal feeding cost. However, nutrient requirements vary greatly between pigs of a given population (Brossard *et al.*, 2009; Pomar, 2007) and for each pig over time following individual patterns. To maximize the desired population response, which is usually body weight gain, population requirements are associated with those of the most demanding pigs, with the result that most of the pigs receive more nutrients than they need and the efficiency of nutrient utilization is reduced (Hauschild *et al.*, 2010; Jean dit Bailleul *et al.*, 2000). This is because, for most nutrients, underfed pigs will exhibit reduced growth performance and overfed ones will exhibit near optimal performance. Given that unutilized nutrients (other than Carbone) are excreted via the urine or faeces, feeding pigs to maximize population responses is associated with high feeding costs and high levels of nutrient excretion.

Reducing feeding cost, the excretion of excess nutrients such as nitrogen (N) and phosphorus (P), and the use of non-renewable resources is essential to the development of sustainable pig production systems (Honeyman, 1996; Jondreville and Dourmad, 2005; Rotz, 2004). The excretion of N and P is affected mainly by the amount of N and P ingested, the metabolic availability of those nutrients, and the balance between dietary nutrient supply and the animals' requirements (Jongbloed and Lenis, 1992). If feeding cost and nutrient excretion are to be minimized, it is essential that the composition of the available feed ingredients, their nutritional potential and the animals' requirements be properly characterized and that the supply of dietary nutrients be accurately adjusted to match the requirements of the animals. For the feeding of a population, however, optimal feed composition is difficult to estimate, as the response of the population to rising concentrations of nutrients is affected by many factors, including genetics, gender and the environment as well as the variability between the individuals of the population to be fed. Optimal nutrient concentrations should be estimated, given that feeds are provided to heterogeneous populations over long periods (Leclercq and Beaumont, 2000; Pomar *et al.*, 2003).

Precision feeding is an agricultural concept that relies on the existence of between-animal variation and involves the use of feeding techniques that allow the right amount of feed with the right composition to be provided at the right time to each pig in the herd. In commercial production systems, animals within a herd differ from each other in terms of age, weight and production potential and then each have different nutrient requirements. Therefore, precision feeding may be a powerful approach to reducing feeding cost and improving nutrient efficiency by reducing excesses of the most economically and environmentally detrimental nutrients without jeopardizing animal performance. The objective of this study was to evaluate the potential impact of using precision feeding techniques on feeding cost and N and P excretion in commercial production systems.

Material and methods

Animal data

Data from a population of growing-finishing female pigs from 25 to 105 kg of body weight were used in this study to characterize feed intake, body weight gain and nutrient requirement patterns of individual pigs of a population (Pomar, 2007). These pigs were used in a project comparing growth performance, body composition, and nitrogen and phosphorus excretion between a three-phase and a daily multiphase feeding program. Pigs were offered ad libitum complete diets obtained by combining two premixes for which composition was calculated to meet or exceed the animals' requirements throughout the experiment (Letourneau Montminy *et al.*, 2005). Feed consumption was measured throughout the experiment using an automated recording system (IVOG®-station, Insentec, Marknesse, Netherlands). The animals were weighed at least every two weeks. At the beginning and end of the experiment, total body fat and body fat-free lean tissues were estimated by dual-energy X-ray absorptiometry (DXA) using a densitometer (DPX-L, Lunar Corp., Madison, WI, USA). Total body protein and lipids were obtained by converting the muscle and fat values obtained with DXA into their chemical equivalents (Pomar and Rivest, 1996). Data from 68 animals with regular feed intake and growth patterns were used in the present study. The data set used includes measures of daily net energy intake, two-week interval body weight, and total body lipids and protein at the beginning and end of the experiment. The growth period lasted 83 days and was divided in three equivalent feeding periods.

Pig growth modeling

The growing pig module of InraPorc® (Van Milgen *et al.*, 2008) was slightly modified and used in the present study to estimate the lysine requirements and simulate growth and of individual pigs and the overall population. A detailed description of model modifications has been given previously (Hauschild *et al.*, 2010). Because this model estimate animal responses by simulating the utilization of nutrients based on concepts used in net energy and ideal protein systems, comparisons between simulation alternatives cannot be attributed to the model itself.

The model was slightly modified to optimize the utilization of the available animal data. Individual pigs were characterized by their voluntary feed intake and growth potential, which is defined respectively by the pig's appetite expressed as net energy (NE) intake and its potential for protein deposition (PD). The model is based on the transformation of dietary nutrients into body protein (PT) and lipids (LT) (state variables), which are then used to estimate body weight (BW). Euler's integration method is used to solve the differential equations with an integration step (dt) of 1 day. Rate variables are expressed on a daily basis, energy is in megajoules, mass in kilograms, and nutrient concentrations are expressed on a kilogram basis when not otherwise specified in the text. Because the animals' appetite is expressed as net energy intake, standardised ileal lysine (Lys) requirements estimated by either method are expressed in relation to this element (Lys:NE).

As in a previous study (Hauschild *et al.*, 2010), weight changes and daily net energy intake of each animal over time was simulated with a second-order polynomial function because individual growth does not always follow the typical Gompertz growth pattern observed in pig populations (Knap, 2000; Van Milgen *et al.*, 2008). Although the parameters in this polynomial model have little biological meaning, they allow proper representation of the observed individual BW and feed intake data over the 83 days during which animals were fed in the original trial. All results presented in this paper were obtained with the modified InraPorc model. Thus, body weight is estimated in the model by a quadratic function of time (t) fitted to data from the four weightings of each pig. The first derivative of this function is used to estimate daily weight gain ($\partial BW/\partial t$), while PD is estimated as $PD/\partial t = \partial BW/\partial t \times (ADPG/ADG)$ (Schinckel *et al.*, 2007), where $PD/\partial t$ is the simulated daily protein deposition and ADPG and ADG are the average daily PD and weight gain calculated from initial and final BW and protein masses measured on individual pigs. $PD/\partial t$ and $\partial BW/\partial t$ are therefore always equidistant. Energy intake was estimated based on the daily measures of NE intake in relation to BW using a quadratic function. Body weight was chosen to drive NE intake to maintain the same driving forces as in the InraPorc model.

Initial LT and PT masses for each simulated pig were estimated from DXA measurements, from which simulated initial BW was estimated using the relationships proposed in InraPorc. However, initial BW was not normally distributed and showed low variability between animals (s.d. = 1.2 kg; CV <5%). Thus, a new population was generated randomly having normally distributed initial BW, the same average BW and an s.d. of 3.8 kg. The generated initial population variability was chosen to represent the observed variability of commercial conditions (Patience *et al.*, 2004; N. Lafond, Aliments Breton Inc., Québec, personal communication). The body composition of the newly generated population was established by giving each new pig the identification and body protein and lipid proportions of the pig having the same weight rank in the original population (Hauschild *et al.*, 2010).

Population and individual nutrient allowances

In this simulation study the generated population of pigs were fed *ad libitum* either according to a typical three-phase feeding program or a daily tailored feed to each pig as could be provided using precision feeding techniques (Pomar *et al.*, 2009). In the three-phase feeding program all pigs of the population received a common feed which composition was previously estimated to maximise population growth. In the daily tailored feeding program, each pig received every day the feed that satisfied its nutrients requirements.

Lysine requirement for three-phase feeding program

The Lys:NE requirements were estimated by the modified growth model for each of the three 28-day feeding phases for which the initial BW averaged 24.5±4.1, 54.4±4.9 and 81.6±5.4 kg, respectively. The optimal population dietary Lys:NE concentration for each feeding period was estimated by simulating individual pig growth based on feeds containing 10 MJ NE/kg but graded levels of lysine. In this study, optimal population levels of lysine for each feeding phase were those maximising ADG. Maximal population ADG was reached at 1.09, 0.80 and 0.65 g/MJ of Lys:NE ratio in periods one to three, respectively (Hauschild *et al.*, 2010).

Lysine requirement for daily tailored feeding program

The modified pig growth model was inverted to estimate the individual daily lysine requirement according to the animal's current state and its potential for protein deposition (van Milgen *et al.*, 2008). The optimum daily Lys:NE ratio was obtained by dividing the daily lysine requirement by the estimated NE appetite for this day.

Comparing feeding programs

The feeding programs were compared based on average animal performances, feeding cost and N and P balance in the overall growth period (83 days). Animal performance was assessed on the basis of simulated individual feed intake, ADG, feed conversion (FCR) and body PT and LT composition. Feed costs in the three-phase feeding program was estimated as Feed cost ($/kg) = 0.3272 + 0.0791 × Lys:NE, matching the cost of the most common commercial pig feeds sold in Québec from March to June 2008 to its Lys:NE content (N. Lafond, Aliments Breton Inc., Québec, personal communication). Feed cost in the individually tailored feeding program was estimated assuming that the reduction on crude protein and phosphorus was obtained by replacing the soybean meal by corn and by reducing the amount of phosphates in the original diets. The cost of soybean meal, corn and phosphates was 532, 201, and 748 $/t. The results from feeding programs were compared using GLM procedure of the SAS (SAS Institute Inc., 1999, NC, USA).

Results and discussion

Simulated and observed values were similar for both average daily feed intake (ADFI; 2.44 vs. 2.43 kg, P=0.99) and ADG (79.40 vs. 78.30 kg, P=0.70). Model accuracy as estimated by the mean square prediction error was 0.01 kg/d for ADFI and 3 kg for BW. Deviations between observed and simulated performance values were small, which is consistent with the fact that model parameters were estimated for each pig in the population. However, the slope between the predicted and observed ADFI was 1.15, which is higher than 1 (P<0.001), indicating that the model underestimated ADFI during the first feeding phase and overestimated this variable in older pigs. In fact, observed feed intake seemed to approach a plateau at the end of the last feeding period, a trend that was probably not captured by the quadratic function used in this study (Hauschild *et al.*, 2010).

The effect of feeding a population of pigs using a traditional three-phase (3P) or individually daily tailored diets (PF) as would be provided by precision feeding devices on growth performance, body composition, and N and P excretion was studied. Population protein and P requirements were established in this study to maximise population average daily gain and did not include any safety margins as normally used in commercial conditions. In today feeding systems all pigs of the population are fed with a unique feed during relatively long periods of time which are established in relation to the number of feeding phases. In the present study each feeding phases lasted 28 days. Because pigs were fed in the traditional 3P feeding program to maximise population growth, most pigs received more nutrients that required. Determining the optimal composition of this complete feed is complex, and when population responses are optimized, most of the pigs in the population will receive more nutrients than they need, and a small portion of the population will be fed above requirements (Figure 1). In the present study, maximal population ADG was 12% higher than the requirement of the average pig and corresponded to the requirement of pigs in the 82% percentile of the population (Hauschild *et al.*, 2010). Feeding pigs under these requirements may contribute to reduce feeding cost but will also reduce population growth. In fact, the most performing pigs of the population are those that will be the most penalised with this strategy.

Most of the pigs fed in the 3P and all of the pigs in the PF program received every day the amount of nutrients or more required to satisfy their maintenance and growth needs. For most nutrients such as protein and P, overfed pigs will exhibit near optimal performance and therefore, average responses from populations fed according to both feeding programs obtained similar growth performances (Table 1). Furthermore, simulated protein and P retention was not affected by the feeding method (Table 2). However, feeding pigs with daily tailored diets reduced N and P intake respectively by 25% and 29%, and the corresponding excretions were reduced both by more than 38%. Feeding costs was estimated to be reduced by more than 8.7 $/pig which represents a feeding cost reduction of 10%. It should be noted, however, that the most performing pigs of the herd received more protein and P when fed the daily tailored diets than when fed in the three-phase feeding program. Protein and P requirements estimated in this study did not include any safety margin and therefore, the estimated N and P reductions are probably underestimated. Phosphorus estimations have however

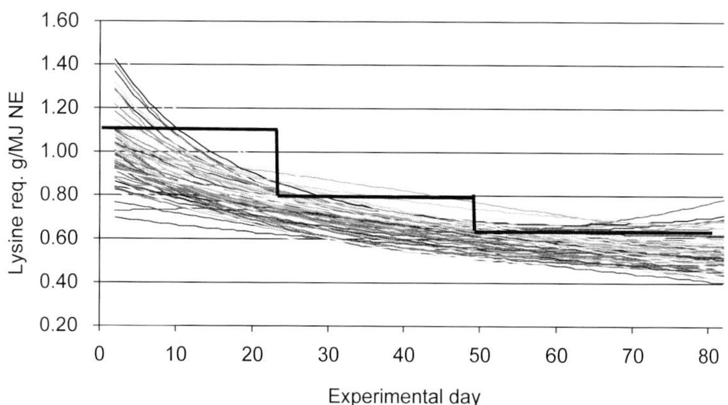

*Figure 1. Estimated individual (—) and population (**—**) optimal lysine concentration in diets fed to a heterogeneous population.*

Table 1. Initial and final animal body conditions and composition and overall growth performance of pigs fed according to a three-phase feeding program or individually daily tailored diets.

	Feeding method		SEM	*P*-value
	Traditional three-phase feeding programme	Individually daily tailored diets		
Initial conditions				
Body weight (kg)	27.2	27.2	0.47	0.989
Body protein (kg)	4.0	4.0	0.08	0.985
Body fat (kg)	4.0	4.0	0.06	0.985
Final conditions				
Body weight (kg)	107.9	107.9	0.83	0.997
Body protein (kg)	17.3	17.3	0.12	0.996
Body fat (kg)	22.2	22.2	0.50	1.000
Overall performance				
Average daily feed intake (kg/d)	2.49	2.49	0.02	0.995
Average daily gain (g/d)	973	972	7.31	0.987
Feed conversion (kg/kg)	2.56	2.56	0.02	0.997

Average values obtained from 68 pigs in an 83 days simulated experiment.

Table 2. Feeding costs and simulated nitrogen (N) and phosphorus (P) balance in pigs fed according to a commercial three-phase feeding programme or individually daily tailored diets.

	Feeding method		SEM	*P*-value
	Traditional three-phase feeding programme	Individually daily tailored diets		
Feed costs/ADG ($/kg)	1.022	0.974	0.01	0.001
N intake (kg)	5.69	4.29	0.05	0.001
N excretion (kg)	3.61	2.21	0.04	0.001
N retention (kg)	2.08	2.08	0.02	0.909
P intake (kg)	0.91	0.65	0.01	0.001
P excretion (kg)	0.49	0.30	0.01	0.001
P retention (kg)	0.35	0.35	0.00	0.909

Average values obtained from 68 pigs in a 83 days simulated experiment.

be interpreted with caution because actual models simulating P retention seldom takes into account the effect of P intake in bone mineralisation.

Phase-feeding involves feeding a number of successive diets, each differing in its protein, energy or amino acid balance to match the evolving nutritional requirements of the growing pigs. It should be expected that increasing the number of feeding phases in group-feeding facilities will reducc fceding costs, decrease nutrient excretion and improve feed efficiency. The economic and environmental benefits of this concomitant nutrient adjustment increase with the number of feeding phases (Bourdon *et al.*, 1995; Van der Peet-Schwering *et al.*, 1999). Thus, it has been demonstrated that daily multiphase feeding can reduce N excretion by 12% in comparison to a traditional 3P feeding program (Pomar *et al.*, 2007). The application of daily multiphase feeding techniques in group-feeding facilities can hardly decrease further this 12% N excretion because the between-animal variation seems to be more important than the over-time variation of the population requirements. The utilisation of precision feeding techniques able to provide to each pig daily tailored diets is therefore a promising alternative to capture most of the observed between-animal variation on nutrient requirements and thus significantly improve nutrient efficiency.

The expected reduction in feeding costs and N and P excretion which can be obtained when feeding pigs with daily tailored diets will be affected by the cost of the available ingredients, the composition of the premixes and the animal variation. Feeds formulated with large safety margins and large amounts of excess nutrients for optimal response in heterogeneous populations (e.g. genetic potential, sex and weight) increase the excretion of nutrients. In these situations, it should be expected that the impact of feeding pigs individually with daily tailored diets will reduce feeding costs and improve dietary nutrient efficiency more than observed in the present study.

This study has demonstrated that in growing-finishing pig facilities precision feeding may be an effective approach for reducing feeding costs and improving nutrient efficiency by reducing excesses of the most economically and environmentally detrimental nutrients without jeopardizing animal performance. However, the proper implementation of precision feeding in livestock production systems ought to include (1) the proper evaluation of the nutritional potential of feed ingredients, (2) the precise determination of nutrient requirements, (3) the formulation of premixes that limit the amount of excess nutrients, and (4) the concomitant adjustment of the dietary supply and concentration of nutrients to match the evaluated requirements of each pig in the herd.

References

Bourdon, D., Dourmad, J.Y. and Henry, Y., 1995. Réduction des rejets azotés chez les porcs en croissance par la mise en oeuvre de l'alimentation multiphase, associée a l'abaissement du taux azoté. Journées de la Recherche Porcine 27:269-278.

Brossard, L., Dourmad, J.-Y., Rivest, J. and Van Milgen, J., 2009. Modelling the variation in performance of a population of growing pig as affected by lysine supply and feeding strategy. Animal 3:1114-1123.

Hauschild, L., Pomar, C. and Lovatto, P.A., 2010. Systematic comparison of the empirical and factorial methods used to estimate nutrient requirements of growing pigs. Animal 4: 714-723.

Honeyman, M.S., 1996. Sustainability issues of U.S. swine production. Journal of Animal Science 74:1410-1417.

Jean dit Bailleul, P., Bernier, J.F., van Milgen, J., Sauvant, D. and Pomar, C., 2000. The utilization of prediction models to optimize farm animal production systems: the case of a growing pig model. In: McNamara, J.P., France, J. and Beever, D. (eds.) Modelling Nutrient Utilization in Farm Animals. CAB International 2000, Wallingford, UK., pp.379-392.

Jondreville, C. and Dourmad, J.Y., 2005. Le phosphore dans la nutrition des porcs. Productions Animales 18:183-192.

Jongbloed, A.W. and Lenis, N.P., 1992. Alteration of nutrition as a means to reduce environmental pollution by pigs. Livestock Production Science 31:75-94.

Knap, P.W., 2000. Time trends of Gompertz growth parameters in 'meat-type' pigs. Animal Science 70:39-49.

Leclercq, B. and Beaumont, C., 2000. Etude par simulation de la réponse des troupeaux de volailles aux apports d'acides aminés et de protéines. Productions Animales 13:47-59.

Letourneau Montminy, M.-P., Boucher, C., Pomar, C., Dubeau, F. and Dussault, J.-P., 2005. Impact de la méthode de formulation et du nombre de phases d'alimentation sur le coût d'alimentation et les rejets d'azote et de phosphore chez le porc charcutier. Journées de la Recherche Porcine 37:25-32.

Patience, J.F., Engele, K., Beaulieu, A.D., Gonyou, H.W. and Zijlstra, R.T., 2004. Variation: costs and consequences. In: Ball, R.O. (ed.), Adv. Pork Prod. University of Alberta, Edmonton, Canada, pp.257-266.

Pomar, C., 2007. Predicting responses and nutrient requirements in growing animal populations: the case of the growing-finishing pig. In: Hanigan, M.D., Novotny, J.A. and Marstaller, C.L. (eds.), Mathematical modeling in nutrition and agriculture. Virginia Polytechnic and State University, Blacksburg, VA., USA, pp.309-330.

Pomar, C., Hauschild, L., Zhang, G.H., Pomar, J. and Lovatto, P.A., 2009. Applying precision feeding techniques in growing-finishing pig operations Revista Brasileira de Zootecnia 38:226-237 (Supl. special).

Pomar, C., Kyriazakis, I., Emmans, G.C. and Knap, P.W., 2003. Modeling stochasticity: Dealing with populations rather than individual pigs. Journal of Animal Science (E. Suppl. 2):E178-E186.

Pomar, C., Pomar, J., Babot, D. and Dubeau, F., 2007. Effet d'une alimentation multiphase quotidienne sur les performances zootechniques, la composition corporelle et les rejets d'azote et de phosphore du porc charcutier. Journées de la Recherche Porcine 39:23-30.

Pomar, C. and Rivest, J., 1996. The effect of body position and data analysis on the estimation of body composition of pigs by dual energy x-ray absorptiometry (DEXA). Proceedings of the 46th annual conference of the Canadian Society of Animal Science (Abstr.), Lethbridge, Alberta, Canada, p. 26.

Rotz, C.A., 2004. Management to reduce nitrogen losses in animal production. Journal of Animal Science 82:E119-137.

Statistical Analysis Systems Institute, 1999. SAS/STAT Software, version 8. SAS Institute Incorporated., Cary, NC, USA.

Schinckel, A.P., Einstein, M.E., Foster, K. and Craig, B.A., 2007. Evaluation of the impact of errors in the measurement of backfat depth on the prediction of fat-free lean mass. Journal of Animal Science 85:2031-2042.

Van der Peet-Schwering, C.M.C., Jongbloed, A.W. and Aarnink, A.J.A., 1999. Nitrogen and phosphorus consumption, utilisation and losses in pig production: the Netherlands. Livestock Production Science 58:213-224.

Van Milgen, J., Valancogne, A., Dubois, S., Dourmad, J.-Y., Sève, B. and Noblet, J., 2008. InraPorc: A model and decision support tool for the nutrition of growing pigs. Animal Feed Science and Technology 143:387-405.

A herd modelling approach to determine the economically and environmentally most interesting dietary amino acid level during the fattening period

L. Brossard[1,2], N. Quiniou[3], J.-Y. Dourmad[1,2] and J. van Milgen[1,2]
[1]INRA, UMR1079 SENAH, 35590 Saint-Gilles, France; ludovic.brossard@rennes.inra.fr
[2]Agrocampus Ouest, UMR1079 SENAH, 35590 Saint-Gilles, France
[3]IFIP – Institut du Porc, B.P. 35104, 35651 Le Rheu cedex, France

Abstract

For a sustainable pork production, emission of pollutants from pig farms has to be reduced as much as possible whilst economical results have to be maximised. To evaluate different feeding strategies in terms of economic performance and environmental impact (through nitrogen excretion), we simulated herd performance using the prediction model InraPorc. A population of 1000 virtual pigs was generated. Performance of these pigs was simulated using different feeding strategies varying in digestible lysine to NE ratio in the diet (85 to 115% of the mean population requirement) and number of diets used (i.e. a 1-phase feeding strategy, a 2-phase strategy or a continuous multiphase strategy where the lysine/NE ratio was changed daily according to requirement). Diets were formulated on a least-cost basis using two contexts of ingredient prices: September 2007 (P1) or March 2009 (P2). Simulations were ended at a mean population body weight of 112 kg. Average daily gain increased while feed-to-gain ratio and duration of growth decreased with increasing lysine content. Maximum performance was achieved for a supply between 105 and 115% of the mean population requirement for lysine/NE. Increasing the lysine/NE ratio from 85 to 100% for the 1-phase strategy and to 105% for multiphase strategies reduced total feed cost in both economic contexts. The financial return (carcass payment minus feed and labour costs) increased with increasing lysine/NE ratio and was maximal with the multiphase strategy and a lysine/NE ratio corresponding to 110% of the mean population requirement. When using a lysine/NE ratio greater than the mean population requirement, multiphase feeding strategies reduced nitrogen excretion by 3 to 16% compared with 1-phase strategy. The relationship between financial return and nitrogen excretion depended on the feeding strategy. Using multiphase strategies allows to better match the nutrient supply to the requirement. Based on feed cost alone, it allows optimizing economic return while reducing nitrogen excretion. Apart from the interest of including the variation between animals, stochastic simulation modelling can be helpful in multiple criteria evaluation of feeding practices.

Keywords: pig, group, model, feeding strategy, multi-criteria evaluation

Introduction

Over the last decade, animal production had to face the challenge of sustainability in terms of social acceptance, economic viability and environmental impact. Concerning this last aspect, animal production, and especially pig production, has an impact on the environment through (amongst others) the excretion of nitrogenous compounds, phosphorus and trace elements, and the emission of methane. This impact can be modulated through nutrition by adapting the nutrient supply (e.g. amino acids such lysine (Lys), net energy (NE)) to the requirement by accounting for dynamic changes in requirement or for variability of requirements between individuals. However, this environmental concern has also financial implications through the price of feed and the impact of the chosen nutritional strategy on performance (both in terms of level of performance and variability in performance). Evaluating simultaneously animal performance, financial return and the environmental impact of a production system is difficult in field conditions because of the number of factors to include and the duration of study. Modelling can be an alternative method to evaluate these different criteria and this has been used to define the economically most-optimal feeding and management strategies for growing pigs (e.g. Boys *et al.*, 2007; Morel *et al.*, 2008; Quiniou *et al.*, 2007). It has been used also in order to study environmental impact (Morel and Wood, 2005; van Milgen *et al.*, 2009; Zhu *et al.*, 2009).

The objective of the present study was to use a herd modelling approach with InraPorc (2006) to evaluate the response of pigs to different nutritional strategies in terms of economic performance and environmental impact.

Material and methods

Pig herd generation

The first step of our study was to generate a population of pigs having a realistic variability in terms of feed intake and growth. Data from a real population of pigs (n = 104; Quiniou *et al.*, 2007) were used to characterise each animal in that population for the five growth and intake parameters used by InraPorc (2006) as described by Brossard *et al.* (2009). These parameters are required in InraPorc to define an animal profile, i.e. the phenotypic potential of a pig in terms of feed intake capacity and growth (Van Milgen *et al.*, 2008). The vector of means and the variance-covariance matrix of the parameters were calculated using the MEANS and the CORR procedures of the SAS (2000), respectively. Based on this matrix and this vector, a population of 1000 virtual pigs was generated through the generation of 1000 sets of five parameters using the IML procedure (VNORMAL subroutine generating multivariate normal random series) of the SAS (2000). The use of this procedure ensured that the generated population had the same means and variance-covariance matrix, and thus the same realistic variability for the five descriptive parameters as the pigs in the original population.

Definition of feeding strategies

The sets of individual parameters were included in InraPorc (2006) to define 1000 pig profiles. The model was then used to predict the daily requirement for the standardised ileal digestible Lys to NE ratio (Lys/NE ratio) for each pig (Van Milgen *et al.*, 2008). The mean daily Lys/NE requirement of the population was calculated from the individual requirements.

Different feeding strategies were then defined in InraPorc (2006). These strategies differed first by the number of diets used, i.e. a 1-phase strategy, a 2-phase strategy with a diet change at 112 days of age, or a continuous multiphase strategy where the Lys/NE ratio was changed daily according to requirements. The strategies differed also by the Lys/NE content in the diet and corresponded to 85, 90, 95, 100, 105, 110 or 115% of a reference level. For the diet of the 1-phase strategy (diet 1), the reference level for dietary Lys/NE content was defined as the mean population requirement at the beginning of the growth period, fixed for each pig at 65 days of age (Figure 1). For the 2-phase strategy, the first diet was identical to the diet 1 described above. The Lys/NE reference level of the second diet (diet 2) was equal to the mean population requirement at 112 days of age. The diets used for the continuous multiphase strategy were obtained by mixing a high Lys/NE content diet, equal to diet 1, and a low Lys/NE content diet (diet 3) with a Lys/NE reference level equal to the mean requirement of the population at the end of the growth period. The Lys/NE requirement (in g/MJ) of a pig typically declines with increasing age because voluntary feed intake (kg/day) increases more rapidly than the Lys requirement (g/day). Consequently, the reference Lys/NE levels in diets

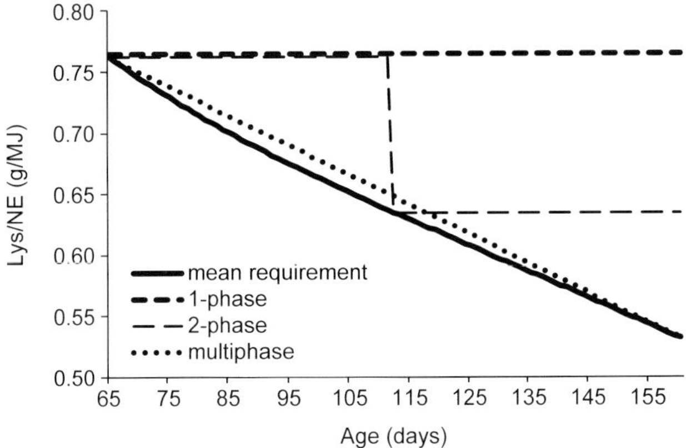

Figure 1. Change in mean lysine requirement as a function of age for the simulated population of pigs (n = 1000) and lysine content in the diet for the reference level for the three feeding strategies.

1 and 2 corresponded to the mean requirement at the beginning of the period of distribution, but exceeded the mean requirement for the remainder of the period. The Lys/NE reference levels are indicated in the Table 1.

Diets used in the different feeding strategies (level of Lys/NE content × number of diets) were formulated on a least-cost basis using a set of constraints (Table 1) and a set of feed ingredients (Table 2). All diets were formulated to contain 9.71 NE MJ/kg and amino acids other than lysine were supplied according to the ideal protein profile (Henry, 1993; Sève and Le Floc'h, 1998). Moreover,

Table 1. Feed formulation constraints.

Item	Value
Dry matter (%)	82.0-95.0
Net energy (MJ/kg)	fixed to 9.71
Crude protein (%)	13.0-16.5
Reference level of lysine (g/MJ NE)[1]	
Diet 1	0.762
Diet 2	0.635
Diet 3	0.531
Crude fibre (%)	2.0-5.0
Calcium (g/kg)	5.8-6.0
Total phosphorus (g/kg)	<4.4
Digestible phosphorus (g/kg)	>2.2
Fatty acid C18:2 (g/kg)	<14.9

[1] The reference level of lysine corresponds to 100% of the mean requirement of the population. Diet 1 was used in the 1-phase strategy and in the first phase of the 2-phase strategy. Diet 2 was used in the second phase of the 2-phase strategy. Diets 1 and 3 were mixed to adjust daily the dietary lysine content to requirement in the continuous multiphase strategy.

Table 2. Prices of feed ingredients used in feed formulation and incorporation rates.

Feed ingredients[1]	Price context P1 (€/t) September 2007[2]	Price context P2 (€/t) Dec. 2008-Feb. 2009[2]	Incorporation rates (g/kg)
Wheat	263	135	
Barley	265	118	
Maize	267	128	
Soybean meal	285	302	
Sunflower meal	216	130	<50
Peas	285	186	
Rape seed meal	244	184	<150
Soy oil	717	591	<15
Wheat middlings	223	113	<150
Wheat bran	204	87	
Cane molasses	137	168	<50
Rape seed	374	307	<50
L-Lysine HCl	1,550	1350	
L-Threonine	2,600	1,717	
DL-Methionine	2,400	4,550	
L-Tryptophan	28,500	19,333	
Phytase	1,100	9,500	0.1
Salt	90	90	3
Calcium carbonate	50	50	
Dicalcium phosphate	330	720	
Vitamin and mineral mixture	384	384	5

[1] Feed formulation was performed using the composition of feed ingredients indicated in INRA-AFZ tables (Sauvant *et al.*, 2004).
[2] Prices of feed ingredient are given by IFIP (http://www.ifip.asso.fr/newsletters/sommaire_lettres.htm).

two scenarios of ingredient prices were used in the least-cost feed formulation: September 2007 (P1, high prices), or December 2008 to March 2009 (P2, moderate prices) (Table 2).

Simulations and calculations

Daily feed intake, growth and N excretion were simulated individually using the InraPorc model from 65 days of age to a mean population BW of 112 kg BW and for each feeding strategy and dietary Lys/NE content, with diets offered *ad libitum*. Individual average daily gain (ADG), average daily feed intake (ADFI), feed-to-gain ratio (F:G), duration of simulation and total feed intake were calculated. The individual feed cost was calculated from the total feed intake and the feed price; an additional cost of 0.02 €/kg of feed was applied by default, corresponding to the manufacturing cost and margin; no additional cost was considered for the multiphase feeding system (Quiniou *et al.*, 2007). A labour cost per pig produced was calculated by adding the fixed cost of labour per pig (0.088 hours per pig) and the cost of daily tasks related to the time spent in the housing (0.001381 hours per day per pig); the hourly income was set at 16.54 € (Quiniou *et al.*, 2007). The individual carcass weight was calculated from the final BW and an individual dressing percentage estimated as follows: the dressing percentage is calculated from the mean dressing percentage of the original population (79.2%) corrected by the difference between the mean slaughter BW of the population (112 kg BW) and the individual BW at the simulation end (Equation 1; Aubry, 2004):

$$\text{Dressing percentage} = 79.2 + 0.015 \text{ (end simulation BW} - 112) \quad (1)$$

The individual carcass payment was calculated according to the French payment grid for lean meat content of the carcass (March 2009). The lean mean content was estimated through the InraPorc model (2006). The financial return for each pig was calculated as the difference between the carcass payment and feed and labour costs. As the economic evaluation did not include all charges and external costs considered as constant per pig, carcass payment, feed cost and financial return were expressed relative to that of the 100% Lys/NE level of the 1-phase strategy for each of the two price contexts.

Statistical analysis

Data were analysed using the MIXED procedure of the SAS (2000). Feeding strategy, level of dietary Lys/NE content and feed price context were included in the model as fixed effects as well as their interactions, and animal was considered as a random effect. The same analyse was performed within each feed price context. When effects were significant ($P<0.05$), least square means were separated using the PDIFF function with the Tukey-Kramer adjustment.

Results and discussion

The average price of diets for the 100% Lys/NE level was 253.2 and 143.0 €/t in P1 and P2, respectively (Table 3). The CP content of these diets averaged 14.1 and 13.3%, respectively. Feed price and CP content increased with increasing dietary Lys/NE content. The CP content was lower in P2 than in P1, which was partly due to a higher inclusion rate of crystalline amino acids (due to lower price in P2, except for DL-Methionine).

In both price contexts, ADG increased while ADFI, total feed intake, F:G and duration of growth decreased with increasing dietary Lys/NE content according to a curvilinear-plateau relationship ($P<0.001$; Table 4). Performance was slightly higher in P1 than in P2 ($P<0.001$). Maximum performance was observed for a dietary Lys/NE content that ranged from 105 to 115% of the reference level. It has been shown previously, that a Lys content (combined with a proportional increase of the other essential amino acids) 10 to 15% higher than the mean population requirement maximises performance (Brossard *et al.*, 2009; Morel *et al.*, 2008; Quiniou *et al.*, 2007). This can be explained by a greater part of the population for which the amino acid requirements for growth is covered, resulting in a reduction of F:G and growth duration to reach market weight.

For dietary Lys/NE contents higher than 100% of the reference level, performance was similar between feeding strategies. For lower Lys/NE contents, ADG decreased and ADFI, F:G, growth duration and total feed intake increased with an increasing number of diets used in the feeding strategy ($P<0.001$). The reduction in performance due to a decrease in Lys/NE content was the greatest with the multiphase strategies. In such condition, a dietary Lys/NE content of 85% of the reference level decreased ADG by 10% and increased F:G by 12%, compared with the 1-phase strategy. In both price contexts, reducing the dietary Lys/NE content from 100 to 85% of the reference level reduced ADG by 4% for the 1-phase strategy and by 11% for the multiphase strategy. The impact on F:G was similar for both strategies.

Table 3. Price and CP content of feeds formulated in two different price contexts while varying the lysine to net energy content in the diet.

Feed price context[1]		Diet[2]	Dietary Lys/NE content (% of reference level)						
			85	90	95	100	105	110	115
P1									
	Price (€/t)	Diet 1	256.0	257.0	258.0	259.1	260.2	261.8	263.9
		Diet 2	253.8	254.2	254.8	255.7	256.5	257.3	258.2
		Diet 3	252.9	253.0	253.3	253.7	254.0	254.4	255.0
	CP (%)	Diet 1	13.8	14.4	15.0	15.6	16.2	16.5	16.5
		Diet 2	13.0	13.0	13.1	13.6	14.1	14.6	15.1
		Diet 3	13.0	13.0	13.0	13.0	13.0	13.0	13.2
P2									
	Price (€/t)	Diet 1	141.4	143.7	146.7	149.9	153.5	157.2	160.9
		Diet 2	138.1	138.7	139.7	141.0	142.3	144.4	147.2
		Diet 3	137.3	137.5	137.8	138.0	138.5	138.8	140.0
	CP (%)	Diet 1	13.0	13.0	13.2	14.0	14.0	14.4	15.1
		Diet 2	13.0	13.0	13.0	13.0	13.0	13.0	13.3
		Diet 3	13.0	13.0	13.0	13.0	13.0	13.0	13.0

[1] P1: ingredients prices in September 2007; P2: ingredients prices from December 2008 to March 2009.
[2] Diet 1 was used in the 1-phase strategy and in the first phase of the 2-phase strategy. Diet 2 was used in the second phase of the 2-phase strategy. Diets 1 and 3 were mixed to adjust daily the dietary lysine content to requirement in the continuous multiphase strategy.

As for Lys/NE level, the effect of feeding strategy on performance can be explained by the adequacy of Lys/NE supply with the requirements of individual pigs. In the present study, the reference level of the dietary Lys/NE content corresponded to the highest mean population requirement within a phase, which typically corresponds to the requirement at the beginning of a phase. It can thus be assumed that the Lys/NE requirement is covered for approximately 50% of the pigs in the population at the beginning of a phase. During growth, as the Lys/NE requirement typically decreases, the proportion of pigs whose requirement was met will increase. With the same definition of the reference level for dietary Lys content, Brossard *et al.* (2009) showed that the proportion of pigs for which the requirement was covered was reduced when adjusting more frequently the dietary Lys supply to the mean requirement (by increasing the number of diets during the growth period). Subsequently, the mean performance of the population was reduced. This impact of the increased number of phases was amplified by the reduction of Lys/NE content below the reference level, as observed in the present study.

Increasing the dietary Lys/NE content from 85 to 100% of the reference level for the 1-phase strategy and to 105-110% for 2-phase and multiphase strategies was associated with a reduction in total feed cost in both price contexts ($P<0.001$; Table 5). Depending on the context, feed costs were lower for 2-phase or multiphase strategies than for 1-phase strategy for dietary Lys/NE content higher or equal to 95 or 105% of the reference level. The carcass payment increased in both price contexts with an increase in Lys/NE content ($P<0.001$; Table 5). Moreover, the carcass payment was higher for the 1-phase strategy than for multiphase strategies for Lys/NE contents lower than

Table 4. Effect of the feed price, feeding strategy and lysine to net energy content in the diet on growth and feed intake in a simulated population of pigs (n = 1000).

Feed price context[1]	Feeding strategy	Dietary Lys/NE content (% of reference level)							SE[2]	Stat.[3]
		85	90	95	100	105	110	115		
P1										
ADG (g/d)	1-phase	859	876	887	893	897	897	898	23.8	L***
	2-phase	825	853	872	886	892	896	897		S***
	Multiphase	777	812	843	866	881	890	895		L×S***
ADFI (kg/d)	1-phase	2.37	2.37	2.38	2.38	2.39	2.38	2.38	0.020	L***
	2-phase	2.38	2.38	2.38	2.38	2.38	2.39	2.38		S***
	Multiphase	2.39	2.39	2.39	2.39	2.39	2.38	2.39		L×S***
F:G (g/g)	1-phase	2.75	2.70	2.67	2.66	2.66	2.65	2.65	0.062	L***
	2-phase	2.88	2.79	2.72	2.69	2.67	2.66	2.65		S***
	Multiphase	3.09	2.94	2.83	2.75	2.70	2.67	2.66		L×S***
Growth duration (days)	1-phase	99	97	96	95	95	94	94	0.2	L***
	2-phase	103	100	97	96	95	95	94		S***
	Multiphase	110	105	101	98	96	95	95		L×S***
Total feed intake (kg)	1-phase	237	232	231	228	229	226	226	3.0	L***
	2-phase	247	241	233	231	229	229	226		S***
	Multiphase	266	253	244	236	231	229	229		L×S***
P2										
ADG (g/d)	1-phase	860	871	878	890	894	897	899	22.2	L***
	2-phase	826	848	864	883	889	894	897		S***
	Multiphase	777	812	841	865	879	889	896		L×S***

Table 4. Continued.

Feed price context[1]	Feeding strategy	Dietary Lys/NE content (% of reference level)							SE[2]	Stat.[3]
		85	90	95	100	105	110	115		
P2 (continued)										
ADFI (kg/d)	1-phase	2.37	2.36	2.37	2.38	2.38	2.38	2.38	0.018	L***
	2-phase	2.38	2.38	2.38	2.38	2.38	2.38	2.38		S***
	Multiphase	2.39	2.39	2.39	2.39	2.38	2.38	2.39		L×S***
F:G (g/g)	1-phase	2.75	2.71	2.70	2.66	2.66	2.64	2.64	0.059	L***
	2-phase	2.88	2.80	2.74	2.69	2.67	2.66	2.65		S***
	Multiphase	3.08	2.94	2.84	2.75	2.70	2.67	2.66		L×S***
Growth duration (days)	1-phase	99	97	97	95	95	94	94	0.2	L***
	2-phase	103	100	98	96	95	95	94		S***
	Multiphase	110	105	101	98	96	95	95		L×S***
Total feed intake (kg)	1-phase	237	232	233	228	228	226	226	2.8	L***
	2-phase	248	240	235	231	228	229	226		S***
	Multiphase	265	253	243	236	231	229	229		L×S***

[1] P1: ingredients prices of September 2007; P2: ingredients prices from December 2008 to March 2009.
[2] SE: residual standard error of the mean.
[3] Statistics with effects of level of dietary Lys/NE content (L), feeding strategy (S) and their interaction (L×S) within feed price context; ***$P<0.001$. The effect of feed price context was highly significant ($P<0.001$) for all criteria.

Table 5. Effect of the feed price, feeding strategy and lysine to net energy content in the diet on the variation of costs and return[1] in a simulated population of pigs (n = 1000).

Feed price context[2]	Feeding strategy	Dietary Lys/NE content (% of reference level)							SE[3]	Stat.[4]
		85	90	95	100	105	110	115		
P1										
Feed cost (Δ €/pig)	1-phase	1.5	0.6	0.4	0.0	0.4	-0.1	0.4	0.77	L***
	2-phase	4.2	2.5	0.7	0.3	-0.1	0.2	-0.3		S***
	Multiphase	9.0	5.8	3.3	1.5	0.3	-0.2	0.2		L×S***
Carcass payment (Δ €/pig)	1-phase	-2.2	-1.1	-0.2	0.0	0.5	-0.1	0.0	3.28	L***
	2-phase	-5.4	-3.1	-2.0	-0.7	-0.3	0.4	-0.2		S***
	Multiphase	-9.4	-6.6	-4.2	-2.5	-1.5	-0.7	0.1		L×S***
Financial return (Δ €/pig)	1-phase	-3.8	-1.8	-0.6	0.0	0.1	0.0	-0.4	2.92	L***
	2-phase	-9.8	-5.7	-2.7	-1.1	-0.2	0.2	0.2		S***
	Multiphase	-18.7	-12.7	-7.7	-4.0	-1.8	-0.5	0.0		L×S***
P2										
Feed cost (Δ €/pig)	1-phase	-0.6	-0.8	0.0	0.0	0.9	1.3	2.2	0.36	L***
	2-phase	0.7	-0.2	-0.5	-0.7	-0.7	0.1	0.3		S***
	Multiphase	3.4	1.8	0.5	-0.2	-0.6	-0.6	0.0		L×S***
Carcass payment (Δ €/pig)	1-phase	-1.6	-1.3	-0.4	0.0	0.4	0.1	0.5	3.10	L***
	2-phase	-4.9	-3.2	-2.0	-0.7	-0.5	0.3	0.1		S***
	Multiphase	-8.8	-6.2	-4.0	-2.0	-1.2	-0.5	0.6		L×S***
Financial return (Δ €/pig)	1-phase	-1.1	-0.5	-0.5	0.0	-0.5	-1.1	-1.6	2.93	L***
	2-phase	-5.7	-3.1	-1.5	0.1	0.2	0.2	-0.2		S***
	Multiphase	-12.6	-8.2	-4.7	-1.9	-0.6	0.1	0.6		L×S***

[1] Difference relative to the 100% Lys/NE level of the 1-phase strategy.

[2] P1: ingredients prices of September 2007; P2: ingredients prices of December 2008 to March 2009.

[3] SE: residual standard error of the mean.

[4] Statistics were performed from absolute with effects of level of dietary Lys/NE content (L), feeding strategy (S) and their interaction (LxS) within feed price context; only difference between the different levels and the reference are presented; ***$P<0.001$. The effect of feed price context was highly significant ($P<0.001$) for all criteria.

105% of the reference level. The changes in financial return were similar to those observed for carcass payment (Table 5).

Due to lower feed prices, financial return was higher in P2 than in P1 (data not shown). The improvement in financial return with the increase in Lys/NE content and the better results for the 2-phase and multiphase strategies with high Lys/NE content are due to the improvement of performance (reduction of F:G and total feed intake) and lower feed prices for multiphase strategies. In a simulation study, Morel *et al.* (2008) observed also that a Lys/NE content higher than the mean requirement of the population allowed to improve gross margin. The required Lys/NE content (in % of the reference level) that allows maximising performance and/or financial return depends on the variability of pigs in the population, as observed by Morel *et al.* (2008). The more the population is variable (e.g. in terms of growth potential, weight), the higher the Lys/NE content has to be to maximise performance. Moreover, the response may further depend on the context for feed prices.

Response patterns of N excretion to the dietary Lys/NE content differed depending on the feeding strategy and price context (Table 6). Due to the higher CP content in P1 (Table 3), N excretion was higher in P1 than in P2 (P<0.001). When using a 1-phase strategy, increasing Lys/NE content resulted in an increased N excretion within both feed price contexts. For the 2-phase strategy, N excretion decreased when the Lys/NE content increased from 85 to 95% of the reference level and increased for higher Lys/NE contents. For the multiphase strategy, the decrease in N excretion was observed for Lys/NE contents up to 110% of the reference level. The N excretion was higher for the 1-phase strategy than for multiphase strategy for Lys/NE contents higher than 95 or 100% of the reference level (P<0.001). When using a Lys/NE content greater than the reference level, multiphase feeding strategy reduced N excretion by 3 to 16% compared with the 1-phase strategy. The reduction in N excretion between P1 and P2 and between 1-phase and multiphase strategies is due both to a reduction in dietary CP content and to the increased efficacy for protein deposition, shorter duration of growth, and reduction in total feed intake. A reduction in the dietary CP content can be used to minimize N excretion without decreasing growth performance (Dourmad *et al.*, 1993). The reduction in CP content under P2 compared with P1 context was achieved mainly by increased incorporation rates of crystalline amino acids in the diets. The use of such amino acids

Table 6. Effect of the feed price context[1], feeding strategy and lysine to net energy content in the diet on nitrogen (N) excretion in a simulated population of pigs (n = 1000).

N excretion (kg/pig)	Feeding strategy	Dietary Lys/NE content (% of reference level)							SE[2]	Stat.[3]
		85	90	95	100	105	110	115		
Context P1	1-phase	3.14	3.27	3.44	3.61	3.84	3.88	3.88	0.068	L***
	2-phase	3.23	3.15	3.10	3.24	3.39	3.53	3.59		S***
	Multiphase	3.66	3.47	3.34	3.27	3.26	3.24	3.27		L×S***
Context P2	1-phase	2.85	2.75	2.83	3.01	3.03	3.12	3.35	0.050	L***
	2-phase	3.12	2.95	2.86	2.87	2.82	2.88	3.00		S***
	Multiphase	3.54	3.25	3.06	3.01	2.90	2.90	3.00		L×S***

[1] P1: ingredients prices of September 2007; P2: average ingredients prices from December 2008 to March 2009.

[2] SE: residual standard error of the mean.

[3] Statistics with effects of level of Lys/NE content (L), feeding strategy (S) and their interaction (L×S) within feed price context; ***P<0.001. The effect of feed price context was highly significant (P<0.001).

increases feed efficiency and reduces N excretion (Dourmad *et al.*, 1999). Depending on ingredient prices and availability of crystalline amino acids, a further reduction in CP content would allow a further reduction in N excretion, while growth and financial return should be maintained. Using simulation studies, it has been demonstrated that the optimization of dietary efficiency, depending on factors such as genotype and variability in the population, can improve profit while reducing N excretion (Morel and Wood, 2005; Morel *et al.*, 2008; Zhu *et al.*, 2009). In the present study, the improvement of financial return and the reduction in N excretion depended on feeding strategy and economic context, and were also probably related to the inherent variability of the population. Consequently, the determination of an optimal Lys/NE content that allows maximising economic performance while minimising the environmental impact is not a static process. This raises the issue of the multiple definitions of the 'optimal level' of a nutrient (here Lys). This value may differ whether the objective is to maximise performance or financial return, to minimize environmental impact or to optimise a combination of these. The sustainability of pig production probably calls for the latter and thus for a multi-criteria evaluation of 'requirement' and feeding strategies.

Conclusion

Our simulation study showed that feeding strategies optimizing financial return while reducing N excretion can be obtained with different combinations of number of phases and level of Lys/NE content in the diet. The choice between theses strategies depends on the desired balance between economical result and environmental impact. Moreover, the optimal Lys/NE content in the diet (relative to the mean requirement of the population) depends on the economic context and is related to the inherent variability in the population. Apart from the conceptual interest to model the performance on a herd basis (i.e. to include the variation among animals), stochastic simulation modelling can be helpful in multi-criteria evaluation of feeding strategies.

References

Aubry, A., 2004. Y a-t-il un intérêt économique à alourdir les carcasses? TechniPorc 27:33-36.

Boys, K.A., Li, N., Preckel, P.V., Schinckel, A.P. and Foster, K.A., 2007. Economic replacement of a heterogeneous herd. American Journal of Agricultural Economics 89:24-35.

Brossard, L., Dourmad, J.Y., Rivest, J. and Van Milgen, J., 2009. Modelling the variation in performance of a population of growing pig as affected by lysine supply and feeding strategy. Animal 3:1114-1123.

Dourmad, J.Y., Henry, Y., Bourdon, D., Quiniou, N. and Guillou, D., 1993. Effect of growth potential and dietary protein input on growth performance, carcass characteristics and nitrogen output in growing-finishing pigs. In: Versegen, M.W.A., den Hartog, L.A., van Kempen, G.J.M. and Metz, J.H.M. (eds.), Nitrogen flow in pig production and environmental consequences. Pudoc, Wageningen, the Netherlands, pp.206-211.

Dourmad, J.Y., Sève, B., Latimier, P., Boisen, S., Fernández, J., Van der Peet-Schwering, C. and Jongbloed, A.W., 1999. Nitrogen consumption, utilisation and losses in pig production in France, the Netherlands and Denmark. Livestock Production Science 58:261-264.

Henry, Y., 1993. Affinement du concept de la protéine idéale pour le porc en croissance. INRA Productions Animales 6:199-212.

InraPorc® 2006. Un outil pour évaluer des stratégies alimentaires chez le porc. Version 1.5.2.0. INRA-UMR SENAH. Available at: http://www.rennes.inra.fr/inraporc. Accessed January 2010.

Morel, P. and Wood, G., 2005. Optimisation of nutrient use to maximise profitability and minimise nitrogen excretion in pig meat production systems. Acta Horticulturae 674:269-275.

Morel, P., Wood, G. and Sirisatien, D., 2008. Effect of genotype, population size and genotype variation on optimal diet determination for growing pigs. Acta Horticulturae 802:287-292.

Quiniou, N., Brossard, L., Gaudre, D., Van Milgen, J. and Salaun, Y., 2007. Optimum économique du niveau en acides aminés dans les aliments pour porcs charcutiers: impact du contexte de prix des matières premières et de la conduite d'élevage. Techni-Porc 30:25-36.

Sauvant, D., Perez, J.M. and Tran, G. (eds), 2004. INRA AFZ. Tables de composition et de valeur nutritive des matières premières destinées aux animaux d'élevage: porcs, volailles, bovins, ovins, caprins, lapins, chevaux, poissons. Inra Editions, France, 301 pp.

Sève, B. and Le Floc'h, N., 1998. Valorisation mutuelle du L-tryptophane et de la L-thréonine supplémentaires dans l'aliment deuxième âge du porcelet. Rôle de la thréonine déshydrogénase hépatique, Journées de la Recherche Porcine en France 30:209-216.

Statistical Analysis Systems Institute, 2000. SAS/STAT users guide, version 8.01. SAS Institute, Cary, NC, USA.

Van Milgen, J., Brossard, L., Valancogne, A. and Dourmad, J., 2009. Using InraPorc to reduce nitrogen and phosphorus excretion. In: Garnsworthy, P.C. and Wiseman, J. (eds.), Recent Advances in Animal Nutrition. Nottingham University Press, Notthingham, UK, pp.179-194.

Van Milgen, J., Valancogne, A., Dubois, S., Dourmad, J.Y., Seve, B. and Noblet, J., 2008. InraPorc: A model and decision support tool for the nutrition of growing pigs. Animal Feed Science and Technology 143:387-405.

Zhu, C.L., Vander Voort, G., Squire, J., Rheaume, J. and De Lange, C.F.M., 2009. It pays to fine-tune feeding programs for individual growing-finishing pig units. Canadian Journal of Animal Science 89:175.

Evaluation of two feeding strategies with a herd model integrating individual variability

L. Puillet[1,2,3,4], O. Martin[3,4], M. Tichit[1,2] and D. Sauvant[3,4]
[1]*INRA, UMR 1048 Activités Produits Territoires, 16 rue C. Bernard, 75231 Paris cedex 5, France; laurence.puillet@agroparistech.fr*
[2]*AgroParisTech, UMR 1048 Activités Produits Territoires, 16 rue C. Bernard, 75231 Paris cedex 5, France*
[3]*INRA, UMR 791 Modélisation Systémique Appliquée aux Ruminants, 16 rue C. Bernard, 75231 Paris cedex 5, France*
[4]*AgroParisTech, UMR 791 Modélisation Systémique Appliquée aux Ruminants, 16 rue C. Bernard, 75231 Paris cedex 5, France*

Abstract

Individual variability is a key element to understand and predict herd response to management practices. To study how individual variability is generated within the herd, an individual-based herd model was developed. It was applied to intensive dairy goat systems. The model combines a decisional with a biological sub-model. The decisional sub-model represents technical decisions relative to reproduction, feeding and replacement management. The biological sub-model represents each individual within the herd and is based on a dairy goat model. This latter simulates body weight and milk production dynamics throughout each goat's productive life. Individual goat performance depends on its production potential, physiological regulations and responses to diet. As each goat model has its own individual performance scaling and its own operation driven by management decisions, individual variability is an output of simulation. In the present work, the herd model was used to simulate two feeding strategies involving two-step (2S) or five-step (5S) feeding sequences. The 5S feeding sequence is based on a regular adjustment of the quantity of concentrate feedstuff throughout lactation whereas the 2S feeding sequence is based on a single adjustment in late lactation. Results show that increasing the number of steps within the feeding sequence did not lead to a major difference in terms of milk production and feed cost. When focusing at individual level, results show that the 2S feeding sequence led to a greater proportion of goats whose body weight was lower than their potential body weight. For the 2S feeding sequence, production and efficiency relied on goats which tended to lose more weight than those fed with the 5S sequence. Hence, the same level of herd production and efficiency can be achieved with different biological underlying processes. This result stresses the importance of understanding the mechanisms behind overall performance to achieve a sound evaluation of management effects.

Keywords: dairy goat, productive life, management practices, body weight, milk production

Introduction

The evaluation and design of sustainable livestock farming systems requires the ability to understand and predict herd responses to management practices. Animal scientists have produced knowledge about the individual processes giving rise to biological responses. Such knowledge is used to predict the average animal response to management. However, scaling up this average response to the herd level requires taking into account the individual variability of biological responses (Pomar *et al.*, 2003, Villalba *et al.*, 2006). Such variability is generated by three main components. Firstly, it is related to the variability of production potential of individuals within the herd. Secondly, it is related to the regulations of the biological dynamics of individuals throughout their productive life.

Finally, it is related to the management decisions. These latter target different organizational levels (herd, group and individual) and drive each animal's productive life. For instancc, rcproductive management determines the physiological state sequence whereas replacement management determines the length of productive life. To study the processes giving rise to biological response variability, an individual-based model was developed. It integrates both the management and the biological basis of individual variability. This model was built for intensive dairy goat systems. The objectives of this paper are twofold: first we give an overview of the model; secondly we illustrate the model's ability to study individual variability generated by two simulated feeding strategies.

Materials and methods

Model description

The herd model is made up of a decision sub-model and of a biological sub-model (Puillet *et al.*, 2008). The herd operation throughout time results from the interaction between these two components. The decisional system accounts for the farmer's decision-making process, where the farmer's project is translated into key input parameters reflecting management strategy. These parameters define technical operations relative to reproduction, feeding and replacement. These operations are formalized by discrete events applying decision rules in the form 'if condition is true then action'. Some actions, for instance diet change, are applied on the biological system at a group level. Others, such as culling, are applied at the individual goat level, thus providing feedback information for the decision system.

The biological system is made up of a set of individual goat models. Each model simulates milk production and body weight throughout the goat's productive life (Puillet *et al.*, 2010). The goat model is based on two sub-systems: a regulating one and an operating one. The regulating sub-system accounts for homeorhetic regulations. These regulations define priorities among physiological functions regarding energy allocation (Martin and Sauvant, this issue). They are represented by a set of compartments whose dynamics represent theoretical meta-hormones. These dynamics are transmitted to the operating sub-system with a scaling effect of genetic parameters (mature body weight and milk production potential, expressed as milk production in kg at the peak of the third lactation). The regulating sub-system defines a reference pattern for energy flows associated with growth, maintenance, gestation, body reserves repletion and mobilization, milk energy export and energy intake. A central zero pool compartment accounts for the constant equilibrium between energy inputs and outputs. The actual energy intake from the diet is defined by the feeding management strategy. Compared with the intake defined by the production potential, the actual intake generates an energy differential. This differential is divided between body reserves and milk according to a set of partitioning coefficients. These coefficients are determined by the relative priorities of physiological functions and also by the proportion of body reserves. They enable the calculation of the actual body reserve dynamics and milk exportation in response to the diet. These responses are globally consistent with the INRA feeding system (Sauvant *et al.*, 2007). The whole set of energy flows are converted to material flows and thus body weight and milk production. Hence, the model simulates goat performance throughout its lifespan, from birth to herd exit. The dairy goat model is the elementary component of the herd model. It has its own production potential parameter, stochastically allocated from Gaussian distributions at birth. It also has its own operation within the herd depending on management decisions. With such a structure, the herd model generates individual variability which is an output of simulations.

Simulations

Production potential distribution (mean and standard deviation) and management options relative to reproduction, feeding and replacement are input parameters of the herd model. The model simulates outputs over 20 years at both the herd and goat productive life levels. Due to stochastic processes, each simulation was replicated 15 times. Two experiments were considered in this study by simulating two feeding strategies: two-step (2S) and five-step (5S) feeding sequences. A feeding sequence is defined by the quantity of concentrate distributed to the herd throughout lactation as shown by Figure 1. For the 5S feeding sequence, the quantity of concentrate was regularly adjusted (every two months) whereas for the 2S feeding sequence, the quantity of concentrate was only adjusted in late lactation (after 7 months). The two feeding strategies differ only in the number of feedstuff quantity adjustments throughout lactation. They lead to the same total quantity of concentrate distributed to the herd throughout 300 days of lactation (255 kg of dry matter/goat). This quantity was determined to satisfy the requirements of a goat producing 4 kg of milk at lactation peak. The two management options were tested for a herd of 300 goats mated during the natural breeding season. Production potential distribution and management options relative to reproduction and replacement were the same between the two feeding strategies.

Results and discussion

Herd level

Figure 2 illustrates the simulated dynamics of herd milk production over four years of simulation. The simulated patterns of milk production were globally the same for the two feeding options. However, the 5S sequence led to a slightly higher milk peak whereas the 2S sequence led to higher persistency of milk production. At the beginning of the herd lactation period, the 5S sequence made it possible to better express individual production potential compared with the 2S sequence. Conversely, at the end of the herd lactation period, the 2S sequence led to a higher herd production because individuals were overfed.

Mean herd results over the last 10 years of simulation are presented in Figure 3. The left graphic shows the mean milk production of the herd. There was a slight difference between feeding

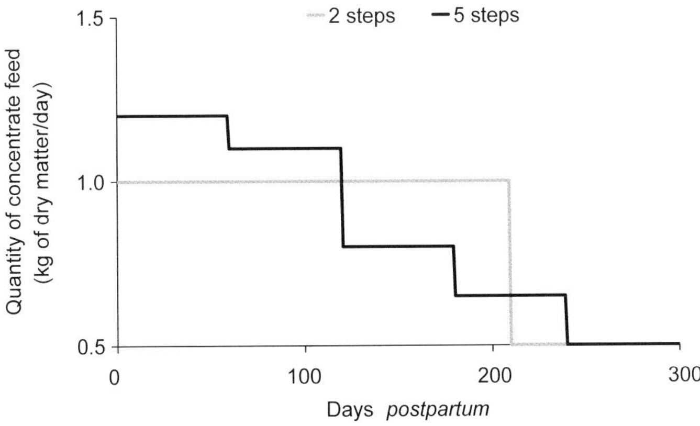

Figure 1. Description of the two feeding management strategies simulated: the two-step and the five-step feeding sequences. The number of steps within the sequence defines the adjustment of the quantity of concentrate (in kg) distributed each day.

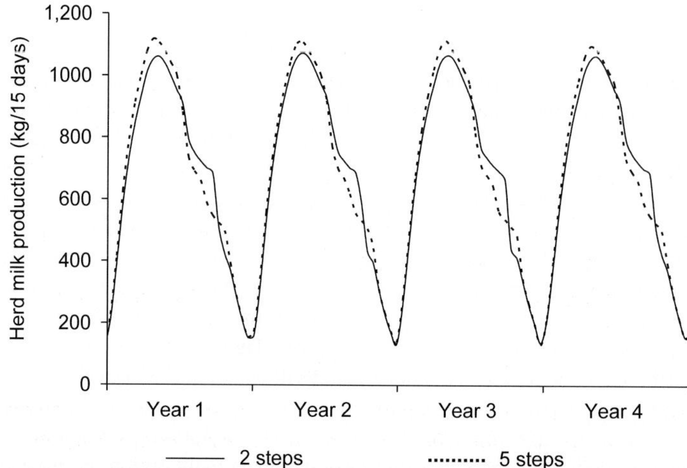

Figure 2. Dynamics of herd milk production (kg/15 days) for the two-step and the five-step feeding sequences over four years of simulation.

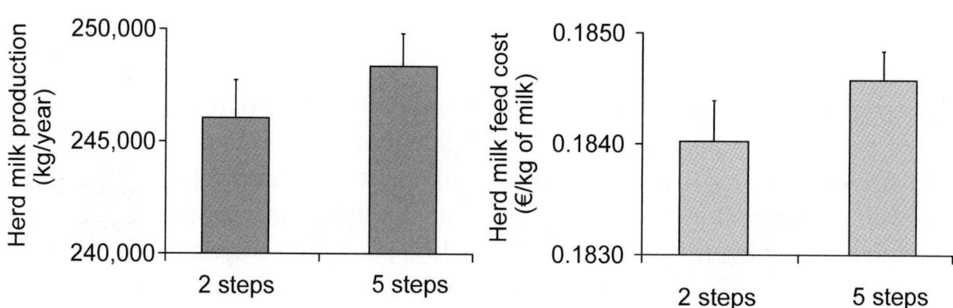

Figure 3. Mean results over 10 years of simulation for herd milk production (in kg) and for herd milk feed cost (in €/kg of milk).

management strategies with an increase of production level when the number of steps within the feeding sequences was higher ($P<0.001$). The right graphic shows the mean milk feed cost of the herd. Increasing the number of steps within the feeding sequence led to a slight increase in herd feed cost ($P<0.001$). Hence, the five-step feeding sequence led to a slight increase in production but with a higher marginal cost.

Individual level

Figure 4 illustrates the body weight and the milk production simulated by the dairy goat model parameterized with a production potential equal to 4 kg and fed with the 2S and the 5S sequences. Results for milk production were globally the same for the two feeding strategies. However, the body weight mobilization for the 2S sequence was greater than for the 5S.

To focus on the goats that generated herd performances, a level plot representation was used to analyze herd outputs at goat level. In Figure 5, each pixel corresponds to a pair of values of milk per day of lactation (kg/day of life) and milk feed cost (€/kg of milk). Both of these outputs were calculated at the scale of individual productive life. Each pixel was shaded depending on the

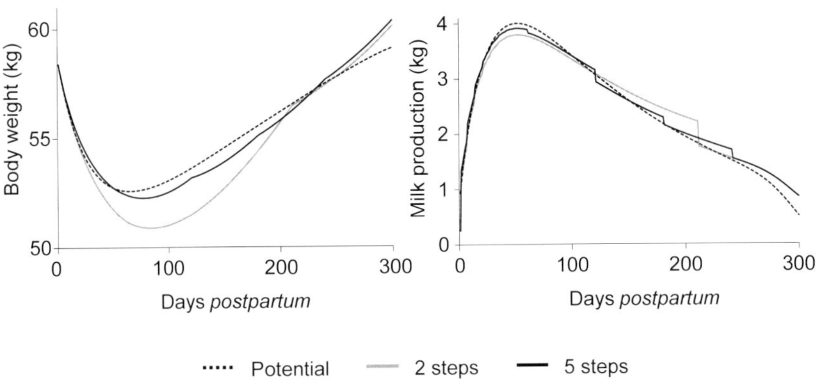

Figure 4. Body weight (kg) and milk production (kg) simulated for a goat with a production potential of 4 kg and fed throughout lactation with the two feeding management strategies.

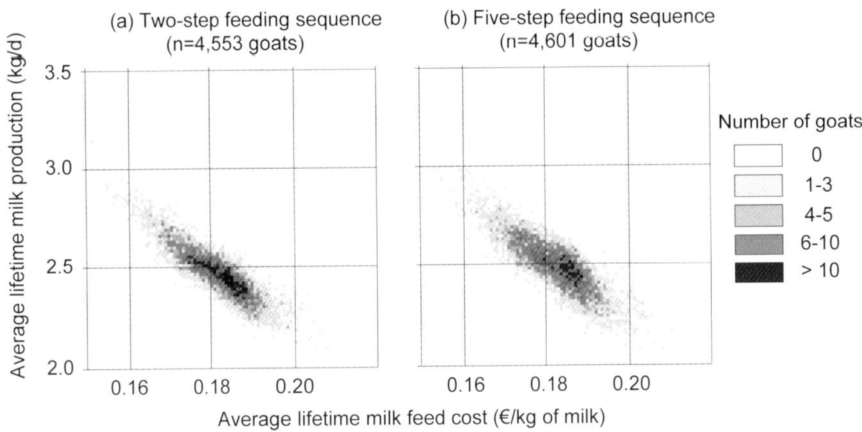

Figure 5. Level plot representation of goat lifetime milk production and milk feed cost for the two-step feeding sequence (a) and the five-step feeding sequence (b).

number of goats concerned. As found for the herd level results, feeding strategies did not give rise to large differences. The distributions had globally the same shape reflecting that higher lifetime production was associated with lower feed cost. This trend represented the dilution of production cost throughout productive life. However, it can be noticed that the 5S strategy led to a slightly more spread distribution than the 2S strategy. As there was a more regular adjustment of concentrate quantity throughout lactation in the 5S strategy, individuals were less nutrient-limited and they were able to express their potential. The 2S strategy tended to reduce this variability as the constant level of concentrate had a smoothing effect on the individual production potential expression.

To further understand the effect of feeding management, the level plot representation was used to analyze individual results at the level of the second lactation. Figure 6 displays the distribution of goats regarding the total milk production over the second lactation and the difference between the actual goat body weight and the body weight defined by its production potential. A positive difference means that a goat gained weight compared with its potential and a negative difference means that a

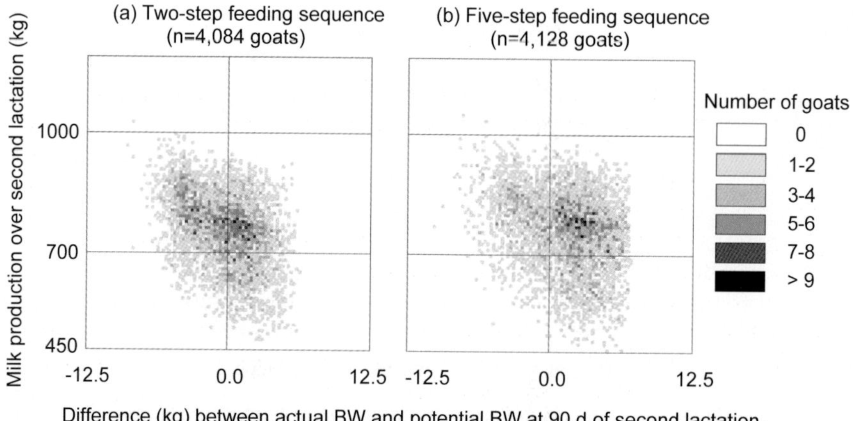

Figure 6. Level plot representation of goat milk production over the second lactation and difference (in kg) between actual and potential body weight (BW) at 90 d of the second lactation for (a) the two-step feeding sequence and (b) the five-step feeding sequence.

goat lost weight compared with its potential body weight. Goat distributions had roughly the same shape. However, when focusing on the negative area of the level plot, 58% of goats lost weight with the 2S feeding sequence and only 38% of goats lost weight with the 5S feeding sequence. Hence, the 2S feeding sequence led to 20% more goats which lost weight compared to the 5S feeding sequence. This result explains why the 5S sequence led to lower herd efficiency than the 2S despite the total of concentrate distributed remained the same in the two feeding strategies. In the 2S, there was a greater proportion of milk production which was obtained with the energy provided by goat body reserves. This higher contribution of goat body reserves to milk production in the 2S strategy led to a better herd efficiency than for the 5S strategy, where production was based to a lesser extent on body reserves. However, the better herd efficiency obtained with the 2S strategy should be interpreted with caution. The dairy goat model does not take into account the delayed effect of a low level of body reserves on reproductive performance and on dry matter intake. Hence, the simulated lifetime milk production of a goat which had a high level of mobilization is likely to be overestimated. Further, if greater use of body reserves has a negative impact on probability of re-breeding then, at herd level, average lifespan may also be negatively affected.

Conclusion

Our results show that the number of steps within a feeding sequence did not lead to major differences in terms of the production level and the production efficiency of both herd and goat productive life levels. However, when focusing at the scale of the second lactation, our results show that production and efficiency do not rely on the same biological processes. With the two-step feeding sequence, herd milk production is based on a greater proportion of individuals mobilizing body reserves. Our results show the importance of integrating individual variability to assess the effects of management strategies on herd performances.

References

Martin, O. and Sauvant, D., 2010. Turning a cow into a goat with a teleonomic model of lifetime performance. In: Friggens, N., Faverdin, P., Sauvant, D. and Van Milgen, J. (eds.) Modelling nutrient digestion and utilisation in farm animal. Wageningen Academic Publishers, Wageningen, the Netherlands, pp.49-53.

Pomar, C., Kyriazakis, I., Emmans, G.C. and Knap, P.W., 2003. Modelling stochasticity: dealing with populations rather than individual pigs. Journal of Animal Science 81:178-186.

Puillet, L., Martin, O., Tichit, M. and Sauvant, D., 2008. Simple representation of physiological regulations in a model of lactating female: application to the dairy goat. Animal 2:235-246.

Puillet, L., Martin, O., Sauvant, D. and Tichit, M., 2010. An individual-based model simulating goat individual variability and long term herd performance. Animal, in press, doi:10.1017/S1751731110001059.

Sauvant, D., Giger-Reverdin, S. and Meschy, F., 2007. Alimentation des caprins. In: Alimentation des bovins, ovins et caprins. Besoins des animaux – Valeurs des aliments. Tables INRA 2007. Editions Quae, Versailles, France, pp.137-148.

Villalba, D., Casasus, I., Sanz, A., Bernues, A., Estany, J. and Revilla, R., 2006. Stochastic simulation of mountain beef cattle systems. Agricultural Systems 89:414-434.

SIMBAL: a herd simulator for beef cattle

L. Pérochon[1], S. Ingrand[2], C. Force[3], B. Dedieu[2], F. Blanc[4] and J. Agabriel[1]
[1]URH1213 Recherches sur les Herbivores, INRA Theix, 63122 Saint Genès Champanelle, France; laurent.perochon@clermont.inra.fr
[2]UMR 1273 Metafort, INRA Theix, 63122 Saint Genès Champanelle, France
[3]UMR 6158 LIMOS, Complexe scientifique des Cézeaux, 63177 Aubière cedex, France
[4]VetAgro Sup, 89 Avenue de l'Europe, B.P. 35, 63370 Lempdes, France

Abstract

A beef cattle herd simulator was designed and developed as a tool for testing the consequences of farm management practices on herd performance. This discrete-event simulator is stochastic and individually based. It integrates a detailed reproduction model together with parameter-settable practices. This article details the various core components of the simulator. We then used it on virtual experiments with a herd of 70 cows. The first virtual experiment compared an autumn calving system with a herd management system based on winter calving. Adapting the herd management rules made it possible for the two systems to give similar outputs. In order to show the importance of the number of mated cows and of the reproductive period length on the performance of the herd, two different strategies were tested in with two new virtual experiments. The first one consisted in reducing the number of mated cows while the second one deals with reducing the reproduction period when the bull is present. The first of these scenarios led to drop in herd size in proportion with the female reduction over the years, whereas under the second scenario, a gap is simulated and the herd was wiped out after 10 years. We also focused an in-depth analysis on the variability of the results. The model shows that an average result not only masks variations from replication to replication but also masks variations in performances from year to year.

Keywords: beef production herd, simulator, individual-based

Introduction

Farmers in the beef cattle sector manage their livestock with different strategies facing the various undergoing constraints (economical, environmental…), which prompts the question of how to organize the production system (e.g. replacement rate, calving period, proportion of fattened livestock…). This question needs to be addressed in terms of both short-term and long-term strategies, but adapting herd management requires time and investments. It could be useful to carry out an *ex ante* evaluation of the technical and management practices options that could be involved (Veysset *et al.*, 2005).

One possibility consists in modeling the herd's functioning, to build-up a simulator solution capable of testing the effects of different strategy-change and/or management-rule scenarios on the technical performance of the herd. This can be approached using a simulation model that interplays decisions and performance outcomes (Cros *et al.*, 2004; Romera *et al.*, 2004). An individual-based simulator (Grimm *et al.*, 1999) is able to express the effect of the actions simulated on the performance simulated for each animal so it is able to express the simulated variability of the performance. These individual simulation-runs can then be pooled to build the fate of the batch. Along these lines, Force *et al.* (2002) developed a simulator modeling the incidence of mastitis in a dairy herd, while Cournut and Dedieu (2004) built a simulator for testing different reproduction configurations in a sheep herd.

These studies use a systemic frame of a herd which consists of (1) a decision-making subsystem representing the farmer whose decisions express its production (2) a 'biological' subsystem representing the livestock dynamic integrating the individual dynamic for each animal. For this, each animal is shaped by deterministic or stochastic biological laws that account for factors such as change in physiological stage, condition score, or level of milk production, and/or, for young cattle, rate of growth. The performance indicators are generally used as inputs by the decision-making subsystem. This is called the information subsystem. We applied this global systemic framework in our modeling process.

Objective and modeling strategy

The 'Simbal' simulator was designed to test the consequences of herd management practices on the status and time-course dynamics of a Charolais herd under French farm conditions. Simbal is a tool for testing different reproduction management strategies according to the characteristics of the component herd-batch animals. It uses an individual-based approach, as animals' characteristics can be used as information by the decision-making process, or can directly shape the biological function.

Analysis of the simulator raised several hypotheses previously described and discussed (Agabriel and Ingrand, 2004).

Stochasticity

Reproductive performances of the cows include a great part of variability. Farmers take it into account and provide safety margins (i.e. room for manoeuvre) to be sure to reach the average calving date they want for the herd. One main lever is the choice of the length of the mating period, i.e. the duration of the bull presence with the cows. The variability of the duration of the mating period is then an input of the model. This allows comparing the consequences of several management strategies not only with a mean result, but also with the range of its variability. We considered that variability mainly occurred at two levels: (1) individual animal characteristics; (2) interplay between biological relations.

Events

Farmer's actions or biological events occur at different points in time, changing the status of the herd system. There may be variability in the time-course occurrence of these events, such as calving dates. We opted for a discrete-event approach (Coquillard and Hill, 1997) in order to account for this other variability characteristic in our model.

Biological functions

The two functions represented for a given animal are growth and reproduction. These functions are governed by feeding levels that are an input parameter in our model (the intake function that determines feed level is not represented).

System boundaries

To simplify the model, we considered our simulated system as closed-loop, i.e. next to no interactions with the outside world. This means, for example, that no purchase of animals can be done. The focus of our model is therefore on the practices related to the interplay between animal renewal and culling in order to maintain a stable herd size.

Entities and outline of the simulator

At the beginning of a run sequence, characteristics of the herd and management practices are specified. Then, no longer change can be done as the replications are executed.

We used a highly specific modeling method (Pérochon 2009) and the formalism chosen is Unified Modeling Language (Rumbaugh *et al.*, 2004) except for the description of biological functions for which we preferred Forrester diagram formalism (Forrester, 1969) for modeling flows. Applications development was performed in Java on the Eclipse development platform. Versioning was managed using Eclipse-platform subclipse plugin. The current version runs on Windows XP or Vista, but it should be compatible with Linux.

The simulator comprises four core interacting model-sets. Three describe dynamic, the farmer decisioning/action rule-set, the animal growth and reproduction functions (Blanc and Agabriel, 2008), and the animal characteristics. The fourth one describes the livestock batches the farmer will use in applying a number of management practice components. These batches are constant over the time.

Animal characteristics

Three types of animal are considered: reproductive cows, heifers and growing or finishing males. Following variables are attributes for each animal type: age, weight and body condition score, parity number of lactation and reproductive status (females), location (cowshed or pasture).

Biological functions reproduction and growth

The more detailed biological function is reproduction. It is based on Blanc and Agabriel (2008) dynamic model, in which reproduction process is modeled as a sequential series of events, as previously proposed by Oltenacu *et al.* (1980). Each cow's anoestrus phase is followed by a period of cyclicity. When cycling, if a bull is present, mating and fertilization may occur. In this model a special focus was given to post-partum anoestrus, which is a decisive factor in determining calving interval in naturally-mating beef cows. It was modeled mechanistically according to parity, taking into account three major effects: body condition at calving, exposure to a bull, and the calving date itself (calendar date), which in primiparous cows has been linked to durational variation in the day of the month preceding calving. This anoestrus sub model has been specifically and independently validated (Blanc and Agabriel, 2008).

After bull insemination, the probability of fertilization depends on number of oestrus cycles and number of times sired. The following pregnancy then culminates in a new calving, when no abortions is taken into account.

Calf gender, as a component of the reproduction function, results from random selection according to a uniform distribution calculated with a 50% probability for each condition.

Animal weight varies throughout time. For various time sequences, average daily gain are initially determined as parameters for each batch, together with a level of variability according to animal type: bull, cow or heifer.

The live weight of cows in reproductive batches depends on both parity number and body condition score (Agabriel *et al.,* 1986). BCS and live weight are linked together (1 BCS= 45 kg LW).

Cow condition score plays a role in the reproduction model or in the culling decision. Change in condition score is calculated at each site-change (cowshed or pasture) as well as at each calving. The model considers that a feed level translates as varying degrees of variation in body condition scores. Winter and summer phases also have an influence.

Setting batches

The farmer reshuffles their animals between batches according to production strategy, which is implemented as actions led at farmer-defined calendar dates, and herd-number objectives differentiated across batches. All the animals in a given batch undergo the same batch-wide management practice.

There are five predefined animal batch-sets: (1) birth; (2) renewing; (3) finishing; (4) culling and (5) reproductive batch (Ingrand *et al.*, 2002). The farmer can create as many sub batches as they want for each of these batches-sets. Batch-set type determines what animal or whole-batch actionable are authorized or prohibited. The user then details input and output rules and intra-batch practices.

At birth, all calves are necessarily channeled into a 'birth' batch-type (Figure 1). They can then be allocated into a 'renewing' batch-type if they are female candidates for reproduction or into a 'finishing' batch-type if they need to be put through a fattening phase. Output from a finishing batch corresponds to animal sales.

Females leaving a renewing batch can then be reassigned to a finishing batch or to a reproductive batch.

After the reproduction cycles when the females are scheduled for culling, they are placed in a culling batch and will only leave that batch to get sold. Once in a culling batch-type, these females can be put through one fattening phase which will last for a highly variable duration determined by the farmer.

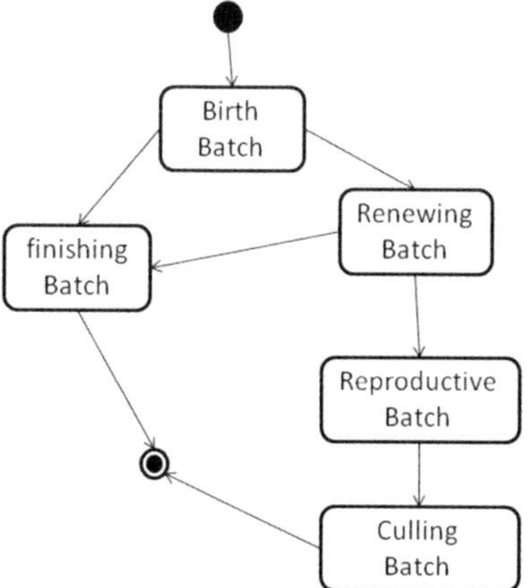

Figure 1. UML states diagram of the batch types.

Run of simulation

The simulated husbandry actions

There are different rule-sets applicable for the following husbandry actions:

- Reproduction: for each 'Renewing' batch-type, the farmer sets the date on which the bull joins the batch and the date it leaves. The duration between the first and last calving for heifers (primiparous) is in accordance with the duration of the corresponding mating period.
- Renewing: the farmer keeps all or part of the female calves he gained over the previous calving season. Using selection criteria, such as live weight or age, the farmer picks a certain number to be allotted to one or more 'Renewing-type' sub batch. The number of cows in these batches will determine the future number of births.
- Culling: farmer removes animals from a renewing batch-type and channeling them into a culling batch-type.
- Sales are made when an animal is taken out of a culling or finishing batch. Specific rules are then applied: thresholds on live weight, or age for example.
- Weaning occurs when an animal is removed from a birth batch, either at defined age or weight or date.
- Fattening is defined by feed level and site (cowshed or pasture). It occurs before selling and is characterized by a specific finishing and culling batch-types.

These rules are set before each simulation run and stored for each simulated year.

Figure 2 illustrates a series of management practice types.

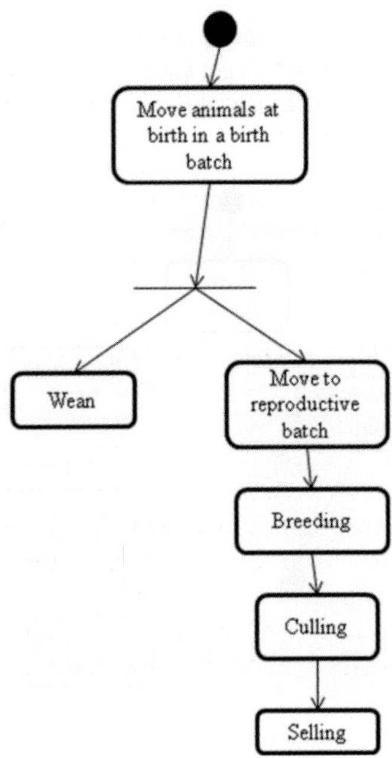

Figure 2. UML activity diagram. Series of management practices led by the farm manager.

Simulator outputs

The stochasticity of the simulator requires several replications before an acceptable variability can be obtained, not just to represent the variability in the target herd system but also to be able to run standard statistical analyses on the output variables.

Simulator outputs are sequenced by replications, and each replication gives per-batch and per-animal outputs. For each animal, the model gives the events it went through: entering or leaving a batch, return to cyclicity, estrus, fertilization, number of times sired and outcome, weaning, culling and sale, weight. Animal inputs and outputs are flagged per batch, with bull inputs and outputs for reproductive batch-types. For each replication, outcome reports are calculated on condition scores, culling, reproduction, calvings, sales, fertilizations, in-out batch moves, and the age-sex pyramid.

Virtual experiments

Objective

The aim is to compare the consequences of different calving period options, i.e. in autumn or winter. The comparison is initially run with all the females assigned to reproduction (exposure to a bull) for a sufficiently long period to enable each female to be sired. We then go on to study the effect of a first limitation by restricting the number of females assigned to reproduction, before studying the effect of shortening the reproductive period. The variability in simulator outputs caused by stochasticity in the reproduction model is given in-depth analysis.

Starting herd

The starting herd comprises 13 heifers in the 'reproductive heifer' batch, 35 cows in the 'reproductive cow' batch, and 22 heifers in the renewing batch, i.e. a total of 70 females. The starting herd had the same profile (age-sex pyramid, weight, condition score) in every trial. Differences come from the estimated first calving date of each cow, taking into account the condition-set governing the reproduction period start-date.

Rule-sets for each run

A first virtual experiment run compares one beef system with winter calving to another with autumn calving.

For the 'winter' condition, a bull is placed in the 'reproductive heifer' batch from 1 March to 1 June (3 months) and into the 'reproductive cow' batch from 1 April to 1 August (4 months). The stockbreeder wants to keep a constant number of cows in the 'reproductive cows' sub batch similar to initial value. He replaces culled cows with heifers from the 'reproductive heifer' sub batch. The remaining heifers in the sub batch are fattened and then culled. Finally, in order to keep a constant overall herd size, the manager does not use the 'growth heifer' sub batch: all the animals leaving the 'renewing' batch join the 'reproductive heifer' batch. The model used two sales dates: 1 February and 1 August.

For the 'autumn' condition, the same types of decisions are used, but with different dates. The 'reproductive heifers' enter mating between 1 November and 1 February, while 'reproductive cows' start on 1 December and finish on 1 April. The other decision rule-sets integrate this time shift in relation to the winter condition. Any difference comes then from the reproduction model, which takes into account the calving date to estimate the anoestrus-period length (Blanc and Agabriel 2008).

A second virtual experiment reintegrates these first two condition-sets, but the livestock manager places some of its animals in the 'heifer growth' sub batch. This has some impacts on herd size if the farmer opts to sell off more animals than before.

A third and last experiment reintegrates the first two condition-sets again, but reproductive period is two months shorter for each batch, lasting respectively one month and two months for heifers and cows.

Batches

The batches used are identical for each virtual experiment (Figure 3).

At birth, young heifers are placed in the birth batch. At weaning, they integrate the renew batch. At the following reselection step, they will be channeled into either the heifer growth batch to fatten or into the heifer reproductive batch to begin reproduction. Certain females will then be shortlisted for the following reproduction cycle by integrating the cow reproductive batch, and the gaps will be filled by culling in the culled heifer batch. Each year, a handful of females from the cow reproductive batch will be culled, following the culling rules, into the culled cow batch. In both culling batch-types, the animals are fattened for growth. Animals leaving the heifer growth, culled heifer and culled cow sub batches are counted as sales.

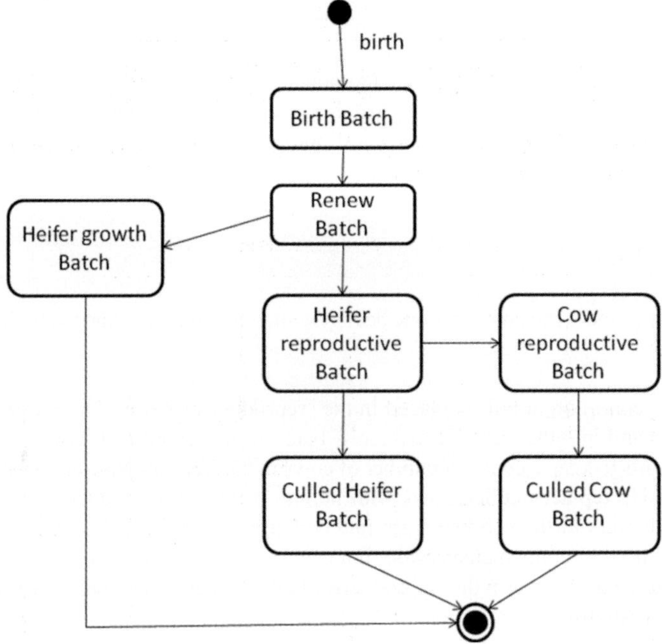

Figure 3. Batches used for each virtual experiment.

Results

The first experiment highlighted a clear difference in calving to cyclicity interval (Figure 4) between autumn and winter. This difference was mainly due to the reproduction model as expected. Intervals are greater under the autumn condition, particularly for heifers. We also recorded significantly large variability for winter heifers, which was caused by the effect of the higher day lengths in winter included in the reproduction model.

However, when looking at calving to calving interval (Figure 5), these differences between winter and autumn condition-sets disappear.

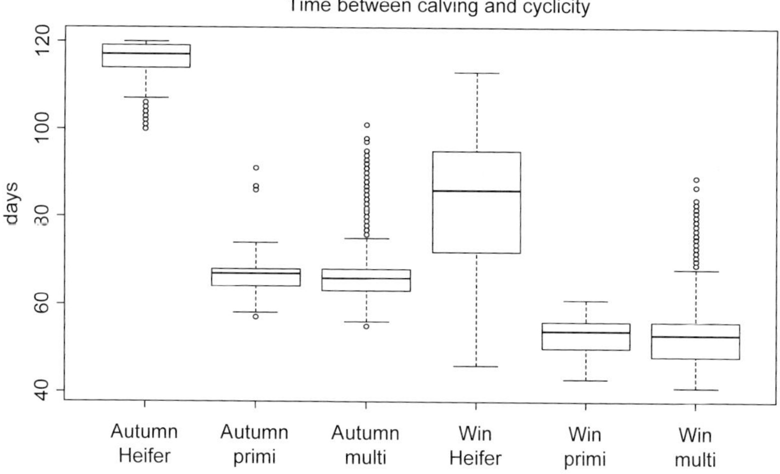

Figure 4. Calving to cyclicity interval per parity and per calving season. Win is for winter, primi for primiparous and multi for multiparous.

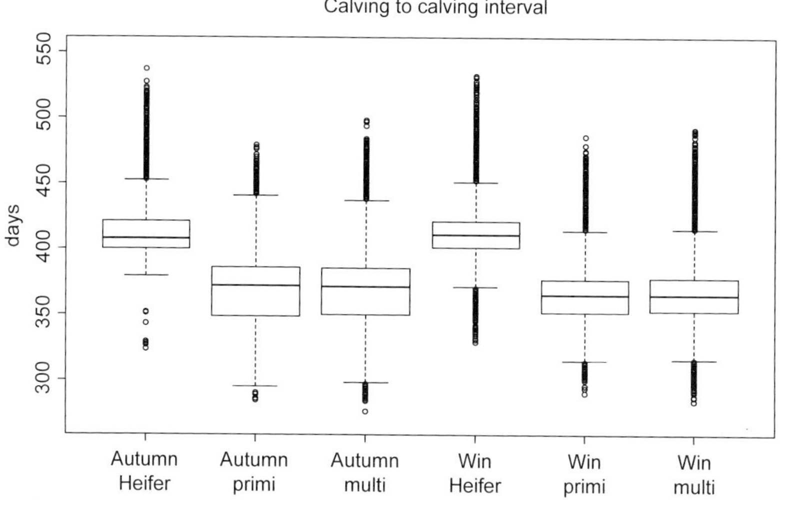

Figure 5. Calving to calving interval per parity and per calving season.

Variability of the calving to calving interval was comparable between winter and autumn heifers. This could be explained by bull exposure. The bull is introduced into a reproductive batch-type at a given date and for a given time-window. For autumn heifers, the rule-set dictates that they are exposed to the bull as soon as the first returns to cyclicity begin, whereas for the winter herd the earliest cycles begin up to 50 days before the bull is introduced. Consequently, these spring calving heifers will not be inseminated during their first cycle but during the following cycles. Insemination dates are delayed for early returns to cyclicity, which makes it possible to synchronize the future calving dates of the herd.

Average sales of animals turn out similar under both scenarios, at 42 per-years. Cumulated total sales at 20 years vary from replication to replication in the two scenarios from 767 to 888, i.e. a non-negligible variation of 121 animals over the 20 years. This cumulated total sales ranged from 821 to 865 for 50% of the replications. These variations cannot be explained by culling policy, which was identical from replication to replication, but are instead due to variation in the sizes of reproductive-batch sub-herds.

The most important batch is the reproductive cows batch, not just because it has the largest herd size but also because its population is kept constantly balanced by the reproductive heifers sub batch. The reproductive cows batch averaged 55 cows under both winter and autumn condition-sets, but masking significant variability, as winter counts fell to 39 and autumn counts fell to 35 (36% less than the average).

The reproductive heifers sub batch reproduced the same pattern. This had a direct impact on the number of calves produced, and consequently on the number of female calves in the herd. Assuming that no cows went unfertilized, which is not the case in the simulator, a population of 55 cows could give a population of 55 female calves. With a population of 39, as previously shown for winter, the female calves figure drops to 19, while 35 cows produce 17 female calves, and the figures could be even lower in years where births of female calves are low. Since our system is closed-loop and no female calves were bought in, this variability becomes a critical factor. We recorded an average female calf population of 29, but year-to-year variation ran from 16 to 44. For half of the years, this population ranged from 26 to 32. This number depends not only on sex ratio at birth but also on the number of females inseminated (reproductive heifers or reproductive cows).

For the second experiment, 10 of the heaviest calves were removed each year from the birth batch to be fattened in the heifer growth sub batch. The consequence was a drop of ten or so cows in the reproductive cows sub batch. This population also tended to decrease over the years, reaching a significant decline in some replications. This is due to the drop in the number of female calves, which despite averaging 21 in autumn and 22 in winter could nosedive to just 5 in autumn and 9 in winter.

For the third experiment, the reproductive period was limited to two months. This had a radical effect, resulting in the number of reproductive cows approaching zero within the space of just 10 years.

Conclusion

Initial experiments with this simulator resulted in the completion of a first evaluation phase covering the overall behavior of the model. Employing a closed-loop system highlighted the importance of conserving the number of reproductive females each year. The simulator can be used to test different policies designed to keep herd population constant over time. It can also predict the risk of running into a deficit of reproductive females depending on herd management practices. From this standpoint, the simulator can prove useful as a tool for informing policy change or deciding on the risk of having to buy-in female calves or heifers in subsequent years. The simulator experiments also highlighted the importance of having an output that gives a global vision of results variability, and the importance to have a validated reproduction model in various conditions. In order to improve the predictions, measures are planned to improve the feed module and to extend the options for herd management conditions.

References

Agabriel J. and Ingrand, S., 2004. Modelling the performance of the beef cow to build a herd functioning simulator. Animal Research 53:347-361.

Agabriel J., Giraud, J.M. and Petit M., 1986. Détermination et utilisation de la note d'état d'engraissement en élevage allaitant. Bull. Tech. CRZV Theix INRA 66:43-50.

Agabriel J. and Petit M., 1987. Recommandations alimentaires pour les vaches allaitantes. Bull. Tech. CRZV Theix INRA 70:153-166.

Blanc F. and Agabriel, J., 2008. Modelling the reproductive efficiency in a beef cow herd: effect of calving date, bull exposure and body condition at calving on the calving-conception interval and calving distribution. Journal of Agricultural Science Cambridge 146:143-161.

Coquillard P. and Hill, D.R.C., 1997. Modélisation et simulation d'écosystèmes. Masson, Recherche en Écologie, Paris, France, 273 pp.

Cournut S. and Dedieu B., 2004. A discrete events simulation of flock dynamics: a management application to three lambings in two years. Animal Research 53:383-403.

Cros M.J., Duru, M. and Garcia, F., 2004 Simulating management strategies: the rotational grazing example Agricultural Systems 80:23-42.

Force C., Perochon, L. and Hill, D.R.C., 2002. Design of a multimodel of a dairy cows herd attacked by mastitis. Simulation Modelling Practice and Theory 10:543-554.

Forrester, J.W., 1969. Urban Dynamics. MIT Press, Cambridge, MA, USA, 285 pp.

Grimm, V., Wyszomirski, T., Aikman, D. and Uschmanski, J., 1999. Individual-based modelling and ecological theory: synthesis of a workshop. Ecological modeling 115:275-282.

Ingrand S., Dedieu, B., Agabriel, J. and Pérochon, L., 2002. Modélisation du fonctionnement d'un troupeau bovin allaitant selon la combinaison des règles de conduite. Premiers résultats de la construction du simulateur SIMBALL, Recontres Recherches Ruminants 9:64.

Oltenacu, P.A., Milligan, R.A., Rousaville, T.R. and Foote, R.H., 1980. Modelling reproduction in a herd of dairy cattle. Agricultural Systems 5:193-205.

Pérochon, L., 2009. MLPS: A method for modeling livestock production systems. ESM'2009, October 26-28, Leicester, UK.

Romera, A.J., Morris, S.T., Hodgson, J., Stirling, W.D. and Woodward, S.J.R., 2004. A model for simulating rule-based management of cow-calf systems. Computers and Electronics in Agriculture 42:67-86.

Rumbaugh, J., Jacobson, I. and Booch, G., 2004. UML 2.0 Guide de référence. Campus Press, Paris, France, 800 pp.

Veysset, P., Bebin, D. and Lherm, M., 2005. Adaptation to Agenda 2000 (CAP reform) and optimisation of the farming system of French suckler cattle farms in the Charolais area: a model-based study. Agricultural Systems 83:179-202.

Modelling the impacts of climate change on suckling grass-based systems with the Pasture Simulation Model

A.-I. Graux[1], *M. Gaurut*[1], *J. Agabriel*[2], *J.-F. Soussana*[1] *and R. Baumont*[2]
[1]*INRA, UR 874 Grassland Ecosystem Research, 63100 Clermont-Ferrand, France; aigraux@clermont.inra.fr*
[2]*INRA, UR1213 Herbivores, 63122 St Genès de Champanelle, France*

Abstract

To simulate changes and possible feedbacks in grazing animal performance in a . context, we have improved a process-based biogeochemical pasture model, PASIM, so that it can simulate performance of suckler cows with their calf, in response to climate and management, and possible feedbacks, namely through methane emissions by ruminants. In this model, an average cow is simulated, which herbage ingestion is predicted according to the animal profile, diet digestibility and herbage availability. Sward depletion by animal intake at pasture affects the herbage growth and quality. The net energy balance determines the daily liveweight and body condition variations, and influences milk production. These changes have feedbacks on energy requirements and ingestion capacity. Net energy intake is used to assess enteric methane energy through a conversion factor which depends on the energetic level of the diet. The model was validated against experimental data for cow performance at pasture: predictions show good agreement with observations. The comparison of previous and new version of PASIM demonstrates the necessity to account for animal performance in projections of climate change impacts on grassland and of feedbacks to the atmosphere. The originality and the validity domain of the model are discussed.

Keywords: grazing, climate change, energy balance, methane, ecosystem services.

Introduction

In the context of the current global food and economic crisis and as world's population is projected to increase from six to nine billion by 2050, providing food to the world population will be a challenge for this century. Around one sixth of global population is food insecure (FAO, 2009). By affecting biotic and abiotic drivers on agricultural production, climate change could exacerbate food insecurity (Tubiello *et al*., 2007). Among the multiplicity of their functions (Hopkins *et al*., 2006; Lemaire *et al*., 2005), grasslands provide animal products including meat which is a major source of proteins. Climatic factors, either alone or in interaction with each other, often limit animal performance (Rémond *et al*., 1982; D'hour and Coulon, 1994). Climate change may thus impact directly and indirectly bovine performance by disturbing physiology and behavior of animals that have to maintain their body temperature (Morand-Fehr and Doreau, 2001), and by affecting the productivity, seasonality and quality of pasture production (Tubiello *et al*., 2007). Severe droughts such as yr. 2003 in Europe and 2007 in Australia may happen more frequently in future (IPCC, 2007a), and thus, may jeopardize the grassland production function. However, the agriculture sector contributes for around 13% of worldwide greenhouse gases (GHG) emissions (IPCC, 2007b). Enteric fermentation and emission from animal manure account for 35-40% of the total annual anthropogenic methane emissions (Koneswaran and Nierenberg, 2008). Reducing methane emissions from animal production systems is among the key priorities for GHG mitigation in the agriculture sector (IPCC, 2007b). Factors relative to the animal (breed, kind and level of production) and to the diet type (amount, chemical composition, interactions between diet components) are well-known to influence animal methane production. To model enteric methane emissions, we used a recent method proposed by Vermorel *et al.* (2008) based on the amount and the energetic level of the diet.

Little is known about the interactions of climate and increasing climate variability with other drivers of change of livestock systems (Thornton, 2009). How far will climate changes affect livestock system performances and in return, what could be the feedbacks in terms of methane emissions on the atmosphere? To our knowledge, only few pasture models are able to simulate accurately both biogeochemical fluxes exchanged in grassland ecosystems and animal performance (See Bryant *et al.*, 2008). In order to simulate changes and possible feedbacks in grazing animal performance in a global warming context, we have improved a biogeochemical pasture model, PASIM (Riedo *et al.*, 1998), so that it can simulate performance of suckler cattle in response to climate, and possible feedbacks, through enteric methane emissions.

Model description

Key processes

PASIM key processes are well described in Vuichard *et al.* (2007a). PASIM is a process-based grassland biogeochemical model derived by Riedo *et al.* (1998) from the Hurley Pasture Model (Thornley, 1998). Grassland processes are simulated on a time step of about 30 minutes. Simulations are limited to the plot scale when animals are grazing at pasture (PASIM does not consider indoors periods). Simulations last at least one year but the model can run on several years. The code consists in around 50,000 lines and is written in Fortran 90. As with other advanced biogeochemical models, PASIM simulates water, carbon (C) and nitrogen (N) cycles, the latter having been improved by Schmid *et al.* (2001). In PASIM, the C assimilated by photosynthesis is either respired or allocated dynamically to one root and to three shoot compartments (each of those consisting of four age classes). Accumulated aboveground biomass is either used by cutting or grazing, or enters a litter pool. The N cycle in PASIM considers three different types of N inputs to the soil via atmospheric N deposition, fertilizer N addition, and symbiotic N fixation by legumes. The inorganic soil N is available for root uptake and may be lost through leaching, volatilization and nitrification/denitrification, the latter processes leading to N_2O gas emissions to the atmosphere. Management includes N fertilization, mowing and grazing and can either be set by the user or optimized by the model (Vuichard *et al.*, 2007b). The same parameterization of the PASIM model as described in Riedo (1998) and Schmid (2001) was used in all simulations, except for some key vegetation parameters which were adjusted according to the vegetation description.

The animal module in PASIM

Previous version. The original animal module of PASIM was first implemented by Riedo *et al.* (2000) to simulate cow or sheep herbage ingestion and milk production (MP) and methane (CH_4) emissions at pasture. Vuichard *et al.* (2007a) included animal diet selection among herbage age classes characterised by their digestibility and abundance. CH_4 emissions were therefore calculated according to Pinarès-Patino (2003) from a linear regression of the amount of digestible fibres in the intake (DNDFI). The animal module remained quite simple as liveweight (LW) and ingestion capacity (IC) were assumed to be constant and diet was limited to grazed herbage. Moreover, the prediction of intake considered neither the kind of production (milk, beef) nor net energy requirements for maintenance (NEM) and production (NEP). Animals were removed from the paddock when aboveground plant biomass (BM) was lower than a threshold of 300kg DM/ha.

New version. In order to simulate changes in grazing animal performance, we have coupled PASIM with the animal production model SEBIEN, developed by Jouven *et al.* (2008) for the simulation of suckler systems. In this model it is assumed that (1) cows calve only once a year, (2) gestation lasts 9 months and lactation 10 months, following by two months of drying up and (3) calving interval lasts one year. The threshold for removing animals is set to 1,100kg DM/ha in this new version.

Simulation of grazing-based suckling system performance. The SEBIEN model is well described in Jouven *et al.* (2008). SEBIEN is a model of intake and production of an average suckler cow with its calf. This model is composed of two modules: one for intake and the other for performance. The intake module calculates (1) herbage organic matter digestibility (OMD) at pasture as a result of selective intake of sward structural components differing in quality and abundance, and (2) the amount of herbage ingested by herbivores, based on the animal's LW, MP and body condition score (BCS). The performance module calculates (1) daily MP, according to Wood (1967), based on theoretical daily milk production (MP_{pot}) and the stage of lactation, and influenced by the net energy balance (NEB) after 3 months of lactation, and (2) daily LW and BCS variations, based on the animal NEB between energy from net energy intake (NEI) and NEM plus NEP. Our model does not simulate LW loss for calves.

To simulate grass-based suckling systems, we have implemented in PASIM the equations of the SEBIEN model for animals grazing continuously on pastures, excepted for (1) the calculation of the net energy content (NEL) of the grazed herbage that has been adapted on all kind of green forage data, and for (2) selective intake of sward structural components, that already exists in PASIM (Vuichard *et al.*, 2007a). NEL is calculated from the OMD of the intake, according to INRA feed tables (2007). In PASIM, the two modules interact on a time step of about 30 minutes: the intake and requirements at time t depend on animal's profile at t, which in turn depends on the NEB on time (t-1) (see Appendix A).

In order to better simulate climate effects on livestock systems, we have added high temperatures (HT) limitation on ingestion, as proposed in Grazefeed by Freer *et al.* (1997). As reviewed by Delagarde *et al.* (2005), Grazefeed is the only feeding system model that takes into account ingestion limitation by HT. In this model, potential dry matter intake (DMI) is depressed when the daily average ambient temperature exceeds 25 °C and when the daily minimum temperature is greater than 22 °C, through the calculation of a temperature factor that modulates DMI.

Simulation of enteric methane emissions from grazing-based suckling systems. As PASIM now simulates NEL, we have improved the PASIM simulation of CH_4 emissions by animals, with the method proposed by Vermorel *et al.* (2008). In this method, metabolizable energy intake (MEI) is used to assess enteric methane energy (E_{CH4}) through a conversion factor (Y_m', expressed in kcal of CH_4 for 100kcal of MEI), which depends on the kind of production, the diet type (for animals fed at pasture with grazed herbage or fed at barn) and the animal performance. CH_4 is finally calculated as the ratio of E_{CH4} to the energy content of methane (55.65 MJ/kg CH_4). To calculate MEI with PASIM, we multiply NEI by the average rate of use of MEI for lactation (k_l=0.6, Vermorel *et al.*, 2008). This method accounts for the kind and the level of production, the amount and the type of diet. However it does not consider the effects on methane production due to animal breed and to interactions between diet components. For suckler cows grazing at pasture, Y'_m is assumed to be constant (Y'_m=12).

Model implementation

The new animal module of PASIM was first tested with Berkeley Madonna software (developed by Robert Macey and George Oster of the University of California at Berkeley) then developed in Fortran 90. Model inputs for animals are (1) the grazing periods (maximum 10) and the corresponding stocking density (SD), (2) the calving date, (3) the average maximum MP_{pot} ($MP_{pot,max}$), (4) the initial average LW and BCS of cows for each grazing period, (5) the proportion of young cows ($\leq$ 4 years old) in the cattle, (6) the average birth LW of calves and the average LW of calves at the beginning of the grazing period, (7) the calf weaning period, and (8) the average age of calves at sale. A sensitivity analysis with input values of the SEBIEN model was performed by Jouven *et al.* (2008), highlighting the major role of herbage digestibility and availability for the animal performance.

Model validation against experimental data

We compared daily model predictions with data from an experiment measuring the ingestion and the performance of suckler cows and their calves at pasture. To assess the accuracy of the model's predictions, we calculated root mean square error (RMSE). To know how much RMSE is systematic in nature and what portion is unsystematic, it has been split in specific $RMSE_s$ and unspecific $RMSE_u$, as defined by Willmott (1981). $RMSE_s$ scales with the linear bias of the model, whereas $RMSE_u$ may be interpreted as a measure of precision. To compare the capacity of model to predict different variables, we also calculated normalized RMSE, $RMSE_s$ and $RMSE_u$: NRMSE, $NRMSE_s$, $NRMSE_u$ respectively. The latter variables are comprised between 0 and 1. The best fit is obtained when they are closed to zero.

We simulated a grassland-based suckler system at the Laqueuille site (upland grassland in the French Massif-Central, 1,050 m.a.s.l), corresponding to experiment III in Jouven *et al.*, 2008. In this experiment, a group of 6 young and 3 mature suckler Charolais cows with their calves, calving in early January, were grazing continuously at low SD from 23 May to 15 September in 2005 on permanent pastures dominated by *Festuca rubra* and *Agrostis tenuis* and containing around 6% of legumes. Every 3 weeks, sward height was characterised by stick measurements and sward biomass was measured. Sward OMD was assessed by pepsine-cellulase digestibility. Cow LW and MP were measured every 2 weeks. MP was estimated by double-weighting of calves (Le Neindre, 1973). Cows were bodyscored every three weeks (Agabriel *et al.*, 1986). We simulated the experimental treatments for the average cow of the group from the 1st of January 2005.

Model predictions of suckler cow and calf performance are good (RMSE represents 3 to 10% of the averaged observed value; Table 1), especially for liveweight, and well reproduce the dynamics of observations (Figure 1).Vegetation availability and digestibility are well reproduced by the model. Even if simulated values of biomass remain in the standard deviation interval, average biomass tends to be overestimated during the growing season (RMSE and $RMSE_s$ represent respectively 30% and 21% of the averaged observed value, see Table 1). This overestimation may be explained by the fact that the model does not integrate the spatial heterogeneity of vegetation due to continuous grazing.

Importance of animal performance in projections of climate changes impacts on grassland services

With both previous and new versions of PASIM, we simulated climate change impacts on the performance of the studied Charolais suckler system during the grazing period at the Theix site (upland grassland in the French Massif-Central, 890 m.a.s.l.) using the A2 IPCC SRES scenario

Table 1. Statistical indicators of model performance[1].

Variable	No. of points compared	RMSE (value)	NRMSE (%)	$RMSE_S$ (value)	$NRMSE_S$ (%)	$RMSE_U$ (value)	$NRMSE_U$ (%)
BCS_{young}	7	0.06	3	0.04	2	0.04	2
BCS_{mature}	7	0.13	6	0.07	3	0.11	5
BM	7	0.06	30	0.04	21	0.04	21
OMD	8	0.05	9	0.05	8	0.03	5
LW_{calf}	8	29.66	4	25.17	3	15.68	2
LW_{young}	10	35.13	4	29.90	4	18.44	2
LW_{mature}	10	23.73	9	20.14	8	12.55	5
MP	8	0.65	10	0.35	5	0.55	8

[1] Root mean-squared error (RMSE value and % relative to the average observed value), and contribution to specific and unspecific errors, for six major variables: standing biomass (BM, kg/m^2), organic matter digestibility (OMD,g/g) of intake, young and mature cow liveweight (LW_{young} and LW_{mature}, kg/animal), young and mature cow body condition score (BCS_{young} and BCS_{mature}), cow milk production (MP, kg/animal/d), and calf live weight ($LW_{calf,}$ kg/animal).

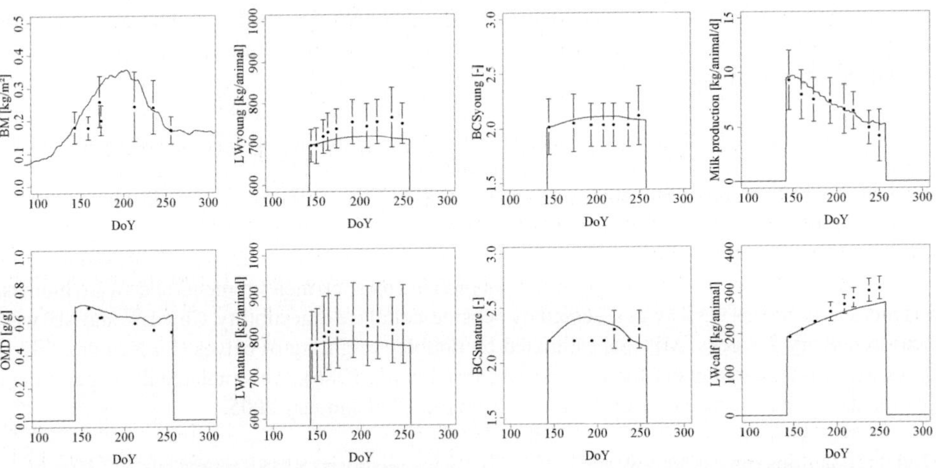

Figure 1. Comparison of simulations (lines) vs. observations (dots) during the grazing period for standing biomass (BM), organic matter digestibility of the herbage (OMD), liveweight of young (LW_{young}) and mature cows (LW_{mature}), body condition of young (BCS_{young}) and mature cows (BCS_{mature}), average milk production (MP) and calf liveweight (LW_{calf}). Time is expressed in day of year (DoY). Standard deviations of measurements are shown as vertical bars.

and the ARPEGE climate model. Soil organic matter (SOM) was initialized at equilibrium with the climate in the 1950's. With fixed agricultural practices, projections for 30 yrs time slices centered (Figure 2) show that (1) previous and new model predictions of annual soil organic carbon (SOC) variation and global warming potential (GWP) are not significantly different, (2) the daily LW gain of cows and calves could be reduced in far future, probably due to the decrease in summer gross primary production (GPP) and to high temperature effect on animal intake, (3) CH_4 emissions

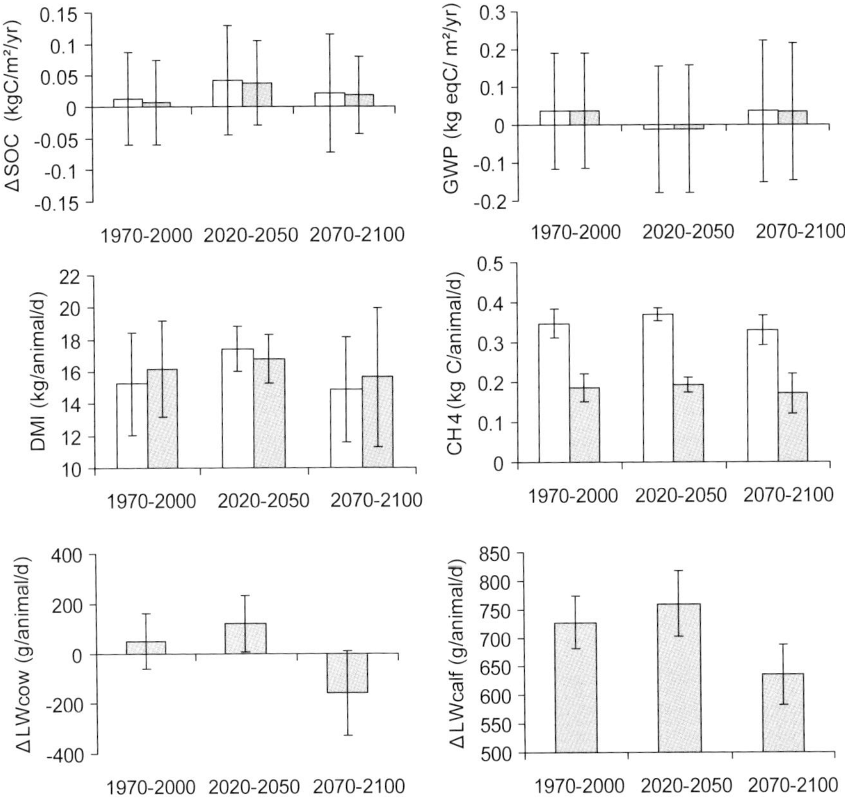

Figure 2. Projected climate change impacts over the grazing period for annual soil organic carbon variation (ΔSOC), global warming potential (GWP), dry matter intake (DMI), enteric methane (CH_4), daily liveweight variations of cows (ΔLWcow) and calves (ΔLWcalf) for the A2 IPCC SRES scenario. Error bars show the interannual variability during the period. White bars, previous model version; grey bars, new model version. Note that ΔLW was not calculated by the previous model version. A negative GWP indicates a net sink of GHG for the atmosphere in CO_2 equivalents.

calculated with previous and new methods give significantly different results, with new results which are more representative of daily CH_4 emissions from suckler cows (about 0.19 kg/animal/d according to Vermorel *et al.*, 1995).

Discussion

The objective of this work was to develop a grassland biogeochemical model that could predict, at pasture, the impacts of climate and management on performance of suckler cows with their calves, and the feedbacks to the atmosphere.

PASIM has been already used to make predictions of GHG fluxes from grasslands at European scale (Vuichard *et al.*, 2007b). It also simulates forage production, dry matter intake through selective grazing and influenced by high temperatures, suckler cow and calf performance from the forecast of herbage, and enteric CH_4 emissions according to the energy content of the intake. Management, including cutting, grazing and N fertilization can be either determined by the user or optimized by the model for grass-based systems self-sufficient in terms of animal feed production. Model predictions of suckler cow performance from net energy balance (NEB) already exist, but to our knowledge, few of them also simulate biogeochemical cycles (Bryant *et al.*, 2008). The novelty of the new model version stems from its ability to predict mechanistically suckler cow performance at grazing, pasture biogeochemical cycles, and their interactions.

Model predictions of suckler cow and calf performance show good agreement with measurements, but pasture aboveground biomass tends to be overestimated during summer. Such error is difficult to interpret as it could be attributed to: (1) the model assumption of a homogeneous distribution of sward height and biomass in the paddock continuously grazed; indeed, under such conditions, the development of sward heterogeneity can affect intake and hence biomass accumulation (Dumont *et al.*, 1996), and/or (2) to the capacity of the model to simulate extensive and low growing swards, and/or (3) to the underestimation of dry matter intake, that has already been observed in summer by Jouven *et al.* (2008). Moreover, the model was able to simulate the seasonal dynamics of animal performance throughout the grazing season, but was unable to reproduce the increase in cow liveweight observed in spring. When forced with biomass and organic matter digestibility measurements, the model tends to underestimate animal performance, showing that the intake limitation by biomass used for intake calculation may be overestimated, especially at low levels of biomass availability. At higher levels, intake is less limited by biomass and therefore less underestimated. As a feedback, the underestimation of intake may contribute partly to biomass overestimation during the same period.

The validity domain of the model corresponds to European temporary and permanent pastures and to suckling grass-based systems. The model can be used to simulate a range of soil-climate situations and grazing managements: more or less intensive in terms of N fertilization and stocking density, and for continuous or rotational grazing, but for only one grazed paddock. Animals are simulated at pasture but not in the barn, and are fed with grazed herbage but received no forage or concentrate. As a consequence, PASIM is well suited to cattle fed with balanced herbage diet. Furthermore, the assumptions of fixed composition of gain, linear relationship between LW and body condition score variations and effect of the cow NEB on milk production are valid for cows moderately overfed or underfed. The INRA feed evaluation systems, on which the model is based, is adapted to French breeds (Charolais and Salers). In its current version, the model does not consider LW loss for calves. Since our model does not integrate all the existing constraints on selective intake due to vegetation structure and to animal capacities (Dumont *et al.*, 1996), the model should perform best at moderate to high stocking rates. As the model is highly sensible to forage characteristics, its prediction of intake and animal performance will depend on its vegetation calibration and its ability to predict accurately the dynamics of the sward profile.

Conclusion

The originality of the pasture model PASIM results from its ability to predict mechanistically cow ingestion and performance at pasture, biogeochemical cycles, and their interactions. Our approach, which has been designed for grass-based suckling systems based on temporary or permanent pastures, can be applied to short or long-term simulations. The model predicts forage production and performance of suckler cows and their calves at pasture with good accuracy. Previous vs. new model projections of climate change impacts on grasslands do not show significantly different results, except for methane production, ensuring that the functioning of the model was not affected by the addition of the new module. Furthermore, the model is now able to make projections of climate change impacts on the grassland production function by affecting animal performance. The validity domain of the model could be extended and its accuracy could be improved by (1) adding a module simulating mechanistically dairy cows at pasture, (2) adding the possibility to supplement animals (3) modulating energy requirements depending on herbage availability, paddock size and climate (4) assessing dry matter intake and methane emissions against experimental data.

Acknowledgements

This work was supported by the Auvergne Region and by the ANR-07-VULN 'Vulnérabilité: Milieux et Climat', VALIDATE project.

References

Agabriel, J., Giraud, J.M. and Petit, M., 1986. Détermination et utilisation de la note d'état d'engraissement en élevage allaitant. Bulletin technique du centre de recherches zootechniques et vétérinaires de Theix 66:43-50.

Bryant, J.R., and Snow, V.O., 2008. Modelling pastoral farm agro-ecosystems: a review. New Zealand Journal of Agricultural Research 51:349-363.

Delagarde, R. and O'Donovan, M., 2005. Les modèles de prévision de l'ingestion journalière d'herbe et de la production laitière des vaches au pâturage. INRA Productions Animales18:241-253.

Dumont B., 1996. Préférences et sélection alimentaire au pâturage. INRA Productions Animales 9:359-366.

D'hour, P., and Coulon, J.B., 1994. Variations de la production et de la composition du lait au pâturage en fonction des conditions climatiques. Annales Zootechniques 43:105-109.

FAO (Food and Agricultural Organization of the United Nations), 2009. The State of Food Insecurity in the World – Economic crises: impacts and lessons learned. Electronic Publishing Policy and Support Branch Communication Division FAO, Rome, Italy, 61 pp.

Freer, M., Moore, A.D. and Donnelly, J.R., 1997. GRAZPLAN: Decision Support Systems for Australian Grazing Enterprises-II. The Animal Biology Model for Feed Intake, Production and Reproduction and the GrazFeed DSS. Agricultural Systems 54:17-126.

Hopkins, A. and Holz, B., 2006. Grassland for agriculture and nature conservation: production, quality and multi-functionality. Agronomy Research 4:3-20.

IPCC (Intergovernmental Panel on Climate Change), 2007a. Climate change 2007: the physical science basis. Contribution of Working Group I to the Fourth Assessment Report of the Intergovernmental Panel on Climate Change. Cambridge University Press, Cambridge, UK.

IPCC (Intergovernmental Panel on Climate Change), 2007b. Climate Change: Impacts Adaptation and Vulnerability. Contribution of WG II to the Fourth Assessment Report of the Intergovernmental Panel on Climate Change. Cambridge University Press, Cambridge, UK.

Jouven, M., Agabriel J. and Baumont R., 2008. A model predicting the seasonal dynamics of intake and production for suckler cows and their calves fed indoors or at pasture. Animal Feed Science and Technology 143:256-279.

Koneswaran, G. and Nierenberg, D., 2008. Global Farm Animal Production and Global Warming: Impacting and Mitigating Climate Change. Environmental Health Perspectives 116:578-582.

Lemaire, G., Wilkins, R. and Hodgson, J., 2005. Challenges for grassland science: managing research priorities Agriculture, Ecosystems and Environment 108:99-108.

Le Neindre, P., 1973. Observations sur l'estimation de la production laitière des vaches allaitantes par la pesée du veau avant et après tétée. Annales Zootechniques 22:413-422.

Morand-Fehr, P. and Doreau, M., 2001. Ingestion et digestion chez les ruminants soumis à un stress de chaleur. INRA Productions Animales 14:15-27.

Pinares-Patino, C.S., Baumont R., Martin, C. 2003. Methane emissions by Charolais cows grazing a monospecific pasture of timothy at four stages of maturity. Canadian Journal of Animal Science 83:769-777.

Rémond, B. and Vermorel M., 1982. Influence du climat et de la saison sur la production laitière au pâturage. In: Actions du climat sur l'animal au pâturage. INRA Editions, Paris, France, pp.115-129.

Riedo, M., Grub, A., Rosset, M. and Fuhrer, J., 1998. A pasture simulation model for dry matter production and fluxes of carbon, nitrogen, water and energy. Ecological Modelling 105:141-183.

Riedo, M., Gyalistras, D. and Fuhrer, J., 2000. Net primary production and carbon stocks in differently managed grasslands: simulation of site-specific sensitivity to an increase in atmospheric CO2 and to climate change. Ecological Modelling 134:207-227.

Schmid, M., Neftel, A., Riedo, M. and Fuhrer, J., 2001. Process-based modelling of nitrous oxide emissions from different nitrogen sources in mown grassland. Nutrient Cycling in Agroecosystems 60:177-187.

INRA, 2007. Alimentation des bovins, ovins et caprins – besoin des animaux – valeurs des aliments. Quae éditions, France, 308 pp.

Thornley, J.H.M., 1998. Grassland dynamics. An ecosystem simulation model. CAB International. Wallingford, UK, 256 pp.

Thornton, P.K., Van de Steeg, J., Notenbaert A. and Herrero M., 2009. The impacts of climate change on livestock and livestock systems in developing countries: A review of what we know and what we should know. Agricultural Systems 101:113-127.

Tubiello, F.N., Soussana, J.F. and Howden S.M., 2007. Crop and pasture response to climate change. Proceedings of the Nataional Academy of Sciences of the United States of America 104:19686-19690.

Vermorel, M., 1995. Emissions annuelles de méthane d'origine digestive par les bovins en France. Variations selon le type d'animal et le niveau de production. INRA Productions Animales 8:265-272.

Vermorel, M., Jouany, J.P., Eugène, M., Sauvant, D., Noblet, J. and Dourmad, J.Y., 2008. Evaluation quantitative des émissions de méthane entérique par les animaux d'élevage en 2007 en France. INRA Productions Animales 21:403-418

Vuichard, N., Ciais, P., Viovy, N., Calanca, P. and Soussana, J.F., 2007a. Estimating the greenhouse gas fluxes of European grasslands with a process-based model: 1. Model evaluation from *in situ* measurements. Global Biogeochemical Cycles 21 GB1004, doi:10.1029/2005GB002611.

Vuichard, N., Ciais, P., Viovy, N., Calanca, P. and Soussana, J.F., 2007b. Estimating the greenhouse gas fluxes of European grasslands with a process-based model: 2. Simulations at the continental level, Global Biogeochemical Cycles 21, GB1005, doi:10.1029/2005GB002612.

Wood, P.D.P., 1967. Algebraic model of the lactation curve in cattle. Nature 216:164-165.

Willmott, C.J., 1981. On the validation of models. Physical Geography 2:184-194.

Appendix A. Diagram illustrating the animal model (grey box) applied to a cow at pasture and integrated into PASIM for the simulation of suckler cow ingestion and performance.

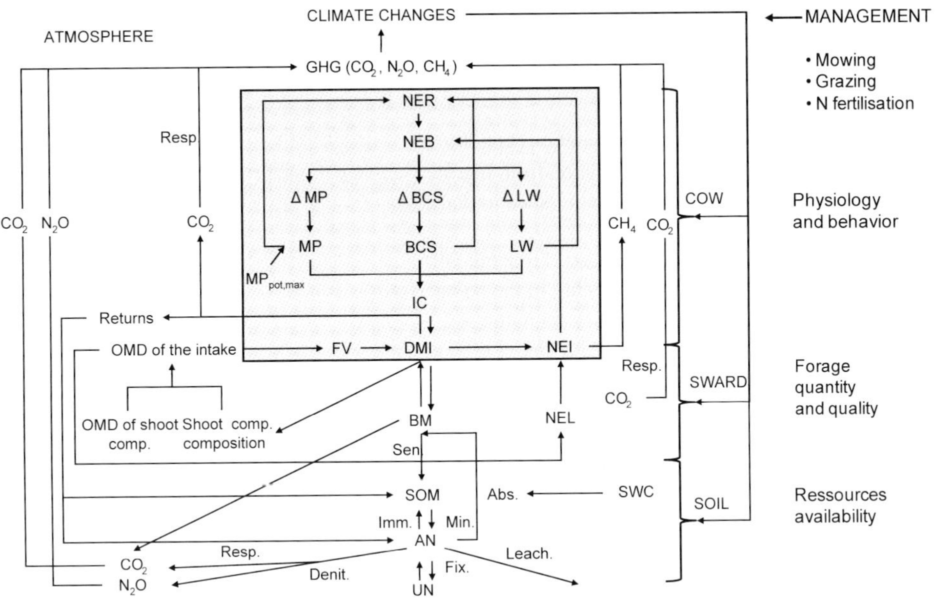

Arrows show how soil, vegetation, herbivores, climate and management interact. Grasslands contribute to net GHG emissions through their net CO_2 balance, through enteric CH_4 production and N_2O emissions from nitrification and denitrification processes. In return, climate changes affects grassland functioning by limiting cow ingestion, by modifying soil water content and nutrients available for sward growth and thus, standing biomass for animals, and by influencing plant processes such as photosynthesis, transpiration and senescence.

Appendix B. Abbreviations

Abs.	Root absorption
AN	Available nutrients in soil for vegetation growth
BCS	Body condition score (0-5)
ΔBCS	Daily variation of BCS
BM	Pasture aboveground biomass (kg DM/ha)
CH_4	Enteric methane production from ingested herbage (kg C/m^2)
Comp.	Shoot compartments: lams, sheaths and ears
Denit.	Denitrification
DMI	Daily dry matter ingestion (kg/animal/d)
OMD	Organic matter digestibility of the herbage (g/g)
Fix.	Nitrogen fixation by bacteria in the soil
FV	Fill value of the herbage (FU/kg DM)
GHG	Greenhouse gases
IC	ingestion capacity (FU)
Imm.	Immobilisation
Lea.	Nitrates leaching
LW	Cow liveweight (kg/animal)
ΔLW	Daily variation of LW (kg/animal/d)
Min.	Mineralisation
MP	Milk production (kg/animal/d)
$MP_{pot,max}$	Maximum of theoretical daily milk production (kg/animal/d)
NEB	Net energy balance (MJ)
NEI	Net energy Intake (MJ)
NEL	Net energy content of the grazed herbage (MJ)
NEM	Net energy requirement for maintenance (MJ)
NEP	Net energy requirement for production (MJ)
NER	Total net energy requirement (maintenance and production) (MJ)
SD	Stocking density (animal/m^2)
Sen.	Senescence
SOM	Soil organic matter
UN	Unavailable nutrients in soil for vegetation growth
SWC	Soil water content

Part 6 – Modelling the environmental impact of animal production

The fate of dietary phosphorus in the digestive tract of growing pigs and broilers

M.P. Létourneau-Montminy[1,5], A. Narcy[1], M. Magnin[2], P. Lescoat[1], J.F. Bernier[4], C. Pomar[5], D. Sauvant[3] and C. Jondreville[6]
[1]INRA, UR83, 37380 Nouzilly, France
[2]BASF Nutrition animale, 53200 Château-Gontier, France
[3]INRA, UMR791, AgroParisTech, 75231 Paris, France
[4]Université Laval, Département des Sciences Animales, Québec, QC, G1V 0A6, Canada
[5]Agriculture and Agri-Food Canada, Sherbrooke, QC, J1M 1Z3, Canada; Marie-Pierre.Letourneau@agr.gc.ca
[6]INRA, USC 340, Nancy Université, 54500 Vandoeuvre-lès-Nancy, France

Abstract

The reduction of the phosphorus (P) content in the diet without affecting performance is an environmental and economic concern for sustainable pig and poultry production that requires precise knowledge of P requirement and its availability. A prerequisite is thus a good understanding of the fate of ingested P in the digestive tract, according to its origin (phytic, non phytic, mineral), its interaction with dietary calcium (Ca) and exogenous phytase supply. With this aim, a mathematical model was constructed using *in vitro* or *in vivo* literature data. The core of the model is based on a compartmental structure which distinguishes three sections in which successively occur: (1) P solubilisation and phytic P hydrolysis, (2) P absorption and (3) P insolubilisation. To better understand the physicochemical and transit characteristics that could be key factors in P utilisation, parameterisation of equations was conducted in both growing pigs and broilers. Sensitivity analysis indicated that P solubilisation/insolubilisation and especially phytic P solubilisation prior to hydrolysis by phytase were sensitive to pH and dietary Ca supply. Transit time also modulated the extent of P hydrolysis. Behaviour analysis indicated that the model was able to predict the lower efficacy of plant phytase compared with microbial phytase in both species. The negative impact of dietary Ca on digestive P utilisation was higher in pigs than in broilers due to physicochemical differences in the gastric section. The model accurately predicted apparent digestibility of P in pigs. External validation could not be performed in broilers due to lack of digestibility data. Before the model can be used as a predictive tool, additional information is needed to validate some parameters, particularly for broilers. The model offers the potential to be combined with a metabolic model that integrates regulation of P and Ca metabolism.

Keywords: mathematical model, mineral, monogastric, environment

Introduction

Reducing the dietary phosphorus (P) given to pigs and broiler, without negatively affecting their health or productivity, is a way to reduce releases of this potentially environmentally harmful element. The approach is based on more accurately adjusting dietary P intake to meet the animals' requirements and on seeking ways to improve P availability in feed. Implementing such a strategy requires, among other things, a good understanding of the fate of P in the digestive tract depending on its origin, particularly in interaction with dietary calcium (Ca) and exogenous phytase intake. With the objective of improving the understanding of P and Ca metabolism and their integration, a deterministic and mechanistic compartmental modelling approach was chosen. To better understand the phenomena under contrasting conditions (pH and rate of transit), the work was carried out both in pigs and broilers. To date, few models exist of P and Ca metabolism in pigs (Fernandez, 1995) and broilers (Hurwitz *et al.*, 1983; Kebreab *et al.*, 2009). To our knowledge, no models have addressed digestion of P and Ca in detail. The objective of this study is to model the quantities of P and Ca absorbable per day that represent the inputs of a metabolic sub-module integrating P and Ca fluxes at the animal level. In this paper, we will emphasis on P and on the comparison between the two species.

Materials and methods

General structure

The model is compartmental and divides the digestive tract on sections that are involved in P and Ca digestion: (1) the gastric area (i.e. the stomach in pigs and the crop and proventriculus and gizzard in broilers), where partial solubilisation of the ingested compounds and, eventually, hydrolysis of phytates by exogenous phytase occur, (2) the proximal small intestine (pSI), where a portion of the soluble compounds are absorbed, and (3) the distal small intestine (dSI), where certain compounds are insolubilised because of an increase in pH. In each compartment, P is divided into its phytic (PP) and non-phytic (NPP) forms, which can be solubilised (PPs, NPPs) or non-solubilised (PPns, NPPns). Ca is divided into its non-solubilised plant form (CaPns) and its solubilised (Cas) or non-solubilised (Cans) non-plant forms (Figure 1).

There are thus seven state variables per compartment, with three compartments in pigs and four in broilers (Figure 2).The model integration step is one hour and the digestive phenomena can be measured at that scale. Model outputs, however, are expressed on daily basis, which is the usual time for representing livestock feeding and performance. The model was parameterised using the ModelMaker software (ModelMaker, Wallingford, United Kingdom). Parameter values were estimated from *in vivo* or *in vitro* data and adjusted until they provided an adequate fit of the data.

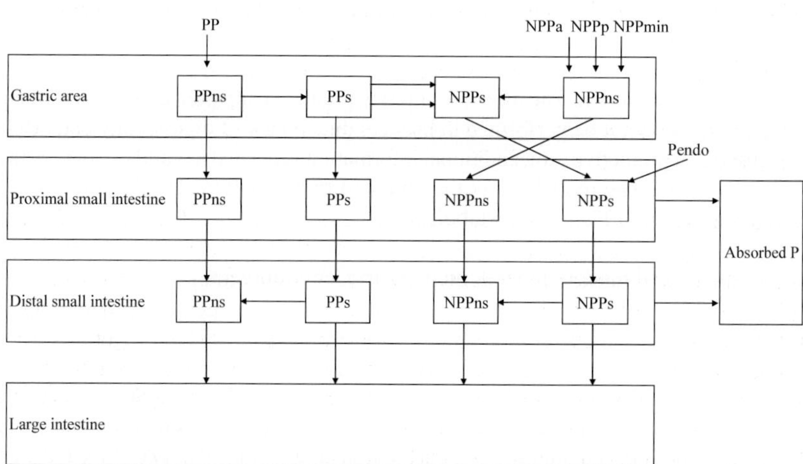

Figure 1. Diagram of the model. Inputs are: phytic P (PP), non-phytic P (NPP), of animal (NPPa), plant (NPPp) and mineral (NPPmin) origin and endogenous P (Pendo). Three anatomical sections: gastric area, proximal and distal small intestine. Four pools within each sections: PP non-solubilised (PPns), PP solubilised (PPs), NPP non-solubilised (NPPns), and NPP solubilised (NPPs). Two outputs: large intestine and absorbed P.

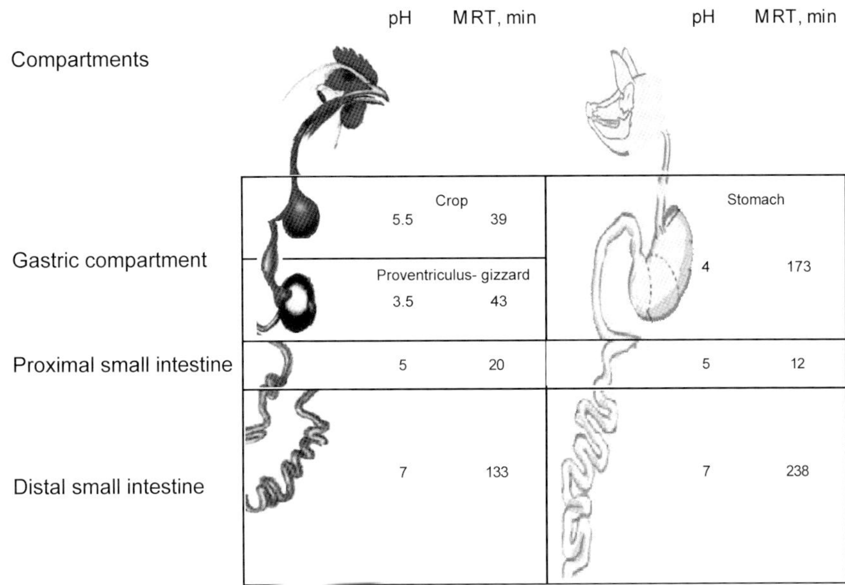

Figure 2. pH and mean retention time condition (MRT, min) of the model compartments.

Inputs and outputs

Model inputs are dietary sources of P, in PP and NPP forms, which is of animal (NPPa), plant (NPPp) or mineral (NPPmin) origin, and Ca of plant (Cap) or mineral (Camin) origin; these inputs enter the stomach. The model inputs also include endogenous sources that enter the duodenum. Feed intake, considered to be continuous feeding, is represented by 24 meals a day, corresponding to the 24 time-steps of the model. The model outputs represent P and Ca absorbed in the intestinal segments, depending on active or passive mechanisms, and unabsorbed P and Ca that enter the large intestine.

Characterisation and representation of the different digestive segments involved in the fate of dietary phosphorus and calcium

The model represents a pig weighing about 50 kg and a broiler of 21 days of age. These characteristics correspond to the majority of data on P utilisation reported in the literature. Given the key role played by pH in the digestive utilisation of P and Ca, the model compartments were divided to isolate different pH zones. The gastric area is the digestive area where solubilisation and hydrolysis occur and consists of one compartment in pigs and two compartments in broilers (Figure 2). Absorption and insolubilisation occur in the small intestine and is represented by two sub-segments to account for the intense absorption and low pH in the proximal portion (pSI), and for the insolubilisation of P and Ca and higher pH in the distal portion. Each compartment is characterised by the prevailing pH conditions and the mean retention time (MRT) of the digesta. The pH and MRT values are averages of the data in the literature.

Fate of phosphorus in the gastric segment

A portion of PPns may be solubilised (PPs), the step prior to its hydrolysis by the exogenous phytase that changes it to the NPPs form. Solubilisation was represented by the following equation:

$$R_{sol,PPns} = \max(((K_{sol,PPns} - (PPs/(PPs + PPns))) \times PPns),0) \quad (1)$$

Thus, when the phytase hydrolyses a portion of the PPs, a similar portion of the PPns is solubilised, a mechanism that was previously proposed by Kemme *et al.* (2006). The solubilisation coefficient ($K_{sol,PPns}$) depends on the pH and was parameterised using data from Scheurmann *et al.* (1989). The solubilisation coefficient also depends on the dietary Ca level, because high Ca intakes reduce the solubilisation of PPns. Hydrolysis of PPs followed a conventional Michaelis-Menten law:

$$R_{hydro,PPns} = V_{max}\ S/K_m + S \quad (2)$$

where the maximum velocity (V_{max}) of hydrolysis comes from *in vitro* studies (Ullah, 1987), and the mean quantity of substrate to achieve 50% V_{max} (K_m) was adjusted using *in vivo* hydrolysis data in pigs (e.g. Rapp *et al.*, 2001). Because of a lack of similar *in vivo* data in broilers, the same hydrolysis flux parameters were used in both species. Hydrolysis depends on the origin of the phytase, the gastric pH (Eeckhout and de Paepe, 1992), and the presence of proteases (Frapin, 1996; Philippy *et al.*, 1999; Rapp *et al.*, 2001). The P ingested in the non-phytic form joins the pool of NPPns and can then be partially solubilised:

$$R_{sol,NPPns} = K_{sol,NPPns} \times NPPns \quad (3)$$

where the value of $K_{sol,NPPns}$ depends on the form in which the NPPns is provided, namely NPPa, NPPp or NPPmin, with NPPmin divided into different categories based on the type of phosphate provided (e.g. monocalcium phosphate, dicalcium phosphate; Jongbloed *et al.*, 2002).

Fate of phosphorus in the proximal small intestine

In the pSI, the P present in the non-phytic solubilised form can be absorbed through two mechanisms: an active saturable absorption described by a Michaelis-Menten equation similar to Equation 1; and a passive absorption proportional to the NPPs pool with a linear equation similar to Equation 2, using a passive absorption coefficient ($K_{abs,pass,NPPs}$). The values of the parameters of these two absorption fluxes were estimated from the work of Fox *et al.* (1978) using isolated jejunal loops in pigs. Because of a lack of similar data in broilers, the values obtained in pigs were used for both species.

Fate of phosphorus in the distal small intestine

In the dSI, the P present in the non-phytic solubilised form can also be absorbed passively and actively, fluxes for which the parameters have been estimated, as in the proximal portion, from the work of Fox *et al.* (1978). A portion of the NPPs is also insolubilised (NPPns) because of the pH of that compartment (pH 7), which promotes the formation of insoluble complexes with Ca (Hurwitz and Bar, 1971; Cromwell, 1996). This phenomenon was parameterised on the basis of the relative distribution of the ionic species of phosphates as a function of pH (Vanden Bossche, 1999).

Results and discussion

A sensitivity analysis was first performed to determine the factors that influence P and Ca utilisation the most. This was followed by a two-step validation process. First, an external validation by simulating several diets and comparing the results of apparent digestibility coefficient of P simulated by the model with measured values was performed. Second, to compare the response of both species, specifics simulations were performed.

Sensitivity analysis

A conventional univariate sensitivity analysis was performed by iteratively varying major input variables according to its standard deviation, while keeping all other input variables at their central value. The calculated univariate sensitivity indices are expressed relative to the results obtained with original standard values. When transit and absorption parameters were considered, values of corresponding parameters in pSI and dSI were both altered at the same time. A reference situation, namely a feed containing (per kg feed): 2.5 g PP, 1.1 g NPPp, 6 g Ca and 500 FTU microbial phytase was used. The results of the sensitivity analysis of the apparent digestibility coefficient of P at variations of the main parameters were tested.

The parameters of the hydrolysis flux are those that affect the apparent total tract digestibility (ATTD) of P the most, with similar amplitude in pigs and broilers (Table 1). That similarity between the two species is not surprising, given that the parameters obtained in pigs were directly applied to broilers because of a lack of data. The hydrolysis parameters were parameterised using *in vitro* and *in vivo* data in pigs. Given the significant variability of these parameters in broilers, it will be essential to further validate these. Because of the regulating effect of rate of transit on the contact time between the enzyme and its substrate, that rate also appears to be a key parameter of PP hydrolysis. This is particularly true in the crop of chickens and the stomach of pigs, whereas a pH change in the proventriculus-gizzard has less impact on P ATTD. Gastric emptying has been the subject of many studies in pigs and appears to be variable between literature sources. In contrast, data on transit time in the crop, proventriculus and gizzard of chickens are scarce. The importance of transit in the gastric area in pigs and broilers means that additional studies are needed to validate the current values. A change in the rates of transit in the segments of the small intestine also modifies P ATTD, an effect that is slightly more marked in broilers. Dietary Ca intake also has a significant effect on P digestibility, with greater magnitude in pigs. The higher sensitivity of P ATTD to dietary Ca intake in pigs originates from the fact that Ca affects the digestive utilisation of P in both the stomach and the dSI, as opposed to an effect only in the dSI in broilers.

Table 1. Univariate sensitivity indices for apparent total tract digestibility of phosphorus.

	Pigs			Broilers		
Variation, STD	-1	0	1	-1	0	1
Rate of transit						
Stomach	0.91	1.00	1.08			
Crop				0.94	1.00	1.05
Proventriculus-Gizzard				0.98	1.00	1.02
Small intestine	0.89	1.00	1.06	0.83	1.00	1.10
pH						
Stomach	0.99	1.00	1.04			
Crop				0.97	1.00	0.99
Proventriculus-Gizzard				1.00	1.00	1.01
Dietary Ca	1.09	1.00	0.90	1.04	1.00	0.96
V_{max}phyt	0.90	1.00	1.07	0.91	1.00	1.07
K_mPhyt	1.08	1.00	0.94	1.09	1.00	0.94

V_{max}phyt: maximum velocity constant of phytase, $K_m$phyt: affinity constant of phytase.

Behaviour of the model

The present study was carried in pigs and broilers to understand the phenomena in contrasting conditions, mainly caused by differences in the physicochemical conditions in the digestive tract. In particular, the effectiveness of exogenous phytase differed between both species due to differences in pH and the residence time of digesta. It is therefore important to test the behaviour of the model for those specific points.

In a feed containing (per kg of feed) 2.3 g PP, 6.0 g Ca and 500 FTU microbial phytase, microbial phytase provided 0.69 and 0.64 g digestible P for pigs and broilers, respectively, representing PP hydrolysis of 33% and 26%. This result is in line with that reported by other (e.g.Tamin *et al.*, 2004; Kornegay *et al.*, 2001). The positive effect of phytase is thus similar in both species despite differences in pH and MRT. A comparison of the three gastric segments shows that hydrolysis is 33, 20 and 6% in the stomach, crop and proventriculus-gizzard, respectively. Given the MRT, which is 43, 39 and 173 minutes, respectively, the crop provides the most favourable conditions for PP hydrolysis by phytase. The low pH in the crop (5.5) allows the microbial phytase to achieve maximum activity. Thus, an increase in the residence time in the crop could result in even greater hydrolysis. The model provides a good representation of PP hydrolysis as a function of microbial and plant phytase supply. The model simulates hydrolysis that is about twice as low in the presence of plant phytase compared to microbial phytase in pigs. That value is in line with results obtained for pigs by Eeckhout and De Paepe (1992) and Zimmermann *et al.* (2002). In broilers, the model simulates PP hydrolysis by microbial phytase that is about 1.4 times less effective than hydrolysis by plant phytase, in line with previous studies (Frapin, 1996). The fact that plant phytase is less affected than microbial phytase in broilers is in large part due to the absence of proteases in the crop, resulting in greater effectiveness compared with a degradation of plant phytase of 40 and 65% in the stomach of pigs and the proventriculus-gizzard of broilers, respectively (Frapin, 1996; Rapp *et al.*, 2001).

Validation

The apparent digestibility of P is mainly the result of the hydrolysis of PP by phytases, of the insolubilisation of NPP, and of the insolubilisation of absorbable P by Ca. In practice, digestibility is commonly used to determine the nutritional value in terms of P attributed to a feed or a raw material. It is therefore important to study the prediction quality of the model for these two digestibility variables. Given the difficulty of separating broilers urine and faeces, a problem that greatly limits the number of digestibility data, external validation could not be performed in broilers.

It is difficult to directly compare the apparent digestibility coefficients simulated with the model and those measured in the literature because of high inter-experiment variability compared to the variability measurable within a single experiment (Sauvant *et al.*, 2008). For example, the study effect includes the effects of the analysis method, live weight, breed, etc. A model that includes the study effect was therefore created, as proposed previously by Offner and Sauvant (2004) and more recently used by Strathe *et al.* (2007). The GLM procedure of the SAS software program (1990) was used to consider the study effect in a fixed manner, as proposed by Sauvant *et al.* (2008).

In the external validation, the intercept and slope did not differ from 0 and 1, respectively, showing that the model correctly predicted the digestive utilisation of P within tests, independent of its construction (Figure 3). Nevertheless, the model slightly underestimated P ATTD, particularly for the high digestibility values. Not all experiments were representative of practical conditions and in some studies high levels of plant phytase or PP were used. The underestimation at high P ATTD may

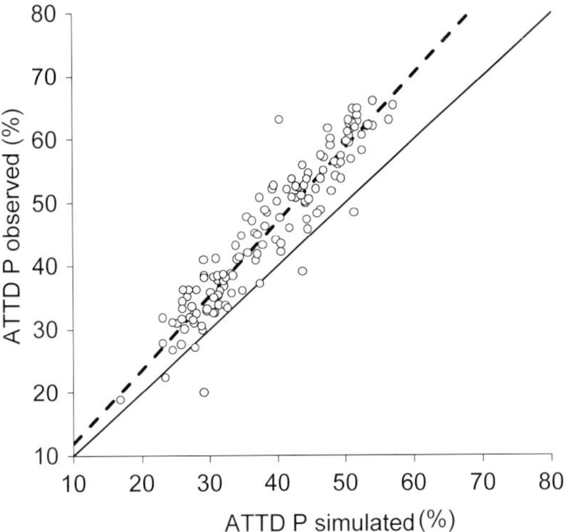

Figure 3. Apparent total tract digestibility (ATTD) of P observed, adjusted for study effect, and simulated by the model. (Dashed line: P ATTD observed = -0.005 + 1.18 P ATTD simulated, RMSE = 0.05, R^2 = 0.94, intercept and slope did not differ to 0 and 1 respectively; Continuous line = Y = X; each point represents one observation).

also be caused by an adaptation of the animal, leading to increased absorption, which is currently not accounted for in the model. The underestimation may also originate from an overestimation of the portion of P that is actively absorbed, which does not allow increased absorption when P intake levels are high.

Conclusions and perspectives

The objective of the model was to simulate the fate of P and Ca in the digestive tract of growing pigs and broilers as a function of diet form, the presence of exogenous phytase, and interactions between P and Ca. The ultimate goal was to improve understanding of the components of the P and Ca metabolism, including their interaction. To the authors' knowledge, neither the hydrolysis of phytates by phytase nor the digestive interactions of P and Ca have been considered in mathematical models in growing pigs or broilers. In pigs, the model adequately predicts P digestibility, a validation that could not be performed in broilers because of a lack of data. Nevertheless, the model has certain limitations.

Given the sensitivity of the fate of P to the MRT of the digesta in the different segments of the digestive tract and the high variability of the MRT values measured in the literature, more research could be helpful. However, no digestion model created to date (e.g. Bastianelli *et al.*, 1996; Strathe *et al.*, 2007) has taken into account the regulation of transit time. The digestibility of P is also sensitive to the pH in the digestive tract, especially in the gastric area. For simplification purposes, the pH is currently fixed within each segment. The representation of feed intake through meals, and the associated variations in pH, would make it possible to test the impact of fluctuations in pH, in relation to solubilisation of phytates and hydrolysis by phytases. It would also be important to validate *in vivo* the hypotheses concerning phytate solubility as a function of pH that currently come from *in vitro* studies.

It was difficult to evaluate the absorption of P, due to the lack of *in vivo* data. The absorption of P and Ca is, however, not very sensitive to variations in the associated parameters. The further development of the model will probably need to account for the metabolic regulation of absorption. The same is true for the endogenous secretions, which can be approached more appropriately when the metabolic portion is addressed. Consequently, the current model cannot be used (yet) as a prediction tool to determine the fate of dietary P and Ca.

References

Bastianelli, D., Sauvant, D. and Rérat, A., 1996. Mathematical modeling of digestion and nutrient absorption in pigs. Journal of Animal Science 74:1873-1887.

Cromwell, G.L., 1996. Metabolism and role of phosphorus, calcium, and vitamin d3 in swine nutrition. In: Coelho, M.B. and Kornegay, E.T. (eds.), Phytase in animal nutrition and waste management. BASF Corporation, Mount Olive, NJ, USA, pp.101-110.

Eeckhout, W. and De Paepe, M., 1992. Phytase de blé, phytase microbienne et digestibilité apparente du phosphore d'un aliment simple pour porcelets. Revue de l'Agriculture 45:195-207.

Fernandez, J.A., 1995. Calcium and phosphorus metabolism in growing pigs. III. A model resolution. Livestock Production Science 41:255-261.

Fox J. and Care, A.D., 1978. The use of a thiry-vella loop of jejunum to study the intestinal absorption of calcium and inorganic phosphate in the conscious pig. British Journal of Nutrition 39:431-439.

Frapin, D., 1996. Valorisation du phosphore phytique végétal chez l'oiseau: Intérêt et mode d'action des phytases végétales et microbiennes. École Nationale Supérieure Agronomique de Rennes, Rennes, France, 133 pp.

Hurwitz, S. and Bar, A., 1971. Calcium and phosphorus interrelationships in the intestine of the fowl. Journal of Nutrition 101:677-686.

Hurwitz, S., Fishman, S., Bar, A., Pines, M., Riesenfeld, G. and Talpaz, H., 1983. Simulation of calcium homeostasis: Modeling and parameter estimation. American Journal of Physiology 245:R664-R672.

Jongbloed, A.W., Kemme, P.A., De Groote, G., Lippens, M. and Meschy., F., 2002. Bioavailability of major and trace minerals. EMFEMA International Association of the European (EU) Manufacturers of Major, Trace and Specific Feed Mineral Materials, Brussels, 112 pp.

Kebreab, E., France, J., Kwakkel, R.P., Leeson, S., Darmani Kuhi, H. and Dijkstra, J., 2009. Development and evaluation of a dynamic model of calcium and phosphorus flows in layers. Poultry Science 88:680-689.

Kemme, P.A., Schlemmer, U., Mroz, Z. and Jongbloed, A.W., 2006. Monitoring the stepwise phytate degradation in the upper gastrointestinal tract of pigs. Journal of the Science of Food and Agriculture 86:612-622.

Kornegay, E.T., 2001. Digestion of phosphorus and other nutrients: The role of phytases and factors influencing their activity. In: Bedford, M.R. and Partridge, G.G. (eds.), Enzymes in farm animal nutrition, CAB International, Wallingford, UK, pp.237-271.

Offner, A. and Sauvant, D., 2004. Comparative evaluation of the Molly, CNCPS, and LES rumen models. Animal Feed Science and Technology 112:107-130.

Phillippy, B.Q., 1999. Susceptibility of wheat and *Aspergillus niger* phytases to inactivation by gastrointestinal enzymes. Journal of Agricultural and Food Chemistry 47:1385-1388.

Rapp, C., Lantzsch, H.J. and Drochner, W., 2001. Hydrolysis of phytic acid by intrinsic plant or supplemented microbial phytase (*Aspergillus niger*) in the stomach and small intestine of minipigs fitted with re-entrant cannulas. Journal of Animal Physiology and Animal Nutrition 85:406-413.

SAS Institute, 1990. Statistical analysis of software package version 8.1 SAS Institute Incorporated., Cary, N.C., USA.

Sauvant, D., Schmidely, P., Daudin, J.J. and St-Pierre, N., 2008. Meta-analyses of experimental data: Application in animal nutrition. Animal 2:1203-1214.

Scheuermann, V.S.E., Lantzch, H.-J. and Menke, K.H., 1989. *In vitro* und *in vivo* Untersuchungen zur Hydrolyse von Phytat II. Aktivitat Pflanzlicher Phytase. Journal of Animal Physiology and Animal Nutrition 60:64-75.

Strathe, A.B., Danfaer, A. and Chwalibog, A., 2007. A dynamic model of digestion and absorption in pigs. Animal Feed Science and Technology 143:328-371.
Tamim, N.M., Angel, R. and Christman, M., 2004. Influence of dietary calcium and phytase on phytate phosphorus hydrolysis in broiler chickens. Poultry Science 83:1358-1367.
Ullah, A.H. and Gibson, D.M., 1987. Extracellular phytase (e.C. 3.1.3.8) from *Aspergillus ficuum* nrrl 3135: Purification and characterization. Preparative Biochemistry 17:63-91.
Van den Bossche, H., 1999. Devenir du phosphore apporté sur les sols et risques de contamination des eaux de surface. Cas des boues de stations d'épuration. PhD Thesis, Université de Rennes 1, Rennes, France, 338 pp.
Zimmermann, B., Lantzsch, H.J., Mosenthin, R., Schöner, F.J., Biesalski, H.K. and Drochner, W., 2002. Comparative evaluation of the efficacy of cereal and microbial phytases in growing pigs fed diets with marginal phosphorus supply. Journal of the Science of Food and Agriculture 82:1298-1304.

Modelling the profile of growth in monogastric animals

E. Kebreab[1], A.B. Strathe[2], C.M. Nyachoti[3], J. Dijkstra[4], S. López[5] and J. France[6]
[1]Department of Animal Science, University of California, Davis 95156, USA; ekebreab@ucdavis.edu
[2]Section of Nutrition, Department of Basic Animal and Veterinary Sciences, Faculty of Life Sciences, University of Copenhagen, 1870, Frederiksberg, Denmark
[3]Department of Animal Science, University of Manitoba, Winnipeg MB R3T 2N2, Canada
[4]Animal Nutrition Group, Wageningen Institute of Animal Sciences, Wageningen University, Marijkeweg 40, 6709 PG Wageningen, the Netherlands
[5]Instituto de Ganadería de Montaña (IGM), Universidad de León, Consejo Superior de Investigaciones Científicas (CSIC), Departamento de Producción Animal, 24071, León, Spain
[6]Centre for Nutrition Modelling, Department of Animal and Poultry Science, University of Guelph, Guelph, Ontario N1G 2W1, Canada

Abstract

Growth functions have been used in animal science for over a century to summarize time course data on the growth of an organism. The chapter aims to review growth functions that have been used in monogastric animal nutrition. Although the Gompertz equation (1825) has been used extensively, particularly to describe growth in pigs, alternative functions have been shown to perform better. The Gompertz and the logistic equations have a fixed point of inflexion while the Richards, López (Morgan) and von Bertanalffy equations have a flexible point of inflexion. This might give the latter an advantage over simpler equations, despite an additional parameter required to obtain flexibility. In almost all studies that compared growth equations, the López (Morgan) equation has consistently shown to be superior to the other equations in describing pig and poultry growth data. An additional important point is that time course data by nature are repeated measures on the same animal or group of animals, therefore, consideration of correlation structure is needed when using any type of growth function. All studies reviewed showed that an autoregressive correlation structure improved equation prediction significantly and should be part of the analysis. Most growth functions struggle to come up with a good estimate of initial body weight regardless of fitting procedure. However, because the objective of using a growth function is primarily to estimate rate of . and time to reach market weight, this limitation does not preclude their use in animal science.

Keywords: growth functions, nonlinear mixed models, pigs, poultry

Introduction

Swine and poultry industries face various decisions in the production cycle that include nutrient and mineral supply to animals, cost and type of feed and a range of animal health, welfare and environmental issues that affect the profitability of an operation. Prediction of growth and identifying the times of maximum growth rate and when the animals are ready for sale are important factors that contribute to the profitability of swine and poultry operations. Traditionally, growth functions have been used to relate body weight (W) to age of the animal or cumulative nutrient intake. It is useful to examine intake in order to assess genetic improvements (Bermejo *et al.*, 2003) or to increase return over feed costs.

A useful growth function should describe data well and contain biologically and physically meaningful parameters (France *et al.*, 1996a). Some of the attributes quantified using growth functions include protein deposition (Strathe *et al.*, 2010a), body weight (Kebreab *et al.*, 2007; Porter *et al.*, 2010), and mineral deposition (Kebreab *et al.*, 2010). López (2008) reviewed nonlinear functions used in animal nutrition to represent time-dependent processes and events, and examined current and potential use of these functions to describe response to nutrients. Growth functions can be broadly classified into three categories: those that describe diminishing returns behaviour (e.g. monomolecular), sigmoidal behaviour with a fixed point of inflexion (e.g. Schumacher, Gompertz, logistic), and sigmoidal behaviour with a flexible point of inflexion (e.g. von Bertalanffy, Richards, López (sometimes referred to as Morgan), Bridges, Weibull), as described in Chapter 5 of Thornley and France (2007). The flexible functions are often generalized models that encompass simpler models for particular values of certain additional parameters.

Growth functions can also be used to determine efficiency of nutrient utilization (Darmani Kuhi *et al.*, 2003), which is the derivative of the relationship between W and cumulative nutrient intake. Kebreab *et al.* (2007) and Schulin-Zeuthen *et al.* (2007) used growth functions to describe utilization of total and available phosphorus for gain.

The objective of this chapter is to review mathematical models that are applicable for analyzing growth profiles in monogastric animals.

Mathematical models

The functional form of the mathematical models that have been used for growth analysis are summarized in Table 1. The monomolecular equation is the simplest nonlinear equation used in most studies reviewed and contains three parameters to be estimated. The monomolecular equation describes the progress of a simple, irreversible first-order reaction. Among the growth functions with sigmoidal behaviour, the Gompertz equation (Gompertz, 1825) can be derived by assuming substrate (feed) is non-limiting, the quantity of growth machinery is proportional to W, and effectiveness of the growth machinery decays exponentially with time according to a decay constant (Table 1). Inflexion in this sigmoidal growth function is fixed and occurs at $W = W_f/\mathrm{e}$, where W_f is theoretical final weight of the animal. The assumptions in the logistic equation are that the quantity of growth machinery is proportional to W, the growth machinery works at a rate proportional to amount of substrate and growth is irreversible. The inflexion point is fixed at exactly half of W_f. Schulin-Zeuthen *et al.* (2008) introduced another function, the Schumacher equation, to describe growth in pigs. The function which exhibits sigmoidal behaviour with a fixed point of inflexion ($W = W_f/\mathrm{e}^2$) is analysed in detail by Thornley and France (2007) and deviates considerably from its original empirical construct described in Schumacher (1939). Like the Gompertz and logistic, the Schumacher can also be derived as a two-state variable problem in which the specific growth rate decays according to a 3/2 power function rather than log-linearly (as in the Gompertz) (Thornley and France, 2007). In the von Bertalanffy equation, the assumptions are that substrate is non-limiting, and the growth process is the difference between anabolism and catabolism. It has an inflexion point that occurs at $W = (1 - n)^{1/n} W_f$. The Richards equation is an empirical construct and therefore does not have the underlying biological basis of the von Bertalanffy. However, it belongs in the same group of classic growth functions, and its flexibility, due to its shape parameter n (dimensionless), makes it a generalized alternative to other equations (e.g. monomolecular, Gompertz, logistic) (Thornley and France, 2007). France *et al.* (1996b, Table 1) introduced a new growth function capable of describing a range of diminishing returns and sigmoidal growth patterns. It has the advantages over the Richards of being able to describe a wider variety of possibilities, and of having a more mechanistic derivation. The López equation is a generalized Michaelis-Menten equation which was derived as a growth function by López *et al.* (2000) and proposed earlier as a response function by Morgan *et al.* (1975).

Table 1. The functional forms used to describe the relationship between body weight, W(t), and age or time, t.

Function[1]	$W(t)$	References[2]
Straight line	$W_f t - W_0$	Schulin-Zeuthen *et al.* (2007)
Monomolecular	$W_f - (W_f - W_0)e^{-ct}$	Kebreab *et al.* (2007), Schulin-Zeuthen *et al.* (2007)
Gompertz	$W_0 \exp\left[(\ln \frac{W_f}{W_0})(1 - e^{-ct})\right]$	Kebreab *et al.* (2007), Schulin-Zeuthen *et al.* (2007, 2008), Darmani Kuhi *et al.* (2003), Strathe *et al.* (2010a)
Logistic	$\frac{W_f W_0}{W_0 - (W_f - W_0)\exp(-ct)}$	Strathe *et al.* (2010b), Darmani Kuhi *et al.* (2003)
Von Bertalanffy	$[W_f^n - (W_f^n - W_0^n)e^{-ct}]^{1/n}$	Kebreab *et al.* (2007), Darmani Kuhi *et al.* (2003)
Richards	$\frac{W_0 W_f}{(W_0^n + (W_f^n - W_0^n)e^{-ct})^{1/n}}$	Kebreab *et al.* (2007) Schulin-Zeuthen *et al.* (2007), Darmani Kuhi *et al.* (2003)
Weibull	$W_f - (W_f - W_0)\exp\left[-(ct)^n\right]$	Schulin-Zeuthen *et al.* (2008)
Bridges	$W = W_0 + W_f(1 - \exp(-ct^n))$	Strathe *et al.* (2010b), Craig and Schinckel (2001)
López (Morgan)	$W = \frac{W_0 K^n + W_f t^n}{K^n + t^n}$	Porter *et al.* (2010), Strathe *et al.* (2010b), Darmani Kuhi *et al.* (2003)
Schumacher	$W_0 \exp\left[\frac{\mu_0 t_0 t}{t + t_0}\right]$	Schulin-Zeuthen *et al.* (2008)
France	$W = W_0,\ t < T,$ $W_f - (W_f - W_0)\exp\left[-c(t - T) + 2d(\sqrt{t} - \sqrt{T})\right],\ t \geq T$	France *et al.* (1996), Darmani Kuhi *et al.* (2003)

[1] W_0 is initial weight, W_f is final weight, the parameters μ_0, t_0, c d and K are positive entities, $n \geq -1$.
[2] Example of studies in monogastric animals where the equation has been considered as a growth function.

The differential form of the Weibull equation is non-autonomous (i.e. explicitly time-dependent), so the equation cannot be derived as a state variable problem and therefore has little credible mechanistic basis in terms of a simple model. Nonetheless, it is a well behaved and useful function. A special case of the Weibull (with $W_0 = 0$) was proposed by Bridges *et al.* (1992) by measuring time from conception rather than from birth. The initial condition $W_0 = 0$ of the Bridges equation might limit its applicability in the study of animal growth curves (Schulin-Zeuthen *et al.*, 2008).

Statistical considerations

For growth analysis, first a database is constructed containing information collected from several animals. The animal is often the experimental unit and its response is observed at multiple time points. The observations may or may not be spaced equally, which might affect the variance-covariance structure of the data during statistical analysis. The hieratical structure of the model may be extended if some of the animals have a common background, e.g. litter mates, animals originating from the same herd. The common statistical term is multilevel model, which refers to the tree-type structure of the data. For example, in growing pigs, Strathe *et al.* (2010b) used multilevel model representation and organized data according to clusters, i.e. some observations originate from the same litter and within that litter data are clustered longitudinally according to each animal. It makes biological sense to assume that individual random effects are correlated because an animal with larger birth weight than the population might also have a larger mature weight. This assumes that the random effects related to birth and final weight are correlated positively. Breed, genotype, and sex may be considered as fixed-effect variables depending on the objective of the study.

In most cases growth profiles are generated from multiple measurements taken from the same animal, therefore correlated errors need to be considered. To define the appropriate correlation structure for the within-animal residuals, it is important to determine whether W measurements were made at equal time intervals. In such cases, a first-order autoregressive correlation structure can be implemented and added to the basic model (e.g. Porter *et al.*, 2010). If W measurements are unequally spaced in time and differ across animals, a continuous time autoregressive process of first order (CAR(1)) that has the properties for dealing with unequally spaced observations needs to be included in the model. The CAR(1) process is a generalization of an autoregressive process of first order (Littell *et al.*, 2006). The correlation structure accounts for local fluctuations in growth over time, which are not described by the expected values. Local fluctuations are natural properties of such data because environmental disturbances are always present (Strathe, 2009).

Implementation of growth functions in standard statistical packages can be summarized as follows:

$$W_{ij} = f(\beta_i, age_{ij}) + e_{ij},$$

where W_{ij} denotes the W (kg) of the ith animal ($1 \leq i \leq n_i$) at the jth day of age_{ij} ($1 \leq j \leq n_i$), n_i is number of observations within each animal, f is a known nonlinear function of age_{ij} and the parameter vector β_i. The parameter vector may include additional fixed effects, such as treatment, gender, and other covariants specific to the study objective. The e_{ij} at first instance can be assumed as identical, independent and normally distributed, but this may be a simplistic assumption if the modelling exercise deals with repeated measures. A model that includes serial correlation within animal errors should be fitted first. Proper utilization of diagnostic plots (e.g. residual and autocorrelation plots) can pinpoint model deficiencies (Strathe *et al.*, 2010b). For comparison between different models, the best performing model is normally identified using the following goodness-of-fit indicators: Akaike information criterion (AIC), Bayesian information criterion (BIC) and residual standard deviation. Bayesian information criterion and AIC are model-order selection criteria based on parsimony and impose a penalty on more complicated models for inclusion of additional parameters. Both AIC and BIC combine the maximum likelihood (data fitting) and the choice of model by penalizing the (log) maximum likelihood with a term related to model complexity. For example Leonard and Hsu (2001) define BIC as:

$$BIC = -2 \log(\hat{J}) + K \log(N),$$

where $\hat{J}$ is maximum likelihood, K is number of independent parameters in the model and N is sample size. A smaller numerical value of BIC indicates a better fit when comparing models.

The growth models can be implemented in several statistical languages. Non-linear mixed effect (NLME) modelling of growth in monogastric animals has previously been approached in SAS (SAS Inst. Inc., Cary, NC) using the NLMIXED procedure (e.g. Craig and Schinckel, 2001; Kebreab *et al.*, 2007). However, the NLMIXED-procedure only offers a single RANDOM and no REPEATED statement (Littell *et al.*, 2006). Thus, specification of multilevel NLME-models with correlation structures, which is required to give a complete description of growth data, is not possible within the framework of the NLMIXED-procedure. Implementation of multilevel NLME growth models with repeated measurements in SAS is possible by means of the %NLINMIX macro using the latest version (nlmm8.sas) (Moser, 2004; Littell *et al.*, 2006). Similar analysis can be done using open-source software such as the R language. Strathe *et al.* (2010b) showed that SAS syntax was more technical compared to the NLME function in R. Complete SAS and R codes for fitting nonlinear mixed models with specification of variance power and a continuous first-order autoregressive process using the López equation in pigs are given by Strathe *et al.* (2010b), and a similar code using a first-order autoregressive process in poultry is given by Porter *et al.* (2010).

Application of growth functions in monogastric nutrition

Comparison of three and four parameter growth functions shows that in many cases a four parameter growth function is required to describe growth adequately. For example, Kebreab *et al.* (2007) reported that using a more complex (four parameter) function resulted in lower BIC values (therefore, better) than using three parameter functions and the additional parameter was justified because it gave flexibility to a growth function in pigs. The authors argue that although the Gompertz has been used extensively, four parameter growth functions such as the Richards were superior in describing growth over time and are recommended for use in growth data analysis in monogastric animals. However, using the Richards equation can sometimes lead to a difficult optimization problem and the process can fail to converge, primarily due to difficulty in estimating initial W (W_0) values. If W_0 is known and does not need to be estimated, fixing W_0 might solve the problem of non-convergence with the Richards. On the other hand, if W_0 has already been estimated using the equation, fixing it might worsen the fit (France *et al.*, 1996a). Among the four parameter growth functions, Schinckel *et al.* (2006) showed that the López equation produced a marginally better fit for pig growth data than the Weibull and the Bridges equations. Strathe *et al.* (2010b) compared four growth functions to describe growth profiles from barrows, boars, and gilts and reported that based on AIC the ranking was López > Bridges > Gompertz > logistic. The López equation gave good predictions when it was fitted with variance power and continuous autoregressive correlation structure. One of the reasons for the logistic performing poorly when describing growth is the inflexion point occurs at a much earlier stage than half mature body size. The Gompertz has a lower inflexion point compared to the logistic and usually does better in describing growth. Schulin-Zeuthen *et al.* (2008) reported that the Schumacher equation did better than the Gompertz based on BIC values. Part of the reason could be that in the Schumacher the inflexion point occurs at an earlier age compared to the Gompertz and the logistic. The authors also reported that the Weibull was better than the Gompertz based on residual mean squares but some of the parameters were not significant. Kebreab *et al.* (2010) used the Gompertz and the López to compare 2 diets (standard and a diet with reduced inorganic P content supplemented with phytase) as to how they affect the growth profile of pigs. Both equations showed that parameter estimates of the growth profiles were very close for the 2 diets, indicating that for the average pig, reducing P in the diet and supplementing it with phytase did not alter the growth profile significantly. Figure 1 shows the population curve was indistinguishable for pigs

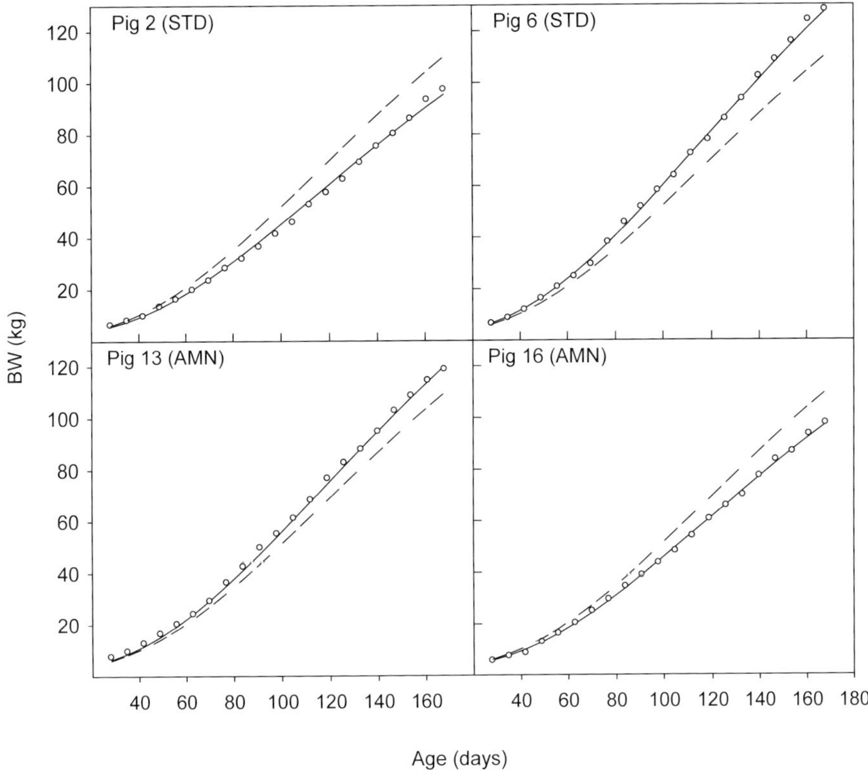

Figure 1. Selected growth profiles of 4 pigs fed standard (STD) or amended (AMN) diets (symbol). The López equation was used to produce population (broken lines) and pig specific (solid lines) curves.

on standard and amended diets, although individual pigs had variable growth profiles and random mixed models were needed to describe their growth profile accurately.

In poultry, Darmani Kuhi *et al.* (2003) compared the Gompertz, logistic, López, Richards, von Bertalanffy and France equations for describing growth in meat and egg strains of chicken. In general the authors found the four parameter equations gave a better fit but all parameter estimates were not always significant. The Richards was found to be the best fitting model and, similar to pig growth, the Gompertz fitted better than the logistic due to the lower inflexion point observed in the data. Therefore, consideration of flexible growth functions as an alternative to simpler equations (with a fixed point of inflexion) for describing the relationship between live weight and age are recommended in chickens (Darmani Kuhi *et al.*, 2003). Porter *et al.* (2010) compared three flexible growth functions (Von Bertalanffy, Richards, and López (Morgan)) and evaluated them with regard to their ability to describe the relationship between W and age in growing turkey hens from commercial flocks. They also compared the flexible functions with the Gompertz. The authors reported that the fixed inflexion point of the Gompertz was a limitation, and that the relationship between W and age in turkeys was best described using flexible growth functions. However, they ran into optimization problems with the Richards and reported that the López was best in describing growth in turkeys.

Conclusions

For describing growth in monogastric animals, a fixed point of inflexion can be a limitation with equations such as the Gompertz and logistic. In general, the point of inflexion in pigs and poultry happens at weights less than half of final weight and varies depending on age, sex, breed and type of animal. Therefore, it has been reported in many cases that four parameter equations with a flexible inflexion point fit growth data better. In some cases optimization problems can occur with equations such as the Richards but consideration is needed on a case by case basis. The López has frequently been shown to be appropriate for growth profile analysis in pigs and poultry, particularly within a non-linear mixed-effect model framework. Consideration of autoregressive correlation structure is also recommended regardless of the growth function selected.

References

Bermejo, J.L., Roehe, R., Rave, G. and Kalm, E., 2003. Comparison of linear and nonlinear functions and covariance structures to estimate feed intake pattern in growing pigs. Livestock Production Science 82:15-26.

Bridges, T.C., Turner, L.W., Stahly, T.S., Usry, J.L. and Loewer O.J., 1992. Modeling the physiological growth of swine part I: Model logic and growth concepts. Transactions of ASAE 35:1019-1028.

Craig, B.A. and Schinckel, A.P., 2001. Nonlinear mixed effects model for swine growth. Professional Animal Scientist 17:256-260.

Darmani Kuhi, H., Kebreab, E. López, S. and France, J., 2003. An evaluation of different growth functions for describing the profile of live weight with time (age) in meat and egg strains of chicken. Poultry Science 82:1536-1543.

France, J., Dijkstra, J. and Dhanoa, M.S., 1996a. Growth functions and their application in animal science, Annales de Zootechnie 45:165-174.

France, J., Dijkstra, J., Thornley, J.H.M. and Dhanoa, M.S., 1996b. A simple but flexible growth function, Growth, Development and Aging 60:71-83.

Gompertz, B., 1825. On nature of the function expressive of the law of human mortality, and on a new mode of determining the value of life contingencies. Philosophical Transactions of the Royal Society 115:513-585.

Kebreab, E., Schulin-Zeuthen, M., López, S., Dias, R.S., de Lange, C.F.M. and France, J., 2007. Comparative evaluation of mathematical functions to describe growth and efficiency of phosphorus utilization in growing pigs. Journal of Animal Science 85:2498-2507.

Kebreab, E., Strathe, A.B., Yitbarek, A., Nyachoti, C.M., Dijkstra, J., López, S. and France, J., 2010. Modelling the efficiency of phosphorus utilization in growing pigs. Journal of Animal Science, in press.

Leonard, T. and Hsu, J.S.J., 2001. Bayesian methods. Cambridge University Press, Cambridge, UK.

Littell, R.C., Milliken, G.A., Stroup, W.W., Wolfinger, R.D. and Schabenberger, O., 2006. SAS System for Mixed Models, 2nd ed. SAS Institute Incorporated, Cary, NC, USA.

López, S., France, J., Gerrits, W.J.J., Dhanoa, M.S., Humphries, D.J. and Dijkstra, J., 2000. A generalized Michaelis-Menten equation for the analysis of growth. Journal of Animal Science 78:1816-1828.

López, S., 2008. Non-linear functions in animal nutrition. In: France, J. and Kebreab, E. (eds.), Mathematical modelling in animal nutrition. CAB International, Wallingford, UK, pp.47-88.

Morgan, P.H., Mercer, L.P. and Flodin, N.W., 1975. General model for nutritional responses of higher organisms. Proceedings of the National Academy of Science 72:4327-4331.

Moser, E.B., 2004. Repeated measures modeling with PROC MIXED. Paper 188-29. Proceedings of the 29th annual SAS users group international conference. SAS Institute Incorporated, Cary, NC, USA.

Porter, T., Kebreab, E. Darmani Kuhi, H., López, S., Strathe, A.B. and France, J., 2010. Flexible alternatives to the Gompertz equation for describing growth with age in turkey hens. Poultry Science 89:371-378.

Schinckel, A.P., Pence, S. Einstein, M.E. Hinson, R. Preckel, P.V. Radcliffe, J.S. and Richert B.T., 2006. Evaluation of different mixed model nonlinear functions on pigs fed low-nutrient excretion diets. Professional Animal Scientist 22:401-408.

Schulin-Zeuthen, M., Kebreab, E., Dijkstra, J., López, S., Bannink, A., Darmani Kuhi, H., Thornley, J.H.M. and France J., 2008. A comparison of the Schumacher with other functions for describing growth in pigs. Animal Feed Science and Technology 143:314-327.

Schulin-Zeuthen, M. Kebreab, E., Gerrits, W.J.J., López, S., Fan, M.Z., Dias, R.S. and France J., 2007. Meta-analysis of phosphorus balance data from growing pigs. Journal of Animal Science 85:1953-1961.

Schumacher, F.X., 1939. A new growth curve and its applicability to timber yield studies. Journal of Forestry Research. 37:819-820.

Strathe, A.B., 2009. Stochastic modelling of feed intake, growth and body composition in pigs. PhD Thesis. Faculty of Life Sciences, University of Copenhagen, Denmark.

Strathe, A.B., Danfær, A., Chwalibog, A., Sørensen, H. and Kebreab, E., 2010a. A multivariate nonlinear mixed effect method for analyzing energy partition in growing pigs, Journal of Animal Science, in press.

Strathe, A.B., Danfær, A., Sørensen, H. and Kebreab, E., 2010b. A multilevel nonlinear mixed-effects approach to model growth in pigs. Journal of Animal Science 88:638-649.

Thornley, J.H.M. and France, J., 2007. Mathematical models in agriculture, 2nd ed. CABI Publishing, Wallingford, UK, 923 pp.

Effects of nutritional strategies on simulated nitrogen excretion and methane emission in dairy cattle

J. Dijkstra[1], J. France[2], J.L. Ellis[2], E. Kebreab[3], S. López[4], J.W. Reijs[5] and A. Bannink[6]
[1]Animal Nutrition Group, Wageningen University, Marijkeweg 40, 6709 PG Wageningen, the Netherlands; jan.dijkstra@wur.nl
[2]Centre for Nutrition Modelling, Department of Animal and Poultry Science, University of Guelph, Guelph, Ontario N1G 2W1, Canada
[3]Department of Animal Science, University of California, Davis 95616, USA
[4]Instituto de Ganadería de Montaña (IGM), Universidad de León, Consejo Superior de Investigaciones Científicas (CSIC), Departamento de Producción Animal, 24007 León, Spain
[5]Agricultural Economics Research Institute, Wageningen UR, P.O. Box 29703, 2502 LS The Hague, the Netherlands
[6]Livestock Research, Wageningen UR, P.O. Box 65, 8200 AB Lelystad, the Netherlands

Abstract

To assess the relation between emission of methane (CH_4) and faecal and urinary losses of nitrogen (N) in dairy cattle, various dietary strategies were evaluated using a mechanistic model of fermentation and digestion processes. To simulate faecal and urinary composition, an extant dynamic, mechanistic model of rumen function and post-absorptive nutrient supply was extended with static equations that describe intestinal digestion and hindgut fermentation. The extended model predicts organic matter, carbon and N output in faeces and urine. Methane emissions were simulated using the same model including a mechanistic description of methanogenesis in the rumen and in the hindgut. Four different types of grass silage were explored at high and low N fertilization levels and early or late cutting. For each grass silage, 10 supplementation strategies that differed in level and type of supplement (no supplement, maize silage, straw, beet pulp, potatoes) and level of concentrate (20 or 40% of total diet DM) were studied. Simulated total N and CH_4 excretion ranged from 211 to 588 g/d and 334 to 441 g/d, respectively, with a small, positive correlation (r^2=0.15). When expressed per unit fat and protein corrected milk (FPCM), a reduced N excretion (g N/kg FPCM) was associated with increased CH_4 emission (g CH_4 / kg FPCM) although the coefficient of determination was small (r^2=0.22). This relationship varied between different treatments. For example, reducing N fertilization level lowered N excretion per kg FPCM, but increased CH_4 emission per kg FPCM, whereas supplementation with maize silage reduced both N excretion and CH_4 emission per kg FPCM. The ratio of urea-N in urine to total N excretion was negatively related to emission of CH_4 per kg FPCM (r^2=0.54). This is of particular concern since urea in the urine, being quickly converted to ammonia, is susceptible to rapid volatilization. The present simulations indicate that measures to reduce N pollution from dairy cattle may increase CH_4 emission and highlight an important area for experimental research.

Keywords: modelling, manure composition, greenhouse gases, livestock

Introduction

Growing populations and incomes, along with changing food preferences, are rapidly increasing demand for livestock products, particularly in the developing world (Steinfeld *et al.*, 2006). Global production of milk and meat are expected to double over the next five decades. Increased production by any enterprise is likely to have negative consequences for the environment, unless steps are taken to ensure that the natural resource base (land, vegetation, water, air and biodiversity) is sustained while still increasing food production. In intensive dairy systems, major areas of concern are excretion of nitrogen (N), phosphorus (P) and greenhouse gases (GHG) including methane (CH_4) and nitrous oxide (N_2O). Nutritional control and interactions between N, P and CH_4 have been discussed (Dijkstra *et al.*, 2007a). The present paper will focus on interactions between N excretion and CH_4 emission only.

In many intensive dairy production systems, intensification has been accompanied by an increase in N surplus. This has a negative environmental impact on surface water (eutrophication), groundwater (pollution with nitrates) and on the atmosphere (de-nitrification and ammonia volatilisation). Dietary intake or nutrient density at which efficiency of production (output divided by input) is maximized is mostly different from that which maximizes financial profits (VandeHaar and St-Pierre, 2006). Therefore, surpluses of nutrients excreted in faeces and urine have increased dramatically with increased intake and production levels, whereas proportionally less feed N has been transferred to milk or body protein N. Increasing the efficiency of N use in dairy cow feeding is an important factor in decreasing environmental pollution (Kebreab *et al.*, 2002; Rotz, 2004). Nutritional strategies to reduce N excretion by cattle include reduction of N fertilization level of pasture, postponement of herbage cutting moment, and use of low protein, high energy concentrates (Børsting *et al.*, 2003; Peyraud and Astigarraga, 1998; Valk *et al.*, 2000). Diet composition not only affects the utilization of N by the cow, but also affects composition of faeces (Van Vliet *et al.*, 2007), urine and manure (Reijs *et al.*, 2007). Nitrogen utilization from soil-applied dairy cow manure has been shown to be affected by nutrition (Sørensen *et al.*, 2003; Reijs *et al.*, 2007).

Globally, agriculture accounts for 0.18 of the projected anthropogenic GHG effect. On a worldwide basis, the livestock sector produces 0.37 of anthropogenic CH_4 (Steinfeld *et al.*, 2006). Methane is produced predominantly in the rumen (0.87) and to a small extent in the large intestine (0.13) of ruminants (Murray *et al.*, 1976). Diet composition has a significant effect on enteric CH_4 formation. The energy lost through CH_4 production can range from 0.02 to 0.12 of gross energy (GE) intake (Johnson and Johnson, 1995). Ellis *et al.* (2008) recently reviewed the development and improvement of CH_4 prediction models in order to increase overall understanding of the system and to evaluate mitigation strategies for CH_4 reduction.

It is challenging to recommend diets that simultaneously reduce the potential risk of N excretion and CH_4 emission from dairy cows operations, whilst maintaining desired milk production levels and milk composition. This requires consideration of the various factors and interactions involved in a quantitative manner. Nutritional strategies to reduce N excretion may increase CH_4 emission and *vice versa* (Bannink *et al.*, 2010). Thus, it is necessary to develop models that account for pollutants in an integrative manner and evaluate diets for their potential environmental impact, while taking into account effects of dietary changes upon the profile of nutrients available for absorption and production by the animal (Dijkstra *et al.*, 2007b). Strategies to mitigate waste excretions of N and CH_4 at the whole-farm level need to include variation at the animal sublevel. Current evaluations of farm systems appear unsatisfactory in this regard (Ellis *et al.*, 2009).

The objective of this paper is to assess the relation between N excretion and CH_4 emission in dairy cattle associated with various dietary strategies, using a mechanistic model of digestion and fermentation processes that predicts the amount and composition of faecal and urinary N and C and the amount of enteric CH_4.

Material and methods

To simulate faecal and urinary nutrient excretion and composition, an extant dynamic, mechanistic model of rumen function and post-absorptive nutrient supply (Dijkstra *et al.*, 1992, 1996) was extended with static equations that describe intestinal digestion and hindgut fermentation. The extended model predicts organic matter (OM), carbon (C) and N output in faeces and urine, classified in different components representing the availability of N or OM following manure application to crops, and is described fully by Reijs (2007). Potentially degradable fibre, starch or protein not degraded in the rumen is assumed to be partly degraded in the large intestine at a rate determined by various factors including the ratio of k_d (fractional degradation rate in hindgut) and the sum of k_d and k_p (fractional passage rate in hindgut). Separate, small intestine digestion coefficients for feed and microbial protein, starch and fat are adopted. Microbial growth and VFA production in the hindgut are related to the OM fermented there assuming a fixed microbial efficiency of 24 g N/kg degraded OM and adopting the VFA coefficients for protein, starch, and fibre described by Bannink *et al.* (2006) for roughage diets. Milk production is predicted from the supply of absorbed nutrients as described by Dijkstra *et al.* (1996) for a dairy cow in mid lactation and a fixed milk composition of 40, 33 and 46 g/kg milk of fat, protein and lactose, respectively. Nitrogen excreted in urine is calculated assuming a zero N-balance. For each component excreted in faeces or urine, a specific C and N content is adopted (Reijs, 2007). Excretion of N-constituents other than urea in urine is fixed or related to body weight of the cow. The OM excreted in faeces and urine is distinguished into fibre and non-fibre components, to reflect differences in degradability of manure organic material during storage and after application to soil. The N excreted in faeces and urine is divided into three fractions representing availability of N following manure application to crops, *viz.* immediately available N (mainly urea-N), resistant N (N in undigested faecal feed components), easily decomposable N (all other N containing components).

Enteric CH_4 production was simulated using the model of Dijkstra *et al.* (1992) with a description of methanogenesis in the rumen and in the hindgut as fully reported by Mills *et al.* (2001). The VFA stoichiometric coefficients of Bannink *et al.* (2006) are adopted. In this model, the assumption is made that excess hydrogen formed in the rumen or in the hindgut, which is the net balance of hydrogen produced and hydrogen utilized in various reactions, is used completely to form CH_4. The model has been evaluated against independent data with a variety of diets and showed good agreement between observed and predicted results (Mills *et al.*, 2001; Kebreab *et al.*, 2008).

Methane and N excretion were simulated for grass silage based diets. Four different types of grass silage were explored at high and low N fertilization levels (350 and 150 kg N/ha/year, respectively) and early or late cutting (at 3,000 or 4,500 kg DM/ha, respectively). For each grass silage, 10 supplementation strategies that differed in level and type of supplement (no supplement, 50% maize silage, 15% straw, 15% beet pulp, 15% potatoes) and for each supplement in level of concentrate (20 or 40% of total diet DM) are simulated to give a total of 40 simulated diets. A reduction of fertilizer decreases crude protein and increases water soluble carbohydrate levels of grass silage. Postponement of cutting moment increases NDF and decreases crude ash content of grass silage. Full details on other assumptions and calculations for intake, composition and fractional degradation rates can be obtained from Reijs (2007).

Results and discussion

Diet composition, intake, simulated digestion and milk production are presented in Table 1. Simulated excretion of CH_4 and OM and N in faeces and urine are presented in Table 2. Large variation in diet composition, excretion of CH_4, OM and N and milk production was observed

Table 1. Diet composition, feed intake and simulated digestion and milk production for cows fed grass silage based diets (n=40 diets).

	Mean ± s.d.	Minimum	Maximum
Diet composition (g/kg DM)			
OM	903±10.0	888	925
Crude protein	162±27.8	116	225
NDF	463±36.6	404	547
Starch and sugars	190±46.2	99	271
Intake (kg DM/d)	19.6±1.73	16.0	22.4
Apparent digestion (% of intake)			
OM	74.8±3.30	70.4	81.6
Crude protein	68.8±4.70	58.9	77.5
NDF	68.3±7.07	55.8	82.6
Starch and sugars	99.6±0.00	98.7	99.8
Milk production (kg FPCM/d)	27.5±3.55	19.1	33.8

OM: organic matter, NDF: neutral detergent fibre.

Table 2. Simulated excretion of CH_4 and of nutrients in manure for cows fed grass silage based diets (n = 40 diets).

	Mean ± s.d.	Minimum	Maximum
CH_4 emission (g/d)	407±27.0	334	441
CH_4 emission (g/kg DMI)	20.8±0.94	19.0	23.0
CH_4 emission (g/kg FPCM)	14.9±1.25	12.6	17.9
OM excretion (kg/d)	5.1±0.76	3.7	6.3
Faecal OM excretion (kg/d)	4.5±0.77	3.1	5.8
Urinary OM excretion (kg/d)	0.6±0.23	0.2	1.1
N excretion (g/d)	365±77.1	211	558
Faecal N excretion (g/d)	154±12.8	128	177
Urinary N excretion (g/d)	211±83.9	81	388
Manure C:N ratio	6.6±1.76	3.4	10.6
Manure N_{imm} (% total N)[1]	48.2±22.87	29.3	63.5
Manure N_{eas} (% total N)[2]	36.4±21.01	26.9	48.9
Manure N_{res} (% total N)[3]	15.3±7.50	9.6	21.8
N excretion (g/g dietary N)	0.71±0.037	0.63	0.77
N excretion (g/kg FPCM)	13.2±2.41	8.8	17.8

OM: organic matter.
[1]Immediately available N in manure (mainly urea-N).
[2]Easily decomposable N in manure (all N not in immediately available or in resistant N in manure).
[3]Resistant N in manure (N in undigested faecal feed components).

across nutritional strategies. Average N excretions and CH_4 emissions were 365 (range: 211-558) g/d and 407 (range: 334-441) g/d, respectively. There was a weak, positive correlation between N and CH_4 excretion (r^2=0.15) (Figure 1a). A positive correlation is expected, since feed intake level is a major determinant of these excretions.

Emissions were also expressed per unit of fat and protein corrected milk (FPCM). Simulated N and CH_4 excretion were on average 13.2 (range: 8.8-17.8) and 14.9 (range: 12.6-17.9) g/kg FPCM, respectively. When expressed per unit FPCM, N excretion and CH_4 emission were negatively correlated (r^2=0.22) (Figure 1b). This relationship varied between treatments. Reducing N fertilization level lowered N excretion per kg FPCM, but increased CH_4 emission per kg FPCM. Upon an increase of N fertilization of the grass, feed intake and milk production were increased in line with observations by Valk *et al.* (2000) at the lowest and highest N fertilization levels evaluated in their study. Higher intake and production levels result in the well-known dilution of maintenance effect. As a cow eats more feed to support greater milk production, a smaller proportion of feed energy and protein intake is partitioned toward meeting maintenance needs and a greater proportion is transferred to milk (VandeHaar and St-Pierre, 2006). Thus, at lower production levels maintenance requirements of the cow and hence the basal amounts of N and CH_4 excretion are distributed over a smaller amount of product (milk), and N and CH_4 efficiency per unit FPCM at cow level tend to decrease as well. For N excretion, this decrease is more than offset by the lower N intake level (on average 575 and 442 g N/d at high and low N fertilization levels, respectively) and therefore lower N excretion by the cow. Therefore, reducing N fertilization level lowered N excretion per kg FPCM, but increased CH_4 emission per kg FPCM. Similarly, late compared with early cutting of grass silage affects feed intake and production, but also N intake levels, giving rise to lowered N excretion but increased CH_4 emission per kg FPCM produced. Such simulation results are qualitatively in line with the much higher CH_4 emission (as % of GE intake) for cows grazing pastures in summer compared to spring, attributed to grass maturity (Robertson and Waghorn, 2002). In contrast, supplementation with diets low in protein and rich in starch, such as maize silage, reduced both N and CH_4 excretion per kg FPCM. Generally, starch fermentation in the rumen gives rise to higher propionic acid and less acetic and butyric acid molar proportions (Bannink *et al.*, 2006) associated with reduced CH_4 formation in the rumen. The simulations show that strategies to reduce N excretion per unit milk may decrease or increase CH_4 emission per unit milk, depending on the actual dietary changes evaluated.

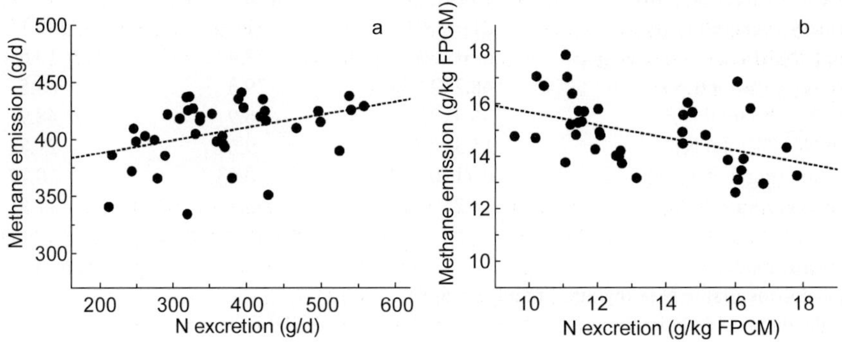

Figure 1. Relationship between simulated N excretion and CH_4 emission in dairy cattle fed grass silage based diets. Dashed line indicates linear regression. (a) CH_4 (g/d) = 366 (±17) + 0.112 (±0.041) × N (g/d), P=*0.015; r^2 = 0.146; (b) CH_4 (g/kg FPCM) = 18.1 (±1.0) – 0.243 (±0.080) × N (g/kg FPCM),* P=*0.002, r^2 = 0.218. FPCM, fat and protein corrected milk.*

The present simulations indicate that variation in response between N excretion and CH_4 emission to dietary changes may be large. However, many current whole farm approaches to evaluate management strategies at the soil-plant-animal interface employ static, empirical representations for the individual subsystems with presumed values for transfer rates between these subsystems. For example, Brink *et al.* (2001) concluded that several ammonia abatement options (including reducing dietary N content) have no or negligible effect on CH_4 emissions. However, fixed Intergovernmental Panel on Climate Change (IPCC) based CH_4 emission factors were adopted in their model, and fixed factors are not in line with the observed large variation in CH_4 emissions in response to dietary changes as simulated (Table 2; Kebreab *et al.*, 2008) or observed (Ellis *et al.*, 2008). Similarly, using a model for nutrient flows on intensive dairy farms, Schils *et al.* (2006) reported that for distinct stages in the implementation of N mitigation policies in the Netherlands (1985, 1997 and 2002) GHG emissions from dairy farms decreased. Reduction of 1 g N surplus per kg of milk (less grazing, less application of artificial fertilizer) was calculated to reduce GHG emissions by 29 g CO_2 equivalents per kg milk. Thus Schils *et al.* (2006) concluded that N surplus at the farm level is a useful indicator of GHG emission. However, Schils *et al.* (2006) used a fixed emission of 102 kg CH_4/cow/year, or in their refined approach used fixed CH_4 emission factors for only three categories of feed (viz., concentrate, grass and maize silage), ignoring the substantial variation present within a class of feedstuffs. For example, in feedlot cattle corn based concentrates result in markedly lower CH_4 emission than barley-based concentrates (2.8 and 4.0% of GE intake, respectively) (Beauchemin and McGinn, 2005). Equally, large differences in CH_4 emission related to grass maturity have been observed (Robertson and Waghorn, 2002). In these situations, clearly a more detailed analysis of variation in CH_4 emission due to dietary changes is needed to evaluate mitigation options for varying farm management conditions.

The type and proportion of N excreted in faeces versus N excreted in urine affects potential volatilization and leaching losses as well as the potential fertilizer value of manure (Sørensen *et al.*, 2003; Reijs *et al.*, 2007). Therefore, the relationship between type of N excreted and CH_4 emission from the cow was evaluated as well (Figure 2a-d). The ratio of urea-N in urine to total N in manure was negatively related to CH_4 emission per kg FPCM ($r^2 = 0.54$). In contrast, excretion of easily decomposable N and in particular resistant N fractions in dairy cattle manure were positively related to CH_4 emission per kg FPCM ($r^2 = 0.22$ and 0.54, respectively). This is of particular concern since urea in the urine, being quickly converted to ammonia, is susceptible to rapid volatilization, and in the field urinary-N may leach significantly to lower depths in the soil (Pakrou and Dillon, 1995). According to our simulations, N-mitigation options aimed at reducing urinary N excretion, may well result in elevated CH_4 emission levels per kg FPCM. In view of the lack of actual observations on the relationship between urinary N excretion and CH_4 emission, the present simulations therefore highlight an important area for experimental research.

Conclusions

Changes in dietary composition to decrease excretion of N and emission of CH_4 are possible. The simulations indicate that excretions of N and CH_4 per unit of milk are slightly negatively related. Since most of the avoidable N losses are in urine and have the most distinct negative environmental effects, the clear negative relation between urea-N and CH_4 excretion is of particular concern. However, these simulated relationships remain to be confirmed by experimental research. Strategies to mitigate waste emissions of N and CH_4 by dairy cattle at the whole-farm level need to include variation at the cow sublevel.

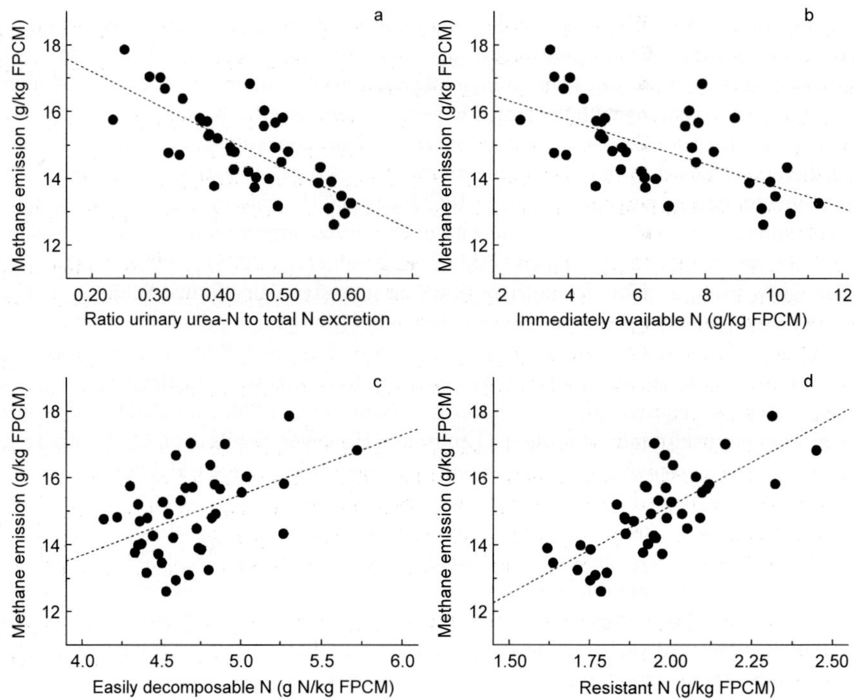

Figure 2. Relationship between simulated N excretion and CH_4 emission in dairy cattle fed grass silage based diets. Dashed line indicates linear regression. (a) CH_4 (g/kg FPCM) = 19.1 (±0.6) – 9.50 (±1.41) × ratio urinary urea-N to total N, P<0.001, *r^2 = 0.544; (b) CH_4 (g/kg FPCM) = 17.1 (±0.5) – 0.330 (±0.070) × immediately available N (g/kg FPCM),* P<0.001, *r^2=0.371; (c) CH_4 (g/kg FPCM) = 6.5 (±2.6) + 1.80 (±0.551) × easily decomposable N (g/kg FPCM),* P=0.002, *r^2=0.219; (d) CH_4 (g/kg FPCM) = 4.6 (±1.5) + 5.27 (±0.767) × resistant N (g/kg FPCM),* P<0.001, *r^2=0.554. FPCM, fat and protein corrected milk.*

References

Bannink, A., Kogut, J., Dijkstra, J., France, J., Kebreab, E., Van Vuuren, A.M. and Tamminga, S., 2006. Estimation of the stoichiometry of volatile fatty acid production in the rumen of lactating cows. Journal of Theoretical Biology 238:36-51.

Bannink, A., Smits, M.C.J., Kebreab, E., Mills, J.A.N., Ellis, J.L., Klop, A., France, J and Dijkstra, J., 2010. Simulating the effects of grassland management and grass ensiling on methane emission from lactating cows. Journal of Agricultural Science 148:55-72.

Beauchemin, K.A. and McGinn, S.M., 2005. Methane emissions from feedlot cattle fed barley or corn diets. Journal of Animal Science 83:653-661.

Børsting, C. F., Kristensen, T., Misciatelli, L., Hvelplund T. and Weisbjerg, M.R., 2003. Reducing nitrogen surplus from dairy farms. Effects of feeding and management. Livestock Production Science 83:165-178.

Brink, C., Kroeze, C. and Klimont, Z. 2001. Ammonia abatement and its impact on emissions of nitrous oxide and methane in Europe – Part 1: method. Atmospheric Environment 35:6299-6312.

Dijkstra, J., Neal, H.D.St.C., Beever, D.E. and France, J., 1992. Simulation of nutrient digestion, absorption and outflow in the rumen: model description. Journal of Nutrition 122:2239-2256.

Dijkstra, J., France, J., Assis, A.G., Neal, H.D.St.C., Campos, O.F. and Aroeira, L.J.M., 1996. Simulation of digestion in cattle fed sugar cane: prediction of nutrient supply for milk production with locally available supplements. Journal of Agricultural Science 127:247-260.

Dijkstra, J., Bannink, A., France, J. and Kebreab, E., 2007a. Nutritional control to reduce environmental impacts of intensive dairy cattle systems. In: Q.X. Meng, L.P. Ren and Z.J. Cao (eds.), Proceedings of the 7th International Symposium on the Nutrition of Herbivores. Herbivore Nutrition for the Development of Efficient, Safe and Sustainable Livestock Production, China Agricultural University Press, Beijing, pp.411-435

Dijkstra, J., Kebreab, E., Mills, J.A.N., Pellikaan, W.F., López, S., Bannink, A. and France, J., 2007b. From nutrient requirement to animal response: predicting the profile of nutrients available for absorption in dairy cattle. Animal 1:99-111.

Ellis, J.L., Dijkstra, J., Kebreab, E., Bannink, A., Odongo, N.E., McBride, B.W. and France, J., 2008. Aspects of rumen microbiology central to mechanistic modelling of methane production in cattle. Journal of Agricultural Science 146:213-233.

Ellis, J.L., Bannink, A., Dijkstra, J., Kebreab, E. and France, J., 2009. Prediction of methane production by cattle in some current whole farm models. In: Y. Chilliard, F. Glasser, Y. Faulconnier, F. Bocquier, I. Veissier and M. Doreau (eds.), Ruminant Physiology. Digestion, Metabolism, and Effects of Nutrition on Reproduction and Welfare, Wageningen Academic Publishers, Wageningen, the Netherlands, pp.168-169.

Johnson, K.A. and Johnson, D.E., 1995. Methane emissions from cattle. Journal of Animal Science 73:2483-2492.

Kebreab, E., France, J., Mills, J.A.N., Allison, R. and Dijkstra, J., 2002. A dynamic model of N metabolism in the lactating dairy cow and an assessment of impact of N excretion on the environment. Journal of Animal Science 80:248-259.

Kebreab, E., Johnson, K. A., Archibeque, S. L., Pape, D. and Wirth, T,. 2008. Model for estimating enteric methane emissions from United States dairy and feedlot cattle. Journal of Animal Science 86:2738-2748.

Mills, J.A.N., Dijkstra, J., Bannink, A., Cammell, S.B., Kebreab, E. and France, J., 2001. A mechanistic model of whole-tract digestion and methanogenesis in the lactating dairy cow: model development, evaluation and application. Journal of Animal Science 79:1584-1597.

Murray, R.M., Bryant, A.M. and Leng, R.A., 1976. Rates of production of methane in the rumen and large intestines of sheep. British Journal of Nutrition 36:1-14.

Pakrou, N. and Dillon, P., 1995. Preferential flow, nitrogen transformations and ^{15}N balance under urine-affected areas of irrigated and non-irrigated clover-based pastures. Journal of Contaminant Hydrology 20:329-347.

Peyraud, J.L. and Astigarraga, L., 1998. Review of the effect of nitrogen fertilization on the chemical composition, intake, digestion and nutritive value of fresh herbage: consequences on animal nutrition and N balance. Animal Feed Science and Technology 72:235-259.

Reijs, J.W., 2007. Improving Slurry by Diet Adjustments. A Novelty to Reduce N Losses from Grassland Based Dairy Farms. PhD Thesis, Wageningen University, the Netherlands.

Reijs, J.W., Sonneveld, M.P.W., Sørensen, P., Schils, R.L.M., Groot, J.C.J. and Lantinga, E.A., 2007. Effects of different diets on utilization of N from cattle slurry applied to grassland on a sandy soil in the Netherlands. Agriculture, Ecosystems and Environment 118:65-79.

Robertson, L.J. and Waghorn, G.C., 2002. Dairy industry perspectives of methane emissions and production from cattle fed pasture or total mixed rations in New Zealand. Proceedings of the New Zealand Society of Animal Production 62:213-218.

Rotz, C.A. 2004. Management to reduce nitrogen losses in animal production. Journal of Animal Science 82:E119-E137.

Schils, R.L.M., Verhagen, A., Aarts, H.F.M., Kuikman, P.J. and Šebek, L.B.J., 2006. Effect of improved nitrogen management on greenhouse gas emissions from intensive dairy systems in the Netherlands. Global Change Biology 12:382-391.

Sørensen, P., Weisbjerg, M.R., and Lund, P., 2003. Dietary effects on the composition and plant utilization of nitrogen in dairy cattle manure. Journal of Agricultural Science 141:79-91.

Steinfeld, H., Gerber, P., Wassenaar, T., Castel, V., Rosales, M. and de Haan, C., 2006. Livestock's Long Shadow: Environmental Issues and Options. FAO, Rome, Italy.

Valk, H., Leusink-Kappers, I.E. and Van Vuuren, A.M., 2000. Effect of reducing nitrogen fertilizer on grassland on grass intake, digestibility and milk production of dairy cows. Livestock Production Science 63:27-38.

VandeHaar, M.J. and St-Pierre, N., 2006. Major advances in nutrition: relevance to the sustainability of the dairy industry. Journal of Dairy Science 89:1280-1291.

Van Vliet, P.C.J., Reijs, J.W., Bloem, J., Dijkstra, J. and De Goede, R.G.M., 2007. Effects of cow diet on the microbial community and organic matter and nitrogen content of feces. Journal of Dairy Science 90:5146-5158.

A whole farm-model to simulate the environmental impacts of animal farming systems: MELODIE

X. Chardon[1,2,3], *C. Rigolot*[4,5,6], *C. Baratte*[1,2], *R. Martin-Clouaire*[7], *J.P. Rellier*[7], *C. Raison*[3], *A. Le Gall*[3], *J.Y. Dourmad*[4,5], *J.C. Poupa*[8,9], *L. Delaby*[1,2], *T. Morvan*[10,11], *P. Leterme*[10,11], *J.M. Paillat*[10,11], *S. Espagnol*[6] *and P. Faverdin*[1,2]
[1]*INRA, UMR1080, Production du Lait, 35590 St-Gilles, France; chardon@agroparistech.fr*
[2]*Agrocampus Ouest, UMR1080 Production du Lait, 35000 Rennes, France*
[3]*Institut de l'Elevage, Monvoisin, 35650 Le Rheu, France*
[4]*INRA, UMR1079, Système d'élevage nutrition animale et humaine, 35590, St-Gilles, France*
[5]*Agrocampus Ouest, UMR1079 Système d'élevage nutrition animale et humaine, 35590, St-Gilles, France*
[6]*IFIP – Institut du porc, 35650 Le Rheu, France*
[7]*INRA, UR875 Biométrie et Intelligence Artificielle, 31326 Castanet-Tolosan, France*
[8]*INRA, UMR1302 Sciences sociales, agriculture et alimentation, espace et environnement, 35000, Rennes, France*
[9]*Agrocampus Ouest, UMR1302 Sciences sociales, agriculture et alimentation, espace et environnement, 35000, Rennes, France*
[10]*INRA, UMR1069 Sol Agro et hydrosystème Spatialisation, 35000 Rennes, France*
[11]*Agrocampus Ouest, UMR1069 Sol Agro et hydrosystème Spatialisation, 35000 Rennes, France*

Abstract

The *ex ante* environmental evaluation of farming systems is increasingly demanded when proposing new developments of animal farming systems. Modelling is a promising approach to reduce the cost and the delay in studying the relationship between farming management and risky emissions. The simulation of decision is essential to better analyze *ex ante* changes in farm management, but is rarely considered in environmental models. MELODIE simulates the flows of carbon, nitrogen, phosphorus, copper, zinc and water within the whole pig and dairy farm over the long term. MELODIE upscales dynamic models developed at the field or animal scale by considering the management of the whole farm system coherently with the livestock farming system. The model is structured according to an ontology of agricultural production systems to represent the interactions between the biotechnical system and the decision system. The biotechnical module simulates the nutrient flows at a daily time step for each entity of the sub-models (soil/crop, animal and manure processes). MELODIE represents decisions at two time scales: every year, for drawing annual activity plans and every day for the context-dependent application of this plan. Thanks to the interactions between the biotechnical system and the decision system at different time scales, MELODIE is able to run consistently under different long-term climate series. The goal is to study the emerging properties of the system. Besides, because the nutrient flows within the farm are dynamically simulated, it is possible to study both the spatial and temporal heterogeneity of the environmental risks. This approach enables a better understanding of variability in the farming system according to climate. MELODIE is intended for use in research, not as a decision support system for farm management. It is a framework for virtual experimentation on animal farming systems, and could be extended to deal with other issues than nutrient flows.

Keywords: farming systems, whole-farm model, environmental impact, decision-making

Introduction

In regions of intensive pig and dairy farming, water quality can be threatened by nutrient losses from these farming systems. Greenhouse gases emissions from animal production contribute to climate change and ammonia emissions are a threat for air quality. Accumulation of trace elements like copper and zinc in soils are also sources of concern. But in mixed farms, manure and pasture can restore the organic content of soils and improve soil structure and fertility. Although farming systems impact their environment in many different ways, both positive or negative, nutrient flows are one of the first concerns, at least in regions of high animal densities. Nutrient flows and the subsequent impacts depend on many factors, including climate and soil type but also farmer decisions, from the strategic level (farming system: stocking rate, type of animal housing, crops grown, etc...) to the operational level (weather during waste application, etc...).

Changes of practices to improve one aspect can have negative consequences on other aspects, creating some dilemmas. An overall view of the system is required to avoid risks of pollution swapping. Therefore, environmental impact evaluations must include complementary criteria concerning the different aspects. Furthermore, decisions are mostly taken at farm scale. Because of the interactions between the different parts of the farm and the tradeoffs between different economic and/or environmental objectives, the environmental impact of a farming system may be different from what was inferred at lower scales. To be able to study these emerging properties, the choice was made to simulate nutrient flows at farm scale. However, at this scale, only emission indicators can be calculated, not impact indicators (Payraudeau and Van der Werf 2005).

Several dynamic models simulating farming systems exist. As far as pig and/or dairy farms are concerned, four dynamic mechanistic models deal with nutrient flows: the Integrated Farm Systems Model: IFSM (Rotz and Coiner 2004), DairyNZ's Whole Farm Model: WFM (Wastney *et al.*, 2002), Farm ASSEsment Tool: FASSET (Jacobsen *et al.*, 1998) and DairyWise (Schils *et al.*, 2007). IFSM and WFM are centred on the technical and economic results of the modelled farm and were not designed to perform long term simulations with decisions adapted to the context of each climatic year. FASSET uses a planning module generating a specific management at the beginning of each year, but few adaptations of the plan are made in the course of the year. DairyWise only performs simulations for an average climatic year. Thus, these models do not make it possible to study the interactions between climate variability, farm management and environmental impacts.

This paper presents the model MELODIE (french acronym for 'object oriented model of animal farms to evaluate their environmental impacts'), which aims to evaluate ex ante the environmental impacts of production strategies in pig and dairy farms. The evaluation of the environmental impacts is centred on nutrient flows and the associated environmental risks. The model is intended to be used in research, to compare different strategies at different time scales. The aim is not to use it directly as a decision support system for farm management. MELODIE can be applied for the main systems encountered in France (where dairy and/or pig farm often also produce annual crops), and probably in many other countries thanks to its very flexible conception.

Model description

Overview

The general organisation of the model is shown in Figure 1. The nutrients taken into account are the ones whose losses are linked with environmental risks: carbon (C), nitrogen (N), phosphorus (P), potassium (K), copper (Cu) and zinc (Zn). Water (H2O) flows are also simulated. In order to take

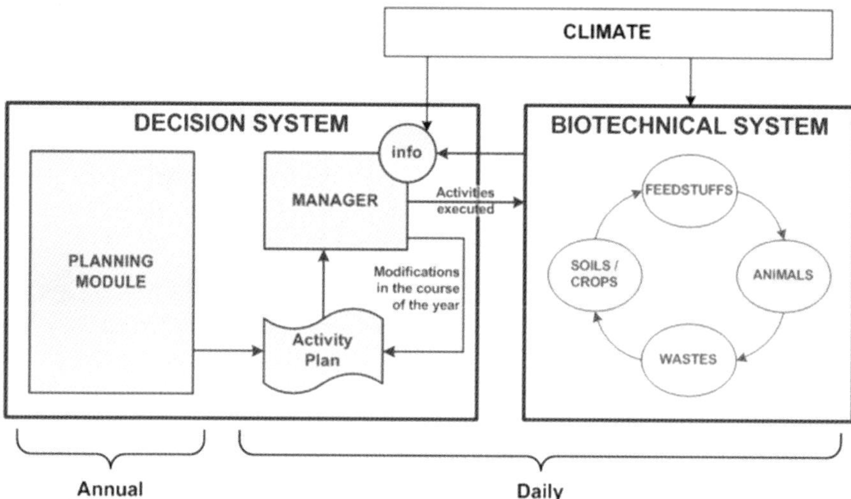

Figure 1. General organisation of MELODIE. The model consists of a biotechnical system interacting with a decision system.

climate variability into account, long term (several decades) simulations are performed, which means that farmer decisions have to be modelled, at least for tactical and operational decisions. Integrating the effect of farmers' practices is a key element to find environmentally friendly production systems.

The main outputs of the model are the different products (crops, milk, meat) and the losses to the environment of different emissions, calculated daily over decades for each animal class, field or waste storage unit. These outputs can be used to calculate indicators of environmental impact, such as those used in Life Cycle Analysis (LCA).

MELODIE is based on the ontology of agricultural production systems (Figure 2) proposed by Martin-Clouaire and Rellier (2003, 2009). In this ontology, a production system is composed of three subsystems: the biotechnical system (or controlled system), the decision system (or manager) and the operating system. The operating system includes the resources used to conduct activities, like labour and machinery. These resources are not taken into account in the present version of MELODIE, thus the operating system is not modelled.

The biotechnical system

The nutrient flows are calculated at a daily time step by the biotechnical system, which is a set of connected sub-models. Four main nutrient pools are considered (Figure 1): animals, agricultural wastes (storage and treatment), soils and crops, and feed stocks. Nutrient losses to air and water are simulated, as well as nutrient flows between and within these pools. Different levels of precision are associated with the nutrients: for example, the N cycle is more detailed than the Cu or Zn flows, for which only balances between pools are calculated. MELODIE uses existing models wherever possible. For every process covered, the existing models or equations were studied, and the most appropriate were chosen. When no suitable model was found, new models were developed.

For soils and crops, each field is represented individually. MELODIE uses Stics (Brisson *et al.*, 2003), a generic model simulating the flows of N, organic matter (i.e. C) and water, as well as

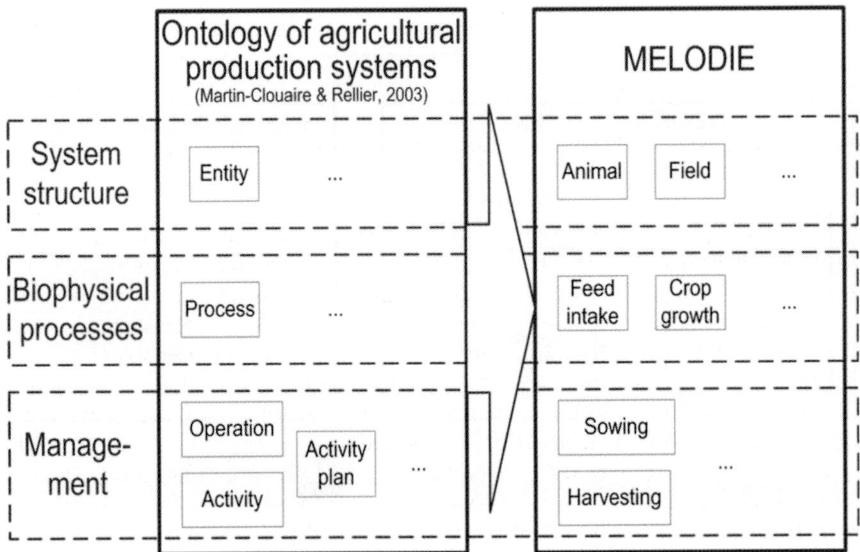

Figure 2. The objects of MELODIE are specialisations of broader concepts defined in an ontology of agricultural production systems.

crop growth and development. As far as P, K, Cu and Zn are concerned, mass balances at field scale are calculated, assuming average contents in crops. The nutrient contents of animal wastes are calculated within the wastes sub-model (see below).

Animals are not simulated individually, but by groups. For dairy cows and heifers, the model GEDEMO (Coquil *et al.*, 2005) dynamically simulates the demography of the herd, i.e. the size of the 21 groups of homogeneous animals in terms of age and/or physiological status. For pigs, the demography model is based on practical references and expert knowledge, and is closely connected with the animal housing system. A group can be constituted either of sows managed together, or of fattening pigs born at the same cycle of a group of sows. For both pigs and cattle, the nutrient flows are calculated for each group, by multiplying an individual result, calculated for an average animal, by the size of the group. For dairy cattle, feed intakes are calculated using the equations of the INRA system (INRA 2007). The model of Maxin (2006) describes the nutrient (N, C, minerals, water) balances of dairy cattle and in particular allows calculating the nutrient content of urine and faeces as well as methane production, using easily available data. For pigs, equations taken in the literature for growth, feed intake and nutrient excretion are used (C. Rigolot, personal communication).

The storage and treatment of all animal wastes are handled by a common module, which calculates the evolution of the wastes and the losses to the air, from the excretion to the land application. This module was built specifically for MELODIE from a set of existing empirical equations and emission factors (C. Rigolot, personal communication).

The decision system

Decisions made by farmers are simulated by the decision system, which interacts with the biotechnical system throughout the simulation (Figure 1). The role of the decision system is to dynamically determine the operations that should be applied to the different entities of the biotechnical system, in order to apply the farmer's management strategy. Decisions are taken at two time scales. Every year, a planning module generates plans. The ontology of agricultural production systems provides a consistent framework to describe explicitly these plans and their flexible application (and modification, if necessary and if an alternative strategy is available).

The general organisation of the planning module is shown in Figure 3. The goal of this module is to generate a cropping plan and a waste application plan for the upcoming year. Crops are allocated to fields by the cropping plan generator Tournesol (Garcia *et al.*, 2005). It considers the feed and straw requirements associated with the feeding strategy and applies agronomic knowledge (potential of the fields and effects of crop sequences) to generate a cropping plan that best satisfies the goals and priorities defined by the model user. Likewise, the model Fumigene (Chardon *et al.*, 2007) is used to generate yearly waste allocation plans, according to the needs of each field and to management rules. The needs of each field are calculated using a balance sheet method which takes into account the type of soil, the history of the field (fertilisation and past cultivation of grass if any), the crop to be grown for the current year and its expected yield. The planning sub-models Tournesol and Fumigene interact with the biotechnical system by using information on the yearly variations of stocks of feed, straw and wastes. For example, if in a given year the quantity of grazed grass is high thanks to favourable conditions, the maize silage stocks will be high at the end of the year. The cropping area devoted to maize the following year will be decreased. Similarly, if the quantity of slurry is higher than expected, more slurry applications are planned. These interactions between planning sub models and the biotechnical system are a key feature of MELODIE, enabling the adaptation of practices to climate variability considered here on an interannual basis.

The activity plan is a set of activities organized by different temporal or programmatic operators that indicate how the plan should unfold (for example sequence or iteration). The plans are examined every day, for context-dependent application. The actual dates when activities should be executed

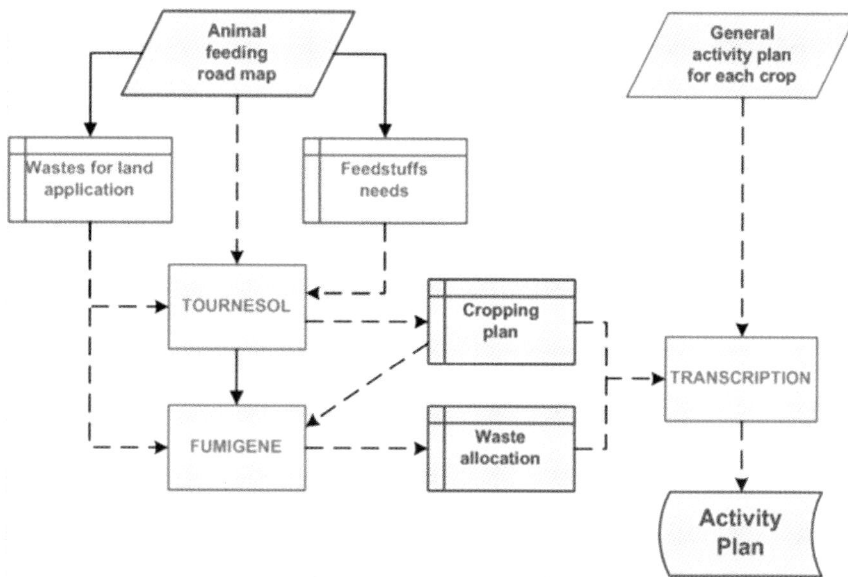

Figure 3. Organisation of the planning module, used to generate an activity plan for crops and waste management every year.

are not included. The plan only contains time windows and 'opening predicates' (conditions relating to the biophysical state) indicating when execution is possible. The operations scheduled are executed only when the conditions are suitable. If necessary, the activity plan can be modified in the course of the events.

In MELODIE, these mechanisms are used, for example, to describe dairy cows feeding and in particular grazing management. Dairy cows feeding is adapted to the resources available on farm. It is thus very variable within years (several diets are used depending on the season) as well as between years (dates of transition between diets and quantities of complementary forage at grazing are variable).

The model user must provide a feeding road map for the animals, i.e. the feed (type and amount) to be provided, for different periods of the year and for different groups of animals (Figure 4). The user must also provide an activity plan containing (1) the activities of transition between diets and (2) the activities of grazing management (paddock changes and diet adjustment).

At the beginning of a simulation, animals are indoors for wintering. In spring, when the quantity of grass available per cow on the whole farm reaches a fixed threshold, the opening predicate of the 'turn out to grazing' activity becomes true, triggering the grazing season. During this period, an activity moving the herd from a paddock to another is iterated (Figure 3). At each paddock change, the state of the system is examined and decisions are taken. The ratio between the quantity of grass available (total on all paddocks except those intended to be cut) and the quantity of grass eaten every day by the cows is compared to a desired interval (model input, in days). If too little grass is available, some paddocks initially intended to be cut are grazed and complementary forage is distributed. Conversely, if too much grass is available, complementary forage distribution is stopped and some paddocks can be cut as silage. The complementary forages to use are specified as an input

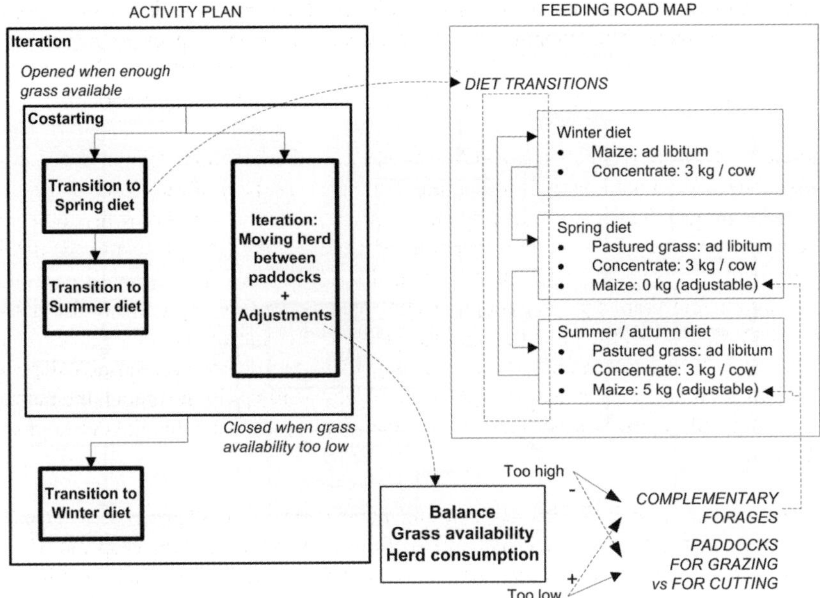

Figure 4. Example of feeding road map and activity plan for feeding. Flexible and dynamic management of feeding is defined by the combination of both.

of the model: each diet contains 'adjustable' elements, which means that the quantities specified of the corresponding feed can be altered if necessary (Figure 4). When not enough grass is available to feed the herd during one day after using 'adjustable' elements, the herd goes back to its winter diet and stays indoors. It is possible to resume grazing during the same year when grass growth allows it. The whole mechanism described here is iterated throughout the simulations. The model user provides a single general description of the management with which long term simulations are performed, including adaptations to climate and grass growth variability. Animal feeding is therefore planned, but the plan leaves room for context-dependent application.

In summary, MELODIE includes models of the decision-making processes of farmers, which is relatively rare for models at this scale. The main strength of the decision system is the deep interactions with the biotechnical system, in the operational management during the year as well as in the planning step taken every year. Thanks to these interactions at two time steps, farm management is permanently adjusted to climate conditions and to the state of the system, which is a key feature for realistic simulations of farming systems on the long term and to study both the management variability and the resilience of farming systems.

Discussion

Consistency of the farming systems simulated

MELODIE was designed to perform *ex ante* evaluations of farming systems. The goal is to evaluate different systems, existing or not, in the same context. A major difficulty is to ensure the consistency of the system simulated, particularly for those which do not yet exist in commercial farms. For example, the cropping plan must match the feeding road map of the animals and the level of feed autonomy targeted by the farmer. In Melodie, the planning module plays a key role in this regard, because it ensures the consistency of the different elements of the plan. To design a new system, it is necessary to provide the animal feeding road map, and the farmer's goals and priorities used by Tournesol and Fumigene. The cropping plan and the waste management plan are then automatically generated each year, so as to best match the goals defined by the model user. The consistency of the generated system depends on the consistency of the goals and their representation.

Another element of consistency in *ex ante* evaluation is the capability of the decision-making process to cope with climate variability and to respond to the state of the biophysical system. In practise, management decisions can vary greatly between years, in terms of dates and parameters of the activities planned, and even in terms of activities executed. The plans must be flexible and leave room for context-dependent adaptation. In MELODIE, climate variability is taken into account at two levels, thanks to the interactions between the decision system and the biotechnical system. On a yearly basis, the planning module integrates the variations of the stocks of feed and animal wastes. In the course of the year, farm management can also be very variable depending on the conditions, as demonstrated earlier in the case of herd feeding. In the model, the management is thus automatically consistent with the state of the system, provided that the decision system is correctly modelled.

The decision system in MELODIE also makes the management spatially consistent. Decisions are taken according to factors specific of each field. Two fields with the same crop can be managed differently, particularly in terms of fertilisation. During a long term simulation, each field follows its own trajectory, and the consistency is enforced at the whole farm scale. Spatial heterogeneity creates constraints and opportunities which justify taking decisions for each field individually. This capacity of the model could be particularly efficient to upscale farm models to a territory with a true spatial dimension.

Model applications

MELODIE is intended to be used in research and development, to compare the environmental impact of different production strategies in several series of yearly climatic scenarios. MELODIE is not intended to be used directly on farm as a decision support system, but should be considered as a virtual experimentation framework. The simulations should be designed like real experiments would be. Simulations can be performed for typical farm configurations in a region, in order to propose the most desirable practises and evolution in each case. MELODIE is complementary with other methods of investigation.

Conclusion

MELODIE is a true whole farm model made possible by the association of biotechnical and decision models. It upscales a set of pre-existing models of animal, plant-soil and biological processes in effluents and new decision models which are combined in a generic structure relying on the object-oriented paradigm. Thanks to this change of level of organisation, MELODIE is a framework for virtual experimentation on animal systems. It enables users to perform multi-criteria *ex ante* evaluations of the environmental pressures resulting from production strategies. Such evaluations are complementary with experimental approaches and MELODIE can be extended to include new knowledge on nutrient flows and the underlying biophysical processes. MELODIE could also be extended to deal with other issues than nutrient flows, for example in order to provide evaluations of sustainability.

Acknowledgements

This work was partially funded by the French national research agency, project SPA/DD (project no. ANR-06-PADD-017).

References

Brisson, N., Gary, C., Justes, E., Roche, R., Mary, B., Ripoche, D., Zimmer, D., Sierra, J., Bertuzzi, P. and Burger, P., 2003. An overview of the crop model STICS. European Journal of Agronomy 18:309-332.

Chardon, X., Le Gall, A., Raison, C., Morvan, T. and Faverdin, P., 2008. FUMIGENE: a model to plan the allocation of agricultural wastes at the farm level. Journal of Agricultural Science Cambridge 146:521-539.

Coquil, X., Faverdin, P. and Garcia, F., 2005. Dynamic modelling of dairy herd demography. Proceedings of Rencontres Recherche Ruminants 12:213.

Garcia, F., Faverdin, P., Delaby, L. and Peyraud, J.-L., 2005. Tournesol: a model to simulate cropping plans in dairy production systems. Proceedings of Rencontres Recherche Ruminants 12:195-198.

INRA, 2007. Alimentation des bovins, ovins et caprins: besoins des animaux – valeurs des aliments. Tables Inra 2007, Eds Quae, Paris, France.

Jacobsen, B. H., Petersen, B. M., Berntsen, J., Boye, C., Sørensen, C.G., Søgaard, H.T. and Hansen, J.P., 1998. An integrated economic and environmental farm simulation model (FASSET). Danish Institute of Agricultural and Fisheries Economics, Copenhagen, Denmark.

Martin-Clouaire, R. and Rellier, J.-P., 2003. A conceptualization of farm management strategies. Proceedings of EFITA-03 conference, July 5-9, Debrecen, Hungary, pp 719-726.

Martin-Clouaire, R. and Rellier, J.-P., 2009. Modelling and simulating work practices in agriculture. International Journal Metadata, Semantics and Ontologies 4:42-53.

Maxin, G., 2006. Modélisation des bilans Entree/Sortie des éléments carbone, azote, eau et minéraux chez la vache laitière. Masters Thesis, ESITPA, Rouen, France.
Payraudeau, S. and Van der Werf, H.M.G., 2005. Environmental impact assessment for a farming region: a review of methods. Agriculture, Ecosystems & Environment 107:1-19.
Rotz, C.A. and Coiner, C.U., 2004. The integrated farm system model: reference manual. Available at: www.ars.usda.gov/Main/docs.htm?docid=8519. Accessed May 2009.
Schils, R.L.M., De Haan, M.H.A., Hemmer, J.G.A., Van den Pol-van Dasselaar, A., De Boer, J.A., Evers, A.G., Holshof, G., Van Middelkoop, J.C. and Zom, R.L.G., 2007. Dairywise, a whole-farm dairy model. Journal of Dairy Science 90:5334-5346.
Wastney, M.E., Palliser, C.C., Lile, J.A., McDonald, K.A., Penno, J.W. and Bright, K.P., 2002. A whole-farm model applied to a dairy system. Proceedings of the NZSAP 62:120-123.

Development and validation of a biophysical model of enteric methane emissions from Australian beef feedlots

S.K. Muir[1], D. Chen[2], D. Rowell[2] and J.Hill[1]
[1]Department of Agriculture and Food Systems, School of Land & Environment, University of Melbourne, Parkville 3010, Australia; muirs@unimelb.edu.au
[2]Department of Resource Management and Geography, School of Land & Environment, University of Melbourne, Parkville 3010, Australia

Abstract

Feedlot producers face considerable pressure to reduce emissions of greenhouse gases and excretion of nitrogen and phosphorus. This paper reports on the development and validation of a biophysical model to predict greenhouse gas emissions from Australian beef feedlots, specifically enteric methane emissions. The developed model was based on the current Australian methodology for greenhouse sources and sinks, with the addition of two recently developed beef cattle specific models. The model was validated using the results of published studies and compared with emissions measured using open path spectroscopy and micrometeorology from two Australian feedlots during two seasons. The best performing equations were Ellis *et al.* (2007) and Moe and Tyrell (1979) with Lins concordance values and 95% confidence intervals of 0.4509 (0.1018) and 0.3696 (0.1362). Average residuals were 118.6 and 98.2 g/head/day for the two best performing equations. The IPCC Tier II equation demonstrated the lowest average residual 0.6 g/head/day but also the poorest concordance (Pc 0.0657, 95% CI -0.024). This study demonstrates that the current Australian methodology for estimating enteric emissions from feedlot cattle is overestimating emissions.

Keywords: greenhouse gas, beef cattle, modeling

Introduction

In Australia, beef cattle are estimated to account for 58% of livestock greenhouse gas (GHG) emissions (ALFA 2008) or a 7% of total national emissions (Chen *et al.*, 2009). Of the national beef cattle population of 28.8 million, around 680,000 are managed in feedlots at any time, generating approximately 3.5% of livestock GHG emissions or 0.4% of total national emissions (ALFA 2008, Chen *et al.*, 2009). The primary sources of GHG in feedlot systems are the animals themselves, via enteric fermentation (CH_4) and gases produced by the decomposition of manure (CH_4, N_2O and NH_3 as an indirect GHG). Typically, feedlots consist of multiple pens with watering and feeding sites, and each pen holding may hold more than 100 animals for several months (Miller and Berry 2005), the resulting build up of manure leads to high concentrations in GHG reflecting the intensity in animal production. Furthermore, the high stocking rates also lead to high concentrations of these GHGs being emitted as a point source rather than a diffuse source. The models currently used in national inventories were developed in the 1960's and 1970's from short duration metabolism experiments, and not for the purpose of greenhouse inventories. More importantly, these equations were developed based on dairy cattle fed diets which differ significantly from today's feedlot cattle rations. This study aims to investigate the accuracy of the current Australian methodology for estimating emissions from feedlot cattle and compare it to alternative equations, through the development and validation of a biophysical model.

Materials and methods

Model structure

The biophysical model is based on the Australian Methodology for greenhouse sources and sinks (NGGIC 2007). It integrates animal and production system data into a range of equations predicting greenhouse gas production from the feedlot system (Figure 1). The model incorporates equations to predict methane from enteric fermentation; Blaxter and Clapperton (1965), Moe and Tyrell (1979), Ellis *et al.* (2007) and (2009) and IPCC Tier I and II (NGGIC 2007), as well as for CH_4 emission from manure and N_2O and NH_3 from the nitrogenous compounds in manure. This discussion will focus on the structure and validation of the model for enteric methane emissions.

Data source-inputs

The model has been designed so that data obtained from commercial feedlot systems can be utilized. Feedlot management software can produce reports which contain detailed information about the current crop of cattle, including numbers, placement weights, estimated current weights, days on feed, class amounts of feed offered and detailed ration information.

Prediction of enteric methane emissions

The IPCC national greenhouse gas inventory guidelines suggests 3 levels of complexity for prediction of enteric emissions (Kebreab *et al.*, 2008), Tier I, which is equivalent to 164 g/head/day for feedlot cattle in Oceania (NGGIC, 2007), Tier II and Tier III, which is country specific (as discussed above). The Tier II model uses the gross energy content of the ration and a standard emission fact of 3(±1)% for feedlot cattle. A important issue surrounding the use of both Blaxter and Clapperton (1965) (B&C) and Moe and Tyrell (1979) (M&T) is that they were both developed based on dairy cattle and have difficulty predicting emissions outside the range on which they were developed (Ellis *et al.*, 2007; Wilkerson *et al.*, 1995). More recently a number of studies have developed models based on beef cattle (Ellis *et al.*, 2007, 2009) and a smaller number have evaluated equations for feedlot cattle (Kebreab *et al.*, 2008).

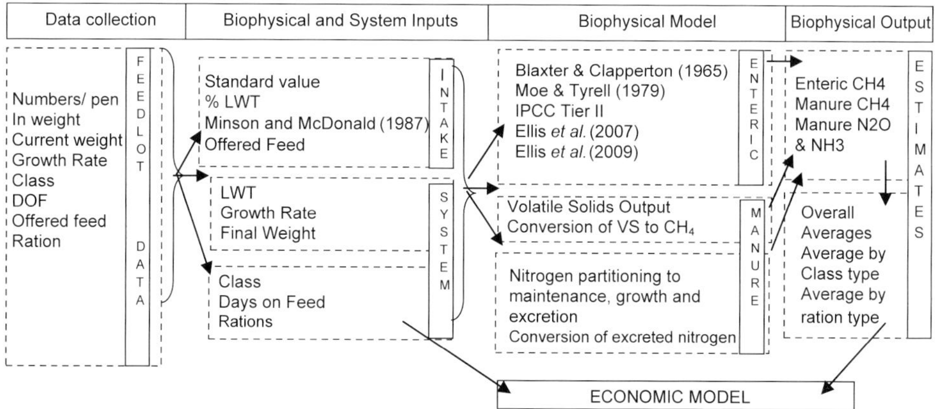

Figure 1. Diagrammatic representation of the basic modelling approach used to estimate greenhouse gas emissions from feedlot systems.

The model described in this paper includes five equations linking nutritive value to enteric methane production Blaxter and Clapperton (1965), Moe and Tyrell (1979), IPCC Tier II, Ellis *et al.,* 2007 and Ellis *et al.* (2009). The current Australian methodology uses Moe and Tyrell (1979) (M&T: Equation 1) as the country specific model for feedlot cattle. It was developed using dairy cattle fed high grain diets and relates CH_4 production to the CH_4 carbohydrate fractions (soluble residue, cellulose and hemicelluloses) in the diet.

$$CH_4\ (MJ/day) = 3.406 + 0.510\ Soluble\ residue\ (kg/day) + 1.736\ Hemicellulose\ (kg/day) + \\ + 2.648\ Cellulose\ (kg/day) \quad (1)$$

The equation of (Blaxter and Clapperton) (B&C: Equation 2) calculates the gross energy content of the diet and estimates how much is converted to methane based on digestibility at maintenance energy requirement and the level of feed intake relative to intake required for maintenance.

$$CH_4\ (MJ/day) = 1.3 + 0.112\ DMD\ (\%) + Relative\ intake\ (2.37 - 0.050\ DMD(\%)) \quad (2)$$

This equation forms the basis of a number of national inventories and is used as the Australian methodology for dairy and grazing beef cattle. However, previous validation studies for dairy systems (Wilkerson *et al.,* 1995) have shown this equation to be inaccurate when used on diets outside the range on which it was developed. Ellis *et al.* (2007) developed a range of Equation 3 based on measured dietary variables to predict CH_4 production. The model reported with the lowest root mean square prediction error (RMSPE) was selected for inclusion into the biophysical model in this study. This equation uses lignin and ADF as well as metabolisable energy (ME) intake.

$$CH_4\ (MJ/day) = 2.94 + 0.59\ MEI\ (MJ/day) + 1.44\ ADFI\ (kg/day) - 4.16\ LIGI\ (kg/day) \quad (3)$$

This equation was developed from analysis of 14 different studies; however only a small number of these studies were feedlot cattle and include cattle fed high levels of pasture and forages, which is likely to produce a different relationship to high grain diets. Ellis *et al.* (2009) developed a further Equation 4 for beef cattle, using 12 different studies, with an increased proportion of feedlot based studies. This equation is similar to M&T in that it utilizes the hemicelluloses and cellulose content of the ration, however it also includes the fat content. This has the potential to significantly improve estimates, as inclusion of lipids in the diet

$$CH_4\ (MJ/day) = 2.72 + 0.0937\ MEI\ (MJ/day) + 4.31\ Cellulose\ (kg/day) - \\ - 6.49\ Hemicellulose\ (kg/day) - 7.44\ fat\ (kg/day) \quad (4)$$

Emissions from manure are estimated (Equation 5) using volatile solids output, and a conversion factor for methane, according to IPCC standards (NGGIC 2007)

$$CH_4\ (kg/head/day) = Volatile\ Solids\ (kg/day) \times Emissions\ potential\ of\ manure\ (CH/kg\ VS) \times \\ \times\ methane\ conversion\ factor\ (\%) \times methane\ density\ (kg/m^3) \quad (5)$$

Assumptions and default values

The primary assumptions used in the model (Table 1) are based on the current Australian methodology for estimating greenhouse sources and sinks (NGGIC, 2007). In addition, there are assumed to be 3 production categories (ref), short fed domestic (<100 days), Short fed export (100-200 days) and Long Fed Export (>200 days). Cattle were grouped into these classes based on data provided by the feedlot. Detailed ration composition (cellulose, hemicelluloses and lignin) is

Table 1. Assumptions of the standard biophysical model (NGGIC, 2007).

Parameter	Set value
Energy content of methane	55.22 MJ/kg
Density of methane	0.662 kg/m^3
Gross energy content of feed	18.4 MJ/kg DM
Dry matter digestibility	80%
Y_m (methane conversion factor)	3%

not commonly reported, or provided in ration information. Where these values were required for model calculations they are calculated based on (Givens and Moss, 1990).

Validation

In order to validate the model, published results were used. The model was evaluated for its ability to reflect the physiological changes associated with changes in diet, primarily changes in digestibility and intake, and changes in the fermentation products produced by different diet types (e.g. propionate dominant in a high grain diet vs. acetate dominant in a high forage diet). Five published studies were selected, Beauchemin and McGinn (2005), Boadi *et al.* (2004), Beauchemin and McGinn (2006), Hegarty *et al.* (2007) and Lovett *et al.* (2003). These studies focused on manipulating dietary factors (eg. forage content, grain type and backgrounding vs. finishing rations) to mitigate CH_4 emissions from feedlot cattle. The primary factor under investigation was the energy density of the diet; high vs. low forage. The studies were mainly conducted in the northern hemisphere with the exception of Hegarty *et al.* (2007).

Emissions from Australian beef feedlots

The model for enteric fermentation was also tested against measurements from two Australian beef feedlots (Chen *et al.*, 2009). Measurements were conducted using open-path spectroscopy (OP-FTIR, University of Wollongong, Australia and GasFinder2.0, Boreal Las Inc, Edmonton Alberta, Canada) and micrometeorology (including a three-dimensional sonic anemometer, CSAT-3, Campbell Scientific and a high-speed data logger CR5000, Campbell Scientific) in combination with atmospheric dispersion modeling (backward Lagrangian Stochastic method, WindTrax, Thunder Beach Scientific, Canada) to measure emissions from beef cattle feedlots.

Two feedlots representative of Australian beef cattle feedlots were selected for the study, located in the south (36°21'41" S, 143°24'5" E) and north (27°8'14" S, 151°26'3" E) of the country. Data was collected during eight 2- week field campaigns, at each site during summer and winter of two consecutive years (Chen *et al.*, 2009). During these measurement periods data was collected from the feedlot operator in the form of standard lot or bunk sheets, providing details on placement and current (at time of measurement) weights, intakes and rations fed at the time of measurements (Table 2). The dry pack management of manure in feedlots is thought to result in very low emissions of methane. Boadi *et al.* (2004) report emissions of only 0.92 and 1.48 g/head/day, whilst recent Australian measurements suggest emissions of 0.65 g/head/day (Muir, unpublished data). Therefore for the purpose of this study, the emissions measured using the open path system were assumed to primarily represent enteric emissions.

Table 2. Animal and ration characteristics used in modelling emissions from Australian feedlots.

	Southern site		Northern site	
	Winter 2007	Summer 2008	Winter 2007	Summer 2008
Number of head	13,100	12,800	10,500	6,100
Live weight (kg)	440.4±91.5	454.9±68.2	529.8±117.8	576.4±132.8
Intake (kg DM/head)	9.4±1.8	9.7±1.9	11.3±2.0	11.4±2.0
Measured emission (g/head/day)	122.8±41.5	91.0±62.3	138.3±78.6	60.2±102.5
Ration[1]				
Grain (%)	70	75	70	78
Forage (%)	15	15	15	6
ME (MJ/kg DM)	13.2	12.8	12.8	12.3
CP (%)	14.4	13.5	13.6	13.5
Oil (%)	1.1	0.5	1.1	0.5

[1] By-product protein feeds (e.g. cottonseed) make up the remainder of the rations at each site.

Results

Validation

For the validation data (Figure 2), IPCC Tier I (set value of 164.4 g/head/day) over estimates the measured emissions by an average of 29.6 g (min-max -28.5, 102.3). IPCC Tier II underestimates emissions by an average of 55.9 g/head/day (min-max -109.9, 5.9). B&C overestimates emissions by an average of 109.9 g/head/day (min-max 56, 229.8) and M&T by 57.5 g/head/day (min-max 32.0, 85.9). The equations developed by Ellis *et al.* (2007) and (2009) overestimated emissions by 33.7 (min-max -26.6, 98.6) and 34.6 (min-max -74.1, 111.5) g/head/day respectively. The variability of measured emissions (95% confidence interval was 27.7 g/head/day. This is similar to B&C (26.8) however variability is considerably reduced in all other estimates, with 95% confidence intervals of 9.3, 15.8, 14.4 and 15.8 g/head/day for Tier II, M&T and Ellis *et al.* (2007) and Ellis *et al.* (2009) respectively.

Australian measurements

When predicted are compared with measured Australian feedlot data (Figure 3), the IPCC Tier I equation over estimates all emissions by 61 g/head/day (min-max 26.1, 104.2), whilst estimates from IPCC Tier II are closer to measured values, only slightly over estimating, an average of 0.6 g/head/day (min-max -29, 52.7). All other equations over estimate measured emissions quite considerably, B&C 203.4, M&T 98.2, Ellis *et al.* (2007) 118.6, Ellis *et al.* (2009) 145.1 g/head/day, with ranges of 151.4- 279.4, 56.1- 165.3, 87.8- 172.9 and 94.9- 186.8 respectively. Variability (95% confidence intervals) associated with measured emissions from these studies was 40.3 g/head/day. The smallest variation was associated with the IPCC Tier II equation at 10.7. 95% confidence intervals associated with the other models were 32.5 (B&C), 16.9 (M&T), 18.0 (Ellis *et al.*, 2007) and 22.4 (Ellis *et al.*, 2009). An analysis of the concordance between equations was conducted using the method of Lin (2000). Concordence (Pc) between measured and IPCC Tier II was 0.0659 (95% CI -0.024), B&C 0.1248 (CI -0.043), M&T 0.3696 (CI 0.1362), Ellis *et al.*, 2007 0.4509 (CI 0.1018) and measured and Ellis *et al.*, 2009 -0.177 (CI -0.1726).

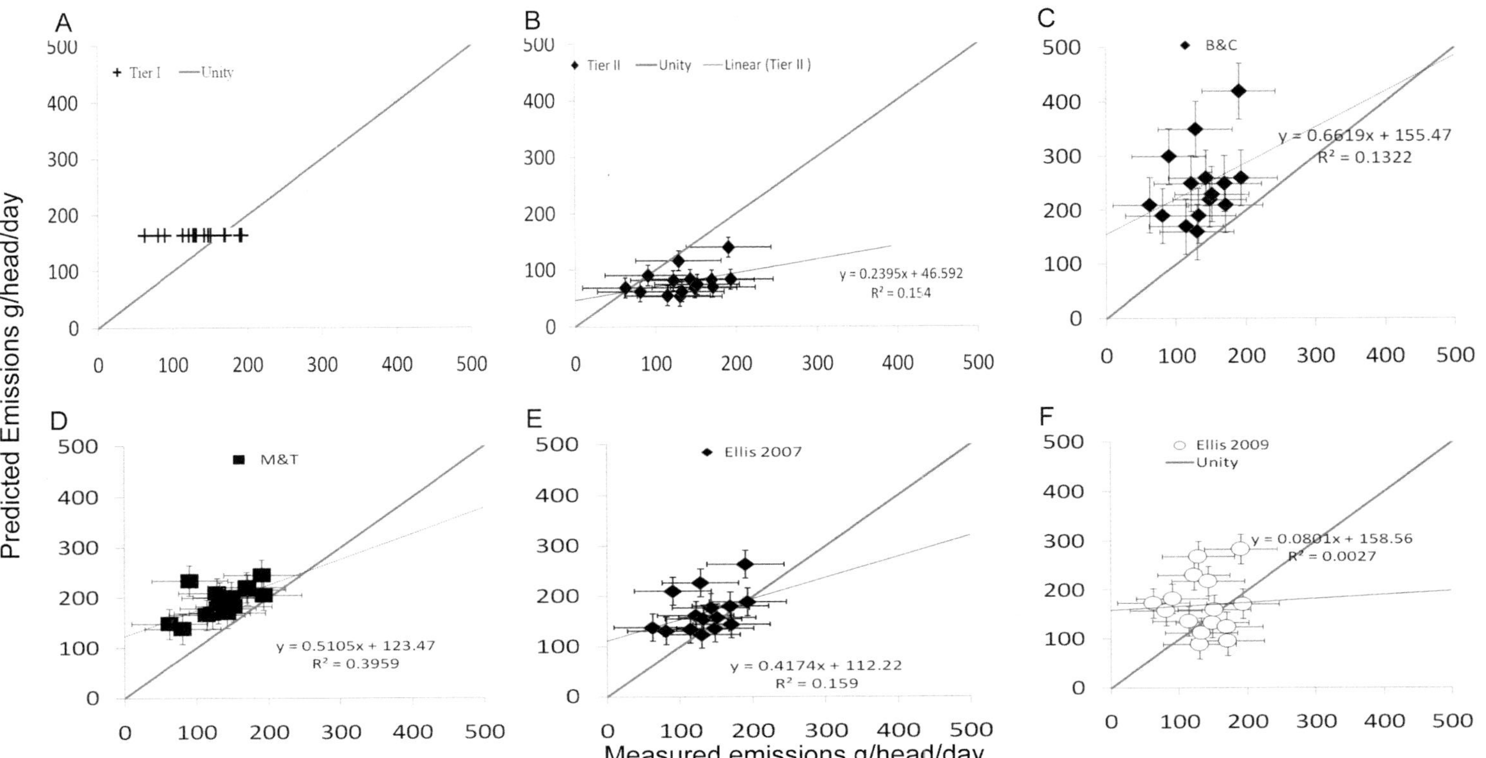

Figure 2. Measured vs. predicted emissions based on a database of published data. A. IPCC Tier I, B. IPCC Tier II, Blaxter and Clapperton (1965), Moe and Tyrell (1979), Ellis et al. *(2007), Ellis* et al. *(2009).*

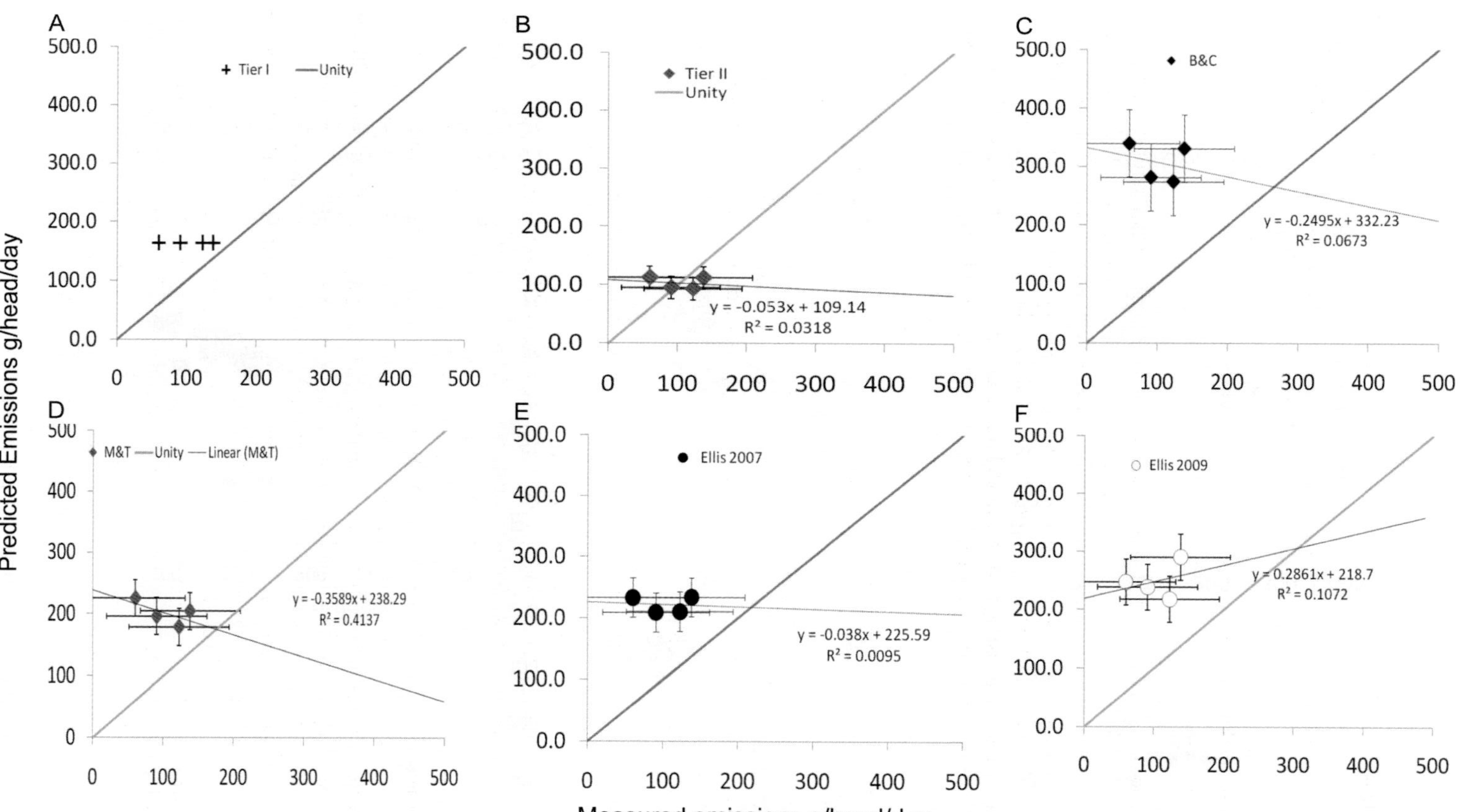

Figure 3. Measured vs. predicted emissions from Australian beef feedlots during summer and winter. A. IPCC Tier I, B. IPCC Tier II, Blaxter and Clapperton (1965), Moe and Tyrell (1979), Ellis et al. *(2007), Ellis* et al. *(2009).*

Discussion

The commitment of the Australian government to introduce a carbon trading scheme has raised questions in the agricultural industries about the inclusion of greenhouse gases emitted from agricultural systems. Concern has also been expressed about the current Australian methodology for estimating emissions from feedlot operations. This study aimed to investigate the current Australian methodology for estimating greenhouse gas emissions from feedlot cattle (Moe and Tyrell 1979) and compare this methodology with other potential equations, in order to assess the most effective methodology for Australian feedlot systems.

The current model has been designed to utilize information available from commercial feedlot management software, and to use inputs which are commonly measured (or for which we can use published data (e.g. Givens and Moss, 1990) with reasonable confidence). However, the desired simplicity does not allow the use of a dynamic model which relates CH_4 output to a detailed representation of rumen function (e.g. Kebreab *et al.*, 2008). The validation component of the study aimed to evaluate the accuracy of the model for feedlot type rations. This study shows that the three levels of methodology provided by the IPCC differ in complicity, but also in accuracy. The Tier I methodology is the simplest method, but also shows high residuals (29.6) being accurate only at ad libitum feeding levels containing 50% forage. In general, feedlot rations contain much less than 50% forage, which would result in an over estimation of emissions using this method. This is consistent with the results of previous studies (Beauchemin and McGinn, 2005) who report that the IPCC Tier I methodology over or under estimates emissions depending on the ration used. The second level of the methodology (Tier II) tends to overestimate emissions from the studies examined (average residual of 55.9 g/head/day), but increases in accuracy at low forage proportions (less than 20%). This indicates some limitations to the Y_m value of 3%. This is consistent with the results of Beauchemin and McGinn (2005) who report Y_m values ranging from 2.81 to 7.55%, supporting the use of diet specific Y_m values.

Comparison of the model with measured values from an Australian feedlot system demonstrate the IPCC Tier I value (164 g/head/day) to overestimate under all feeding conditions studied. IPCC Tier II appears to be a reasonable estimate of emissions, however this equation performed worst in the concordance analysis (Pc 0.0657) indicating a poor relationship between the measured and predicted values. Emissions predicted by Blaxter and Clapperton (1965) are consistently over estimating measure values suggesting the equation is not to be used under feedlot conditions. The best performing equations based on Lins concordance were Ellis *et al.* (2007) (Pc 0.4509) and M&T (Pc 0.3596). The current Australian methodology for feedlot cattle, M&T overestimates emissions, by an average of 98.2 (and 57.5 for the validation study). Despite this, the Lins concordance value (0.3696) indicates reasonable relationship between the measured and predicted emissions. Ellis *et al.* (2007) produced an improved estimation of measured emissions. Average residuals were 33.7g for the validation study and 188.6 for the Australian data. Ellis *et al.* (2009) appears to improve estimates in the validation study, with average residuals of 34.6. However, this is not supported by very high residuals in the Australian data (average 145.1) or the negative value of concordance. Moe and Tyrell (1979) and Ellis *et al.* (2007) were the most accurate in their prediction of CH_4 emission for the winter campaigns, although intakes and animal live weights are very similar for each site (north and south) in each season (summer and winter). The primary difference in the ration composition between summer and winter at each site is the grain composition (wheat vs. wheat/ barley and sorghum vs. sorghum/ barley) suggesting variation in intake of NDF and other non polysaccharide sources that may yield acetate. The variability in measured emissions suggests that other factors (e.g. environmental conditions, behavior) (discussed in detail in Chen *et al.*, 2009) may be affecting emissions as well as ration composition.

Conclusion

This study showed that the use of equations developed based on beef cattle improved the estimates of emissions from feedlot cattle. It also demonstrated the need for further investigation of the IPCC Tier II methodology and the development of improved and diet specific Ym values, to enable this equation to be effective in predicting emissions. The current Australian methodology for estimating emissions from feedlot cattle results in a significant over estimation of emissions, however a diet specific equation appears to be the most effective choice. With the decision to include agricultural industries in the Australian carbon pollution reduction scheme looming emissions estimates based on an alternative method should be considered.

References

ALFA (The Australian Lot Feeder's Association), 2008. Australian Lotfeeders association response to carbon pollution reduction scheme green paper. Public domain submissions to the Federal Government, Canberra, Australia.

Beauchemin, K.A. and McGinn, S.M., 2005. Methane emissions from feedlot cattle fed barley or corn diets. Journal of Animal Science 83:653-661.

Beauchemin, K.A. and McGinn, S.M., 2006. Enteric methane emissions from growing beef cattle as affected by diet and level of intake. Canadian Journal of Animal Science 86:401-408.

Blaxter, K.L. and Clapperton, J.L., 1965. Predicting of the amount of methane produced by ruminants. British Journal of Nutrition 19:511- 522.

Boadi, D.A., Wittenberg, K.M., Scott, S.L., Burton, D., Buckley, K., Small, J.A. and Ominski, K.H., 2004. Effect of low and high forage diet on enteric and manure pack greenhouse gas emissions from a feedlot. Canadian Journal of Animal Science 84:445-453.

Chen, D., Bai, M., Denmead, O.T., Griffith, D.W.T., Hill, J., Loh, Z.M., McGinn, S.M., Muir, S., Naylor, T., Phillips, F. and Rowell, D., 2009. Greenhouse gas emissions from Australian beef feedlots. Meat and Livestock Australia, North Sydney, Australia.

Ellis, J.L., Kebreab, E., Odongo, N.E., Beuchemin, K., McGinn, S., Nkrumah, J.D., Moore, S.S., Christopherson, R. Murdoch, G.K., McBride, B.W., Okine, E.K. and France, J., 2009. Modeling methane production from beef cattle using linear and nonlinear approaches. Journal of Animal Science 87:1334-1345.

Ellis, J.L., Kebreab, E., Odongo, N.E., McBride, B.W., Okine, E.K. and France, J., 2007. Prediction of methane production from dairy and beef cattle. Journal of Dairy Science 90:3456-3467.

Givens, D.I. and Moss, A.R. (eds.), 1990. UK Tables of Nutritive Value and Chemical Composition of Feedingstuffs. Rowett Research Services Ltd, Aberdeen, UK.

Hegarty, R.S., Goopy, J.P., Herd, R.M. and McCorkell, B., 2007. Cattle selected for lower residual feed intake have reduced daily methane production. Journal of Animal Science 85:1479-1486.

Kebreab, E., Johnson, K.A., Archibeque, S.L., Pape, D. and Wirth, T., 2008. Model for estimating enteric methane emissions from United States dairy and feedlot cattle. Journal of Animal Science 86:2738-2748.

Lin, L.I., 2000. A note on the concordance correlation coefficient. Biometrics, 56:324-325.

Lovett, D., Lovell, S., Stack, L., Callan, J., Finlay, M., Conolly, J. and O'Mara, F.P., 2003. Effect of forage/ concentrate ratio and dietary coconut oil level on methane output and performance of finishing beef heifers. Livestock Production Science 84:135-146.

Miller, D.N. and Berry, E.D., 2005. Cattle feedlot soil moisture and manure content: 1. Impacts on Greenhouse Gases, Odor Compounds, Nitrogen Losses and Dust. Journal of Environmental Quality 34:644-655.

Moe, P.W. and Tyrrell, H.F., 1979. Methane production in dairy cows, Journal of Dairy Science 62:1583-1586.

NGGIC (National Greenhouse Gas Inventory Committee), 2007. Australian Methodology for the estimation of greenhouse gas emissions and sinks 2006. Agriculture, Australian Government Department of Climate Change, Canberra, Australia.

Wilkerson, V.A., Casper, D.P. and Mertens, D.R., 1995. The prediction of methane production of Holstein cows by several equations. Journal of Dairy Science 78:2402-2412.

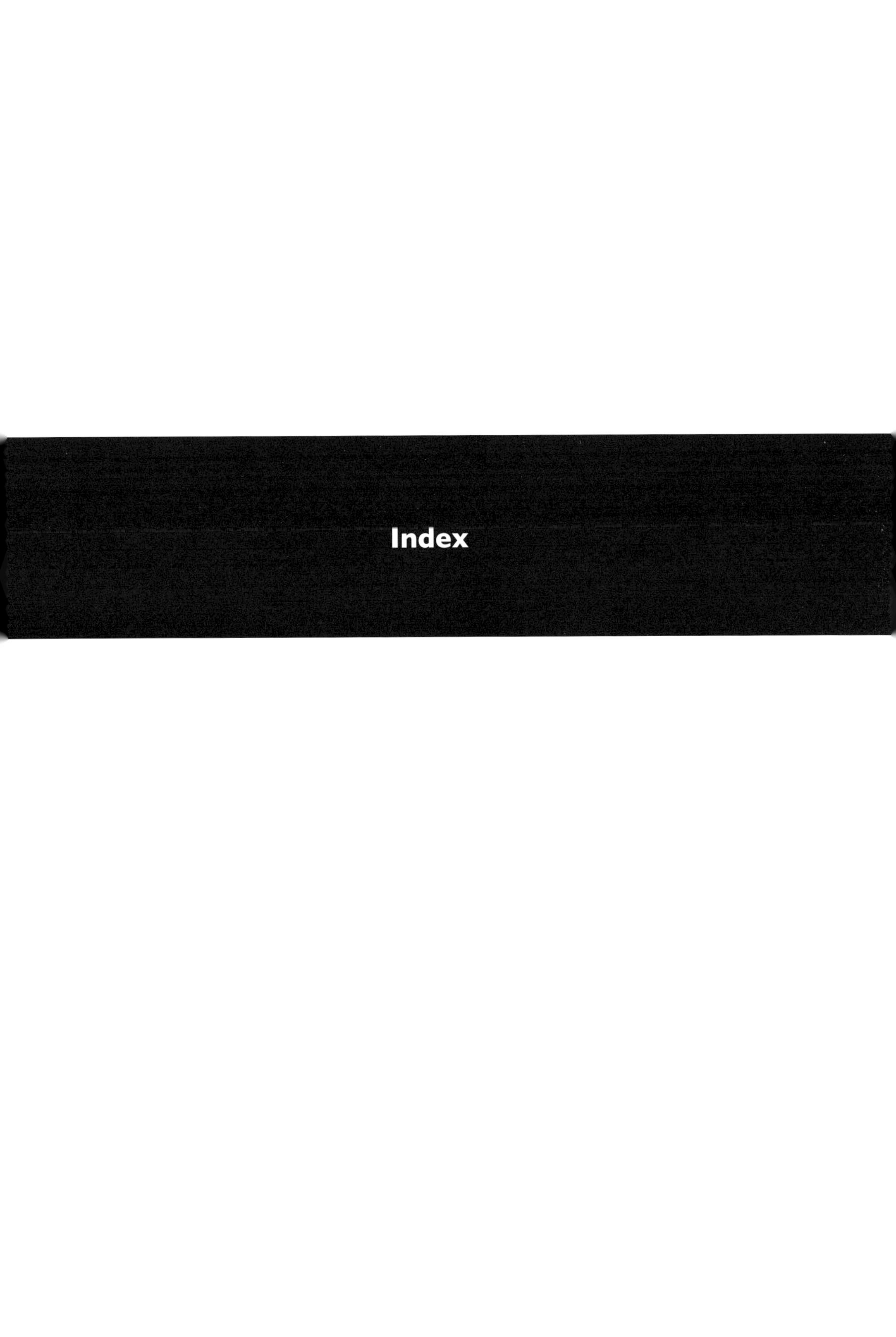

Index